Applied Statistics for the Behavioral Sciences

APPLIED STATISTICS FOR THE BEHAVIORAL SCIENCES

Fourth Edition

Dennis E. Hinkle
TOWSON UNIVERSITY

William Wiersma
UNIVERSITY OF TOLEDO

Stephen G. Jurs
UNIVERSITY OF TOLEDO

HOUGHTON MIFFLIN COMPANY   BOSTON   NEW YORK

Sponsoring Editor: David C. Lee
Senior Associate Editor: Jane Knetzger
Packaging Services Supervisor: Charline Lake
Senior Production/Design Coordinator: Carol Merrigan
Senior Manufacturing Coordinator: Priscilla J. Bailey
Marketing Manager: Pamela J. Laskey

Cover design: Harold Burch, Harold Burch Design, New York City

Computer Examples in selected chapters courtesy SPSS, Inc. All rights reserved.

Printed in the U.S.A.
Library of Congress Catalog Card Number: 97-72486
ISBN: 0-395-87411-4

123456789-DH-01 00 99 98 97

Brief Contents

Contents

7 Probability, Sampling Distributions, and Sampling Procedures   152

8 Hypothesis Testing: One-Sample Case for the Mean   188

9 Estimation: One-Sample Case for the Mean   217

10 Hypothesis Testing: One-Sample Case for Other Statistics

11 Hypothesis Testing: Two-Sample Case for the Mean

12 Hypothesis Testing: Two-Sample Case for Other Statistics

13 Determining Power and Sample Size

17 Linear Regression: Estimation and Hypothesis Testing   461

18 Multiple Linear Regression   479

19 Analysis of Covariance   516

Preface

The goal of the fourth edition of *Applied Statistics for the Behavioral Sciences*, as in the previous editions, is to give the reader both a conceptual understanding of basic statistical procedures used in the behavioral sciences and the computational skills to use them. The book presumes no prior course in statistics—only a basic knowledge of mathematics and algebra. Presentations are designed to give students a clear understanding of the basic theoretical background needed to understand these statistical procedures.

We begin the text with a review of basic mathematical concepts and introductory statistical concepts. We then move into increasingly more sophisticated statistical concepts and procedures, both descriptive and inferential. Where relevant throughout the text, we include discussions of examples that demonstrate the use of contemporary computer software packages in carrying out these procedures.

Changes in the Fourth Edition

Several substantive improvements have been made in the fourth edition based on the suggestions of instructors and reviewers. These changes will serve to enhance the understanding and application of the statistical procedures. They include the following:

- A major change in this edition has been to again separate the discussion of regression into two chapters. The first of these chapters (Chapter 6) follows the chapter on Correlation (Chapter 5) and considers the descriptive components of regression. The second of these chapters (Chapter 17) follows the many chapters on inferential statistics and considers estimation and hypothesis testing. This is the order and format that was used in the second edition of the book. Based on reviewer comments, we have returned to that format.

- The chapter dealing with the χ^2 tests for frequencies (Chapter 21) has been rewritten; it now has better continuity, with examples moving from the one-sample case, through the two-sample case, to the k-sample case.

- In Chapter 10, we have changed the discussion for testing the H_0: $P = a$. Based on a paper presented at the American Education Research Association (AERA), we have used the hypothesized parameter (P) rather than the sample statistic (p) in the computation of the standard error for testing the null hypothesis. However, we have used the sample statistic (p) in the standard error when only the confidence interval is being developed without testing the null hypothesis.

- Computer examples using printouts from SPSS have been added to many chapters. The purpose of these examples is to expose the reader to use of statistical software packages and to ensure their understanding of the entries in the printouts.

- We have provided five data sets on a diskette that accompanies every copy of the text. These data sets, described in full in Appendix B, are:
  1. High School and Beyond ($n = 500$)
  2. High School and Beyond ($n = 75$)
  3. Northwest Ohio School Districts ($n = 94$)
  4. Applicants for Management Positions ($n = 100$)
  5. Children's Protective Services Youth ($n = 50$)

  In addition, computer exercises using these data sets are presented in Appendix B. The computer printouts for these exercises appear in the *Test Bank and Solutions Manual* available to instructors.

- We have added a full glossary to the book (see Appendix A), which we hope will be helpful to both instructors and students.

- We have added one additional exercise per chapter, whose complete solution is provided in the answer key (Appendix D). The answer key provides briefer solutions for all other exercises. In addition, instructors have access to complete solutions to all exercises in the *Test Bank and Solutions Manual*.

Features Retained in the Fourth Edition

The pedagogical approach that has proven to be so effective in the three previous editions has been retained and enhanced. Sections in several chapters have been rewritten for better flow. In addition, careful attention has been given to ensuring that typographical and computational errors have been eliminated.

Several important features that have been retained include:

- A list of key concepts is presented at the beginning of each chapter. These concepts are printed in boldface when they first appear in the chapter, and they are included in the new glossary.

- Important concepts, definitions, and conclusions are highlighted in boxes following their presentation in the text.

- Each chapter concludes with an extensive set of exercises that provide students with practice calculating and interpreting statistics. Many of the exercises involve realistic scenarios representing a range of areas in the behavioral sciences. As noted above, answers are provided in Appendix D and in the *Test Bank and Solutions Manual*.

- The most challenging chapters for begining and intermediate students have been reviewed and revised as needed. We are gratified to learn from those who have adopted the book that these chapters have been particularly effective:
    ◇ Chapter 7 - Probability, Sampling, and Sampling Distributions
    ◇ Chapter 13 - Determining Power and Sample Size
    ◇ Chapter 15 - Multiple Comparison Procedures
    ◇ Chapter 18 - Multiple Regression
    ◇ Chapter 19 - Analysis of Covariance

Organization of the Text

Chapters 1 through 6 introduce basic statistical concepts and cover descriptive statistics. They include discussions of elementary mathematical concepts, the rules of summation, and fundamental statistical concepts (Chapter 1); organizing and graphing data (Chapter 2); describing distributions, measures of central tendency, measures of variation, percentiles, and standard scores (Chapter 3); the normal distribution (Chapter 4); correlation (Chapter 5); and regression (Chapter 6).

The discussion of inferential statistics begins in Chapter 7, which introduces the concepts of probability, sampling, and sampling distributions. This chapter along with Chapters 8 and 9 introduce the logic for hypothesis testing and estimation for the one-sample case for the mean. Chapters 10 through 12 apply this logic to the one-sample case for other statistics and to the two-sample case for the mean and other statistics. Those instructors wishing to study all one-sample tests before going on to the two-sample tests would use the chapters in their current order. For instructors wishing to study the one-sample and two-sample case for the mean before dealing with the other statistics, the appropriate chapter sequence would be 7, 8, 9, 11, 10, and 12.

The recurring question of sufficient sample size is addressed in Chapter 13. This discussion relates the concept of sample size to the power of the test and to the effect size of interest to the researcher.

An introduction to more complex procedures begins in Chapter 14 with a discussion of one-way analysis of variance and simple repeated measures analysis of variance. These discussions are followed in Chapter 15 with a presentation of multiple comparison procedures, including both *a priori* and *post hoc* comparisons. Analysis of variance, two-way classification, is discussed in Chapter 16. This chapter also includes discussions of the meaning of main effects and interaction, simple effects tests, and post hoc multiple comparisons. Sample size determination for both one-way and two-way analysis are also included in Chapters 14 and 16.

The concept of regression introduced in Chapter 6 is expanded in Chapter 17 to contain discussion of estimation and hypothesis testing in linear regression. Testing the regression coefficient for statistical significance is presented and the relationship of this test to the significance test for the bivariate correlation coefficient is shown. The discussions in Chapter 17 are expanded in Chapter 18 to include multiple correlation and multiple regression. The discussion of multiple regression includes a set of steps for conducting a multiple regression analysis, including the significance test for R^2, for the regression coefficients, and for the R^2 increase or decrease. Because of the computation complexity of multiple regression, computer printouts are included for all examples.

Analysis of covariance is discussed in Chapter 19; this discussion relates concepts of analysis of covariance to regression analysis and analysis of variance. The chapter contains a discussion of the assumptions underlying analysis of covariance and the post hoc procedures for the adjusted means.

The final three chapters cover other correlation coefficients and measures of association (Chapter 20) and selected nonparametric statistics (Chapters 21 and 22). The chapter on χ^2 tests (21) has been revised for better continuity and includes a discussion of standardized residuals as a post hoc procedures following the χ^2 test of homogeneity.

Supplements to the Text

Workbook

The text is accompanied by a student workbook prepared by Dennis E. Hinkle, Keith W. Zoski, and John R. Cox. The workbook has been updated to reflect the changes made in the textbook. Each of the chapters includes a programmed chapter review and a set of exercises that will help students assess their comprehension and retention of the text material. The exercises also provide the opportunity for students to hone their computational skills.

Test Bank and Solutions Manual

The *Test Bank and Solutions Manual*, prepared by Dennis E. Hinkle and Stephen G. Jurs, contains multiple-choice questions, detailed solutions to exercises in the textbook, and computer printouts for the computer exercises in Appendix B of the textbook. The multiple-choice questions have been expanded and updated. The detailed solutions outline the intermediate steps that are needed to solve the text exercises. Using these detailed solutions, instructors can help their students discover what portion of the statistical procedure is not understood. Using the computer printouts, instructors can correct students' work or have the students correct their own work.

Computerized Test Bank

The multiple-choice questions are also available on disk. The test bank software allows instructors to add and edit questions and prepare tests. On-line testing and gradebook options are available.

Using SPSS for Windows

A manual entitled *Using SPSS for Windows*, by Charles A. Stangor of the University of Maryland, can be shrinkwrapped with the text at a discount. This brief handbook introduces students to the basics of SPSS, with step-by-step instructions, sample output, and student exercises based on the data sets provided on disk with our text.

Pyschology Website

Houghton Mifflin's Psychology website can be reached by pointing to the Houghton Mifflin homepage at http://www.hmco.com and going to the College Division Psychology page. This location provides access to additional useful and innovative teaching and learning resources that support this book.

Acknowledgments

The publication of a textbook results from the efforts of many people. We are indebted to Debra Buchman, Corinna Ethington, Charles Hinkle, Diane Horm-Wingerd, Steve Kaeuper, James Nimmer, Robert Rachor, Edgar Roland, Alan Sack, Michael Scott, and Keith Zoski for their assistance with the various phases of this project. We are also grateful for the assistance provided by the following colleagues and reviewers who guided us on this edition of the project:

Jackson Barnette, *University of Alabama*
Bernard C. Beins, *Ithaca College*
Mike Cline, *University of Illinois at Chicago/Roosevelt University*
Charles Johnson, *University of Maryland*
Ravinder Nath, *University of Memphis*
Anton Netusil, *Iowa State University*
Brian J. Stone, *Wichita State University*
Holly R. Straub, *University of South Dakota*
William B. Ware, *University of North Carolina at Chapel Hill*
Christa Winter, *Springfield College*

In addition, we would like to acknowledge the excellent professional support and assistance given by the staff at Houghton Mifflin: David C. Lee, Sponsoring Editor, Jane Knetzger, Senior Associate Editor, and Charline Lake, Packaging Services Supervisor.

Finally, we are grateful to the Literary Executor of the late Sir Ronald A. Fisher, F.R.S., to Dr. Frank Yates, F.R.S., and the Longman Group Ltd., London, for permission to reprint Table IIi, part III, part IV, part VII, and part XXIII, from their book *Statistical Tables for Biological, Agricultural and Medical Research* (Sixth Edition, 1974).

D. E. H.
W. W.
S. G. J.

1

Introduction

The word *statistics* means different things to different people. Weathermen report daily weather statistics, such as high and low temperatures and amount of rainfall; sportscasters give halftime statistics that include total yards rushing and total yards passing. Mathematicians and researchers in the behavioral and physical sciences talk about statistics in a much different way. Mathematicians describe statistics as a major area in mathematics, and researchers discuss the appropriate statistics for analyzing the results of a particular investigation.

Are all these people using the word *statistics* in the same way? Not exactly, but in their own contexts they are all using the term correctly. Weathermen and

1

sportscasters use the word *statistics* to mean one or more bits of information that describe weather conditions or the action in a game; mathematicians use the word to define a body of theory and methods that can be used in analyzing data.

In this book, we are interested in what *statistics* means to behavioral scientists. We will begin by looking at the meaning and role of statistics in the behavioral sciences. We will review some very basic mathematical operations that are needed to understand statistics. In the rest of the chapter, we will introduce some of the fundamental terms and concepts used in statistics.

The Meaning and Role of Statistics

What does **statistics** mean to the researcher in the behavioral sciences? Essentially it means the methods or procedures that researchers apply in an attempt to understand data. **Data** consist of information — for the most part, information in numerical form that represents a certain characteristic. For example, if we want to know the anxiety level of a group of individuals, the data might be scores on an anxiety scale, and we would use statistics to describe and understand those scores.

> **Statistics is a collection of theory and methods applied for the purpose of understanding data.**

The behavioral and social sciences could not exist without statistics. Behavioral scientists use statistics to explain the results of research studies and to provide empirical evidence to support or refute theories. Often researchers monitor individuals' behavior by representing that behavior in numerical form; then they use statistics to evaluate, report, and explain the behavior. For example, in several chapters of this book and in the accompanying exercises, we use data from a longitudinal survey of high school seniors.[1] The purpose of the survey was to determine the students' educational and vocational activities, plans, aspirations, and attitudes as well as the relationship between these characteristics and the students' prior education and personal characteristics. The student information is represented by numbers such as scores on a mathematics achievement test, numbers from 1 to 8 to represent letter grades, 0 for male and 1 for female. Then these data are analyzed by using statistics, such as procedures to find average scores and procedures to determine whether mathematics test scores are related to gender, previous education, or other characteristics.

[1] National Opinion Research Center, *High School and Beyond Information for Users: Base Year (1980) Data* (Chicago: National Opinion Research Center, 1980).

Many people who are not researchers also need to be knowledgeable consumers of research results. They, too, rely on statistics. Often statistics provides the basic rationale behind planning and decision making by educators, counselors, and others whose work is tied to the behavioral sciences. A school superintendent, for example, might use statistics to plan and make decisions about an innovative curriculum. A clinical psychologist might decide on the basis of statistics that a particular client is more likely to benefit from group therapy than from intensive psychoanalysis. Whatever the reason for using statistics, researchers and consumers of research alike should understand what information the statistics provide and what conclusions one can draw from them.

Basic Mathematics for Statistics

Statistics involves everything from basic to complex mathematics. In this introductory text no complex mathematics is required. But you should be familiar with some arithmetic, elementary algebra, and basic statistical symbols. A brief review of mathematical concepts and operations follows. You should master them to understand the statistical concepts and procedures used in this text.

Four Basic Arithmetic Operations

Addition, subtraction, multiplication, and division are the four **basic arithmetic operations** performed with numbers.

The algebraic sign for addition is the plus sign (+), and the addition of two numbers — say, X and Y — is written $X + Y$. For example, if $X = 15$ and $Y = 5$, then $X + Y = 15 + 5 = 20$. The result of the addition operation is called the *sum*.

The algebraic symbol for subtraction is the minus sign (−). Subtracting Y from X, we find that $X - Y = 15 - 5 = 10$. The result of the subtraction operation is called the *difference*.

Multiplication of two numbers can be denoted algebraically in three ways: $X \cdot Y$ or $(X)(Y)$ or simply XY. We will use all three ways in this book.

$$X \cdot Y = 15 \cdot 5 = 75$$

$$(X)(Y) = (15)(5) = 75$$

$$XY = (15)(5) = 75$$

The individual numbers that are multiplied are called *factors*; the result of the multiplication operation is called the *product*.

Division of two numbers can also be denoted in three ways: $X \div Y$ or X/Y or $\frac{X}{Y}$.

$$X \div Y = 15 \div 5 = 3$$

$$X/Y = 15/5 = 3$$

$$\frac{X}{Y} = \frac{15}{5} = 3$$

In this last form, the number on top is called the *numerator*, and the number on the bottom is called the *denominator*. The result of the division operation is called the *quotient*.

It is important to note the relationship between the multiplication operation and the division operation. The quotient X/Y can also be expressed as the product $(X)(1/Y)$. For example,

$$X/Y = 15/5 = 3$$

$$(X)(1/Y) = (15)(1/5) = 3$$

Arithmetic Operations with Real Numbers

In elementary arithmetic, we are most familiar with *positive numbers* — that is, all numbers zero (0) or greater. In the study of statistics, however, we must deal with what mathematicians call *real numbers*. **Real numbers** are all positive and negative numbers from negative infinity $(-\infty)$ to positive infinity $(+\infty)$. Thus real numbers include not only the positive and negative integers (whole numbers) but also all fractions and decimals.

When both positive and negative numbers are used in an arithmetic operation, we often call them **signed numbers**. Sometimes we want to ignore the algebraic sign and consider only the **absolute value** of the number — that is, the value of the number without regard to its algebraic sign. For example, the absolute value of -7, which is denoted $|-7|$, is 7. In absolute values, $|-7| = |+7| = 7$.

> The real numbers are all positive and negative numbers from $-\infty$ to $+\infty$. The absolute value of any real number is the value of the number without regard to its algebraic sign.

The algebraic sign in front of a number affects the results of algebraic operations. In the following paragraphs, we will see these effects as we define the rules for the arithmetic operations. We will use the following numbers: $W = -10$, $X = +15$, $Y = +5$, and $Z = -25$.

Addition   When two numbers have the same sign, add the absolute values of the numbers and retain the sign.

$$W + Z = (-10) + (-25) = -35$$

$$X + Y = (+15) + (+5) = +20$$

When adding two numbers that have opposite signs, subtract the absolute value of the smaller number from the absolute value of the larger number. The resulting sum takes the sign of the number having the larger absolute value.

$$W + X = (-10) + (+15) = +5$$

$$Y + Z = (+5) + (-25) = -20$$

Subtraction When subtracting numbers, change the sign of the number being subtracted and then apply the rules for addition.

$$W - Y = (-10) - (+5) = (-10) + (-5) = -15$$

$$W - Z = (-10) - (-25) = (-10) + (+25) = +15$$

Multiplication When we multiply two numbers that have the same sign, the product is positive. When we multiply two numbers that have different signs, the product is negative.

$$(W)(Z) = (-10)(-25) = +250$$

$$(X)(Y) = (15)(5) = +75$$

$$(W)(X) = (-10)(15) = -150$$

$$(Y)(Z) = (5)(-25) = -125$$

Division When we divide two numbers that have the same sign, the quotient is positive. When we divide two numbers that have different signs, the quotient is negative.

$$X/Y = +15/+5 = +3$$

$$W/Z = -10/-25 = +0.40$$

$$X/W = +15/-10 = -1.50$$

When performing arithmetic operations with real numbers, consider the signs of the numbers carefully.

Special Numbers: Zero and One There are two special numbers in the real number system: 0 and 1. When the number 0 is added to or subtracted from any number, the result is that number: $X + 0 = X$ and $X - 0 = X$. When any number is multiplied by 0, the result is 0: $(X)(0) = 0$. Now consider the number 1. When any number is multiplied or divided by 1, the result is that number: $(X)(1) = X$, and $X/1 = X$.

Rounding Numbers

In our use of statistical methods, we often must divide one number by another. For example, when we divide Y by X (given $X = 15$ and $Y = 5$), we have 5/15, or 1/3. If we express this fraction as a decimal, the result is $0.3\bar{3}$. This decimal can be extended indefinitely. However, we must determine (1) where the decimal should be terminated and (2) which value should be assigned to the last number in the decimal.

In a strict mathematical sense, these two concerns are addressed by several rules for determining the number of *significant digits*. In the study of statistics, however, the following rule is usually adopted:

> After every arithmetic operation that yields a fraction, carry the decimal to three places and round to two places more than there were in the original data.[2]

For example, if the original data are in whole numbers, we carry the decimal to three places and round to the second decimal place. If the data are recorded in tenths, we carry the decimal to four places and round to the third decimal place, and so on.

After we decide on the number of decimal places, we must determine the value of the last digit. The following rule is generally adopted:

> If the number beyond the last digit to be reported is equal to or greater than 5, increase the last digit to the next higher number. If the number beyond the last digit is less than 5, the last digit remains the same.

Consider the following examples, in which the decimal is to be carried to three places and then rounded to two places.

> 3.826 rounds to 3.83
> 4.842 rounds to 4.84

To round numbers, carry the decimal to three places and round to two places more than were in the original data.

---

[2]Instructors may have established their own conventions for rounding to significant digits. Any convention that maintains an adequate degree of accuracy is appropriate.

Properties of Real Numbers

Real numbers have three important characteristics, which are often used in statistical formulas and calculations. These characteristics are the *commutative property*, the *associative property*, and the *distributive property*.

Commutative Property The **commutative property** tells us that two numbers (X and Y) may be added (or multiplied) in any order to achieve the same result. For addition:

$$X + Y = Y + X$$
$$15 + 5 = 5 + 15 = 20$$

For multiplication:

$$(X)(Y) = (Y)(X)$$
$$(15)(5) = (5)(15) = 75$$

Associative Property The **associative property** tells us that, when three numbers are added (or multiplied), we can add (or multiply) X and Y first and then W, or we can add (or multiply) W and X first and then Y. The result will be the same. For addition:

$$W + (X + Y) = (W + X) + Y$$
$$-10 + (15 + 5) = (-10 + 15) + 5$$
$$-10 + 20 = +5 + 5$$
$$10 = 10$$

For multiplication:

$$W(X \cdot Y) = (W \cdot X)Y$$
$$(-10)(15 \cdot 5) = (-10 \cdot 15)(5)$$
$$(-10)(75) = (-150)(5)$$
$$-750 = -750$$

Distributive Property The **distributive property** tells us that the product of a number (Y) with the sum of two numbers (W and X) is the same as the sum of the products of Y with W and Y with X.

$$Y(W + X) = (Y)(W) + (Y)(X)$$
$$5(-10 + 15) = (5)(-10) + (5)(15)$$
$$5(5) = -50 + 75$$
$$25 = 25$$

Three properties of the real number system are used in statistical formulas and calculations: the commutative, associative, and distributive properties.

Special Indicators of Arithmetic Operations

Two special indicators of arithmetic operations appear often in statistics. These are the exponent and square root symbols. *Squaring* a number, or multiplying a number by itself, is a common calculation in statistics. For example, to find the square of Y, we multiply Y times Y: $(Y)(Y) = (5)(5) = 25$. Symbolically, the square is denoted Y^2, or Y to the second power. The number 2 is called the **exponent**, the power to which the number is raised. If Y is raised to the fourth power, Y^4, we have $Y^4 = (Y)(Y)(Y)(Y) = (5)(5)(5)(5) = 625$.

Finding the square root of a number is also a common calculation. The **square root** of a number is a real number such that, when you multiply the square root by itself, the product is the original number. The symbol $\sqrt{}$ over the number indicates that the square root is to be taken. For example, $\sqrt{36} = 6$ because $6^2 = (6)(6) = 36$. The square root of a number can also be indicated by the fractional exponent 1/2; that is, $\sqrt{36} = (36)^{1/2} = 6$.

Finding the square root of a number involves a very cumbersome procedure. It is unnecessary to review this process, because we can find square roots readily by consulting a table of square roots or by using a calculator.

Order of Operations

When we combine two or more operations in one solution, the order in which we do the operations is important. For example, for the problem

$$5 \times 8 + 16 \div 4$$

do we multiply 5 times 8 (40), add 16 to obtain 56, and divide 56 by 4 for a final solution of 14? No. The operation indicators mean that we divide 4 only into 16; then we add this quotient to the product of 5 times 8. So the solution is 44.

The rules for the order of operations are

1. First perform any operations in parentheses (or brackets or braces). If there are parentheses within brackets, perform the operations from the inside out.

2. Next perform operations with exponents, such as squares or square roots, if they were not contained within the parentheses.

3. When these operations are complete, clear the expression of parentheses, do the multiplications and divisions, then the additions and subtractions.

Consider another example. Suppose we want to find

$$\sqrt{49} + 3(7 + 8) - (10 + 6) \div 2^2$$

The sequence of operations is

$\sqrt{49} + (3)(15) - 16 \div 2^2$ operations in parentheses

$7 + (3)(15) - 16 \div 4$ operations with exponents
                          (and the square roots)

$7 + 45 - 4$ multiplication and division

48 addition and subtraction

The Summation Operator

One very important mathematical operator used in statistics is the **summation operator**, which is often used to simplify statistical formulas. It is denoted by the Greek capital letter sigma (Σ), which stands for "the sum of." Thus Σ indicates the summing of whatever immediately follows in the expression.

For example, suppose we have the following five numbers:

$$X_1 = 3 \quad X_2 = 7 \quad X_3 = 4 \quad X_4 = 2 \quad X_5 = 8$$

The subscripts 1 through 5 on the X's simply mean that they stand for different numbers. If we want to sum the five numbers, we proceed as follows:

$$\sum_{i=1}^{5} X_i = X_1 + X_2 + X_3 + X_4 + X_5$$

$$= 3 + 7 + 4 + 2 + 8$$

$$= 24$$

X_i (called "cap X, subscript i," or just "X sub i") is the general symbol for the number. The notation under the Σ indicates the first number in the summation (in this case, $X_1 = 3$), and the number above the Σ indicates how far the summation continues (in this case, through $X_5 = 8$). In general, the notation

$$\sum_{i=1}^{N} X_i$$

means that the summation begins with the first number and concludes with the Nth. (In running text, this is written $\Sigma_{i=1}^{N} X_i$.) Often the notations above and below the Σ are omitted. When they are, Σ means summation from the first through the Nth or last number.

Here are three of the rules that govern the summation operator.

Rule 1   Applying the summation operator, Σ, to the products resulting from multiplying a set of numbers by a constant (unchanging value) is equal to multiplying the constant by the sum of the numbers. Symbolically, if C is a constant,

$$\sum_{i=1}^{N} CX_i = C \sum_{i=1}^{N} X_i$$

For example, let 2 be the constant, and let the numbers be $X_1 = 1$, $X_2 = 3$, $X_3 = 4$, and $X_4 = 7$. Then we find

$$\sum_{i=1}^{4}(2)X_i = 2(1) + 2(3) + 2(4) + 2(7)$$
$$= 2 + 6 + 8 + 14$$
$$= 30$$

and

$$2\sum_{i=1}^{4}X_i = 2(1 + 3 + 4 + 7)$$
$$= 2(15)$$
$$= 30$$

Rule 2   Applying the summation operator, Σ, to a series of constant scores is the same as taking the product of N times the constant score. Symbolically,

$$\sum_{i=1}^{N} C = NC$$

Suppose $X_1 = 4$, $X_2 = 4$, and $X_3 = 4$; that is, each of three scores is equal to the constant 4. Then

$$\sum_{i=1}^{3} C = 4 + 4 + 4$$
$$= 12$$

and

$$NC = 3(4)$$
$$= 12$$

Rule 3   Applying the summation operator, Σ, to the algebraic sum of two (or more) scores for a single individual and then summing these sums over the N

individuals is the same as summing each of the two (or more) scores separately over the N individuals and then summing the scores. Symbolically,

$$\sum_{i=1}^{N}(X_i + Y_i) = \sum_{i=1}^{N}X_i + \sum_{i=1}^{N}Y_i$$

Suppose we have four individuals with the following X and Y scores:

Individual	X	Y
1	2	7
2	5	9
3	3	6
4	1	5

We find that

$$\sum_{i=1}^{4}(X_i + Y_i) = (X_1 + Y_1) + (X_2 + Y_2) + (X_3 + Y_3) + (X_4 + Y_4)$$
$$= (2+7) + (5+9) + (3+6) + (1+5)$$
$$= 9 + 14 + 9 + 6$$
$$= 38$$

and

$$\sum_{i=1}^{4}X_i + \sum_{i=1}^{4}Y_i = (X_1 + X_2 + X_3 + X_4) + (Y_1 + Y_2 + Y_3 + Y_4)$$
$$= (2+5+3+1) + (7+9+6+5)$$
$$= 11 + 27$$
$$= 38$$

The summation operator, Σ, is one of the most widely used symbols in statistics. It indicates the summing of the numbers that immediately follow it in an expression.

Variables

We have already used the term *data*, but where do data come from? They are collected on some characteristic of a group of individuals. If a characteristic is the same for every member of the group, then the characteristic is called a **constant**. But if the characteristic can take on different values, it is called a **variable**. Suppose we conduct a study of third-graders. Grade level is *constant* for this group. But a group of third-graders will differ in gender, height, intelligence, attitudes, and many other ways. These characteristics are therefore *variables* for this group.

> A constant is a characteristic that assumes the same value for all members of a group under study. A variable is a characteristic that can take on different values for different members of a group being studied.

The survey of high school seniors that we will use in chapter exercises provides other examples. We will be considering only the data collected from high school seniors, so grade level will be a constant for our data. There were many variables in the study, including the gender of the respondent, the parents' educational level, self-reported grade-point averages, and scores on several standardized tests.

Researchers distinguish independent variables from dependent variables. **Independent variables** are variables that the researcher controls or manipulates in accordance with the purpose of the investigation. They can be either manipulated or classifying variables. For example, in a study to determine the effect of level of drug dosage on the performance of some task, the researcher can manipulate the dosage level; thus dosage is a manipulated variable in the study. If the independent variable is a classifying variable, it simply categorizes the individuals under study. For example, if the researcher in our hypothetical drug study classifies the subjects by gender or age and then compares these groups' performance after giving them a certain dosage, then gender or age would be the classifying variable in the study. Or consider an investigation of the effects of three teaching methods on arithmetic achievement. Teaching method here is the independent variable, and the three methods are *levels* of the independent variable.

A **dependent variable** is a measure of the effect of the independent variable. In the preceding examples, performance on a task and arithmetic achievement are the dependent variables. As the independent variable is changed or varied, the researcher observes the changes in the dependent variable in order to determine how it is associated with changes in the independent variable.

> Independent variables are either manipulated or classifying variables. Dependent variables are measures of the effect of the independent variables.

Scales of Measurement

The data collected on variables are the result of measurement. **Measurement** is a process of assigning numbers to characteristics according to a defined rule. For example, a student's knowledge of chemistry is measured through a written test. Here the test score represents measurement of the characteristic of knowledge. The score is determined using a rule whereby numbers are assigned to test responses. The higher the score, the greater the student's measured knowledge of chemistry.

Clearly, not all measurement is the same. Some measurements are more precise than others. If we say that an individual is tall, that is not as precise as saying that the person is six feet, seven inches. Some measurement scales are more precise than others. And some characteristics are more amenable to precise measurement than others. Given an accurate bathroom scale, we can measure an individual's weight very precisely. But it is more difficult to measure anxiety level or opinions precisely.

The precision of measurement of a variable is important in determining what statistical methods should be used to analyze the data in a study.[3] Measurement scales of variables are classified in a hierarchy based on their degree of precision. This hierarchy includes the nominal, ordinal, interval, and ratio scales.

Nominal Scale

The least precise measurement scale is the nominal scale. In fact, to call it a measurement scale at all is to extend the definition of measurement to its lowest limit. A **nominal scale** classifies objects into categories based on some defined characteristic. Then the number of objects in each category is counted. For example, if we make a nominal measurement of automobiles, we might classify them into various makes and then count the number of cars of each make. Gender, color of eyes, and ethnic background are examples of variables measured on a nominal scale.

Notice that in nominal measurement there is no logical ordering of the categories. For example, when we classify automobiles, one make of automobile is not considered better, higher, or stronger than any other category. Similarly, we measure the religious backgrounds of a group of people by setting up categories such as Catholic, Protestant, and so on, but we cannot order these categories from lowest to highest or in any other kind of numerical ordering.

[3] A comprehensive discussion of the pros and cons of relating appropriate statistical procedures to the measurement scales of variables is beyond the scope of this book. Interested readers are referred to N. H. Anderson, "Scales and Statistics: Parametric and Nonparametric," *Psychological Bulletin* 58 (1961): 305–316; H. F. Kaiser, review of *Measurement and Statistics* by Virginia Senders, *Psychometrika* 25 (1960): 411–413; F. M. Lord, "On the Statistical Treatment of Football Numbers," *American Psychologist* 8 (1953): 750–751; S. S. Stevens, "Measurement, Statistics, and the Schemapiric View," *Science* 101 (1968): 849–856; S. S. Stevens, ed., "Mathematics, Measurement, and Psychophysics," *Handbook of Experimental Psychology* (New York: Wiley, 1951), pp. 1–49.

Thus nominal data have the following properties:

1. Data categories are mutually exclusive; that is, an object can belong to only one category.

2. Data categories have no logical order.

In short, a nominal scale simply classifies, nothing more.

Ordinal Scale

If we not only classify objects or characteristics but also give a logical order to the classification, then we have an **ordinal scale**. The measurement process is essentially the same in ordinal measurement as in nominal measurement: categories are identified, and numbers are assigned to the categories. But when a variable is measured on an ordinal scale, differences in the *amount* of the measured characteristic are discernible, and numbers are assigned according to that amount. If we use an ordinal scale, scores are ranked from highest to lowest and categorized within that order.

The letter-grading system — A, B, C, D, and F — is an example. If these are the grades in a history course, we know that an individual who gets an A has a *higher* level of achievement in history (according to the criteria used for measurement) than one who gets a B. Still, we cannot infer that the difference between an A and a B is the same as the difference between a B and a C. Similarly, we might set up a scale to measure aggressiveness and assign a 5 to indicate violently aggressive, a 4 to indicate very aggressive, a 3 to indicate moderately aggressive, and so on. The scale indicates that a person who receives a 5 shows more aggression than a person who receives a 3, but it has no meaning to say that one person has 2 units more aggression than the other, or exactly how much more aggressive one is than the other.

The properties of ordinal data are as follows:

1. Data categories are mutually exclusive.

2. Data categories have some logical order.

3. Data categories are scaled according to the amount of the particular characteristic they possess.

Interval Scale

The next level in the hierarchy of measurement is the interval scale. Variables measured on an **interval scale** have all the properties of those measured on ordinal scales, plus one additional property. The differences between levels of categories on any part of the scale reflect *equal differences* in the characteristic measured. That is, an equal unit is established in the scale. For this reason, the interval scale is also called an *equal unit scale*.

Temperature is a familiar example of a variable measured on an interval scale. Equal differences between any two temperatures are the same, regardless of the temperatures' positions on the scale. For example, the difference between temperatures of 85°F and 88°F is 3°, and the difference between temperatures of 23°F and 26°F is also 3°. It is important to note that in this example — and in all interval scales — the point 0 is just another point on the scale; it does not reflect the starting point of the scale or the total absence of the characteristic.

The properties of interval data are as follows:

1. Data categories are mutually exclusive.

2. Data categories have a logical order.

3. Data categories are scaled according to the amount of the characteristic they possess.

4. Equal differences in the characteristic are represented by equal differences in the numbers assigned to the categories.

5. The point 0 is just another point on the scale.

Ratio Scale

The highest level in the hierarchy of measurement scales is the **ratio scale**. Variables measured on the ratio scale exhibit the most precise type of measurement. This scale has one property in addition to the properties of the interval scale: a known, or true, zero point that reflects an absence of the characteristic measured. Thus, when we have variables measured on a ratio scale, we can make statements not only about the equality of the differences between any two points on the scale but also about the proportional amounts of the characteristic that two objects possess. For example, a temperature of 50°F is *not* twice as warm as a temperature of 25°F, but a bag of apples weighing 30 pounds does weigh twice as much as one weighing 15 pounds.

The properties of ratio data are as follows:

1. Data categories are mutually exclusive.

2. Data categories have a logical order.

3. Data categories are scaled according to the amount of the characteristic they possess.

4. Equal differences in the characteristic are represented by equal differences in the numbers assigned to the categories.

5. The point 0 reflects an absence of the characteristic.

We can summarize the four levels of measurement scales as follows:

The nominal scale categorizes without ordering the categories.

The ordinal scale categorizes and orders the categories.

The interval scale categorizes, orders, and establishes an equal unit in the scale.

The ratio scale categorizes, orders, establishes an equal unit, and contains a true zero point.

Qualitative and Quantitative Variables

Variables measured on the nominal or ordinal scales are often referred to as **qualitative variables** because the measurement consists of unordered or ordered (ranked) discrete categories, such as religious backgrounds or grades. In contrast, variables measured on the interval or ratio scales are **quantitative variables**. It is assumed that quantitative variables have underlying continuity; that is, they can take on any value on the measurement scale. Height, weight, temperature, and IQ are examples of quantitative, continuous variables.

Often the values reported for quantitative variables give the impression that the measure is discrete rather than continuous. For example, a thermometer can measure temperature only to the nearest degree or tenth of a degree; however, any temperature can theoretically occur. Similarly, we might say that a person has a score of 85 on a creativity measure. Although we use discrete units on the measurement scale, the variable itself — creativity — is assumed to have underlying continuity.

Qualitative variables consist of unordered or ordered (ranked) discrete categories; quantitative variables are assumed to have underlying continuity.

Populations and Samples

The terms *population* and *sample* are frequently used in the study of statistics. By definition, a **population** includes *all* members of a specified group. For example, all residents of New York City make up a population; all surgical patients in a given hospital at a particular time constitute another population, as do all students enrolled at Linkous School during the 1997–1998 academic year. In the survey of high school seniors, for instance, the population is all high school seniors in the United States in the spring of 1980.

Quite often we think of populations as containing large numbers of members. In fact, some populations are large, but others are not. The distinguishing characteristic of a population is *not* that it is large, but that all those who meet the definition for membership are included.

In many research situations, it is not feasible to involve or measure all members of a population. So a subset of the population, called a **sample**, is selected, and only the members of the sample are included in the research study. For example, a sample of 28,240 seniors was selected for participation in the survey of high school seniors.

A measure of a characteristic of a population is called a **parameter**. For example, in Chapter 3 we will compute the measure called a *mean*. If we determine from census data the mean income of all residents of New York City, the mean would be a parameter, or a descriptive measure, of this population. In contrast, a descriptive measure of a sample is called a **statistic**. (This is another, more specific meaning of the term *statistic*.) If we draw a sample of New York City residents and determine the mean income for this sample, this mean would be a statistic.

In order to distinguish between descriptive measures of a population and descriptive measures of a sample, we need two sets of symbols. *Greek* letters are used to denote parameters (population measures); *Roman* letters are used to denote statistics (sample measures). For example, in Chapter 3 we will use μ (mu, pronounced "mew") as the symbol for the mean of a population and $\overline{X}$ ("X bar") as the symbol for the mean of a sample.

A population includes all members of a defined group; parameters are descriptive measures of a population. A sample is a subset of a population; statistics are descriptive measures of a sample.

Descriptive Statistics and Inferential Statistics

The study of statistics is often described by two broad categories: descriptive statistics and inferential statistics. **Descriptive statistics** are used to classify and summarize numerical data; i.e., to describe data. **Inferential statistics** consist of procedures for making generalizations about a population by studying a sample from that population. With data from the sample, we use inferential statistics to draw conclusions about characteristics of the population based on the corresponding characteristics of the sample. In the terminology of the preceding section, we make inferences from the statistics of the sample to parameters of the population.

We do not simply compute statistics and guess at parameters. Rather, we test a hypothesis about the value of the parameter or estimate the value of the parameter based on the corresponding statistic. For example, one purpose of the survey of high school seniors was to estimate the average mathematics achievement. Suppose the directors of the survey are interested in testing the hypothesis that the mean mathematics score in the population is 14. First they determine the mean for the sample of 28,240 seniors ($\overline{X}$). Based on the value of $\overline{X}$, the survey directors would decide whether the hypothesis was tenable. Or they might instead (or subsequently) estimate the population mean (μ) using the sample mean ($\overline{X}$). We will discuss the specifics of this process at length, beginning in Chapter 8.

> Descriptive statistics is a collection of methods for classifying and summarizing numerical data. Inferential statistics is a collection of methods for making inferences about the characteristics of the population from knowledge of the corresponding characteristics of the sample.

It would be incorrect to say that descriptive and inferential statistics are mutually exclusive categories; they are not. Both are necessarily descriptive. In inferential statistics, the measures computed on the sample data are descriptive of the sample, but inferential methods have the additional function of generating inferences from the sample to the population.

Summary

In this chapter, we have explained how statistics can be used in the behavioral sciences. If you have limited mathematical skills, you may have felt swamped by the terminology, but familiarity with the terms explained here will greatly enhance your understanding of subsequent chapters. The mathematics required for understanding statistical concepts is no more complex than the three rules of summation.

Statistics helps us to understand data. The variables examined may be independent or dependent, qualitative or quantitative. And our measurement of them may have varying degrees of precision. If the scale of measurement is nominal, then we are only classifying the data. If we use ordinal measurement, then we are also ordering the classifications, ranking them in order of magnitude. Interval measurement scales have equal units, so that we can measure and compare the differences between points on the scale. Ratio scales have an additional property—a true zero point.

Usually we measure, not a population, but a subset of the population—a sample. Thus we need to use not only descriptive statistics, which describe data, but also inferential statistics, which allow us to draw conclusions about parameters (measures of a population) from statistics (measures of a sample).

Throughout this book, we will consider many examples using many different data. Some data will be hypothetical; others, such as those from the survey of high school seniors, will be actual data. The statistical procedures discussed will provide the tools for organizing and summarizing these data.

Exercises*

1. For each of the following, perform the basic arithmetic operations indicated.
   a. $7 + 4 + 53 =$ **b.** $232 - 87 =$
   c. $(5 + 6)(7 + 2) =$ **d.** $(20 + 5)(4 + 8) =$
   e. $(9)(6) + 32 =$ **f.** $283 + 97 - 106 =$
   g. $156/12 =$ **h.** $36 - (-16) =$

2. Solve the following, using absolute values as indicated.
   a. $36 + |-14| =$ **b.** $|-18| - |-43| =$
   c. $|95|/5 =$ **d.** $|9(-3)| + |-200| =$

3. Solve the following, using the exponents indicated.
   a. $7^3 =$ **b.** $(13^2) =$
   c. $\sqrt{64} =$ **d.** $(81)^{1/2} =$
   e. $5^5 =$ **f.** $(5 + 6)^2 =$
   g. $6 + 3^6 =$

*4. Use the rules for order of operations to solve the following:
   a. $(3 + 7)(8 + 1)$ **b.** $(3)(7) + (128/8)$
   c. $\sqrt{169}(3 - 15)$ **d.** $3^2 + 5(6) - (16 + 3)$

5. Multiply each of the following numbers by 5 and then use the first rule of summation to show that the sum of the products is equal to the sum of the original numbers multiplied by 5.

   4   6   12   2   2

6. Find $\Sigma(X + Y + Z)$ when

a. $X_1 = 4$	$Y_1 = 7$	$Z_1 = 52$
b. $X_2 = 9$	$Y_2 = 6$	$Z_2 = 37$
c. $X_3 = 7$	$Y_3 = 1$	$Z_3 = 22$
d. $X_4 = 5$	$Y_4 = 2$	$Z_4 = 41$
e. $X_5 = 3$	$Y_5 = 5$	$Z_5 = 36$

*Complete solutions for selected exercises, typically one per chapter, are provided in the "Answers to Exercises" section at the end of the book. The exercises with complete solutions are marked with an asterisk.

7. Classify each of the following as a discrete (qualitative) or a continuous (quantitative) variable.

   a. Intelligence

   b. Rank in class

   c. Ethnicity

   d. Marital status

   e. Mathematical ability

8. Identify the measurement scale for each of the following.

   a. Number of children in families

   b. Order of finish in the Boston Marathon

   c. Grading system (A, B, C, D, F)

   d. Level of blood sugar

   e. Time required to complete a maze

   f. Political party affiliation

   g. Amount of gasoline consumed

   h. Major in college

   i. IQ scores

   j. Number of fatal accidents

9. Let $A = -7$, $B = \sqrt{144}$, $C = |-3.42|$, and $D = (2.5)^2$. Carry out the following arithmetic operations, taking into consideration the rules of significant digits discussed in this chapter. (Use the positive square root.)

   a. $A + B =$

   b. $A - B =$

   c. $C(A + D) =$

   d. $(A)(C)(D) =$

   e. $|(A)(B)| =$

   f. $A/D =$

   g. $B/C =$

   h. $|A(C - D)| =$

   i. $|C/D| =$

   j. $(A)(C) =$

   k. $|(A)(D)| =$

10. Using the values for A, B, C, and D given in Exercise 9, illustrate the commutative, associative, and distributive properties. (Use the positive square root.)

    a. $A + (B + D) = (A + B) + D$

    b. $A(B \cdot D) = (A \cdot B)D$

    c. $D(A + B) = (D)(A) + (D)(B)$

11. Let $X_1 = 2$, $X_2 = 7$, $X_3 = 5$, $X_4 = 6$.

    a. Use the rules of summation to find ΣX_i.

    b. Find ΣX_i^2.

    c. Show that $(\Sigma X_i)^2 \neq \Sigma X_i^2$.

    d. If $c = 3$, show that $\Sigma cX = c\Sigma X$.

2

Organizing and Graphing Data

In research studies in the behavioral sciences, data generally consist of numbers. These numbers represent the measurement of selected variables for a group of individuals. For example, 180 college freshmen enrolled in an introductory psychology course undoubtedly differ in gender, age, composite Scholastic Aptitude Test (SAT) scores, and midterm and final examination scores. A researcher studying them would come up with numbers that represented how the students differ in terms of these characteristics, such as how many students are 19 years old, how many received a grade of 50 on the midterm examination, how many received a grade of 50 on the final examination, and so on. But by themselves, these numbers are not very interesting or informative. To learn anything from them, the researcher must first organize the data and then present them in a meaningful way.

In this chapter, we will discuss how to code data. Then we will show how to organize and summarize data using *stem-and-leaf displays* and *frequency distributions*, and how to depict data in *graphs*. The procedures we will discuss are the first steps in providing a meaningful representation of a data set.

Coding Data

The purpose of any statistical analysis is to organize and summarize collected data. Before beginning the analysis, the researcher must prepare the data. Today this process usually involves entering the data directly into a computer file through a computer terminal or microcomputer.

Suppose a professor has collected and recorded data on the age, gender, composite SAT scores, and midterm and final examination scores of 180 freshman psychology students. Table 2.1 shows these data for five students. Now suppose the professor wants to enter the data into a computer file. To do so, the professor must first make two decisions: (1) how to code the data, and (2) where the data will go on the data record.

Coding data involves assigning numerical values to nonnumeric categories. In this example, age, composite SAT scores, and midterm and final examination scores all have numerical values, so no coding system is needed. Instead, the actual values are recorded. But the name and gender of each student must be coded.

How does the professor code the name and gender of each student? For names, the professor might assign Sue an ID number of 001; Mary, 002; and so on. Several coding schemes could be used for the gender of each student. In this example, we have coded female as 0 and male as 1. But we could have used 1 for male and 2 for female. Any coding system is arbitrary.

The process of coding data involves assigning numerical values to nonnumeric categories of a variable.

TABLE 2.1
Data for Five Freshman Psychology Students

Name	Gender	Age	SAT	Midterm	Final
Sue	Female	19	960	22	49
Mary	Female	20	1040	31	62
Joe	Male	19	980	34	58
Amy	Female	19	1130	42	73
Bill	Male	20	1120	38	66

Now consider how the professor enters the data into the computer, constructing a *data file*. The format used is arbitrary. By convention, the data for each student are given on a single line, or row, called a *data record*. Thus each row presents data for a different student, and the columns record data for different variables, such as name, gender, age, SAT, midterm, and final.

With the expansion and improvement of the application of computer technology, especially microcomputer technology, to statistical analysis, different methods can be used to enter data into a computer file. Many statistical software packages have data entry and data management subroutines built into the packages. The user of the software packages need only refer to the instruction manual for determining how data are to be entered.

These software packages can also be used to analyze data that have been stored in computer files external to the packages; these files are *imported* into the system prior to analyzing the data. Most of these packages can use data files that have been stored in **ASCII format**, such as illustrated in the above computer file, or in a spreadsheet. First, consider data stored in an ASCII file. These data could have been entered into the file using a word-processing program. When all the data have been entered, the file must be saved in ASCII format. Using such an approach with the data for the five psychology students, we would use the following format for each data line.

Data	Columns	Data	Columns
ID	1–3	SAT	10–13
Gender	5	Midterm	15–16
Age	7–8	Final	18–19

That is, the student's ID number is entered in columns 1 to 3, gender in column 5, age in columns 7 to 8, and so on. Thus, we enter the data records for these five students into the computer file in the following way:

Columns of data record

```
1    5   10    15    20    25    30    35    40
|____|____|____|____|____|____|____|____|____|

001  0 19   960  22 49
002  0 20  1040  31 62
003  1 19   980  34 58
004  0 19  1130  42 73
005  1 20  1120  38 66
```

In a data file, each row contains the data record for a single subject, and the columns contain the data for the variables.

			Worksheet1					
	A	B	C	D	E	F	G	H
1								
2								
3								
4								
5								
6								
7								
8								
9								
10								
11								
12								
13								
14								
15								
16								
17								
18								
19								
20								
21								
22								
23								
24								
25								
26								
27								

FIGURE 2.1
Example of spreadsheet

The use of spreadsheets for data entry and data management is becoming more common. A **spreadsheet** is a two-dimensional array of cells, as is illustrated in Figure 2.1. For the purpose of data entry, the *rows* of the spreadsheet are used to represent the data lines mentioned above. The *columns* of the spreadsheet represent the variables to be included in the data record. For the above example, we have included the original data from Table 2.1 as well as both an ID number and the coding for gender. Figure 2.2 illustrates how these data have been entered into the spreadsheet.

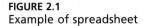

Data – Psychology Students

Name	ID Number	Gender	Gender-Code	Age	SAT	Midterm	Final
Sue	1	Female	0	19	960	22	49
Mary	2	Female	0	20	1040	31	62
Joe	3	Male	1	19	980	34	58
Amy	4	Female	0	19	1130	42	73
Bill	5	Male	1	20	1120	38	66

FIGURE 2.2
Psychology student data entered in a spreadsheet

Once the data have been entered into the computer file and imported into the statistical software package, we can use one of the many subroutines for analyzing them. The specific commands necessary for using the various subroutines are discussed and illustrated in Appendix B. However, in order to be able to analyze data on our own and to understand the printouts from the statistical packages, we need to understand the basic procedures for organizing and summarizing data.

Stem-and-Leaf Displays

The distribution of scores for a particular variable is best illustrated with a graph. Graphing techniques will be discussed in a later section of this chapter, but it is important that we set the stage for these discussions. For example, **stem-and-leaf displays** are one useful strategy for "exploring" distributions of data. These displays are one of many techniques developed by Dr. John W. Tukey in the late 1970s. In his classic book *Exploratory Data Analysis* (1977), Tukey presented a series of relatively simple, but effective, techniques for exploring data.

Consider a study by the Chamber of Commerce of a city in the upper Midwest. The purpose of the study was to illustrate seasonal variation in the high temperatures. The data collected for this study are found in Table 2.2; these data are the high-temperature readings for each Saturday for the previous year. The stem-and-leaf display for these data is found in Figure 2.3.

TABLE 2.2
Chamber of Commerce Study of High Temperatures on 52 Saturdays

January	April	July	October
22	43	82	72
34	47	87	68
28	52	79	59
18	55	91	49
31			53
February	**May**	**August**	**November**
34	54	88	45
28	57	84	41
36	68	93	38
42	62	87	42
	71	84	
March	**June**	**September**	**December**
49	71	81	29
39	67	79	38
51	74	72	24
53	80	84	16

9	1 3
8	0 1 2 4 4 4 7 7 8
7	1 1 2 2 4 9 9
6	2 7 8 8
5	1 2 3 3 4 5 7 9
4	1 2 2 3 5 7 9 9
3	1 4 4 6 8 8 9
2	2 4 8 8 9
1	6 8

FIGURE 2.3
Stem-and-leaf display of a Chamber of Commerce study of high temperatures on 52 Saturdays

The first step in developing stem-and-leaf displays is to determine the **stem**. In this example, we will use the "tens" digit for the stems, i.e., $9 = 90$, $8 = 80$, $7 = 70$, etc. The **leaves** for the display are the "units" digit. In Figure 2.3, the first line (9 | 1 3) indicates the two temperatures 91 and 93. The second line (8 | 0 1 2 4 4 4 7 7 8) indicates the temperatures 80, 81, 82, 84, 84, 84, 87, 87, and 88. Overall, these data tend to indicate that this city had a relatively moderate climate for the previous year. Referring back to the original data in Table 2.2, the stem-and-leaf display confirms the very mild, almost warm, fall.

The above example illustrates a simple stem-and-leaf display with nine stems. When there are fewer stems easily identified in the original data, there are techniques for "stretching" the stem-and-leaf display. Similarly, when there are many more stems, there are techniques for "squeezing" the display. The reader is referred to Tukey's book for an in-depth discussion of these techniques.

Stem-and-leaf displays provide the data analyst with a quick way to illustrate a distribution of scores while maintaining the actual scores in the displays. However, the effectiveness of this technique is limited to distributions with small numbers of observations. These displays are most often used in the preliminary exploration of data and are rarely presented in a final report or presentation of data. The following sections contain discussions of the techniques most often used for presenting data in a final report or published article.

Frequency Distribution

In the last section, which dealt with stem-and-leaf displays, we emphasized the need for organizing and summarizing data for ease of interpretation. However, we indicated that the stem-and-leaf displays would be arduous for large data sets. The more common method for displaying data such as these is to group the data into intervals of scores and then to develop a frequency distribution. A **frequency distribution** is an arrangement of values that shows the number of times a given score or group of scores occurs. It is generally considered a first step in any data analysis.

A frequency distribution is a tabulation that indicates the number of times a given score or group of scores occurs.

Consider the final examination scores listed in Table 2.3. If you are a student in the class and obtain a score of, say, 55, you might wonder how many others did better. A frequency distribution would answer your question.

To construct a frequency distribution from the scores in Table 2.3, we find the highest and lowest scores, list these scores and all the scores in between, and then count and record the number of times each score occurs. Table 2.4 shows the resulting frequency distribution for these scores. It lists all the scores from the final examination and the frequency of each score — that is, the number of times each score occurs. (Notice that frequency is given the symbol f in the table.) There are four scores of 64, two scores of 58, and so on.

Class Intervals

In Table 2.4 we include all of the actual test scores. In other words, we classify the data into as many groups, or *classes*, as there are scores. Constructing a frequency distribution this way can be very long and tedious. We might instead reduce the number of classes by grouping several actual scores into an *interval* of scores. For example, one interval might include all scores between 65 and 69 inclusive; another, the scores between 60 and 64 inclusive; and so on. These intervals are called **class intervals**.

Table 2.5 shows a frequency distribution using class intervals of width 5 — in other words, intervals that include five units on the scale of measurement. Notice

TABLE 2.3
Final Examination Scores for Freshman Psychology Students

68	52	69	51	43	36	44	35	54	57	55	56
55	54	54	53	33	48	32	47	47	57	48	56
65	57	64	49	51	56	50	48	53	56	52	55
42	49	41	48	50	24	49	25	53	55	52	56
64	63	63	64	54	45	53	46	50	40	49	41
45	54	44	55	63	55	62	56	50	46	49	47
56	38	55	37	68	46	67	45	65	48	64	49
59	46	58	47	57	58	56	59	60	62	59	63
56	49	55	50	43	45	42	46	53	40	52	41
42	33	41	34	56	32	55	33	40	45	39	46
38	43	37	44	54	56	53	57	57	46	56	45
50	40	49	39	47	55	46	54	39	56	38	55
37	29	36	30	37	49	36	50	36	44	35	45
42	43	41	42	52	47	51	46	63	48	62	49
53	60	52	61	49	55	48	56	38	48	37	47

TABLE 2.4
Frequency Distribution of Final Examination Scores

Score	f	Score	f	Score	f
69	1	53	7	37	5
68	2	52	6	36	4
67	1	51	3	35	2
66	0	50	7	34	1
65	2	49	11	33	3
64	4	48	8	32	2
63	5	47	7	31	0
62	3	46	9	30	1
61	1	45	7	29	1
60	2	44	4	28	0
59	3	43	4	27	0
58	2	42	5	26	0
57	6	41	5	25	1
56	14	40	4	24	1
55	12	39	3		
54	7	38	4		

that the only difference between the frequency distribution in Table 2.4 and the one in Table 2.5 is the width of the class intervals. In Table 2.4 the interval width is 1, whereas in Table 2.5 the interval width is 5.

There are two general rules for developing frequency distributions that use class intervals other than the intervals of the original scores.

1. For large data sets (100 or more observations) with a wide range of scores, 10 to 20 intervals work well. For smaller data sets (such as those in the exercises

TABLE 2.5
Frequency Distribution of Final Examination Scores Using Class Intervals

Class Interval	f
65–69	6
60–64	15
55–59	37
50–54	30
45–49	42
40–44	22
35–39	18
30–34	7
25–29	2
20–24	1

at the end of the chapter), 6 to 12 intervals work well. Our example uses 10 class intervals of width 5. If we use class intervals of width 3, 16 intervals would be required. By observing this first rule, we obtain a manageable number of intervals without distorting the general distribution of the scores.

2. Whenever possible, the width of the class interval should be an odd, rather than an even, number. This rule makes computations easier, because it ensures that the midpoint of the interval will be a whole number rather than a fraction. The interval **midpoint** is defined as the point halfway through the interval. For example, in Table 2.5 the midpoint of the first interval is 67.

Exact Limits of the Class Interval

We have recorded the final examination scores as discrete values, but the variable final examination scores can be considered to be a continuous, not discrete, variable. (To review the difference between continuous and discrete variables, see Chapter 1.) That is, although we may record a score as a whole number, the recorded value actually represents a value that falls within certain limits, called the **exact limits**. For example, in Tables 2.3 and 2.4, a score of 53 represents a score somewhere between 52.5 and 53.5; a score of 53.7 is recorded as a score of 54. Thus the values 52.5 and 53.5 represent the exact limits of the score 53.

Usually the exact limits of a score extend from one-half unit below to one-half unit above the recorded measure. Suppose we measure and record the examination scores more precisely, to the nearest tenth of a point, recording scores like 53.7. Then the exact limits of the score 53.7 are 53.65 to 53.75.

Similarly, it is necessary to distinguish between the **score limits** of the class interval in a frequency distribution and the exact limits. In Table 2.5, for example, the numbers 50 to 54 represent the score limits of the fourth class interval, but the exact limits are 49.5 to 54.5. The exact limits are one-half unit below and one-half unit above the score limits of the class interval. If our measurement is more precise — say, to the nearest tenth of a point — then the exact limits change accordingly. For example, if the score limits for a class interval are 17.3 to 18.7, the exact limits for the interval are 17.25, to 18.75.

Table 2.6 shows the frequency distribution of the 180 examination scores with the exact limits given for ten class intervals. The table also gives the midpoints of the class intervals. These midpoints are the same whether we use exact limits or score limits.

Exact limits for class intervals are based on the assumption that the variable under study is continuous, even though the measurement may be in whole numbers.

TABLE 2.6
Frequency Distribution of Final Examination Scores Including the Exact Limits and Midpoints

Class Interval	Exact Limits	Midpoint	f
65–69	64.5–69.5	67	6
60–64	59.5–64.5	62	15
55–59	54.5–59.5	57	37
50–54	49.5–54.5	52	30
45–49	44.5–49.5	47	42
40–44	39.5–44.5	42	22
35–39	34.5–39.5	37	18
30–34	29.5–34.5	32	7
25–29	24.5–29.5	27	2
20–24	19.5–24.5	22	1

Assumptions for Class Intervals Greater than One Unit

Reclassifying data into class intervals with a width greater than one unit often makes the data more manageable and easier to interpret. But this reclassification also has an obvious disadvantage: some specific information is lost. For example, the frequency distribution in Table 2.5 no longer specifies the exact number of students with scores of 65, 66, 67, 68, or 69; it tells us only that there are six scores in the interval from 65 to 69. Furthermore, scores that originally differed, such as 65 and 67, are grouped in the same interval. When we represent a frequency distribution graphically or compute certain characteristics of the distribution, we must ignore this loss of information and make two assumptions.

The first assumption is that, in any class interval, the scores are uniformly distributed between the exact limits of the interval. For example, in Table 2.6 there are 37 scores in the interval 55 to 59, which has exact limits 54.5 to 59.5. We assume that these 37 scores are distributed over this interval as follows:

Interval	f
54.5–55.5	7.4
55.5–56.5	7.4
56.5–57.5	7.4
57.5–58.5	7.4
58.5–59.5	7.4
Total	37.0

The second assumption is that, whenever a single score must represent a class interval, the interval midpoint is the representative score. In other words, we assume that all the scores within the interval can be represented adequately by its midpoint. For the class interval just illustrated, we assume that the midpoint, 57, can adequately represent all 37 scores.

> The scores within any class interval are assumed to be uniformly distributed throughout the interval, and all are assumed to be adequately represented by the midpoint.

Graphing Data

So far we have seen how to organize and summarize scores in a frequency distribution. But we have not considered *how* the scores are distributed. For example, are the students' examination scores evenly distributed, or are there heavy concentrations of scores in certain parts of the distribution? Do many scores fall near the middle? Or is there a heavy concentration of very low scores?

Questions like these are easier to answer if we construct a picture of the data — that is, a graph. Seeing the distribution can enhance our understanding of the data. Before discussing how to depict a frequency distribution graphically, we will discuss some general rules and conventions used in graphing and look at two types of graphs.

Constructing a Graph

Practically everyone encounters graphs at one time or another. The word *graph* comes from the Greek word meaning "to draw or write." We define a **graph** as a pictorial representation of a set of data. Common examples are the graphs used in newspapers and weekly newsmagazines to depict changes in the prime interest rate over the past 12 months. These graphs show the relationship between two variables — the prime interest rate and time.

The Axes   The first step in developing a graph that shows the relationship between two variables is to draw the *coordinate axes*, which are two straight lines, one horizontal and one vertical. The scales of measurement for the independent and dependent variables (defined in Chapter 1) are marked off along these two axes. Generally, the scale of measurement for the independent variable is placed on the horizontal axis and that for the dependent variable, on the vertical axis. Using traditional mathematical terminology, we call the horizontal axis the *X axis*, or the *abscissa*. The vertical axis is called the *Y axis*, or the *ordinate*, as Figure 2.4 shows. The intersection of the two axes, called the *origin*, is the point where $X = 0$ and $Y = 0$.

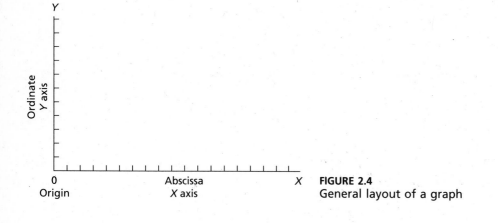

A graph can be used to illustrate the relationship between an independent variable and a dependent variable.

An Example   Suppose we are studying the relationship between annual income and level of education. Level of education is the independent variable; annual income, the dependent variable. First we code the level of education. We could use the following categories for coding:

1   Had fewer than 8 years of formal education

2   Had more than 8 years of formal education, but no high school diploma

3   Finished high school with diploma

4   Completed postsecondary vocational training or associate's degree program

5   Completed bachelor's degree program

6   Completed master's degree program

7   Completed post-master's degree program (M.D., Ph.D.)

These codes are used to mark off the level of education on the X axis in Figure 2.5. Note that this axis represents an ordinal scale of measurement, as defined in

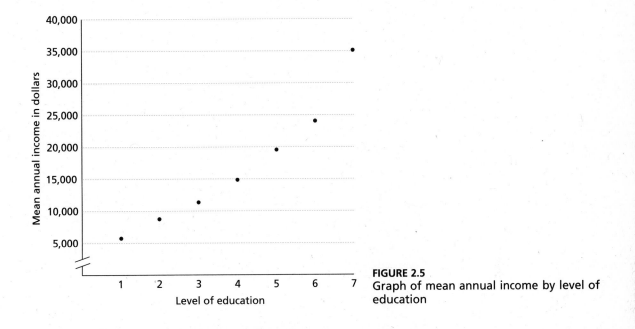

FIGURE 2.5
Graph of mean annual income by level of education

Chapter 1. That is, even though the numbers 1 through 7 are equally spaced along the X axis in Figure 2.5, the numbers do not necessarily reflect equal differences between the levels of education.

Table 2.7 gives the mean annual incomes for different education levels. Since the highest income in the table is $35,525, we mark off a range of mean annual incomes up to $40,000 on the Y axis of Figure 2.5. The measurement of this variable along the Y axis is an example of the ratio scale as defined in Chapter 1. Notice that the zero point is not included along the Y axis and that the axis is

TABLE 2.7
Mean Annual Income for Various Levels of Education

Level of Education	Mean Annual Income
1	$ 7,150
2	8,775
3	12,125
4	15,650
5	20,275
6	24,850
7	35,525

broken just above the origin. Even though the line is broken, it is still assumed that the axis would continue down to the origin. The line is broken only to avoid distorting the data of the graph.

Next we place dots on the graph to represent the data in Table 2.7. For example, according to Table 2.7, those with less than an eighth-grade education (coded as 1) have a mean annual income of $7,150; therefore we place a dot on the graph above number 1, corresponding to the height of $7,150 on the Y axis. Thus the dots on Figure 2.5 illustrate the relationship between level of education and mean annual income. These data indicate that, the more education people in this study had, the greater their mean annual income.

Data Curves   We can connect the data points in Figure 2.5 to illustrate the trend in the data more clearly, as shown in Figure 2.6. By connecting the data points, we portray the data as a **data curve**. This technique makes it easier to get a feeling for the data. However, a data curve is not an actual curve but only a set of data points connected by straight-line segments.

Data curves are formed by connecting the data points of a graph with straight-line segments, illustrating the relationship between two variables.

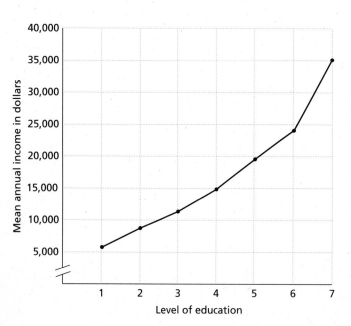

FIGURE 2.6
Data curve for mean annual income (using data from Table 2.7)

Avoiding Distortions   Graphs can exaggerate small differences in data, because the scales of measurement along the axes are quite arbitrary. Using graphs deceptively by exaggerating the scales, analysts with different intentions can enlist the same data to support opposite points of view. For example, by stretching out the measurement scale on the Y axis, we can make a small difference look quite large. Conversely, by condensing the measurement scale, we can make a large difference look quite small.

There is a simple rule that helps ensure consistency and avoid distortion in graphs. It states that the height of the Y axis should be approximately three-quarters the length of the X axis. Using this **three-quarter-high rule** reduces the possibility of misrepresentation and hence the misinterpretation of data.

The three-quarter-high rule, which states that the Y axis should be approximately three-quarters the length of the X axis, should be followed to ensure appropriate representation of data.

Bar Graphs

Suppose a consumer advocate studies the transmission repair rates for five medium-priced domestic automobiles and obtains the data given in Table 2.8. How do we graph these data?

TABLE 2.8
Transmission Repair Rates for Domestic Automobiles

Make of Automobile	Repair Rate (per 1,000 sold)
A	4.2
B	6.8
C	3.3
D	0.4
E	1.2

Notice first that the independent variable — make of automobile — is a qualitative variable measured on a nominal scale. Whenever the independent variable is a nominal variable, a **bar graph** should be used to represent its relationship to other variables. Figure 2.7 is a bar graph of the data given in Table 2.8. The scale for the repair rate is placed on the Y axis, so that the heights of the bars indicate repair rates. The placement of automobile types on the X axis is arbitrary.

Notice that the bars in Figure 2.7 are separated. The bars should be separated by a space whenever nominal data are graphed, because measurements on a nominal scale are categorical. Thus, the graph reflects the fact that there is no continuity among the categories on the scale.

> Bar graphs are used to illustrate the relationship between two variables when the scale of measurement of the independent variable is nominal.

Scatterplots

In the graphs discussed so far, the independent variable has been a qualitative variable measured on either a nominal or an ordinal scale. But suppose we are interested in the relationship between two *quantitative* variables, both measured on either an interval or a ratio scale. The data for two variables can be graphically displayed in a **scatterplot**.

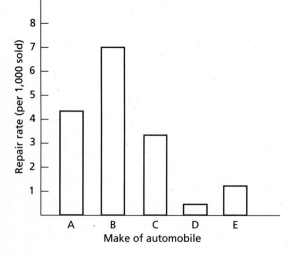

FIGURE 2.7
Bar graph of repair rate by make of automobile

TABLE 2.9
Academic Ability Scores (X) and Semester Hours of Undergraduate Mathematics (Y) for 20 Students

Student	Academic Ability Score, X	Semester Hours of Mathematics, Y	Student	Academic Ability Score, X	Semester Hours of Mathematics, Y
1	54	18	11	39	18
2	29	3	12	42	15
3	42	14	13	55	20
4	60	23	14	47	15
5	33	15	15	50	16
6	28	7	16	29	12
7	56	22	17	34	9
8	48	18	18	48	22
9	55	19	19	56	25
10	59	25	20	46	23

For example, consider the relationship between academic ability and the number of semester hours of mathematics taken as an undergraduate. Hypothetical data for 20 students are given in Table 2.9. Notice that, for these data, neither variable is designated as independent or dependent; we arbitrarily call academic ability variable X and the number of semester hours of undergraduate mathematics variable Y. The data for this example can be graphically displayed in a scatterplot, as shown in Figure 2.8. Each dot in the paired scatterplot represents observations for one

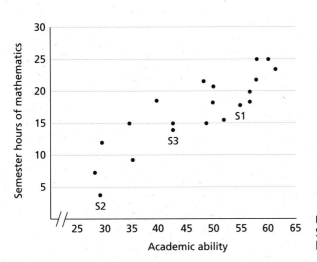

FIGURE 2.8
Scatterplot of academic ability and semester hours in mathematics for 20 students

student; the dots for students 1, 2, and 3 are identified in the figure. For student 1, the dot corresponds to the score of 54 on the X variable (academic ability) and the score of 18 on the Y variable (number of semester hours of mathematics). The dot for student 2 corresponds to scores of 29 and 3 on the X and Y variables, respectively; the dot for student 3, to scores of 42 and 14.

Inspection of the data in Table 2.9 indicates that low academic ability scores tend to be associated with fewer semester hours of undergraduate mathematics. Similarly, high academic ability scores tend to be associated with more semester hours of undergraduate mathematics. Notice that in Figure 2.8 this positive relationship between the two variables is represented by a pattern of data points from the lower left to the upper right of the graph. We will discuss the relationship between two variables in greater detail in Chapter 5.

A scatterplot is a graph that illustrates the relationship between two quantitative variables.

Graphing Frequency Distributions

So far we have discussed procedures for organizing and summarizing data in a frequency distribution and for graphing data. We now combine these procedures to illustrate how to graph frequency distributions. Graphing frequency distributions enhances our understanding of the data by showing the shape of the distribution, allowing us to see how the scores are distributed throughout the range of values.

Histograms

A **histogram** is a type of bar graph that depicts the frequencies of individual scores or scores in class intervals by the length of its bars. The scale of measurement on the Y axis is the frequency of the scores; the scale of measurement on the X axis is the range of scores for the variable under consideration. A histogram of the final examination scores we considered earlier is given in Figure 2.9. Notice that the plot points on the X axis are the exact limits of the class intervals and that the bars are *not* separated, illustrating that the final examination scores represent a continuous quantitative variable.

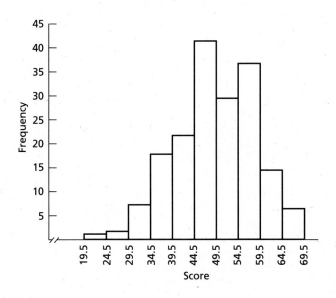

FIGURE 2.9
Histogram of final examination scores

Frequency Polygons

We can also depict a frequency distribution by using a frequency polygon. For the **frequency polygon**, we assume that the scores in the class interval can be represented by the midpoint. For each interval we plot the frequency of scores at the midpoint and then connect the midpoints with straight lines. Figure 2.10,

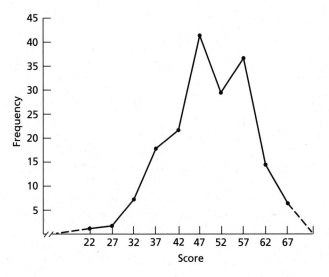

FIGURE 2.10
Frequency polygon of final examination scores

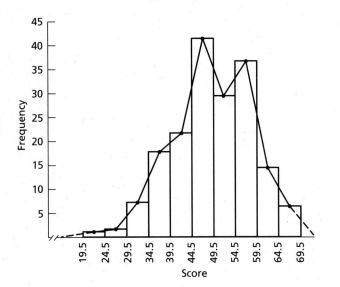

FIGURE 2.11
Frequency polygon superimposed on histogram of final examination scores

which is the frequency polygon for the 180 final examination scores, provides an example. Figure 2.11 shows the relationship of the histogram to the frequency polygon. Notice that we have used the three-quarter-high rule in Figures 2.9, 2.10, and 2.11 to avoid distorting the data.

Frequency distributions can be graphically depicted by using either a histogram or a frequency polygon. For both, it is important to use the three-quarter-high rule in order to avoid distorting the data.

The Cumulative Frequency Distribution and Its Graph

Very often we are interested in knowing how many scores are higher or lower than a particular score. For example, our hypothetical psychology professor might want to know what number or percentage of the final examination scores was higher than 50. To answer questions like this, we use yet another kind of graph — a graph of the cumulative frequency distribution or the cumulative percentage.

The **cumulative frequency distribution** is constructed by adding the frequency of scores in any class interval to the frequencies of all the class intervals below it on the scale of measurement. Table 2.10 provides an example. To construct this cumulative frequency distribution, we begin at the lowest class interval, 20–24, which has a frequency of 1. For the next class interval, 25–29, we add the frequency of the interval, which is 2, to the frequency of the 20–24 class interval,

TABLE 2.10
Frequency Distribution of Final Examination Scores, Including Cumulative Frequency and Cumulative Percentage

Class Interval	Exact Limits	Midpoint	f	cf	%	c%
65–69	64.5–69.5	67	6	180	3.33	100.00
60–64	59.5–64.5	62	15	174	8.33	96.67
55–59	54.5–59.5	57	37	159	20.56	88.34
50–54	49.5–54.5	52	30	122	16.67	67.78
45–49	44.5–49.5	47	42	92	23.33	51.11
40–44	39.5–44.5	42	22	50	12.22	27.78
35–39	34.5–39.5	37	18	28	10.00	15.56
30–34	29.5–34.5	32	7	10	3.89	5.56
25–29	24.5–29.5	27	2	3	1.11	1.67
20–24	19.5–24.5	22	1	1	0.56	0.56

which is 1, to obtain the cumulative frequency (cf) of the interval, 3. We proceed in the same way for each class interval, adding the frequency for that interval to the frequencies for all the preceding class intervals to obtain the cumulative frequency.

Table 2.10 also contains the cumulative percentages ($c\%$) of the distribution of final examination scores. The *cumulative percentage* gives the cumulative frequency in percentage form. It is computed by dividing the cumulative frequency of a class interval by the total frequency of the distribution and then multiplying by 100. For example, for the class interval 30–34 in Table 2.10, the cumulative frequency (10) divided by the total frequency (180) multiplied by 100 equals 5.56, which is the cumulative percentage for the interval.

The graph of a cumulative percentage distribution, the cumulative frequency polygon, is called an **ogive**; the ogive for the distribution in Table 2.10 appears in Figure 2.12. Notice that the upper limits of the respective class intervals are used in the ogive. As we will see in Chapter 3, an ogive can be used to determine various percentile points in a distribution of scores.

The ogive, or cumulative frequency polygon, is the graph of a cumulative frequency distribution. It is useful for determining the various percentile points in a distribution of scores.

Shapes of Frequency Distributions

Frequency distributions have an unlimited number of possible shapes. The shape depends on how the scores are distributed on the scale of measurement. We can

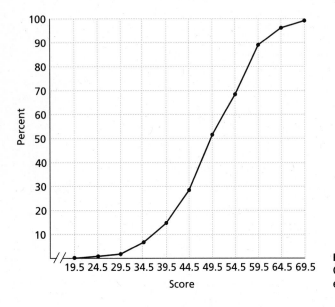

FIGURE 2.12
Ogive (cumulative frequency polygon) of final examination scores

get a general feeling for how frequency distributions vary by comparing the shapes of their graphs.

The shape of a distribution can be classified as uniform or nonuniform, skewed or symmetric. If the scores are evenly distributed throughout a distribution, their distribution is called a **uniform**, or **rectangular**, **distribution** and the frequency polygon will look something like distribution A in Figure 2.13. In contrast, suppose a frequency distribution has many scores toward the lower end of the scale of measurement and progressively fewer scores toward the upper end, as illustrated by distribution B in Figure 2.13. This is a **skewed distribution**, said to be skewed to the right, or *positively skewed*. On the other hand, if there are many scores at the upper end of the distribution and progressively fewer scores at the lower end, as shown in distribution C, the distribution is said to be skewed to the left, or *negatively skewed*. Finally, when the two halves of a graph would coincide if the graph were folded along a central line, we have a **symmetric distribution**.

Distribution D in Figure 2.13 illustrates the most familiar symmetric distribution. It is a *bell-shaped* distribution known as the *normal distribution*, in which there are many scores in the middle of the scale and progressively fewer scores at either extreme. We will discuss the normal distribution in detail in Chapter 4.

Symmetric distributions like those in distributions D, E, and F may vary in **kurtosis** — that is, degree of peakedness. If the vast majority of the scores tend to be located at the center of the distribution, as in distribution E, the distribution is called *leptokurtic*. If the scores are distributed more uniformly, but many scores still cluster at the center (distribution F in Figure 2.13), the distribution is called *platykurtic*. Distributions with a moderate degree of peakedness, such as distribution D, are called *mesokurtic*.

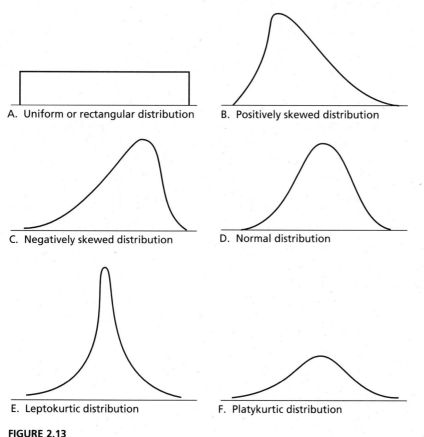

FIGURE 2.13
Various shapes of frequency polygons

The shape of a frequency distribution depends on how the scores are distributed throughout the range of scores. Distributions may be uniform or nonuniform, skewed or symmetric in shape, and distributions may exhibit different degrees of kurtosis.

Computer Example: The Survey of High School Seniors

We introduced a longitudinal survey of high school seniors data set in Chapter 1; we have discussed the data set in more detail in Appendix B. In the Appendix, we defined the variables in the survey of high school seniors data set and the coding scheme used for the variables. The frequency distributions generated for selected variables in the survey of high school seniors data set using the FREQUENCIES procedure in the Statistical Package for the Social Sciences (SPSS) are found in Table 2.11.

Consider the frequency distribution for the variable, GENDER. This frequency distribution indicates that 280 of the 500 high school seniors were Female (56%) and 220 were Male (44%). Note that the frequency distribution contains the frequency within each category the percentage of this frequency, as well as the cumulative frequency and the cumulative percent.

Now consider the frequency distributions for the variables, MAED (Mother's education) and FAED (Father's education). In comparing the two frequency distributions, we note that 114 (22.8%) of the mothers of the high school seniors had less than a high school education compared to 132 (26.4%) of the fathers, i.e., the high school drop out rate for the fathers was higher than for the mothers. These data also indicate that 40.4% of the mothers completed high school while only 25.8% of the fathers were high school graduates. But note that higher percentages of fathers completed college or advanced degrees (15.2%, 7.2%, and 4.4%) compared to the mothers (9.2%, 4.6%, and 1.0%).

In Chapter 14, we will use MAED and FAED as independent variables in the computer exercises. But in order to do so, we have to combine certain categories for both variables. We have included this new frequency distribution below. The recoding scheme for both variables follows along with the new variable names.

MAED/FAED	*MAEDNEW/FAEDNEW*	
2	1	Less than High School
3	2	High School Graduate
4,5,6,7	3	Some Post-Secondary Education
8,9,10	4	College Graduate & Beyond

For the variables, ALG1, ALG2, GEO, TRIG, and CALC, you will note that most of the sudents had taken Algebra 1 (82%) but very few students had taken Trigonometry (28.4%) or Calculus (9.0%). The percentages of the sudents who took Algebra 2 and Geometry were 53.4% and 60.0%, respectively.

Finally, consider the student-reported mathematics grades (MATHGR) and student-reported high school grades (GRADES). Note that for MATHGR, only 49.6% of the sudents indicated that they received "mostly As and Bs." Comparing that percentage with the percentage for total grades, we need to combine the first three categories for GRADES. By doing so, we find that 60.2% (11.8% + 24.2% + 24.2%) of the students received "mostly As and Bs" for the variable GRADES.

TABLE 2.11

```
GENERAL      gender
```

Value Label	Value	Frequency	Percent	Valid Percent	Cum Percent
female	0	280	56.0	56.0	56.0
male	1	220	44.0	44.0	100.0
		-------	-------	-------	
	Total	500	100.0	100.0	

```
Valid cases     500    Missing cases      0
```

```
ALG1       Algebra enroll
```

Value Label	Value	Frequency	Percent	Valid Percent	Cum Percent
not taken	0	90	18.0	18.0	18.0
taken	1	410	82.0	82.0	100.0
		-------	-------	-------	
	Total	500	100.0	100.0	

```
Valid cases     500    Missing cases      0
```

```
ALG2       Algebra2 enroll
```

Value Label	Value	Frequency	Percent	Valid Percent	Cum Percent
not taken	0	233	46.6	46.6	46.6
taken	1	267	53.4	53.4	100.0
		-------	--------	-------	
	Total	500	100.0	100.0	

```
Valid cases     500    Missing cases      0
```

```
GEO        Geometry enroll
```

Value Label	Value	Frequency	Percent	Valid Percent	Cum Percent
not taken	0	200	40.0	40.0	40.0
taken	1	300	60.0	60.0	100.0
		-------	-------	-------	
	Total	500	100.0	100.0	

```
Valid cases     500    Missing cases      0
```

TABLE 2.11
(continued)

```
TRIG         Trigonometry
```

Value Label	Value	Frequency	Percent	Valid Percent	Cum Percent
not taken	0	358	71.6	71.6	71.6
taken	1	142	28.4	28.4	100.0
		-------	-------	-------	
	Total	500	100.0	100.0	

```
Valid cases     500    Missing cases        0
```

```
CALC         Calculus
```

Value Label	Value	Frequency	Percent	Valid Percent	Cum Percent
not taken	0	455	91.0	91.0	91.0
taken	1	45	9.0	9.0	100.0
		-------	-------	-------	
	Total	500	100.0	100.0	

```
Valid cases     500    Missing cases        0
```

```
MAED         Mother's education
```

Value Label	Value	Frequency	Percent	Valid Percent	Cum Percent
less than HS	2	114	22.8	22.8	22.8
HS grad	3	202	40.4	40.4	63.2
less than 2 yr Voc	4	16	3.2	3.2	66.4
more than 2 yr Voc	5	22	4.4	4.4	70.8
less than 2 yr Coll	6	36	7.2	7.2	78.0
more than 2 yr Coll	7	36	7.2	7.2	85.2
Coll grad	8	46	9.2	9.2	94.4
Master's	9	23	4.6	4.6	99.0
MD/PhD	10	5	1.0	1.0	100.0
		-------	-------	-------	
	Total	500	100.0	100.0	

```
Valid Cases     500    Missing Cases        0
```

TABLE 2.11
(continued)

FAED   Father's education

Value Label	Value	Frequency	Percent	Valid Percent	Cum Percent
less than HS	2	132	26.4	26.4	26.4
HS grad	3	129	25.8	25.8	52.2
less than 2 yr Voc	4	14	2.8	2.8	55.0
more than 2 yr Voc	5	38	7.6	7.6	62.6
Less than 2 yr Coll	6	27	5.4	5.4	68.0
more than 2 yr Coll	7	26	5.2	5.2	73.2
Coll grad	8	76	15.2	15.2	88.4
Master's	9	36	7.2	7.2	95.6
MD/PhD	10	22	4.4	4.4	100.0
		-------	-------	-------	
	Total	500	100.0	100.0	

Valid cases   500   Missing Cases   0

MAEDNEW   Mother's education

Value Label	Value	Frequency	Percent	Valid Percent	Cum Percent
less than HS	1	114	22.8	22.8	22.8
HS grad	2	202	40.4	40.4	63.2
some Voc or Coll	3	110	22.0	22.0	85.2
Coll grad or more	4	74	14.8	14.8	100.0
		-------	-------	-------	
	Total	500	100.0	100.0	

Valid cases   500   Missing cases   0

FAEDNEW   Father's education

Value Label	Value	Frequency	Percent	Valid Percent	Cum Percent
less than HS	1	132	26.4	26.4	26.4
HS grad	2	129	25.8	25.8	52.2
some Voc or Coll	3	105	21.0	21.0	73.2
Coll grad or more	4	134	26.8	26.8	100.0
		-------	-------	-------	
	Total	500	100.0	100.0	

Valid cases   500   Missing cases   0

**TABLE 2.11
(continued)**

MATHGR   math grades

Value Label	Value	Frequency	Percent	Valid Percent	Cum Percent
less than A-B	0	252	50.4	50.4	50.4
most A-B	1	248	49.6	49.6	100.0
	Total	500	100.0	100.0	

Valid cases   500   Missing cases   0

GRADES   reported grades

Value Label	Value	Frequency	Percent	Valid Percent	Cum Percent
half A&B	2	3	.6	.6	.6
mostly B	3	26	5.2	5.2	5.8
half B&C	4	66	13.2	13.2	19.0
mostly C	5	104	20.8	20.8	39.8
half C&D	6	121	24.2	24.2	64.0
mostly D	7	121	24.2	24.2	88.2
below D	8	59	11.8	11.8	100.0
	Total	500	100.0	100.0	

Valid cases   500   Missing cases   0

Summary

We have seen that, before we can analyze data, we need to code and classify them, often for entry into a computer data file. To organize and summarize the data, we might construct a frequency distribution of the scores, which shows the number of times a score or group of scores occurs.

Next, in order to make it easier to understand and interpret the data, we construct a graph — a pictorial representation of the data that illustrates the relationship between two variables. There are several conventions to follow when graphing data.

The independent variable is usually given on the horizontal axis, or X axis. The dependent variable is shown on the vertical axis, or Y axis. To avoid presenting a distorted picture of the data, the three-quarter-high rule should be followed. If the independent variable is nominal, a bar graph should be used to represent the data. If we have paired measurements on two continuous variables, we can represent the data on a scatterplot.

Frequency distributions can be depicted using a histogram or a frequency polygon. In constructing a frequency polygon, we assume that the scores in each class interval can be represented by the midpoint of the interval. Often it is useful to construct and graph a cumulative frequency distribution.

The shape of a frequency distribution reflects how the scores are distributed on the scale of measurement. We can classify the shape of a distribution as uniform or nonuniform, skewed or symmetric. Distributions vary in kurtosis, that is, degree of peakedness. But the shape of a frequency distribution tells us only a small part of what we need to know to understand the data and their distribution. In Chapter 3 we will discuss the next step in interpreting data: the use of quantitative measures to describe distributions.

Exercises

*1. A group of 31 male adults took a physical performance test and the following are the resulting scores:

35	36	35
28	32	34
36	29	37
36	30	31
33	34	32
32	35	29
37	33	30
29	35	35
30	29	34
30	34	33
31		

    a. Develop a frequency distribution for the observed scores and construct the histogram for that distribution.

    b. Develop a frequency distribution using 28–30 as the lowest class interval; construct the corresponding histogram.

    c. Superimpose the frequency polygon on the histogram constructed in part b.

2. The following list gives the numbers of Sunday newspapers purchased on the past 52 Sundays.

77	73	75	81
75	72	82	86
71	74	73	73
79	76	102	74
75	82	98	75
88	78	77	75
85	81	83	98
66	68	76	87
80	81	71	74
70	88	74	79
85	77	84	83
72	79	67	68
71	85	72	74

   a. Develop a stem-and-leaf display.

   b. Develop a frequency distribution with 65–69 as the lowest class interval.

   c. Construct a histogram and a frequency polygon for this frequency distribution.

   d. Using these same data, develop a frequency distribution with 66–68 as the lowest class interval.

   e. Superimpose the histogram and the frequency polygon that result from this frequency distribution on those you developed in part c.

3. The following are reading-ability scores of 60 third-grade students:

35	79	42	80
33	81	39	78
44	82	46	82
35	78	28	74
37	81	42	74
47	76	36	71
39	65	48	72
52	61	43	55
54	64	67	54
59	44	62	51
64	56	72	42
74	47	68	33
77	38	81	42
46	34	77	83
65	49	38	56

   a. Develop a stem-and-leaf display.

   b. Develop a frequency distribution with 28–32 as the lowest class interval (interval width equal to 5).

   c. Develop a frequency distribution with 20–29 as the lowest class interval (interval width equal to 10).

d. What is the relationship between the stem-and-leaf display of part a and the frequency distribution of part c?

4. There are two general rules for developing frequency distributions when using class intervals other than the intervals of the original scores. These rules concern the number and size of class intervals to be used. Nevertheless, considerable freedom of choice remains. For example, the following class intervals could have been used when developing a frequency distribution for the scores in Table 2.3:

69–73	9	129
64–68	14	170
59–63	41	156
54–58	34	115
49–53	35	81
44–48	21	46
39–43	10	25
34–38	7	9
29–33	2	2
24–28		

What effect would this choice have had on the shapes of the histogram and the frequency polygon? Use these class intervals to construct a frequency polygon and then compare your graph with Figure 2.10.

5. The following frequency distributions of scores reflect the amount of time (number of seconds) taken by two groups of experimental rats to complete a maze.

Class interval	Group I frequency	Group II frequency
150–154	2	3
145–149	3	7
140–144	4	12
135–139	6	17
130–134	9	25
125–129	18	33
120–124	22	38
115–119	34	23
110–114	37	15
105–109	28	10
100–104	15	8
95–99	12	5
90–94	7	3
85–89	3	1

a. Complete the frequency distributions and construct the cumulative frequency distributions using the methods discussed in this chapter.

b. Draw histograms and frequency polygons for both frequency distributions.

c. Describe and compare the two frequency distributions.

3

Describing Distributions: Individual Scores, Central Tendency, and Variation

Key Concepts		
	Shapes	Range
	Central tendency	Box plot
	Variation	Outliers
	Norm-referenced interpretation	Interquartile range (IQR)
	Percentile	Mean deviation
	Percentile rank	Variance
	Mode	Unbiased estimate
	Modal interval	Deviation formula
	Median	Raw (observed) score formula
	Mean	Standard deviation
	Sum of deviations	Deviation score formula
	Deviation score	Standard score
	Sum of squares	z score
	Least squares	Weighted average score
	Weighted mean	Transformed score

In Chapter 2, we described the procedures for entering data into a computer file, for exploring data using stem-and-leaf displays, and for organizing and summarizing data into frequency distributions. A second step in the process of describing distributions of scores is graphing them. There are, however, other procedures by which we can enhance the description of a distribution. These procedures, which are the subject of this chapter, involve computing quantitative measures of the distribution of scores.

Three kinds of information are required to describe distributions adequately: (1) knowledge of their shapes; (2) measures of central tendency, which are indicators of the average score; and (3) measures of variation, which are indicators of the spread of the differences among scores. Thus we could say that to know a distribution is to answer three questions: what is its shape? where is its average score? how varied are the scores?

We discussed answers to the first question in Chapter 2, where we described various shapes of distributions and procedures for illustrating them graphically. However, before we discuss the measures of central tendency and variation, it is necessary to discuss procedures for describing individual scores. Often, raw scores on any variable are virtually meaningless without an adequate frame of reference — that is, without an indication of the relative position of a score in the total distribution of scores. Psychologists and counselors frequently provide a "norm-referenced interpretation" of scores on personality inventories and achievement tests by indicating the number or percentage of scores greater or less than a given score. For example, if a child scores a 98 on a personality inventory, a psychologist explaining this score to the child's parents might tell them the percentage of children who score lower than 98.

In this chapter, we begin with the process of describing distributions by presenting the concepts of percentiles and percentile ranks. These concepts are needed for discussing the measures of central tendency and variation. Three measures of central tendency will be discussed: (1) *mode*, (2) *median*, and (3) *mean*. Before we discuss the measures of variation, we will illustrate the use of *box plots* as a procedure for describing the spread of scores in a distribution. Then, we will discuss four measures of variation: (1) *range*, (2) *mean deviation*, (3) *variance*, and (4) *standard deviation*. Finally, we will return to norm-referenced interpretations of individual scores that are dependent upon knowledge of the central tendency and variation of scores in a distribution.

To describe a distribution is to indicate its shape, its average score, and its variation. Graphs depict shape; measures of central tendency indicate average scores; and measures of variation indicate how widely scores are spread throughout the distribution.

Percentiles

A **percentile** is the point in a distribution at or below which a given percentage of scores is found. For example, the 28th percentile of a distribution of scores is the point at or below which 28 percent of the scores fall. If a student's raw score on the SAT falls at the 28th percentile of the scores of all students taking the test on the same day, then only 28 percent of the students have lower scores. To indicate the 28th percentile, we write P_{28}.

Computing Percentiles

Consider again the distribution of final examination scores for the 180 freshman psychology students. The frequency distribution using the original scores is found in Table 3.1; the frequency distribution using class intervals of width 5 is found in Table 3.2. Suppose the chair of the psychology department is interested in determining the 75th percentile — the point at or below which 75 percent of the students score. How would we find the value of the 75th percentile using the frequency distribution in Table 3.2?

First, we must find the class interval in which this percentile point falls. Since the 75th percentile is defined as the point at or below which 75 percent of the 180 scores fall, we are looking for the class interval that contains the point below which $0.75 \times 180 = 135$ scores fall — in other words, the interval that contains the point with a cumulative frequency of 135. From Table 3.2 we see that 54.5 is the exact upper limit of the class interval at or below which 122 (67.78 percent) of the scores fall and that 59.5 is the point at or below which 159 (88.34 percent) of the scores fall. Therefore, P_{75} is a point in the interval 54.5–59.5.

Next we find the point within this interval that corresponds to the 75th percentile. Recall from Chapter 2 that we assume that observations or cases are uni-

TABLE 3.1
Frequency Distribution of Final Examination Scores

Score	f	Score	f	Score	f
69	1	54	7	39	3
68	2	53	7	38	4
67	1	52	6	37	5
66	0	51	3	36	4
65	2	50	7	35	2
64	4	49	11	34	1
63	5	48	8	33	3
62	3	47	7	32	2
61	1	46	9	31	0
60	2	45	7	30	1
59	3	44	4	29	1
58	2	43	4	28	0
57	6	42	5	27	0
56	14	41	5	26	0
55	12	40	4	25	1
				24	1

TABLE 3.2
Frequency Distribution of Final Examination Scores, Including Cumulative Frequencies and Cumulative Percentage

Class Interval	Exact Limits	Midpoint	f	cf	%	c%
65–69	64.5–69.5	67	6	180	3.33	100.00
60–64	59.5–64.5	62	15	174	8.33	96.67
55–59	54.5–59.5	57	37	159	20.56	88.34
50–54	49.5–54.5	52	30	122	16.67	67.78
45–49	44.5–49.5	47	42	92	23.33	51.11
40–44	39.5–44.5	42	22	50	12.22	27.78
35–39	34.5–39.5	37	18	28	10.00	15.56
30–34	29.5–34.5	32	7	10	3.89	5.56
25–29	24.5–29.5	27	2	3	1.11	1.67
20–24	19.5–24.5	22	1	1	0.56	0.56

formly distributed through an interval. According to Table 3.2 there are $135 - 122 = 13$ cases or scores below the 135th case in the interval, and there is a total of 37 cases in the interval. Therefore we know that 13/37 of the distance through the width of the interval should equal the point at or below which 135 of the scores are located. Since the interval width is 5, the 75th percentile for this distribution is

$$P_{75} = 54.5 + \left(\frac{13}{37}\right)(5)$$

$$= 54.5 + 1.76$$

$$= 56.26$$

Therefore 56.26 is the 75th percentile, because 75 percent of the students have scores of 56.26 or lower.

We can generalize this procedure to compute any percentile with the following formula:

$$Xth \text{ percentile} = P_X = ll + \left(\frac{np - cf}{f_i}\right)(w) \tag{3.1}$$

where

ll = exact lower limit of the interval containing the percentile point

n = total number of scores

p = proportion corresponding to the desired percentile

cf = cumulative frequency of scores below the interval containing the percentile point

f_i = frequency of scores in the interval containing the percentile point

w = width of class interval

Suppose the psychology instructor decides to assign an F to those with scores at or below the 34th percentile. Applying formula 3.1, we find the 34th percentile for the data in Table 3.2 as follows:

$$P_{34} = 44.5 + \left(\frac{180(0.34) - 50}{42} \right)(5)$$

$$= 44.5 + \left(\frac{61.20 - 50}{42} \right)(5)$$

$$= 44.5 + 1.33$$

$$= 45.83$$

Thus the point below which 34 percent of all the scores fall is 45.83.

We can also find percentiles by using the ogive; the ogive for the 180 final examination scores is shown in Figure 3.1. To find P_{34}, draw a horizontal line from the point 34 on the percentage scale (Y axis) to the curve. Where this line intersects the curve, drop a vertical line perpendicular to the score scale (X axis). This point on the score scale will be P_{34}.

A percentile is the point on the scale of measurement for the distribution at or below which a given percentage of scores is located.

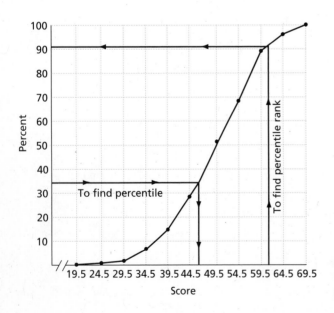

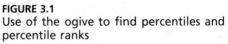

FIGURE 3.1
Use of the ogive to find percentiles and percentile ranks

Percentile Rank

Suppose one of the psychology students has a score of 63 on the final examination and wants to know what percentage of students scored lower. The student is looking for the percentile rank of the score — a concept related to percentiles. The **percentile rank** of a score is the percentage of scores less than or equal to that score. For example, the percentile rank of 63 is the percentage of scores in the distribution that falls at or below a score of 63. It is a point on the percentile scale, whereas a percentile is a score, a point on the original measurement scale.

To find the percentile rank of the score 63 (indicated PR_{63}) consider the data in Table 3.2. The score falls in the class interval 59.5–64.5, and 159 scores are below the exact lower limit, 59.5. As before, we assume that the 15 scores in the interval from 59.5 to 64.5 are uniformly distributed. Therefore the number of scores in the interval that fall below the score 63 is $(63.0 - 59.5)/5 = 3.5/5$, or 7/10 of the distance through the interval. Thus there are $0.7 \times 15 = 10.5$ scores in the interval that are below 63. Adding the 10.5 scores to 159, we find PR_{63} as follows:

$$PR_{63} = \left[\frac{159 + 10.5}{180} \right] \times 100 = 94.17$$

We say that the percentile rank of the score 63 is 94.17.

We can generalize this procedure to compute the percentile rank of any number in a distribution of scores, as follows:

$$PR_X = \left\{ \frac{cf + \frac{(X - ll)}{w} \times f_i}{n} \right\} (100) \qquad (3.2)$$

where

X = score for which the percentile rank is to be determined

cf = cumulative frequency of scores below the interval containing the score X

ll = exact lower limit of the interval containing X

w = width of class interval

f_i = frequency of scores in the interval containing X

n = total number of scores

For example, if we apply formula 3.2 to find the percentile rank of a score of 61 in the set of scores given in Table 3.2, we find that

$$PR_{61} = \left\{ \frac{159 + \left(\frac{61 - 59.5}{5} \right) \times 15}{180} \right\} \times 100$$

$$= \left(\frac{159 + 4.50}{180} \right) \times 100$$

$$= 90.83$$

Thus the score 61 has a percentile rank of 90.83.

The percentile rank can also be found by using the ogive (see Figure 3.1). For example, to find PR_{61}, draw a vertical line from the point 61 on the score scale (X axis) *up* to the curve, and where this line intersects the curve, draw a horizontal line over to the percentage scale (Y axis). This point on the percentage scale is PR_{61}.

The percentile rank of a score is a point on the percentile scale that gives the percentage of scores falling at or below the specified score.

To further illustrate the difference between a percentile and a percentile rank, suppose we have a distribution of scores in which a score of 83 on the measurement scale corresponds to the 72nd percentile. We say that 83 is the 72nd percentile; the percentile rank of a score of 83 is 72.

Use of Percentiles and Percentile Ranks

Percentiles are widely used in reporting the results of standardized tests because they are relatively easy to interpret. For example, suppose a high school counselor is reviewing the results of the SAT with a student's parents. The counselor can tell the parents the student's raw score, but it is more meaningful to tell them the position of the student's score in relation to the scores of other students. If the student's raw score is at the 20th percentile, then only 20 percent of all students taking the SAT scored lower, and the student might be advised to enroll in a technical school rather than a four-year college. In contrast, a student with a raw score at the 85th percentile would probably be advised to enroll in a four-year institution.

The reference group is important when interpreting percentiles. A student's SAT score, for example, might be compared with the scores of all students who took the SAT on a given date, or all students in a particular high school, or all those applying for admission to a specific college, or all those planning to major in engineering. Obviously we interpret a score at the 20th percentile differently if the reference group is National Merit Scholars than if it is all students taking the test in one year.

Still, the usefulness of percentiles is limited. To see why, consider the data from the Differential Aptitude Tests (DAT) shown in Table 3.3. The table gives the percentile norms for twelfth-grade boys on eight subtests of the DAT. Notice that the 50th percentile on the Abstract Reasoning Test is the score 38. Reading up the percentile scale for this subtest, we see that there is a difference of 1 raw score

TABLE 3.3
Percentile Norms for the DAT, Boys—Grade 12 (n = 5, 000+)

					Raw Scores				
Percentile	Verbal Reasoning	Numerical Ability	VR + NA	Abstract Reasoning	Clerical S and A[a]	Mechanical Reasoning	Space Relations	Spelling	Language Usage
99	49–50	40	88–90	49–50	73–100	67–70	58–60	99–100	56–60
97	48	39	86–87	48	66–72	65–66	56–57	97–98	53–55
95	47	38	84–85	47	61–65	64	53–55	94–96	51–52
90	46	36–37	81–83	46	58–60	62–63	51–52	92–93	48–50
85	44–45	35	78–80	45	56–57	60–61	48–50	90–91	45–47
80	43	34	75–77	44	54–55	59	46–47	87–89	43–44
75	41–42	33	72–74	43	52–53	58	44–45	85–86	41–42
70	39–40	31–32	69–71	42	51	57	42–43	82–84	39–40
65	37–38	30	66–68	41	50	55–56	40–41	80–81	38
60	35–36	28–29	63–65	40	49	54	38–39	78–79	36–37
55	33–34	27	60–62	39	47–48	53	36–37	75–77	35
50	31–32	25–26	57–59	38	46	52	34–35	72–74	33–34
45	30	24	54–56	37	44–45	50–51	31–33	70–71	32
40	28–29	22–23	50–53	35–36	43	49	29–30	67–69	30–31
35	25–27	20–21	46–49	34	41–42	47–48	26–28	64–66	28–29
30	23–24	18–19	42–45	33	40	45–46	24–25	61–63	27
25	20–22	16–17	38–41	31–32	38–39	43–44	22–23	57–60	25–26
20	17–19	14–15	33–37	28–30	36–37	40–42	20–21	54–56	22–24
15	14–16	12–13	28–32	23–27	34–35	37–39	18–19	50–53	20–21
10	11–13	10–11	23–27	17–22	31–33	33–36	15–17	45–49	17–19
5	9–10	8–9	18–22	12–16	26–30	29–32	13–14	40–44	14–16
3	7–8	6–7	15–17	9–11	18–25	25–28	11–12	35–39	11–13
1	0–6	0–5	0–14	0–8	0–17	0–24	0–10	0–34	0–10
$\overline{X}$	31.1	24.9	56.0	35.8	45.8	50.6	34.3	71.8	33.8
s	12.2	9.8	20.6	10.1	11.8	10.6	13.0	17.3	11.4

Source: G. K. Bennett, H. G. Seashore, and A. G. Wesman, Manual: Differential Aptitude Tests. 5th ed. (New York: The Psychological Corporation, 1974), p. 51.

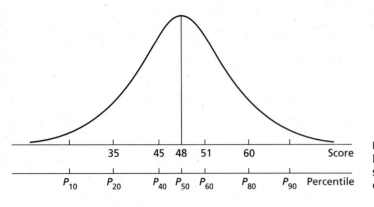

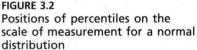

FIGURE 3.2
Positions of percentiles on the scale of measurement for a normal distribution

point for every 5 percentile points. However, below the 50th percentile, there is a difference of 1 or 2 points for every 5 percentile points down to P_{25}. And below the 25th percentile there is a difference of 3 to 5 raw score points for every 5 percentile points. The same characteristic is evident on the other subtests; that is, the percentile points are *nonuniform*. Equal differences in percentiles do not necessarily represent equal differences in the raw scores. In other words, percentiles form merely an *ordinal* scale. This is the major source of limitations on the usefulness of percentiles.

Figure 3.2 provides an example; it shows differences between percentile points that are not uniform throughout the scale of measurement. Notice that the raw score 48 corresponds to P_{50}, whereas the scores 45 and 51 correspond to P_{40} and P_{60}, respectively. Thus a difference of just 6 raw score units in the middle of this normal distribution is equivalent to a difference of 20 percentile points. Because of this piling up of percentile points in the middle of the distribution, we should not make a large distinction between an individual who scores at P_{49} and one who scores at P_{52}. At the *center* of a normal distribution, the use of percentile scores tends to *exaggerate* small, nearly nonexistent differences. On the other hand, in the *tails* of the distribution, the use of percentile scores tends to *underestimate* actual differences.

Because percentiles represent an ordinal scale, with unequal units of measurement, they have another limitation: they should not be manipulated arithmetically. There is no justification for summing or combining them, averaging them or manipulating them in ways which we would manipulate scores that are equally spaced on a scale. Percentiles should be used only for *describing* points in a distribution. If statistical manipulations are to be performed, it is necessary to convert the percentiles into some other kind of score, such as the normalized standard scores discussed in the next chapter, and then perform the manipulations. Anyone using percentiles to report the results of standardized tests should be aware of their ordinal characteristics and the corresponding limitations of these characteristics on their manipulation.

The ordinal nature of percentiles limits the statistical operations appropriate for them.

One result of these limitations is that percentiles are not very useful when we want to make comparisons across distributions. Suppose, for example, a student receives a score of 85 on a comprehensive math test and a score of 75 on a comprehensive history examination. Do these scores imply that the individual is a better student in math or in history? There is no way to answer this question until we know something about how the tests compare and how this student's scores are positioned in the distributions of scores on both tests. We could describe the student's scores as either a raw score or a percentile rank. But the raw scores are not comparable. The average scores on different tests are rarely exactly equal; hence, a single raw score, such as 65, may be above average on one test, average on another test, and well below average on a third test. Thus, using raw scores for comparison across two or more distributions simply does not work if, as is usually the case, the distributions have different scales of measurement. Percentile ranks of scores can be compared across two or more distributions. However, this comparison is limited to an examination of the relative ranking of each score only. Percentiles *cannot* be used to determine differences in relative rank. As mentioned earlier, this is due to the fact that the intervals between adjacent percentiles do not necessarily represent equal raw score intervals.

Because of these limitations of percentiles and percentile ranks, we must look for another method of judging the positions of this student's scores in the two distributions of test scores. We use a procedure that *standardizes* each distribution so that a particular student's scores on the two tests can be both located in the respective distributions *and* compared.

Raw scores and percentile ranks are for the most part inappropriate for making comparisons across distributions.

Measures of Central Tendency

Many studies in the behavioral sciences have distributions with heavy concentrations of scores in the middle and fewer scores trailing out into either extreme (tail). Normal or approximately normal distributions of, say, examination scores fall into this category. The measures of central tendency (average scores) for these distributions lie near the center of the distribution. However, extremely skewed distributions may *not* have measures of central tendency located near the center. As we will see, it is important not only to compute the measures of central tendency for a set of scores but also to interpret them in relation to the overall shape of the distribution.

Mode

The **mode** is the simplest index of central tendency. It is defined as the most frequent score in the distribution. We determine the mode simply by inspecting or counting the data, not by computation. Consider the distribution of final examination scores for our 180 freshman psychology students in Table 3.1. The mode of this distribution is the score 56.

When data are grouped into class intervals, the mode is, in reality, a **modal interval**, and the midpoint of this interval is considered the mode. For example, the modal interval for the frequency distribution in Table 3.2 is 45–49; the mode is the midpoint, 47.

When there is only one mode, the distribution is called *unimodal*. However, when two or more nonadjacent scores (or intervals for grouped data) occur more frequently than adjacent scores (or intervals), the distribution is called *multimodal*. When such a distribution has exactly two modes, it is said to be *bimodal*. For example, consider again the final examination scores for the freshman psychology students (Table 2.6 and Figure 2.9). The interval 55–59 contains 37 cases, whereas the interval 45–49 contains 42 cases. By strict definition, this distribution has one mode (47), but since 57 is the midpoint of a nonadjacent interval that has nearly as many cases, the distribution could be considered bimodal.

The mode is a rather unsophisticated measure of central tendency. It provides little information except when we consider qualitative variables measured on a nominal scale. Although the mode identifies the most frequent score or class interval, it does not lend itself readily to mathematical manipulation and thus has limited value as a statistic. However, when the number of scores is large, the mode can be used in conjunction with other measures to describe the distribution.

The mode is the most frequent score in a distribution.

Median

A second measure of central tendency is the **median**; it is the 50th percentile or the point on the scale of measurement below which 50 percent of the scores fall. When the number of observations is relatively small and when we are using the observed scores (not grouped data), we can compute the median by using the following steps:

1. Arrange the scores in ascending order (from lowest to highest).

2. If the number of scores is odd, the median is the middle score. Consider the following distribution of scores: 3, 6, 12, 18, 19, 21, 23. Since there are seven scores ($n = 7$), the median is the middle, or fourth, score, which is 18.

3. If the number of scores is even, the median is halfway between the two middle scores. Consider the following distribution of scores: 18, 23, 27, 28, 29, 40, 44, 46. Since there are eight scores ($n = 8$), the median is halfway between 28 and 29, or 28.5.

When the number of observations is large and when we group the data into class intervals, we use a variation of formula 3.1 to compute the median. First, we must do the following:

1. Develop the cumulative frequency distribution (see Table 3.2).

2. Find the total number of scores, which we label n, and multiply it by 0.50. This calculation gives the middle score.

3. Find the exact limits of the interval in the cumulative frequency distribution that contains the $n(0.50)$ score.

Next, we apply the following formula:

$$\text{Mdn} = ll + \left(\frac{n(0.50) - cf}{f_i} \right)(w) \qquad (3.3)$$

where

ll = lower exact limit of the interval containing the $n(0.50)$ score

n = total number of scores

cf = cumulative frequency of scores below the interval containing the $n(0.50)$ score

f_i = frequency of scores in the interval containing the $N(0.50)$ score

w = width of class interval

As an example, we can apply this formula to the data in Table 3.2. Notice that, for this distribution, n is 180, so $n(0.50)$ is 90. According to the cumulative frequencies in Table 3.2, this middle score occurs in the class interval 45–49 with a lower exact limit of 44.5. Thus the median for this distribution is

$$\text{Mdn} = 44.5 + \left(\frac{(180)(0.50) - 50}{42} \right)(5)$$

$$= 44.5 + 4.76$$

$$= 49.26$$

The median is the point below which 50 percent of scores fall.

Mean

The **mean** is the arithmetic average of the scores in a distribution. The symbol for the mean of a population is μ; the symbol for the mean of a sample is $\overline{X}$. When we consider inferential statistics in later chapters, we will have to distinguish between the mean of a population and the corresponding mean of a sample. However, at this point we will deal with samples and use $\overline{X}$ as the symbol of the mean.

The mean is determined by adding the scores and dividing by the total number of scores. Symbolically,

$$\overline{X} = \frac{\Sigma X_i}{n} \tag{3.4}$$

where

X_i = each of the scores

n = total number of scores

For the distribution of final examination scores in Table 3.1, we compute the mean as follows:

$$\overline{X} = \frac{69 + 68 + 68 + 67 + \cdots + 24}{180}$$

$$= \frac{8860}{180}$$

$$= 49.22$$

The mean is the arithmetic average of the scores in a distribution.

Properties of the Mean   The mean is the most frequently used measure of central tendency, in part because of the following two properties:

1. The **sum of deviations** of all scores from the mean is zero.

A **deviation score** is the difference between a given score and the mean. Symbolically, if X_i is a given score, x_i is the deviation of that score from the mean; that is,

$$x_i = (X_i - \overline{X})$$

Stated symbolically, this first property of the mean is

$$\Sigma(X_i - \overline{X}) = \Sigma(x_i) = 0$$

Table 3.4 provides a simple example illustrating the property. This property is important in deriving the more complex statistical formulas used in later chapters.

2. The **sum of squares** of the deviations from the mean is smaller than the sum of squares of the deviations from any other value in the distribution.

TABLE 3.4
Data Illustrating the Properties of the Mean

X_i	$x_i = (X_i - \overline{X})$	$x_i^2 = (X_i - \overline{X})^2$	$(X_i - 8)^2$
9	3	9	1
12	6	36	16
7	1	1	1
5	−1	1	9
2	−4	16	36
3	−3	9	25
4	−2	4	16
Σ 42	0	76	104

$n = 7$
$\overline{X} = 6$

This property is also illustrated in Table 3.4. The last column of the table contains the square of the deviations from the value 8. Notice that the sum of the squares of these deviations (104) is greater than the sum of squared deviations from the mean (76). This holds true regardless of the value selected. Thus the mean is the measure of central tendency in the **least squares** sense; that is, the sum of squared deviations from the mean is a minimum. This property will be used in defining the variance and standard deviation later in this chapter.

Two important properties of the mean are
1. The sum of deviations from the mean is zero.
2. The sum of squared deviations from the mean is a minimum.

The Mean of Combined Groups Suppose that, of the 180 freshman psychology students, 106 are female and 74 are male and that the mean final examination score for females ($\overline{X}_F$) is 45.26 and the mean for males ($\overline{X}_M$) is 54.89. To find the mean of the two groups combined, we *cannot* simply compute the "mean" of the means, since the groups have a different number of observations. In order to find the mean for the combined groups, we must compute the **weighted mean**. The formula for the weighted mean for this example is as follows:

$$\overline{X} = \frac{n_F \overline{X}_F + n_M \overline{X}_M}{n_F + n_M} = \frac{\Sigma X_F + \Sigma X_M}{n_F + n_M}$$

where

$\overline{X}_F$ and $\overline{X}_M$ = means for females and males

n_F and n_M = number of females and males

For these data,

$$\overline{X} = \frac{(106)(45.26) + (74)(54.89)}{106 + 74}$$

$$= \frac{4798 + 4062}{180}$$

$$= 49.22$$

The weighted mean is exactly the same as that computed on page 64.

The weighted mean can also be computed when there are more than two groups. To compute the mean of the combined groups, we multiply the number of observations in each group (n_i) by the individual group means ($\overline{X}_i$) and then divide this sum by the total number of observations in all groups (N). Symbolically,

$$\overline{X} = \frac{\Sigma n_i \overline{X}_i}{N} \tag{3.5}$$

where

$\overline{X} =$ individual group means

$n_i =$ number of observations in the individual groups

$N = \Sigma n_i =$ total number of observations in all groups

> **The weighted mean is computed when determining the mean of combined groups in which the number of observations in each group differs.**

Comparison of the Mode, Median, and Mean

What is the best measure of central tendency? To some extent, the answer depends on whether the variable is qualitative or quantitative. If the data are *nominal*, only the mode is appropriate. If the data are *ordinal*, both the median and the mode may be appropriate. All three measures of central tendency may be used if the data are either *interval* or *ratio*.

A second consideration in choosing the most appropriate measure of central tendency is how the measure is to be used. If we wish to infer from samples to populations, the mean usually has a distinct advantage. It can be manipulated mathematically in ways that are inappropriate for the median or the mode, as we will see beginning in Chapter 7. But if the purpose is primarily descriptive, then the measure that best describes the data should be used.

We can compare the three measures of central tendency by looking at frequency distributions with different shapes. Figure 3.3 illustrates four examples. If the distribution is symmetrical and unimodal, the mean, the median, and the

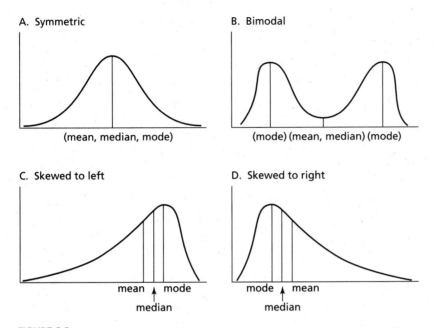

A. Symmetric

(mean, median, mode)

B. Bimodal

(mode) (mean, median) (mode)

C. Skewed to left

mean ⊥ mode
median

D. Skewed to right

mode ⊥ mean
median

FIGURE 3.3
Comparisons of the mode, median, and mean in four distributions

mode coincide, as shown in Figure 3.3A. If the distribution is symmetrical and bimodal, as in Figure 3.3B, the mean and median coincide, but there are two modes. Asymmetrical distributions are illustrated in Figures 3.3C and 3.3D. Notice that, in distribution C, which is skewed to the left (negatively skewed), the mean is less than the median, which is in turn less than the mode. The opposite is true for distribution D, which is positively skewed.

As Figure 3.3 shows, both the mode and the median are generally unaffected by an extreme score. However, the mean is greatly influenced by extreme scores. For example, consider the following distribution of salaries for employees in a small manufacturing company:

Position	Number of employees	Salary	
President	1	$180,000	
Executive vice president	1	60,000	
Vice presidents	2	40,000	
Controller	1	22,800	(Mean)
Senior salespeople	3	20,000	
Junior salespeople	4	14,800	
Foreman	1	12,000	(Median)
Machinists	12	8,000	(Mode)

This distribution is skewed to the right, and the mean is far from the median and the mode. It is greatly influenced by one very high score — the president's salary. If you were the chairman of the local machinists' union negotiating a new wage agreement, which average would you use? In contrast, which would you use if you represented management's position in the negotiations? In general, reporting all three measures of central tendency may provide the most accurate description of the distribution.

The mean is greatly influenced by extreme scores.

Measures of Variation

The shape and central tendency are two of the three characteristics that describe a distribution; the third characteristic is the variation, or spread, of scores in the distribution. In contrast to measures of central tendency, which are points, measures of variation are lengths of intervals that indicate how the scores are spread throughout the distribution. Common measures of variation are the range, mean deviation, variance, and standard deviation.

Measures of variation are lengths of intervals on the scale of measurement that indicate the variation, or spread, of scores in a distribution.

Range

The **range** is the simplest measure of variation. It is the number of units on the scale of measurement that include the highest and lowest values; that is,

$$(\text{Highest score} - \text{lowest score}) + 1$$

The difference between the highest score and the lowest score is increased by 1 so that both extreme values are considered. The range is sometimes defined without +1 added to the difference; however, we will use the former definition throughout this book.

Consider the following distributions:

Distribution 1	11	16	18	23	29	31	37
Distribution 2	18	19	21	23	24	26	29

The medians of both distributions are the same (23), but the ranges differ. The range for distribution 1 ($37 - 11 + 1 = 27$) is greater than the range for distribution 2 ($29 - 18 + 1 = 12$). Thus distribution 1 is more "varied."

Although the range is easy to compute, it may not adequately describe the variation of scores in a distribution. Its most serious limitation is the influence of extreme scores. For example, the range of distribution 1 would double if the highest score were 64 rather than 37. In addition, the range varies with group size; that is, larger groups tend to have larger ranges of scores. Therefore, the ranges of two distributions composed of different numbers of observations are not directly comparable.

> **The range is defined as the number of units on the scale of measurement that include the highest and lowest values.**

Before we define and discuss the remaining measures of variation, we will describe an exploratory technique for graphically illustrating the dispersion of scores in a distribution.

Box Plots

Tukey (*Exploratory Data Analysis*, 1977) devised a simple but highly informative graphical method for displaying the spread of scores in a distribution. This graphical summary, called a **box plot**, can illustrate both the central tendency and the dispersion of scores. The box plot can also be used to identify any unusual scores in the distribution, called **outliers**, that may warrant special consideration. The measure of central tendency used in the **box** of the plot is the median; the measure of dispersion, which is illustrated by the length of the box, is the interquartile range. The **interquartile range (IQR)** is the difference between the 75th percentile (3rd quartile) and the 25th percentile (1st quartile).

Five-Number Summary

In order to graph a box plot for a distribution of scores, a five-number summary is needed; these five numbers are:

1. Max = maximum score
2. Q_3 = third quartile (75th percentile)
3. Mdn = median
4. Q_1 = first quartile (25th percentile)
5. Min = minimum score

Consider the 180 final examination scores for the freshman psychology students; the original data are found in Tables 3.1 and 3.2. For these data, Max = 69 and Min = 24; the third quartile (P_{75}) and the median (Mdn) were computed earlier — $Q_3 = P_{75} = 56.26$ and Mdn = 49.26. The calculation of $Q_1 = P_{25}$, using formula 3.1, is illustrated below.

$$Q_1 = P_{25} = 39.5 + \left(\frac{n(.25) - 28}{22}\right)(5)$$

$$= 39.5 + \left(\frac{17}{22}\right)(5)$$

$$= 43.36$$

Therefore, for the distribution of final examination scores, the five-number summary is

Max $= 69$

$Q_3 = 56.26$

Mdn $= 49.26$

$Q_1 = 43.36$

Min $= 24$

The box plot for these data was graphed using the methods outlined in Chapter 2 and is found in Figure 3.4. As can be seen, the box extends from Q_1 to Q_3, with the Mdn represented by the horizontal bar within the box. The Min and Max values are represented by short horizontal bars connected to the box by vertical lines. Referring back to Figures 2.9 and 2.12, as well as to Figure 3.4, a slight negative skew (skewed to the left) is seen in this distribution of scores.

Box plots can be used to provide a graphical summary of both the central tendency and variation of a distribution of scores.

Outliers

A modification of the box plot illustrated in Figure 3.4 can be used to identify outliers in the distribution. Outliers are defined as unusual scores in the data that are often considered extreme and require special consideration. For example, in the previous section, in which the different measures of central tendency were compared, we illustrated the effect of extreme scores on the calculation and interpretation of the mean.

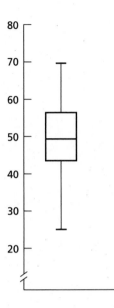

FIGURE 3.4
Box plot of final examination scores

For identifying outliers, it is necessary to modify the box plot illustrated in Figure 3.4. The first step is to compute upper and lower boundaries of *reasonable* values of scores in the distribution. These boundaries are based upon the IQR, which was defined above; algebraically, the IQR is defined as follows:

$$IQR = Q_3 - Q_1 \qquad (3.6)$$

The rules of thumb for computing the "reasonable upper boundary" (RUB) and the "reasonable lower boundary" (RLB) are

$$RUB = Q_3 + 1.5(IQR) \qquad (3.7)$$

$$RLB = Q_1 - 1.5(IQR)$$

For the 180 final examination scores,

$$RUB = 56.26 + 1.5(56.26 - 43.36)$$

$$= 56.26 + 1.5(12.90)$$

$$= 75.61$$

$$RLB = 43.36 - 1.5(56.26 - 43.36)$$

$$= 43.36 - 1.5(12.90)$$

$$= 24.01$$

In the modified box plot, the RUB and RLB replace Max and Min in the five-number summary; the modified box plot is found in Figure 3.5.

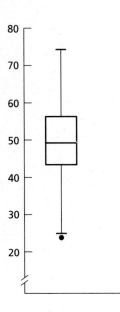

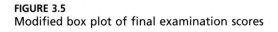

FIGURE 3.5
Modified box plot of final examination scores

$$RUB = 75.61$$

$$Q_3 = 56.26$$

$$Mdn = 49.26$$

$$Q_1 = 43.36$$

$$RLB = 24.01$$

Note that the score 24 is outside the RLB, identifying it as a potential outlier. This score is represented in the modified box plot by a dot below the RLB. It is important to note that using the above rule of thumb does not replace good judgment. For this example, this score is very close to the RLB and would *not* be considered an outlier but a reasonable score in a distribution with a slight negative skew. In the exercises at the end of the chapter, examples with definite outliers have been included.

An outlier is defined as an unusual score in a distribution that may warrant special consideration.

Like stem-and-leaf displays and frequency distributions, box plots are useful for comparing distributions of scores from different groups on the same variable. This usefulness will be discussed in greater detail in Chapters 11 and 14, where we will test the significance of the differences between two groups and more than two groups, respectively. In the meantime, an exercise has been provided at the end of the chapter to introduce the use of box plots for comparing two or more groups.

Other Measures of Variation

Mean Deviation

Earlier in the chapter, we noted that the mean is influenced by extreme scores in a distribution. Thus, in determining the variation of scores in a distribution, we need to incorporate a sensitivity to *all* scores in the distribution by considering the deviation of each score from the mean. Recall that subtracting each score from the mean gives us a *deviation score*; that is,

$$x_i = (X_i - \overline{X})$$

It seems reasonable that if we add the deviations from the mean and divide by the number of scores, n, we will have a measure of the average deviation of all scores from the mean. However, from the first property of the mean, we know that the sum of the deviations equals 0; that is, $\Sigma(X_i - \overline{X}) = \Sigma(x_i) = 0$. Thus, dividing $\Sigma(x_i)$ by n would equal 0 regardless of the distribution. However, the process we have described makes sense intuitively when we consider the *absolute values* (see Chapter 1) of these deviation scores.

Consider the distribution of scores in Table 3.4. The mean of these scores is 6, so the score of 9 has a deviation score of 3 score units ($9 - 6 = 3$) above the mean. Similarly, the score of 3 is also 3 score units from the mean ($3 - 6 = -3$), although the minus sign indicates that this score is below the mean. If we ignore the algebraic sign of the deviation score and use the absolute values of the deviation scores, we can define the **mean deviation** (MD) as the sum of the absolute values of the deviation scores divided by the total number of scores. Symbolically,

$$\text{MD} = \frac{\Sigma|X_i - \overline{X}|}{n} = \frac{\Sigma|x_i|}{n} \tag{3.8}$$

For the data in Table 3.4,

$$\text{MD} = \frac{|3| + |6| + |1| + |-1| + |-4| + |-3| + |-2|}{7}$$

$$= \frac{3 + 6 + 1 + 1 + 4 + 3 + 2}{7}$$

$$= 2.86$$

We can use the mean deviation to compare the variation of several distributions; distributions with larger mean deviations have greater variations. However, the utility of the mean deviation in describing the variation of scores in a distribution is limited because absolute values are used in the computation. More advanced statistical analyses require algebraic manipulations that are difficult, if not impossible, to carry out with absolute values. This problem is overcome by using two measures that are conceptually related to the mean deviation: the variance and the standard deviation.

The mean deviation is the average of the absolute values of the deviation scores.

Variance

Squaring the deviation scores, instead of using absolute values, is an alternative way to eliminate the minus signs before summing the deviation scores. Unlike using absolute values, squaring the deviation scores does not restrict the further algebraic manipulations that are usually required in more complex analyses. Thus to obtain a measure of variation without mathematical restrictions, we first square the deviation scores and then sum them. What we obtain is the sum of squared deviations around the mean, or sum of squares (SS). Symbolically,

$$SS = \Sigma(X_i - \overline{X})^2 = \Sigma(x_i)^2 \tag{3.9}$$

Now, if we divide the sum of squares by the total number of scores, we will have the average or mean of the sum of squares, which is called the **variance**.

The variance is defined as the average of the sum of squared deviations around the mean.

Because of the difference in computation, it is necessary to distinguish between the variance of a population (σ^2) and the variance of a sample (s^2). (To review the difference between a parameter — a descriptive measure of a population — and a statistic — a descriptive measure of a sample — see Chapter 1.) The formula for the variance of a population is

$$\sigma^2 = \frac{SS}{N} = \frac{\Sigma(X_i - \mu)^2}{N} = \frac{\Sigma(x_i)^2}{N} \tag{3.10}$$

where

μ = mean of the population

N = number of scores in the population

Statistical procedures used in the behavioral sciences are usually concerned with the variance of a sample (s^2), which is then used to estimate the variance in the population (σ^2). To compute the variance of a sample, we must alter formula 3.10 in two ways. First, we use the mean of the sample ($\overline{X}$) instead of the population mean (μ) in calculating SS. Second, we divide SS by $n - 1$ (number of observations in the sample minus 1) rather than by N. This results in a sample variance that is an unbiased estimate of the variance in the population. An estimate is an **unbiased estimate** if the mean of all possible sample values, for a given sample size, equals the parameter being estimated. (We will use the concept of unbiased estimates later, in the chapters on inferential statistics.) Thus, the formula for computing the variance of a sample is

$$s^2 = \frac{SS}{n-1} = \frac{\Sigma(X_i - \overline{X})^2}{n-1} = \frac{\Sigma(x_i)^2}{n-1} \tag{3.11}$$

where

$\overline{X} =$ mean of the sample

$n =$ number of scores in the sample

Using formula 3.11 to compute the sample variance for the data in Table 3.4,

$$s^2 = \frac{(9-6)^2 + (12-6)^2 + \cdots + (4-6)^2}{(7-1)}$$

$$= \frac{76}{6}$$

$$= 12.67$$

This formula for computing s^2 is called the **deviation score formula** because the deviation scores are used in the calculation. The deviation formula is convenient to use when the number of cases is small *and* the mean is an integer. If the mean contains a decimal, then squaring the deviation scores requires considerable rounding of numbers. When the number of cases is large, this can become tedious.

A more useful formula for the variance is the **raw score** (or **observed score**) **formula**, which uses the actual scores rather than the deviation scores. The raw score formula for the sample variance[1] is

$$s^2 = \frac{SS}{n-1} = \frac{\Sigma X_i^2 - \left[(\Sigma X_i)^2/n\right]}{n-1} \tag{3.12}$$

---

[1] The corresponding raw score formula for the variance of a population follows; the only difference from formula 3.12 is that the sum of squares is divided by N rather than by $n - 1$.

$$\sigma^2 = \frac{SS}{N} = \frac{\Sigma X_i^2 - [(\Sigma X_i)^2/N]}{N}$$

The summary data for computing the variance of the distribution of scores in Table 3.4 are

$$\Sigma X_i^2 = (9)^2 + (12)^2 + (7)^2 + (5)^2 + (2)^2 + (3)^2 + (4)^2$$
$$= 328$$
$$\Sigma X_i = 9 + 12 + 7 + 5 + 2 + 3 + 4$$
$$= 42$$
$$n = 7$$

Therefore, using formula 3.12, the variance is

$$s^2 = \frac{328 - [(42)^2/7]}{(7-1)}$$
$$= \frac{328 - 252}{6}$$
$$= \frac{76}{6}$$
$$= 12.67$$

For the 180 final examination scores, the variance is computed as follows:

$$\Sigma X_i^2 = (69)^2 + (68)^2 + (68)^2 + (67)^2 + \cdots + (24)^2$$
$$= 450,624$$
$$\Sigma X_i = 69 + 68 + 68 + 67 + \cdots + 24$$
$$= 8860$$
$$n = 180$$

Therefore,

$$s^2 = \frac{450,624 - [(8860)^2/180]}{(180-1)}$$
$$= \frac{450,624 - 436,108.89}{179}$$
$$= \frac{14,515.11}{179}$$
$$= 81.09$$

Standard Deviation

The variance is expressed in *squared units* of the measurement, because the deviation scores (x_i) are squared in order to eliminate the negative signs before they are summed (see formula 3.11). For example, in a report of number of hours worked per week by a sample of people, the variance of scores is expressed in hours squared. What we need, however, is a measure of variation that has the same unit of measurement as the original data. The **standard deviation**, which is the square root of the variance, satisfies this criterion. Symbolically, the standard deviation of the sample[2] is

$$s = \sqrt{s^2} = \sqrt{\frac{SS}{(n-1)}} = \sqrt{\frac{\Sigma(X_i - \overline{X})^2}{(n-1)}} \tag{3.13}$$

Notice that formula 3.13 is the **deviation score formula** for the standard deviation; the corresponding raw score formula is

$$s = \sqrt{s^2} = \sqrt{\frac{SS}{(n-1)}} = \sqrt{\frac{\Sigma X_i^2 - [(\Sigma X_i)^2/n]}{(n-1)}} \tag{3.14}$$

Using formula 3.14, we find that the standard deviation for the 180 final examination scores is

$$s = \sqrt{\frac{450,624 - [(8860)^2/180]}{(180-1)}}$$

$$= \sqrt{\frac{450,624 - 436,108.89}{179}}$$

$$= \sqrt{\frac{14,515.11}{179}}$$

$$= \sqrt{81.09}$$

$$= 9.00$$

If we are interested only in describing the variability of a distribution of scores, both the variance and the standard deviation can be used. That is, a large standard deviation suggests a large amount of variability of scores around the mean, whereas a small standard deviation indicates little variability. However, the standard deviation has certain advantages over the variance because it is expressed in the original

---

[2] The corresponding deviation score formula for the standard deviation of a population is

$$\sigma = \sqrt{\sigma^2} = \sqrt{\frac{SS}{N}} = \sqrt{\frac{\Sigma(X_i - \mu)^2}{N}}$$

units of measurement. For example, consider the interpretation of a score of 115 on an IQ test with a mean of 100 and a standard deviation of 15. The score of 115 can be interpreted as being 1 standard deviation above the mean. Since the variance is $15^2 = 225$, it is not a useful statistic in this situation. We will discuss the use of the standard deviation in the interpretation of individual scores in more detail in the next two chapters.

> The standard deviation is the square root of the variance. It is expressed in the same units as the original measurement of the variable.

Computer Example: The Survey of High School Seniors Data Set

In Chapter 2, we presented frequency distributions for selected variables in the survey of high school seniors data set with the exception of the variables MATHACH (mathematics achievement score), VISUAL (visualization test), and MOSAIC (detecting patterns test). These variables are considered interval scale measures. All of the other variables from the survey of high school seniors data set are considered nominal scale or ordinal scale measures (see Chapter 1).

"Descriptive statistics" for the variables, MATHACH, VISUAL, and MOSAIC, generated by using the DESCRIPTIVES procedures within SPSS are found in Table 3.5. Note we have included the "means," "standard deviations," and "minimum"/"maximum" values for these three variables. As we will see in later chapters, these descriptive statistics are also generated for all variables before additional analyses are performed.

Review these descriptive statistics; specifically, consider the means in relation to the minimum and maximum values. For each of the variables, the means are nearly in the middle of the range of values. For example, the mean for MATHACH is 13.098 and the minimum value is −1.667 and the maximum value is 25.000.

TABLE 3.5
SPSS Printout—Descriptive Statistics for MATHACH, VISUAL, and MOSAIC

Number of valid observations (listwise) =   500.00

Variable	Mean	Std Dev	Variance	Minimum	Maximum	Valid N	Label
VISUAL	5.70	3.89	15.11	-2.750	16.000	500	visualizat
MATHACH	13.10	6.60	43.62	-1.667	25.000	500	math test
MOSAIC	26.89	9.29	86.27	-6.500	56.000	500	pattern te

Standard Scores

Now that we have defined the measures of central tendency and variation, we can illustrate how to overcome the problems in using raw scores and percentile ranks for comparing scores. One way is to transform the scores into scores on an *equal-interval* scale. **Standard scores** do this by using the standard deviation as the unit of measure; they describe the relative position of a single score in the entire distribution of scores in terms of the mean and the standard deviation.

Computing z Scores

The standard score, also called a *z* **score**, is computed by subtracting the mean from the raw score and dividing the result by the standard deviation.

$$\text{Standard score} = \frac{\text{Raw score} - \text{Mean}}{\text{Standard deviation}}$$

or

$$z = \frac{X - \overline{X}}{s} \tag{3.15}$$

For example, suppose a distribution has a mean ($\overline{X}$) equal to 40 and a standard deviation (s) equal to 8. Then the *z* scores that correspond to raw scores of 36, 42, and 50 are

$$z_{36} = \frac{36 - 40}{8} = -0.50$$

$$z_{42} = \frac{42 - 40}{8} = 0.25$$

$$z_{50} = \frac{50 - 40}{8} = 1.25$$

What do these scores mean? The *z* score corresponding to a given raw score indicates how many standard deviations the raw score falls either above or below the mean. A *negative z* score indicates that the raw score is *below* the mean; a *positive z* score indicates that the raw score is *above* the mean; a *z* score of 0 indicates that the raw score is *equal to* the mean. Thus, in our sample calculations, because the raw score of 36 has a *z* score of −0.50, we know that it is 0.50 standard deviation below the mean. Because the raw score 42 has a *z* score of 0.25, we know that this raw score is 0.25 standard deviation above the mean. Similarly, the raw score 50 falls 1.25 standard deviations above the mean.

> The *z* score indicates the number of standard deviations a corresponding raw score is above or below the mean.

TABLE 3.6
Distribution of Raw Scores and z Scores

Subject	Raw Score X	Standard Score z
A	10	1.26
B	9	0.94
C	3	−0.94
D	10	1.26
E	9	0.94
F	2	−1.26
G	2	−1.26
H	10	1.26
I	5	−0.31
J	5	−0.31
K	1	−1.57
L	6	0.00
M	8	0.63
N	6	0.00
O	6	0.00
P	1	−1.57
Q	3	−0.94
R	6	0.00
S	10	1.26
T	8	0.63

$n = 20$
Mean ($\overline{X}$)	6.0	0
Standard Deviation (s)	3.18	1.00

Properties of z Scores

We can calculate a z score for each raw score in a distribution. Consider the data in Table 3.6. Each of the 20 raw scores has been converted to a z score by using formula 3.15. What happens to the distribution of scores when each of the raw scores is converted into a z score?

1. The distribution of standard scores preserves a shape similar to that of the original distribution of raw scores.

2. The mean of the distribution of z scores will always equal 0 regardless of the value of the mean in the raw score distribution. This property is illustrated in Table 3.6.

3. The variance of the distribution of z scores always equals 1. Since the standard deviation is the square root of the variance, the standard deviation will also equal 1. This property is also illustrated in Table 3.6.

Thus, by calculating the z score for each raw score in a distribution, we *transform* the original distribution of scores into a distribution with an identical shape but a mean of 0 and a standard deviation of 1.

> A distribution of z scores (1) retains the shape of the distribution of the original scores, (2) has a mean of 0, and (3) has a variance and standard deviation of 1.

To illustrate how z scores can be used to compare scores across two or more distributions, consider the midterm examination scores for one of the freshman psychology students. Suppose this student has the following scores in three classes: 68 in psychology, 77 in college mathematics, and 83 in history. In which class did the student perform best? To answer this question, we have to determine the mean and standard deviation for each of the three tests and then calculate the z score for each of the student's examination scores. The means and standard deviations for the three tests, as well as the student's z scores, are

Subject	X	$\overline{X}$	s	z
Psychology	68	65	6	0.50
Mathematics	77	77	9	0.00
History	83	89	8	−0.75

As we can see, this student has the highest z score on the psychology midterm examination (0.50) and the lowest on the history midterm (–0.75). Relative to the other students taking these examinations, this student performed best on the psychology midterm.

Weighted Averages

Frequently test scores are combined into a total score that represents some set of weightings established earlier. For example, an employer might give job applicants two short technical tests and one long personality test. The employer might choose to weigh the three scores equally or to give them different weights. Combining percentiles into such an average is inappropriate because of the ordinal nature of percentiles. Using raw scores is not advisable because the three tests probably have different means and, especially, different standard deviations. Averaging the raw scores gives more weight or influence to whichever test has the greatest standard deviation. Instead, the averaging should be done by using standard scores so that the distributions all have the same mean (0) and the same standard deviation (1). Then any set of weights can be easily applied to the scores.

To find a **weighted average score** for an individual, we multiply the person's standard score on each test by the weight of that test and add these products. Then we divide this sum by the sum of the weights of the test. Thus, the general formula for finding a weighted average score is

$$\text{Weighted score}_j = \frac{\Sigma W_i z_{ij}}{\Sigma W_i} \qquad (3.16)$$

where

W_i = weight of each test

z_{ij} = standard score for person j on test i

Suppose the employer in the preceding example wishes to weigh the long personality test twice as heavily as the two short technical tests. If a particular applicant's z scores on the three tests are -0.20, 0.25, and -0.50, the weighted score for this person is

$$\frac{2(-0.20) + 1(0.25) + 1(-0.50)}{2 + 1 + 1} = -0.16$$

When developing a composite score from two or more individual scores, first transform the individual scores into standard scores. Then apply the weights to generate the weighted score.

Transformed Standard Scores

Attractive as z scores are for purposes of comparison, they have some disadvantages. They can be misleading and difficult to manage and to report. For example, consider the z score for subject I in Table 3.6 ($z = -0.31$). To the student or the parents, this score might look considerably worse than it is. First, the minus sign may carry a negative connotation. Laypeople might interpret a z score of 0 as a score of zero correct, rather than the mean value, and might overreact to a negative z score. In addition, the accidental omission of the minus sign drastically changes the meaning of the z score. A second concern is the number of decimal places to retain. Often the usual practice of reporting to two decimal places implies a precision of measurement that raw scores may not possess. Thus, the publisher of an aptitude test who wants to report test scores in a standardized format that can easily be interpreted by both students and parents would prefer some other method of expressing these scores than as z scores.

For these reasons, z scores are commonly transformed into a different distribution of scores, a distribution that has a predetermined mean and standard deviation. This new scale of measurement may be chosen arbitrarily. Suppose we transform a set of z scores into a distribution with a mean of 50 and a standard deviation of 10. The transformation requires two steps: (1) transform the original distribution of raw scores into a distribution of z scores and (2) multiply each z score by 10 (the standard deviation) and add 50 (the mean).

$$\textbf{Transformed score} = (10)(z) + 50 \tag{3.17}$$

Consider the data in Table 3.6. If we want to transform the raw score for subject A, we first compute the z score.

$$z = \frac{10.00 - 6.00}{3.18} = 1.26$$

Then, using formula 3.17, we find subject A's transformed score as follows:

$$\text{Transformed score} = (10)(1.26) + 50$$

$$= 62.60$$

By transforming scores to a distribution with a mean of 50 and a standard deviation of 10, we eliminate negative scores and reduce the need to worry about decimal places.

Behavioral scientists frequently transform a distribution of raw scores into a standardized distribution to report test scores. Many standardized distributions are used. For example, most IQ test scores are reported on the basis of a distribution with a mean of 100 and a standard deviation of 15. The College Entrance Examination Board (CEEB) uses a mean of 500 and a standard deviation of 100. It must be emphasized that these standardized distributions do not emerge magically from the distribution of raw scores. The raw scores must first be converted to z scores; then they are transformed into a distribution with the desired mean and standard deviation by the general formula

$$X' = (s')(z) + \overline{X}' \tag{3.18}$$

where

X' = new or transformed score for a particular individual

s' = desired standard deviation of the distribution

z = standard score for a particular individual

$\overline{X}'$ = desired mean of the distribution

To transform a distribution of scores into a distribution with a desired mean and standard deviation, multiply the z scores by the desired standard deviation and add the desired mean.

Summary

To describe a distribution of scores adequately, we need to know its shape, central tendency, and variation. Different shapes of distributions were described in Chapter 2. In this chapter, we have concentrated on measures of central tendency and variation. We began by focusing on percentiles for providing a norm-referenced interpretation of individual scores in the distribution. We then presented the most common measures of central tendency (mode, median, and mean), which quantitatively define central values in the distribution between the extreme values. Each of these measures of central tendency is useful in describing distributions, but the mean is the most widely used because of its mathematical properties. However, since the mean is sensitive to extreme scores, the median is often used in conjunction with the mean when describing the central tendency of a distribution that contains extreme scores.

Prior to discussing the measures of variation, we illustrated how the box plot can be used to display the spread of scores in a distribution and to identify potential outliers in the data. The measure of central tendency used in the box plot is the median; the spread or variation of scores discussed was the interquartile range. Unlike measures of central tendency, which are points on the score scale, measures of variation (range, mean deviation, and standard deviation) are intervals on the scale. The range is the simplest measure, but tends to vary with the size of the group. The variance and standard deviation are the most widely used measures of variation because of their mathematical properties. Both are based upon squared deviations from the mean. The standard deviation has the advantage of being expressed in the same units as the scores themselves.

In the final sections of the chapter, we illustrated how to use the mean and standard deviation in determining standard (z) scores. Standard scores transform raw scores from a distribution of scores with a specific mean and standard deviation into a distribution of scores with a mean equal to 0, and standard deviation equal to 1. To maximize the interpretation of standard scores, it is also possible to transform z scores into yet another distribution of scores with both a specified mean and a specified standard deviation.

Exercises

1. A college dean examines the cumulative grade-point averages of 120 sophomore students. The following frequency distribution is derived, using class intervals of width 0.3.

Class interval	f
3.8–4.0	4
3.5–3.7	8
3.2–3.4	15
2.9–3.1	18
2.6–2.8	20
2.3–2.5	17
2.0–2.2	12
1.7–1.9	12
1.4–1.6	10
1.1–1.3	4
0.8–1.0	0

   a. Draw a histogram and a frequency polygon for the distribution.

   b. Find P_{10}, P_{45}, P_{60}, and P_{95}.

   c. Find the percentile ranks of cumulative grade-point averages of 2.40, 2.75, 3.25, and 3.60.

2. Find the percentiles and percentile ranks listed in Exercise 1 using the ogive for this frequency distribution.

3. Consider the frequency distribution given in Exercise 4, Chapter 2.

   a. Determine P_{15}, P_{45}, and P_{80}.

   b. Determine PR_{67}, PR_{40}, and PR_{34}.

4. Consider the frequency distribution in Exercise 1 above:

   a. Determine the five-number summary.

   b. Construct the box plot from the five-number summary.

5. Consider the frequency distributions in Exercise 5 of Chapter 2:

   a. Determine the five-number summary for each distribution.

   b. Construct the box plot for each distribution.

   c. Compare the two box plots.

6. Consider the frequency distribution in part d of Exercise 2 of Chapter 2:

   a. Determine the five-number summary.

   b. Construct the box plot from the five-number summary.

   c. Determine the "reasonable upper boundary" (RUB) and the "reasonable lower boundary" (RLB).

   d. Construct the modified box plot using the RUB and RLB instead of Max and Min. Are there any potential outliers?

7. The following are the scores of 25 students who participated in a psychology experiment. The scores represent the number of trials required to complete a memorization test. (Consider the students to be a sample.)

12	10	12	11	6
15	14	17	9	12
13	8	7	15	14
15	18	19	14	10
14	14	16	8	9

a. Determine the mean, median, and mode.

b. Determine the range, variance, and standard deviation.

8. Consider the following scores from a sample: 11, 13, 7, 10, 15, 3, 12, 11, 4, 14.

a. Determine the mean.

b. Using these data, show that $\Sigma(X_i - \overline{X}) = 0$.

9. Suppose we add 4 to each of the scores in Exercise 8. What effect will this have on the mean and standard deviation of these scores?

10. The following are the final examination scores of 40 students in a basic statistics class. These scores were randomly selected from the records of all students who have taken the course over the past 10 years and have taken the standardized final examination.

58	86	70	80	82
88	60	80	72	75
89	61	72	76	80
63	73	82	81	89
75	65	82	86	90
75	63	65	84	82
76	68	82	91	94
68	74	79	84	96

a. Determine the mean and median.

b. Determine the variance and standard deviation.

11. Provide graphical illustrations for the frequency distributions from which the following measures of central tendency were derived. Describe the distributions in terms of symmetry, indicating the direction of skewness of asymmetric distributions.

Mean = 46   Median = 43   Mode = 40
Mean = 43   Median = 43   Mode = 43
Mean = 40   Median = 43   Mode = 46

12. A marital satisfaction inventory was given to a sample of married persons with and without children. The following data were obtained.

	n	$\overline{X}$
Male, no children	48	84.3
Female, no children	63	76.8
Male, with children	56	58.8
Female, with children	67	62.6

   a. Find the mean for the total group.

   b. Find the mean for the males and the mean for the females.

   c. Find the mean for the married persons with children and the mean for those without children.

13. Consider the data given in Exercise 10. What is the effect of removing the five highest and the five lowest scores? Describe this new distribution in terms of the range, mean, and standard deviation.

14. For the following set of scores, show that the sum of squares of deviations about the mean ($\overline{X} = 7.4$) is smaller than the sum of squares of deviations about the value $X = 7$.

    5   8   11   3   10   9   6   7   9   6

15. The mean of a set of eight scores is 37. The first seven scores are 40, 29, 33, 43, 39, 35, and 40. What is the eighth score?

16. Determine the transformed scores for the following z scores for a distribution with mean equal to 50 and standard deviation equal to 10: 2.32, 1.84, 0.00, and –0.37.

17. Consider the following set of scores (assume they represent a sample).

    24   18   20   28   15
    25   24   12   26   18
    14   20   24   17   16

   a. Determine the mean ($\overline{X}$) and the standard deviation (s).

   b. Convert each raw score to a standard score.

   c. Show that this set of standard scores has a mean equal to 0 and a standard deviation equal to 1.

18. For the data given in Exercise 17, multiply each of the standard scores by 10 and add 50. Show that the mean of this transformed distribution has a mean of 50 and a standard deviation of 10.

19. Add 5 to each of the scores in Exercise 17. Compute the mean ($\overline{X}$) and standard deviation (s) for this new distribution. Compute z scores for each of these new scores. How do these results differ from those in Exercise 18?

20. Suppose a distribution has a mean of 18 and a standard deviation of 4 and you want to transform this distribution to one that has a mean of 100 and a standard deviation of 10. Describe the process. What raw score in the original distribution corresponds to a transformed score of 115?

*21. John took two proficiency tests: one in mathematics, and the other in the humanities. His score on the mathematics test was 40; his score on the humanities test was 115. John's reference group for the two tests has the following results:

Mathematics: *Humanities:*

$\overline{X}_M = 35.2$ $\overline{X}_H = 107.8$

$s_M = 4.5$ $s_H = 9.6$

Relative to the reference group, on which test did John have the highest performance?

22. John took three standardized tests with the following results:

Test	Score	$\overline{X}$	s
Chemistry	85	82	10
Mathematics	80	75	8
English	80	90	12

  a. Compute a z score for each of John's test scores.

  b. What appears to be John's strongest subject area?

23. As one of his decision-making criteria, a college admissions director uses a composite score based on SAT verbal performance, SAT mathematics performance, and high school grade-point average. The three variables are weighted 2:1:3 in developing the composite score. Assume that high school grade-point averages are normally distributed with a mean of 2.50 and a standard deviation of 0.50 and that SAT scores are normally distributed with a mean of 500 and a standard deviation of 100. Find the composite z score for an applicant with an SAT verbal score of 450, an SAT mathematics score of 575, and a high school grade-point average of 2.35. Do likewise for an applicant with an SAT verbal score of 650, an SAT mathematics score of 625, and a high school grade-point average of 3.18.

4

The Normal Distribution

| Normal distribution | Normalized standard scores |
| Standard (unit) normal distribution | Normal curve equivalent (NCE) scores |

I n previous chapters, we have noted that many variables in the physical and behavioral sciences are normally distributed. The image that comes to mind is a bell-shaped curve representing the distribution of scores. In this chapter, we will take a closer look at the properties of a normal distribution. We will also see how the concept of the normal distribution and the standard scores discussed in Chapter 3 can be combined to yield a tool for analyzing the distributions of a vast number of variables. And we will see how these concepts are used to generate another type of score, normalized standard scores, which have the advantages of both percentile ranks and standard scores.

The Nature of the Normal Distribution

To begin, we must distinguish between data that are normally distributed and the normal distribution itself. When we say that certain variables are "normally distributed," we do *not* mean that these variables have identical distributions or that their distributions exactly match the normal distribution. We do mean that the distributions of these variables approximate the normal distribution. The normal distribution itself is only a model described by a mathematical equation.

This equation is not determined by any specific event in nature, and it does not reflect a specific law of nature. Nevertheless, it provides a good description of the distribution of many sets of data, such as measures of aptitude and achievement, as well as personality and attitude inventories. In later chapters, we will see that many procedures in inferential statistics depend on the assumption that a distribution is normal.

The Family of Normal Distributions

French mathematician Abraham Demoivre (1667–1754) developed the general equation for the normal distribution in the eighteenth century. He based it on his observations of games of chance, and the equation can be used to determine the probability of certain outcomes. We can think of *probability* as the number of possible successes divided by the total number of possible outcomes. Probabilities can range from 0, or no chance of success, to 1, or certainty of success; the sum of the probabilities for all possible outcomes is 1. We will examine probabilities in Chapter 7; for now, we will just take a brief look at the kind of situation Demoivre studied when he developed the equation for the normal distribution.

Suppose we flip a fair coin. This is a situation with two possible outcomes — heads or tails — and each outcome has an equal probability of occurring. So, if we define getting tails on a single flip of a coin as a "success," then the probability of success when we flip the coin is 1/2 — one possibility of getting tails divided by two possible outcomes. If we flip five coins simultaneously, we might get 0 tails, 1 tail, 2 tails, 3 tails, 4 tails, or 5 tails. Figure 4.1A shows the probability of each of these outcomes. Notice that the distribution is symmetric. Now suppose we flip ten coins simultaneously; Figure 4.1B shows the probability of each possible outcome. Again the distribution is symmetric.

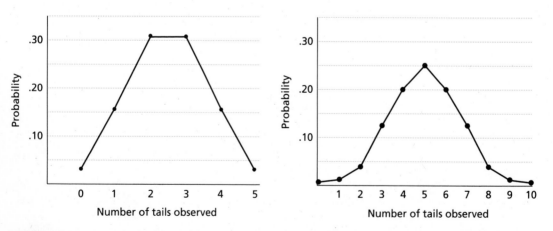

FIGURE 4.1
A. Probability of getting tails flipping five coins simultaneously   B. Probability of getting tails flipping ten coins simultaneously

Demoivre developed a mathematical equation that approximates the frequency polygons shown in Figure 4.1, as well as all possible outcomes of flipping any number of coins. It also defines what is called the **normal distribution**. This equation is

$$Y = \frac{1}{\sigma\sqrt{2\pi}}e^{-(X-\mu)^2/2\sigma^2}$$ (4.1)

where

$Y =$ height of the curve (ordinate) for any given value of X in the distribution of scores

$\pi =$ mathematical value of the ratio of the circumference of a circle to its diameter (3.1416)

$e =$ base of the system of natural logarithms, approximately equal to 2.7183

$\mu =$ mean of the distribution of scores

$\sigma =$ standard deviation of the distribution of scores

This formula is somewhat complicated, but we include it to show that there is a general formula that defines the normal distribution.

Notice that the formula does not specify a particular mean and standard deviation. Instead, for every possible mean (μ) and standard deviation (σ), there is a unique normal distribution. Thus we have, not one normal distribution, but a *family of distributions*. Figure 4.2 illustrates this point. It shows three pairs of normal distributions. The distributions of pair A have the same standard deviation but different means, whereas for pair B the mean is the same but the standard deviations differ. The distributions of pair C have neither mean nor standard deviation in common. All of these curves are members of the family of normal distributions because they are all defined by formula 4.1, but the shape of each individual curve is determined by its particular mean and standard deviation.

The normal distribution is not a single distribution but a family of distributions, each of which is determined by its mean and standard deviation.

Properties of the Normal Distribution

The properties shared by members of the family of normal distributions are as follows:

1. A normal distribution is unimodal (having one mode), symmetrical (that is, the left and right halves are mirror images), and bell shaped, with its maximum height at the mean.

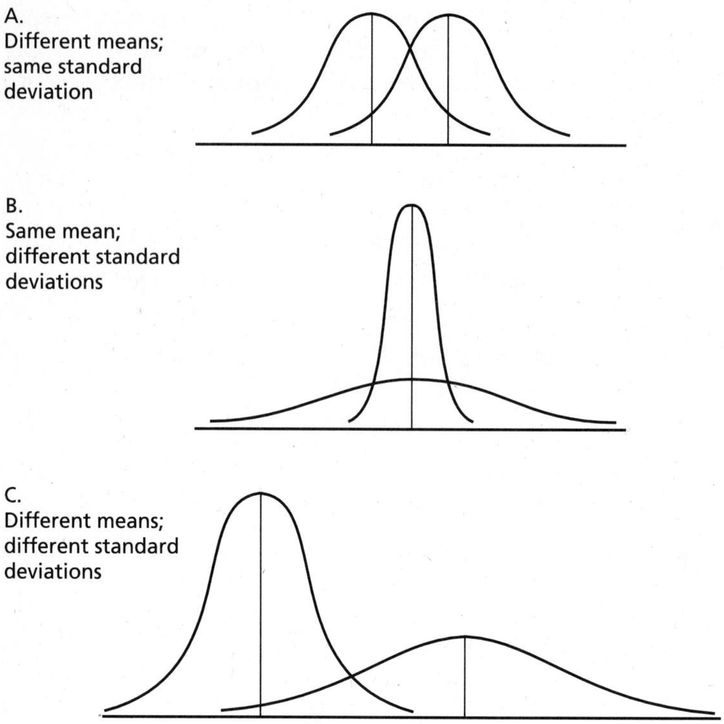

A.
Different means;
same standard
deviation

B.
Same mean;
different standard
deviations

C.
Different means;
different standard
deviations

FIGURE 4.2
Normal distributions with the same
and different means and the same
and different standard deviations

2. A normal distribution is continuous. There is a value of Y (the height) for every value of X where X is assumed to be a continuous rather than a discrete variable. (The scale of measurement for X is the horizontal axis.)

3. A normal distribution is *asymptotic* to the X axis. This means that the farther the curve goes from the mean, the closer it gets to the X axis; but the curve never touches the X axis, no matter how far a particular score is from the mean of the distribution. (This is basically a theoretical property. It allows for the theoretical, but extremely unlikely, possibility that there will be a score more than 3 standard deviations away from the mean.)

The Standard Normal Distribution

The concept of the family of normal distributions gives us a model that can be applied to any number of variables that are normally distributed, regardless of their means and standard deviations. But the number of possible combinations of means and standard deviations is infinite, so the number of different normal distributions is also infinite. To apply the concept of the normal distribution effectively to the many variables that are normally distributed, we need a *standardized* normal distribution that can be used to describe any normally distributed variable.

There is such a normal distribution, called the standard normal distribution, and it is based on the concept of standard scores. The **standard normal distribution** is the distribution of normally distributed standard scores. It is sometimes also called the **unit normal distribution**.

Consider a distribution of scores from a psychological inventory that is normally distributed. If we transform each raw score into a z score, we obtain the standard normal distribution. The distribution of the original scores can be represented by formula 4.1. But because the mean of the z scores equals 0 and the standard deviation equals 1, the normal distribution of z scores can be represented by

$$Y = \frac{1}{\sqrt{2\pi}} e^{-z^2/2} \qquad (4.2)$$

where

$Y =$ height of the curve (ordinate) for any given z score in the standard normal distribution

$z =$ the given standard score, which represents the number of standard deviations of the original score from the mean of the original distribution

$\pi =$ mathematical value of the ratio of the circumference of a circle to its diameter (3.1416)

$e =$ base of the system of natural logarithms, approximately equal to 2.7183

> The standard normal distribution or unit normal distribution is the distribution of normally distributed standard scores with a mean equal to 0 and a standard deviation equal to 1.

Using the Standard Normal Distribution

When a variable is normally distributed and we know the mean and standard deviation, we can use the standard normal distribution to describe the distribution and to determine the percentile rank of a certain score in the distribution, or the Xth percentile of the distribution of scores, or the proportion of scores lying between two points in the distribution.

Suppose we want to know the proportion of scores between the mean and a particular score that lies below the mean. The area under the curve is proportional to the frequency of scores. Thus, the proportion of scores between the mean and the particular score is represented by the area under the curve between the mean and the particular score. Determining this proportion mathematically can be done for any normal distribution using integral calculus; however, it is much easier to transform the particular score into a z score (formula 3.15) and then find the area under the curve for the standard normal distribution. Using this strategy, the area under the curve between the mean of the standard normal distribution (0) and the particular z score would be determined. Again, this process requires integral calculus. However, all the work has been done for us. Statisticians have developed a table of values found in Table C.1 in the Appendix. This table contains the values of z from 0.00 to 4.00 (column A), the proportion of the area between the mean and a given z score (column B), the proportion of the area beyond a given z score (column C), and the height of the standard normal distribution ($Y =$ ordinate) for the given z score (column D).

Figure 4.3 illustrates the use of the table for determining the proportion of the area under the curve that lies between the mean and a given standard score. For example, suppose we want to determine the proportion of the area between $z = 0$ (the mean) and $z = +1$ (one standard deviation above the mean). From Table C.1, we find the proportion to be 0.3413. Thus, approximately 34 percent of the total area falls between the mean and 1 standard deviation above the mean. Similarly, we find from the table that 0.4772, or 47.7 percent, of the area is between $z = 0$ and $z = +2$. The proportion of the area between $z = 0$ and $z = +3$ is 0.4987, or approximately 49.9 percent, as Figure 4.3 shows. Usually, when the standard normal curve is discussed, the area under the curve is defined as 1, so we can refer simply to the *area* of the curve rather than to the "proportion of area."

Because the curve is symmetric, the area between $z = 0$ and $z = +1$ is the same as the area between $z = 0$ and $z = -1$; that is, it is 0.3413. Therefore, the area between $z = -1$ and $z = +1$ is 0.6826, or approximately 68 percent

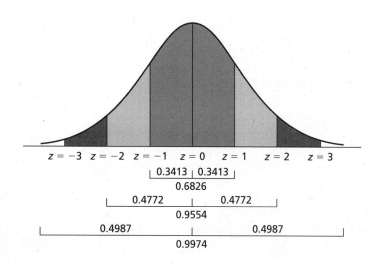

FIGURE 4.3
Areas under the normal curve for
$z = 0, \pm 1, \pm 2, \pm 3$

of the area (see Figure 4.3).

Area between $z = -1.0$ and $z = 0$	$=$	0.3413
Area between $z = 0$ and $z = +1.0$	$=$	0.3413
Area between $z = -1$ and $z = +1.0$	$=$	0.6826

Similarly, the area between $z = -2.0$ and $z = +2.0$ is $0.4772 + 0.4772 = 0.9544$, or approximately 95.4 percent of the area. The area between $z = -3.0$ and $z = +3.0$ is $0.4987 + 0.4987 = 0.9974$, or approximately 99.7 percent of the area.

Notice that the area under the curve for z values greater than $+3.0$ or less than -3.0 is quite small (0.0026). Thus, for all practical purposes, the normal distribution is considered to extend from $z = +3.0$ to $z = -3.0$, or 3 standard deviations on either side of the mean.

Area in the standard normal distribution is given from the mean to the z score or beyond the score. The curve is symmetric, so only areas for positive z scores are given. Approximately 99.7 percent of the area in a normal distribution is contained within 3 standard deviations of the mean—that is, between $z = -3$ and $z = +3$.

Figure 4.4 illustrates use of the table to determine the height (ordinate) of the standard normal curve for several z scores. For example, the height (Y) for the mean ($z = 0$) is 0.3989. For $z = +1$, $Y = 0.2420$; for $z = +2.0$, $Y = 0.0540$. Because the curve is symmetric, the value for $z = -1.0$ is the same as that for

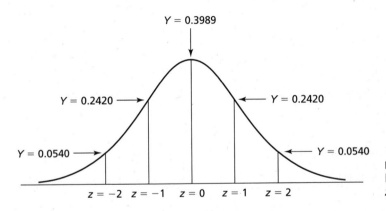

FIGURE 4.4
Heights of the normal curve for
$z = 0, \pm 1, \pm 2$

$z = +1.0$, and the value for $z = -2.0$ equals that for $z = +2.0$. We will use the ordinate in Chapter 19 when computing the biserial correlation coefficient. Notice, too, the asymptotic property of the curve: it approaches the horizontal axis but never reaches it.

Determining Proportions

Consider a distribution of standardized test scores on a psychological inventory for 15,000 students; these scores are assumed to be normally distributed. Suppose we want to determine the proportion and number of scores that are between 0.5 standard deviation *below* the mean and 1.5 standard deviations *above* the mean. In standard score terminology, this means that we want to determine the proportion of the area under the normal curve between $z = -0.5$ and $z = +1.5$. Figure 4.5 illustrates this example. To determine the proportion, we first refer to Table C.1. We find that the area under the standard normal curve between $z = -0.5$ and $z = 0$ is 0.1915. Next we find that the area between $z = 0$ and $z = +1.5$ is 0.4332. Then, since the z scores lie on opposite sides of the mean, the desired proportion is found by adding these two values; that is,

$$
\begin{array}{lcl}
\text{Area between } z = -0.5 & & \\
\quad \text{and } z = 0 & = & 0.1915 \\[1em]
\text{Area between } z = 0 & & \\
\quad \text{and } z = +1.5 & = & \underline{0.4332} \\[1em]
\text{Total area (sum)} & = & 0.6247
\end{array}
$$

Thus 62.47 percent of the test scores are between 0.5 standard deviation below the mean and 1.5 standard deviations above the mean. For this specific example, we

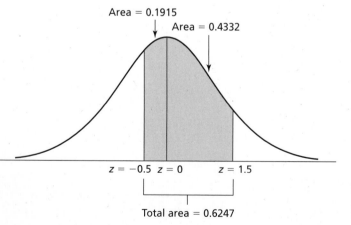

Area = 0.1915

Area = 0.4332

$z = -0.5$ $z = 0$ $z = 1.5$

Total area = 0.6247

FIGURE 4.5
Areas under the normal curve between $z = -0.5$ and $z = +1.5$

can say that

$$0.6247 \times 15,000 = 9,370.5$$

or approximately 9,371 of the scores are in this range.

In the preceding example, the area representing the desired proportion falls on both sides of the mean (see Figure 4.5) so we *add* the two area segments. If the desired area falls between two z scores on the *same* side of the mean, we have to *subtract* to find the proportion. For example, suppose we want to determine the proportion and number of students who have test scores between 1 and 2 standard deviations *above* the mean—that is, between $z = +1.0$ and $z = +2.0$. This example is illustrated in Figure 4.6. Referring to Table C.1, we find that the area under the standard normal curve between $z = 0$ and $z = +1.0$ is 0.3413 and that the area between $z = 0$ and $z = +2.0$ is 0.4772. Because both z scores are on the same side of the mean, we subtract the first area from the second area; that is,

Area between $z = 0$ and $z = +2.0$	= 0.4772
Area between $z = 0$ and $z = +1.0$	= 0.3413
Total area (difference)	= 0.1359

Thus 13.59 percent of the test scores are between 1 and 2 standard deviations above the mean. In other words, $0.1359 \times 15,000 = 2,038.5$, or approximately 2,039 students have test scores in this range.

Consider another example. This time we want to determine the proportion and number of students who have test scores either more than 1.5 standard deviations *below* the mean or more than 2 standard deviations *above* the mean. This example

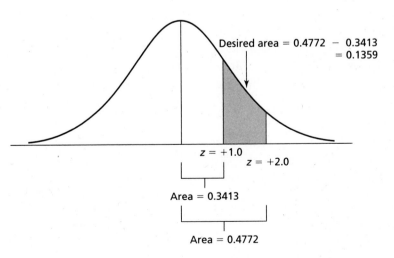

Desired area = 0.4772 − 0.3413
= 0.1359

$z = +1.0$
$z = +2.0$

Area = 0.3413

Area = 0.4772

FIGURE 4.6
Areas under the normal curve between $z = +1.0$ and $z = +2.0$

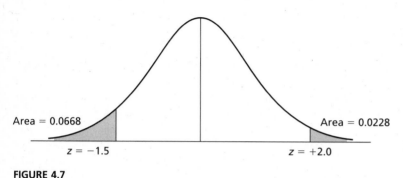

Area = 0.0668

Area = 0.0228

$z = -1.5$

$z = +2.0$

FIGURE 4.7
Areas under the normal curve below $z = -1.5$ and above $z = +2.0$

is illustrated in Figure 4.7. Referring to Table C.1, we find that 0.0228 of the area is beyond (*above*) $z = +2.0$. Thus, $0.0228 \times 15,000 = 342$ of the students have scores in this range. Similarly, 0.0668 of the area is beyond (*below*) $z = -1.5$. Thus, $0.0668 \times 15,000 = 1,002$ of the students have scores in this range. Adding these two values, we find $342 + 1,002 = 1,344$ students with test scores either more than 1.5 standard deviations *below* the mean or more than 2 standard deviations *above* the mean.

> To find areas between z scores in the standard normal distribution, we add the areas if the z scores are on opposite sides of the mean. If the z scores are on the same side of the mean, we subtract the areas.

Determining Percentiles

We can also use the standard normal distribution to find the Xth percentile of a given distribution of scores. (Recall from Chapter 3 that the Xth percentile of a given distribution is the point on the score scale at or below which X percent of the scores fall.) Percentiles below the 50th percentile will have negative z scores; percentiles above the mean will have positive z scores; and the z score of 0 is the 50th percentile or the median.

Suppose we want to determine the 70th percentile (P_{70}) of the distribution of standardized test scores for the 15,000 students. We know that 50 percent of all the scores are below the mean ($z = 0$). From Table C.1, we find that 0.2019 of the area under the standard normal curve is between $z = 0$ and $z = +0.53$. So the area under the curve below $z = +0.53$ is $0.5000 + 0.2019 = 0.7019$. But we want the *precise* z score below which *exactly* 0.7000 of the area falls. Table C.2 can be used to determine this precise z score. Referring to the table, we find that

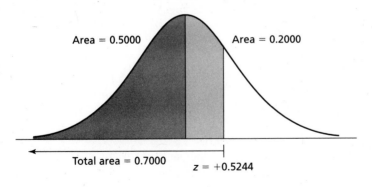

Area = 0.5000   Area = 0.2000

Total area = 0.7000

$z = +0.5244$

FIGURE 4.8
The 70th percentile under the normal
curve—at $z = +0.5244$

0.7000 of the area under the standard normal curve falls below $z = +0.5244$ (see Figure 4.8). Thus the z score of $+0.5244$ is the 70th percentile in the standard normal distribution.

> **To determine percentiles, we find the z score that corresponds to the desired proportion of the area to the left of the z score.**

If we know the percentile in z score form, we can perform a transformation and determine the percentile in raw score form, provided we know the mean and standard deviation of the raw score distribution. Suppose we want to find the 70th percentile on the original scale of measurement for our distribution of 15,000 test scores. We take the z score for the 70th percentile ($z = +0.5244$) and use the general formula for transforming a distribution of standard scores into a distribution of scores with a given mean and standard deviation (see formula 3.18). Assuming that the mean of this distribution is 85 and that the standard deviation is 20, we find the 70th percentile to be

$$X' = (s')(z) + \overline{X}'$$
$$= (20)(0.5244) + 85$$
$$= 95.49$$

Thus, assuming that this distribution of scores is normally distributed, the 70th percentile (P_{70}) is 95.49.

Determining Percentile Ranks

We reverse the process for finding percentiles when we use the standard normal distribution to find the percentile rank of a given score. For example, suppose we want to find the percentile rank of the score 102 on the standardized psychological test we have been discussing—the one with a mean of 85 and a standard deviation

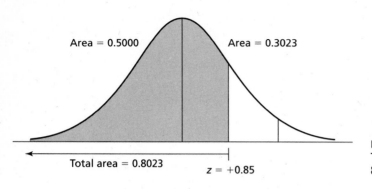

FIGURE 4.9
The percentile rank for $z = +0.85$ is 80.23

of 20. The first step is to determine the z score that corresponds to this raw score. For this example,

$$z = \frac{X - \overline{X}}{s}$$

$$= \frac{102 - 85}{20}$$

$$= +0.85$$

The next step is to determine the proportion of the area that is below this z score. Referring to Table C.1, we find that 0.3023 of the area is between $z = 0$ and $z = +0.85$ (see Figure 4.9). Since 0.5000 of the area is below $z = 0$, $0.5000 + 0.3023 = 0.8023$ of the area is below $z = +0.85$, which means that the percentile rank of the score 102 is 80.23.

In a similar manner, we can find the percentile rank of a raw score below the mean. We know that such a score will have a percentile rank less than 50. Consider the raw score 80. The corresponding z score is

$$z = \frac{80 - 85}{20}$$

$$= -0.25$$

From Table C.1, we find that 0.4013 of the area under the standard normal curve is to the left of $z = -0.25$ (see Figure 4.10). Therefore, we can say that the percentile rank of the raw score 80 is 40.13.

To determine the percentile rank of a raw score, we transform the score into a z score and then determine the proportion of the area below that z score.

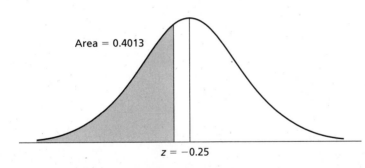

FIGURE 4.10
The percentile rank for $z = -0.25$ is 40.13

Normalized Standard Scores

In Chapter 3, we indicated that the ordinal nature of percentiles and percentile ranks limits their use in mathematical manipulations and statistical analyses, such as summing across subtest scores. This limitation can be overcome, however, by generating normalized standard scores. **Normalized standard scores** are transformed scores that incorporate the advantages of both percentile ranks and standard scores. In the transformation, the percentile rank of the raw score is determined. This percentile rank is then transformed into a z score using Table C.2. As a result, normalized standard scores retain the properties of percentile ranks and also have the advantages of an equal-interval scale.

Examples of normalized standard scores are the **normal curve equivalent (NCE) scores**, which are used to interpret the results of standardized achievement tests. The NCE score scale ranges from 1 to 99 with a mean of 50 and a standard deviation of 21.[1] A score of 1 corresponds to a percentile of 1, a score of 50 to the 50th percentile, and a score of 99 to the 99th percentile. However, for other scores, the NCE and the percentile do not necessarily coincide. Figure 4.11 contains the standard normal distribution and the relative positions of NCE scores, percentiles, and z scores. Notice the inequality of spacing on the percentile scale, which indicates that (as we mentioned earlier) percentiles are not an equal-interval scale. On the other hand, the NCE score scale and the z score scale are equal-interval scales. Because NCE scores have the characteristics of both percentile ranks and an equal-interval scale, we can use them to compare test batteries or subtests within the same battery.

Computing NCE scores involves three steps:

1. Convert the raw score to a percentile rank.

2. Convert the percentile rank to a z score using Table C.2.

3. Convert the z score to the NCE score using the general transformation formula (formula 3.18).

---

[1] The standard deviation is actually 21.38. However, for our purposes, we will use $s = 21$.

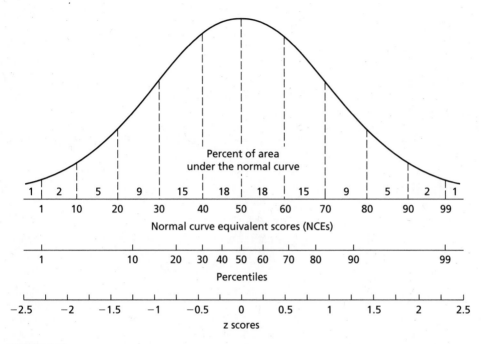

FIGURE 4.11
Positioning of NCEs, percentiles, and z scores for the normal curve

To illustrate this computation, we will use a score from the distribution of final examination scores in Table 3.1. Suppose we want to determine the percentile rank for the raw score of 61. In Chapter 3 we found the percentile rank for the raw score of 61 to be 90.83. (PR_{61} = 90.83), so the first step is completed. Next we use Table C.1 to determine the z score that corresponds to this percentile rank. Rounding this percentile rank to 91 and referring to Table C.1, we find that the corresponding z score is 1.3408. Finally, using the general transformation formula, the NCE score is

$$X' = (s')(z) + \overline{X}'$$
$$\text{NCE}_{61} = (21)(1.3408) + 50$$
$$= 28.16 + 50$$
$$= 78.16$$

The normal curve equivalent (NCE) score is a standard score with mean of 50 and standard deviation of 21. NCE scores range from 1 to 99, and equal intervals are retained in the scale.

Summary

The normal distribution is a very important concept in the behavioral sciences because many variables used in behavioral research are assumed to be normally distributed. Because each variable has a specific mean and standard deviation, there is a family of normal distributions rather than just a single distribution. However, if we know the mean and standard deviation for any normal distribution, we can transform it into the standard normal distribution. The standard normal distribution is the normal distribution in standard score form with mean equal to 0 and standard deviation equal to 1.

The standard normal distribution is presented in Table C.1, with areas and ordinates for the respective z scores. The table contains values for positive z scores only. However, because the distribution is symmetric, corresponding negative z scores have the same values. Areas are given from the mean ($z = 0$) to the specific z scores and beyond those z scores. We can use the table to find areas between z scores, percentiles, and percentile ranks. It is a good idea, at least while learning to use the table, to sketch the areas you are seeking so that you will add or subtract areas properly.

Normalized standard scores provide relative positioning in the normal distribution yet retain an equal-unit measurement in the measurement scale. For example, the normal curve equivalent (NCE) score is essentially a standard score (z score) that is normally distributed with mean equal to 50 and standard deviation equal to 21. NCE scores range from 1 to 99, inclusive.

It is extremely important to know how to use Table C.1, for two reasons. Not only is it necessary for transforming any normal distribution into the standard normal distribution, but the technique of using a statistical table will also be applied to other distributions later in the book.

Exercises

*1. Assume that a distribution of 150 scores is normally distributed with a mean of 92 and a standard deviation of 11.5.

   a. Find the z scores corresponding to observed scores of 87 and 96.

   b. Determine the number of scores between observed scores of 80 and 100; between observed scores of 95 and 100.

   c. Find the 80th percentile, P_{80}.

2. Assume that a set of 200 scores is normally distributed with a mean of 60 and a standard deviation of 12.

   a. What are the z scores corresponding to the raw scores of 76, 38, and 50?

   b. How many scores lie between the values of 48 and 80? 65 and 75? 34 and 52?

   c. How many scores exceed the values of 80, 60, and 40?

   d. How many scores are less than the values of 35, 50, and 75?

   e. Find P_{35}, P_{80}, PR_{55}, and PR_{70}.

3. The norms for a standardized mathematics test, assumed to be normally distributed, are as follows:

   | National norms | $\overline{X} = 75$ | $s = 12$ |
   | Large-city norms | $\overline{X} = 68$ | $s = 15$ |

   John has a score of 80, and Mary has a score of 65. What are their percentile ranks in terms of the national norms? in terms of the large-city norms?

4. A statistics instructor tells the class that grading will be based on the normal distribution. He plans to give 10 percent A's, 20 percent B's, 40 percent C's, 20 percent D's, and 10 percent F's. If the final examination scores have a mean of 75 and a standard deviation of 9.6, what is the range of scores for each grade?

5. For the data given in Table 2.10, convert the following scores to NCE scores: 60, 43, and 33.

6. The dean of a college wishes to know the percentage of students who finish a bachelor's degree program in 3.5 years or less. He knows the average number of years to earn the degree is 4.3 years, with a standard deviation of 0.3 years. Assume a normal distribution in calculating the percentage.

7. The results of an English exam show a mean of 55, with a standard deviation of 6. What percentage of students scored below 46? above 68? Assume a normal distribution.

8. In an ancient culture, the average male life span was 37.6 years, with a standard deviation of 4.8 years. The average female life span was 41.2 years, with a standard deviation of 7.7 years. Use the properties of the normal distribution to find:

   a. What percentage of men died before age 30?

   b. What percentage of women lived to an age of at least 50?

   c. At what age is a female death at the same relative position in the distribution as a male death at age 35?

5

Correlation:
A Measure of Relationship

Key Concepts

Correlation
Correlation coefficient
Slope
Pearson product-moment correlation
  coefficient
Cross-products
Deviation score formula

Raw score formula
Covariance
Linear relationship
Curvilinear relationship
Homogeneity
Coefficient of determination
Spearman rho

T hus far, we have discussed how to describe the distribution of a single variable by using measures of central tendency and of variation. But suppose a researcher has measurements for one group of individuals on two variables — say, SAT scores and final examination scores in introductory psychology — and notices that the scores for the two variables seem to be related. In other words, the researcher sees that some of the students who had high scores on the SAT also had high scores on the psychology final examination. The researcher can describe each variable and each distribution separately, but these descriptions would not indicate the extent to which the two variables are related. Did all of the individuals who scored high on the SAT also score high on the final examination? Did most of those with low SAT scores also have poor examination grades? In more general terms, if performance on one variable is high, does performance on the other variable also tend to be high? If performance on one variable is low, does it tend to be low on the other as well?

Studies of relationships between variables are common in behavioral science. A psychologist might study the relationship between style of child rearing and adult personality, between types of therapy and clients' progress, or between self-esteem and success. An educational researcher might be interested in the relationship between creativity and academic performance or between concept acquisition and types of instruction. A sociologist might be concerned with the relationship between frequency of violent crimes and population density. The number of research questions that can involve the relationship between two variables is almost limitless.

In this chapter, we discuss the standard measure of the relationship between two variables, the Pearson product-moment correlation coefficient. We will see how it is computed, when it can be used, what factors affect its size, and how it can be interpreted. We also introduce the Spearman rho coefficient for use in situations in which the data consist of rankings.

The Meaning of Correlation and the Correlation Coefficient

To illustrate what we mean by a relationship between two variables, let's use the example of final examination scores and SAT scores. Table 5.1 contains the final examination score (Y) and the quantitative SAT score (X) for each of 15 students.

TABLE 5.1
Quantitative SAT Scores and Final Examination Scores for 15 Introductory Psychology Students

Student	Quantitative SAT Score (X)	Final Examination Score (Y)
1	595	68
2	520	55
3	715	65
4	405	42
5	680	64
6	490	45
7	565	56
8	580	59
9	615	56
10	435	42
11	440	38
12	515	50
13	380	37
14	510	42
15	565	53
Σ	8,010	772
	$\overline{X} = 534.00$	$\overline{Y} = 51.47$
	$s_X = 96.53$	$s_Y = 10.11$

What patterns do we find in these data? If we inspect Table 5.1 carefully, we notice that in general those students with high SAT scores tend to have high scores on the final examination; those students with mediocre SAT scores usually have mediocre final examination scores; and those with low SAT scores tend to have low scores on the final examination. In short, in this case there is a tendency for students to score similarly on both variables. Performance on the two variables is related; in other words, there is a **correlation** between the two variables.

Scatterplots and Correlation

Scatterplots, which we introduced in Chapter 2, provide a picture of the relationship between variables. Recall that, in a scatterplot, each point represents paired measurements on two variables for a specific individual. Figure 5.1 is the scatterplot for the data given in Table 5.1. Each student is represented by one point on the scatterplot, which corresponds to the intersection of straight lines drawn through the student's SAT score (measured on the X axis) and the same student's final examination score (measured on the Y axis). The points representing the paired measurements for students 1, 8, and 13 are labeled in Figure 5.1.

Notice that the points in Figure 5.1 produce a lower-left to upper-right pattern. This pattern occurs when there is a *positive* relationship, or *positive correlation*, between two variables. Other patterns occur in a scatterplot when there are other

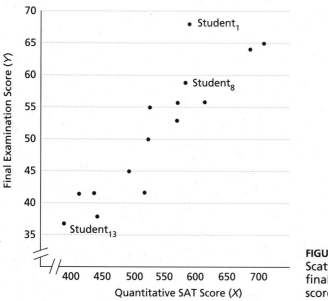

FIGURE 5.1
Scatterplot illustrating the relationship between final examination scores (Y) and quantitative SAT scores (X)

relationships between the variables, as illustrated in Figure 5.2. When there is a *negative correlation* between two variables, the points in the scatterplot form an upper-left to lower-right pattern, as Figure 5.2B illustrates. If there is a perfect correlation between two variables, all the points in the scatterplot lie on a straight line. Figure 5.2C shows a perfect positive correlation and Figure 5.2D, a perfect negative correlation. A scatterplot in which the points do not have an upward trend or a downward trend occurs when there is zero or near-zero correlation, as shown in Figure 5.2E.

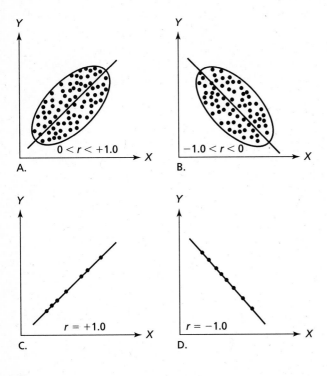

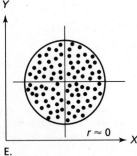

FIGURE 5.2
Scatterplots illustrating varying degrees of relationship between X and Y

Correlation Coefficients

We can obtain some notion of the relationship between two variables by simply inspecting a scatterplot, but this procedure is not sufficiently precise for most statistical purposes. Instead, we must compute a quantitative index of the relationship. The **correlation coefficient** is an index that describes the extent to which two sets of data are related; it is a measure of the relationship between two variables.[1]

A correlation coefficient can take on values between −1.0 and +1.0, inclusive. The sign indicates the direction of the relationship. A plus indicates that the relationship is positive; a minus sign indicates that the relationship is negative. The absolute value of the coefficient indicates the magnitude of the relationship. In other words, a correlation coefficient of +.10 or −.10 indicates that there is little, if any, relationship between the variables, whereas a coefficient of +.90 or −.90 indicates a strong relationship. Thus the value of a perfect positive correlation coefficient is +1.0, and the value of a perfect negative correlation coefficient is −1.0. When there is no relationship between two variables, the correlation coefficient is 0.

The numeric value of a correlation coefficient is a function of the slope of the general pattern of points in the scatterplot and the width of the ellipse that encloses those points. The **slope** indicates the general direction of the relationship. For instance, the direction of the points in Figure 5.1 is from the lower left to the upper right. This illustrates a positive slope, for which we say there is a positive relationship. In other words, the sign of the correlation coefficient is positive. For a negative slope, the direction of the points in the scatterplot is from the upper left to the lower right, and the sign of the correlation coefficient is negative. The width of the ellipse indicates the extent of the relationship and hence the magnitude, or absolute value, of the correlation coefficient. For example, compare the ellipse around the points in Figure 5.2A with that around the points in Figure 5.2E. The narrower ellipse indicates that the degree of relationship is greater and the correlation coefficient is larger.

> The computed value of the correlation coefficient can range from −1.0 to +1.0, inclusive. The sign of the coefficient indicates the direction of the relationship; the absolute value of the coefficient indicates the magnitude of the relationship.

---

[1] In discussing the correlation coefficient in this chapter, we will assume that we are dealing with samples rather than with populations. The formulas for the sample correlation coefficients are presented in the text.

Computing the Pearson r

The correlation coefficient used most often in the behavioral sciences is the **Pearson product-moment correlation coefficient**, symbolized by r. It was developed by the English statistician Karl Pearson (1857–1936).

To understand this coefficient, suppose there is a positive relationship between X and Y. If an individual has a score on variable X that is above the mean of X ($\overline{X}$), this individual is likely to have a score on the Y variable that is above the mean of Y ($\overline{Y}$). An individual with an X score below $\overline{X}$ is likely to have a Y score below $\overline{Y}$. Similarly, if the relationship between the variables is negative, then an individual with an X score above $\overline{X}$ is likely to have a Y score below $\overline{Y}$, and vice versa.

Pearson developed the product-moment correlation coefficient with these relationships in mind. The coefficient involves computing the sum of **cross-products**; that is, multiplying the two scores (X and Y) for each individual and then summing these cross-products across n individuals. This sum is then divided by $n - 1$. In essence, the product-moment correlation coefficient is the mean of the cross-products of scores.

Because of the difference in the measurements for the two variables being correlated, Pearson used standard scores rather than raw scores (see Chapter 3) in developing the product-moment correlation coefficient. He defined it as follows:

$$r_{XY} = \frac{\Sigma z_X z_Y}{n - 1} \tag{5.1}$$

In other words, the coefficient r is expressed as the sum of the cross-products of the standard scores (z_X and z_Y) divided by $n - 1$.

As an example, we can use the data in Table 5.1. If we used formula 5.1 as a computational formula, which would be cumbersome, we would first calculate the mean and standard deviation of the SAT scores, $\overline{X} = 534.00$ and $s_X = 96.53$. Then we calculate the mean and standard deviation of the final examination scores, $\overline{Y} = 51.47$ and $s_Y = 10.11$. Next we convert each raw score to a standard score. This is illustrated below for the first pair of scores:

$$z_X = \frac{595 - 534.00}{96.53} = 0.63 \qquad z_Y = \frac{68 - 51.47}{10.11} = 1.64$$

Then we calculate the cross-products of these standard scores, as shown in Table 5.2. The sum of the cross-products is then divided by $n - 1$, as follows:

$$r_{XY} = \frac{\Sigma z_X z_Y}{n - 1}$$

$$= \frac{12.67}{14}$$

$$= 0.90$$

TABLE 5.2
Data for Calculating the Pearson Product-Moment Correlation Coefficient Using Formula 5.1

X	Y	z_X	z_Y	$z_X z_Y$
595	68	0.63	1.64	1.03
520	55	−0.15	0.35	−0.05
715	65	1.88	1.34	2.52
405	42	−1.34	−0.94	1.26
680	64	1.51	1.24	1.87
490	45	−0.46	−0.64	0.29
565	56	0.32	0.45	0.14
580	59	0.48	0.74	0.36
615	56	0.84	0.45	0.38
435	42	−1.03	−0.94	0.97
440	38	−0.97	−1.33	1.29
515	50	−0.20	−0.15	0.03
380	37	−1.60	−1.43	2.29
510	42	−0.25	−0.94	0.24
565	53	0.32	0.15	0.05
Σ 8,010	772	0.00	0.00	12.67

The Pearson product-moment correlation coefficient is the average cross-product of the standard scores of two variables.

Using formula 5.1 to compute the correlation between two variables is arduous, because each raw score must first be converted to a *z* score. However, by using the definition of the standard score and standard deviations, we can derive more convenient formulas for computing the correlation coefficient. One of these computational formulas is the deviation score formula; the other is the raw score formula. Finally, we will describe how to compute *r* using the covariance.

Deviation Score Formula

Recall from Chapter 3 that the standard score is defined as

$$z = \frac{X - \overline{X}}{s}$$

or, using other notation,

$$z_X = \frac{X - \overline{X}}{s_X} = \frac{x}{s_X} \quad \text{and} \quad z_Y = \frac{Y - \overline{Y}}{s_Y} = \frac{y}{s_Y}$$

In addition, the standard deviation of a set of scores was defined in Chapter 3 as

$$s = \sqrt{\frac{SS}{n-1}} = \sqrt{\frac{\Sigma x^2}{n-1}}$$

Therefore,

$$s_X = \sqrt{\frac{\Sigma x^2}{n-1}} \quad \text{and} \quad s_Y = \sqrt{\frac{\Sigma y^2}{n-1}}$$

Substituting these values for z_X and z_Y in formula 5.1, we get

$$r_{XY} = \frac{\Sigma xy}{\sqrt{\Sigma x^2 \Sigma y^2}} \tag{5.2}$$

Because x and y are the deviations from the means, formula 5.2 is often called the **deviation score formula**.

As an example, we will calculate the correlation between the SAT (X) and final examination scores (Y) cited earlier. These calculations are summarized in Table 5.3.

1. Compute a deviation score for each individual on both variables (x and y).

2. Compute the cross-products of the deviation scores. For example, the cross-product for student 1 is $(61)(16.53) = 1,008.33$.

TABLE 5.3
Data for Calculating the Correlation Coefficient Using the Deviation Formula

	X	Y	x	y	xy	x²	y²
	595	68	61.0	16.53	1,008.33	3,721.0	273.24
	520	55	− 14.0	3.53	− 49.42	196.0	12.46
	715	65	181.0	13.53	2,448.93	32,761.0	183.06
	405	42	− 129.0	− 9.47	1,221.63	16,641.0	89.68
	680	64	146.0	12.53	1,829.38	21,316.0	157.00
	490	45	− 44.0	−6.47	284.68	1,936.0	41.86
	565	56	31.0	4.53	140.43	961.0	20.52
	580	59	46.0	7.53	346.38	2,116.0	56.70
	615	56	81.0	4.53	366.93	6,561.0	20.52
	435	42	− 99.0	− 9.47	937.53	9,801.0	89.68
	440	38	− 94.0	− 13.47	1,266.18	8,836.0	181.44
	515	50	− 19.0	− 1.47	27.93	361.0	2.16
	380	37	− 154.0	− 14.47	2,228.38	23,716.0	209.38
	510	42	− 24.0	− 9.47	227.28	576.0	89.68
	565	53	31.0	1.53	47.43	961.0	2.34
Σ	8,010	772	0.0	0.0	12,332.00	130,460.0	1,429.72

3. Compute the squares of the deviation scores. For student 1, $(61)^2 = 3,721$ and $(16.53)^2 = 273.24$.

4. Sum the cross-products ($\Sigma xy = 12,332$) and the squares ($\Sigma x^2 = 130,460$ and $\Sigma y^2 = 1,429.72$).

Applying formula 5.2 to these data,

$$r = \frac{\Sigma xy}{\sqrt{\Sigma x^2 \Sigma y^2}}$$

$$= \frac{12,332}{\sqrt{(130,460)(1,429.72)}}$$

$$= 0.90$$

Raw Score Formula

We have included formula 5.2 because it is often used in texts on advanced measurement. As illustrated in the above example, it is, like formula 5.1, quite tedious to use. What we would like is a formula that requires fewer calculations and uses raw scores rather than deviation scores. By substituting $(X - \overline{X})$ for x and $(Y - \overline{Y})$ for y and algebraically manipulating formula 5.2, we obtain the following **raw score formula** for the Pearson r:

$$r_{XY} = \frac{n \Sigma XY - \Sigma X \Sigma Y}{\sqrt{[n \Sigma X^2 - (\Sigma X)^2][n \Sigma Y^2 - (\Sigma Y)^2]}} \tag{5.3}$$

The data for calculating the correlation coefficient using the raw score formula are found in Table 5.4. Notice that the preliminary calculations are done on the raw scores rather than on the deviation scores. Applying formula 5.3 to these data,

$$r = \frac{15(424,580) - (8,010)(772)}{\sqrt{[15(4,407,800) - (8,010)^2][15(41,162) - (772)^2]}}$$

$$= 0.90$$

You will probably use the raw score formula most often when computing the Pearson r with a hand calculator.

Using Covariance to Find r

The covariance of X and Y is another way of expressing the relationship between the two variables. Unfortunately, the size of the covariance depends on the units of measurement of each variable. It is not bounded by $+1.0$ and -1.0 as is the correlation coefficient. Even so, there are times when the covariance is a useful

TABLE 5.4
Data for Calculating the Correlation Coefficient Using the Raw Score Formula

X	Y	XY	X^2	Y^2
595	68	40,460	354,025	4,624
520	55	28,600	270,400	3,025
715	65	46,475	511,225	4,225
405	42	17,010	164,025	1,764
680	64	43,520	462,400	4,096
490	45	22,050	240,100	2,025
565	56	31,640	319,225	3,136
580	59	34,220	336,400	3,481
615	56	34,440	378,225	3,136
435	42	18,270	189,225	1,764
440	38	16,720	193,600	1,444
515	50	25,750	265,225	2,500
380	37	14,060	144,400	1,369
510	42	21,420	260,100	1,764
565	53	29,945	319,225	2,809
Σ 8,010	772	424,580	4,407,800	41,162

number. We can use the covariance to find r.

$$\text{covariance} = s_{XY} = \frac{\Sigma(X - \overline{X})(Y - \overline{Y})}{n - 1} = \frac{\Sigma xy}{n - 1} \qquad (5.4)$$

The covariance is the average cross-product of the deviation scores. We again use $n - 1$ in the denominator because we are using data from a sample. For the data in our example taken from Table 5.3:

$$s_{XY} = \frac{12,332}{14} = 880.86$$

We can find the Pearson r by dividing the covariance by the product of the standard deviations of X and Y:

$$r_{XY} = \frac{s_{XY}}{s_X s_Y} \qquad (5.5)$$

Again, substituting the data from the example, we find the Pearson r:

$$r = \frac{880.86}{(96.53)(10.11)} = 0.90$$

Factors Affecting the Size of the Pearson r

If we are interested in determining the relationship between two variables, we cannot simply take any two sets of data and compute r. First, we must satisfy two conditions:

1. The two variables to be correlated must be *paired observations* for the same set of individuals or objects. We cannot determine a correlation coefficient, for example, between anxiety and achievement when one group of thirty students has taken an anxiety inventory and another group has taken an achievement test. The same group must take both tests.

2. Because we use the mean and variance in computing *r*, the variables being correlated must be measured on an interval or ratio scale.

Second, we must be aware of factors that affect the size of the Pearson *r*; these will be discussed below.

Linearity

The Pearson *r* is an index of the linear relationship between two variables. To understand what a linear relationship is, consider the scatterplots in Figure 5.3. The first scatterplot, in Figure 5.3A, illustrates a linear trend in the relationship between variables *X* and *Y*. Thus, when we say that a **linear relationship** exists,

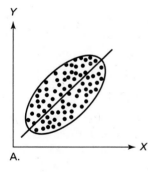

A.

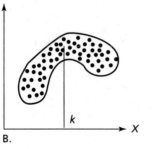

B.

C.

FIGURE 5.3
Scatterplots illustrating linear and nonlinear relationships

we do not mean that the points fall exactly on a straight line but only that the points are located generally along the line. The trend of the data is then said to be *linear*.

In contrast, Figures 5.3B and 5.3C exhibit **curvilinear relationships** between the variables; that is, straight lines do not fit the scatterplots, but some curved lines do. In Figure 5.3B, increasingly greater values of X are associated with increasingly greater values of Y up to some intermediate point; then, as the values of X increase, they are associated with *decreasing* values of Y. The relationship between anxiety and performance might follow such a trend. As subjects' anxiety levels increase, so do their performances, but only up to a point. At that point, an increase in anxiety is no longer associated with a further increase in performance.

Many pairs of variables in the behavioral sciences have linear relationships, and many do not. Some variables have a linear relationship over part of the possible scale of measurement but not over the entire scale. A good example is the correlation between age and psychomotor or cognitive performance. The correlation tends to be highly positive during childhood, then approaches zero during adulthood, and then may even become negative during old age. If a researcher computes the correlation between such variables over all ages, the correlation might well be zero, and the researcher might conclude that there is no relationship between the variables. Instead, there is a relationship, but it is not consistent for all ages.

> **The Pearson r is an index of the *linear* relationship between two variables.**

To illustrate what happens when the Pearson r is applied to data that are not linearly related, consider the data in Table 5.5. A perfect relationship exists between the X scores and the Y scores; the relationship is defined by the mathematical equation $Y = X^2$. Figure 5.4 is the scatterplot for these data. It illustrates the nonlinear, but perfect, relationship in which all the points fall on the curve. However, if we apply formula 5.3 to these data, we find that the Pearson r is 0.0. If the Pearson r is used with curvilinear data, the result will not be an accurate index of the actual relationship between the two variables. The resulting *underestimate* will be substantial when the trend is markedly curvilinear.

> **If the Pearson r is applied to variables that are curvilinearly related, it will underestimate the relationship between the variables.**

The Pearson r is the index of the *linear* relationship between two variables, and it is not an appropriate measure for describing the curvilinear relationships illustrated in Figures 5.3B and 5.3C. There are ways to handle curvilinear data, but

TABLE 5.5
Data Illustrating a Perfect Nonlinear Relationship Using the Mathematical Equation $Y = X^2$

X	Y
4	16
3	9
2	4
1	1
0	0
−1	1
−2	4
−3	9
−4	16

we will discuss them in Chapter 19. For now, if there is any doubt that data exhibit a linear trend, it is a good idea to develop a scatterplot before using the Pearson r.

Homogeneity of the Group

A second factor that affects the size of r is the homogeneity of the group for which the two variables are being correlated. When a group is homogeneous, the range of scores on either or both variables is restricted; as the **homogeneity** of a group increases, the variance decreases. If a group is sufficiently homogeneous on either or both variables, the variance (and hence the standard deviation) tends toward zero. Notice from formula 5.5 that, when this happens, we are dividing by zero, and the formula becomes meaningless. In essence, the variable has been reduced

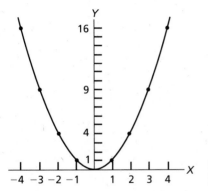

FIGURE 5.4
Data illustrating perfect nonlinear relationship

to a constant. As a group under study becomes increasingly homogeneous, the correlation coefficient decreases.

Consider an example. Suppose a researcher is interested in the relationship between the results of a mathematics aptitude test for advanced placement and first-semester grade-point averages for 75 college freshmen. These data are illustrated in Figure 5.5. The correlation coefficient for these data is 0.69. Suppose that advanced placement is given only to those 34 students who score above 60 on the aptitude test. The correlation coefficient for this homogeneous subgroup is 0.40.

This example illustrates an important concept. When other factors are held constant, the size of the correlation coefficient describing the relationship between two variables will depend on the homogeneity of the group under investigation.

> As the group under study becomes increasingly homogeneous on one or both variables, the absolute value of the correlation coefficient tends to become smaller.

What implications does this have for using the correlation coefficient? One implication, if the researcher is looking for relationships between variables, is that there must be enough variation or heterogeneity in the scores to allow a

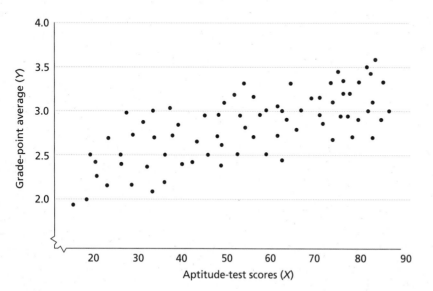

FIGURE 5.5
Scatterplot illustrating the relationship between grade-point averages (*Y*) and aptitude-test scores (*X*)

relationship to manifest itself. If, for example, we are investigating the relationship between IQ scores and performance on a cognitive task, we need to include a wide range of IQ scores. If only individuals with IQs above 140 are included, we are likely to find a very low correlation between IQ and performance on the cognitive task. Does that mean that the two variables are unrelated? Definitely not. It means that the restriction on the IQ scores has limited the magnitude of the correlation coefficient. The coefficient so determined would *not* be representative of the relationship between the variables over a wider range of IQ scores.

If, on the other hand, the research is concerned with a very specific group, such as individuals with IQ scores above 140, then, of course, the data must also come from these individuals. However, the researcher must be careful not to misinterpret the results. A low *r* does not indicate that performance on the cognitive task is not generally related to IQ. It indicates that *for this group*, IQ does very little to differentiate performances on the cognitive task. If the researcher is looking for characteristics within this group that do differentiate, it would be better to consider other characteristics, ones for which the group is more variable.

Size of the Group

Consider again the example of the 34 freshmen in the homogeneous subgroup. Because the smaller correlation coefficient for the subgroup was computed using a smaller number of scores, one might at first propose that the size of the group affected the size of the correlation coefficient. This is *not* the case. The number of observations used in the calculation of the Pearson *r* does *not* influence the value of the coefficient.[2] The number of cases does affect the *accuracy* of the relationship, and this relationship will be discussed in later chapters.

> In general, the size of the group does not affect the size of the correlation coefficient.

Interpreting the Correlation Coefficient

We know that researchers use statistical procedures to enhance their understanding of data, and we also know that the correlation coefficient is a measure of the relationship between two variables. But how large must the coefficient be in order

---

[2]There is one rather trivial exception to this, and that is when $n = 2$. Since, geometrically, a straight line is determined by two points, if $n = 2$ and the two scores on the same variable are not equal, the scatterplot will contain only two points. A straight line can be fitted perfectly, and r will be -1.0, 0, or $+1.0$.

to be noteworthy, and what does r tell us about the data? After the correlation coefficient has been computed, what does it mean?

The Scale of r

Suppose we compute r for three sets of data and find values of .40, .60, and .80. We *cannot* say that the difference between $r = 0.40$ and $r = 0.60$ is the same as the difference between $r = 0.60$ and $r = 0.80$, or that $r = 0.80$ is twice as large as $r = 0.40$. Why? The scale of values for the correlation coefficient is not interval or ratio, but *ordinal*. Therefore, all we can say is that, for example, a correlation coefficient of +.80 indicates a high positive linear relationship and a correlation coefficient of +.40 indicates a somewhat lower positive linear relationship.

> The scale of the correlation coefficient is ordinal.

If we find a correlation coefficient of 0.40 for two variables, does that mean that the relationship between the variables is important? It depends. Deciding what magnitude of r indicates a noteworthy relationship is somewhat arbitrary; the interpretation depends to some extent on the variables under consideration. Nevertheless, there are criteria that we can use as a rule of thumb for interpreting r when we are using it as a descriptive measure; these criteria are listed in Table 5.6. In the following section, we will see why these guidelines are reasonable.

Interpreting r in Terms of Variance

The correlation coefficient provides not only a measure of the relationship between variables but also an index of the *proportion of individual differences* in one variable that can be associated with the individual differences in another variable. It can tell us how much of the total variance of one variable can be associated with the variance of another variable. For example, consider again the data in Figure 5.5 and

TABLE 5.6
Rule of Thumb for Interpreting the Size of a Correlation Coefficient

Size of Correlation	Interpretation
.90 to 1.00 (−.90 to −1.00)	Very high positive (negative) correlation
.70 to .90 (−.70 to −.90)	High positive (negative) correlation
.50 to .70 (−.50 to −.70)	Moderate positive (negative) correlation
.30 to .50 (−.30 to −.50)	Low positive (negative) correlation
.00 to .30 (.00 to −.30)	Little if any correlation

the correlation coefficient of 0.69 between the aptitude-test scores of 75 students and their first-semester grade-point averages (GPAs). This tells us that, for the most part, high scores on the aptitude test were associated with high GPAs and vice versa. The correlation coefficient of 0.69 also suggests that factors other than aptitude, such as health and motivation, could also contribute to individual differences in the GPA. Thus, we can separate the variance in GPA for the total group into two components: the variance in GPA associated with differences in aptitude and the variance in GPA associated with other factors. Symbolically,

$$s_Y^2 = s_A^2 + s_B^2 \tag{5.6}$$

where

$s_Y^2 =$ the total variance in GPA

$s_A^2 =$ the variance in GPA associated with differences in aptitude

$s_B^2 =$ the variance in GPA associated with other factors

The higher the correlation between a group's GPA and its aptitude scores, the larger the portion of the total variance in GPA that can be associated with the variance in aptitude. The specific association is given by the equation

$$r^2 = \frac{s_A^2}{s_Y^2} \tag{5.7}$$

where

$s_A^2 =$ the variance in Y that is associated with the variance of X

$s_Y^2 =$ the total variance of Y

That is, *the square of the correlation coefficient* (r^2) *equals the proportion of the total variance in Y that can be associated with the variance in X.* The square of the correlation coefficient (r^2) is called the **coefficient of determination**. In our example, $r = .69$; so $r^2 = .48$. Thus, we can conclude in this case that 48 percent of the variance in GPA can be associated with the variance in aptitude.

Using the coefficient of determination, we can see the sense behind the guidelines presented in Table 5.6 for interpreting r. For example, consider the second category in the table—correlations between .70 and .90. For these values of r, r^2 is between .49 and .81. In other words, at least 49 percent of the variance in Y is associated with the variance in X, so correlations in the range of .70 to .90 are considered high. Similarly, for correlations of less than .30, r^2 is less than .09, which means that less than 10 percent of the variance in Y is associated with the variance in X. Hence, as a rule of thumb, we can say that correlations of less than .30 indicate little, if any, relationship between the variables.

We can illustrate this interpretation of the coefficient of determination by using overlapping circles, as in Figure 5.6. Each circle in the figure represents the variance of a variable. Because we used standard scores in defining the Pearson product-moment correlation coefficient and (as we saw in Chapter 3) the variance of a

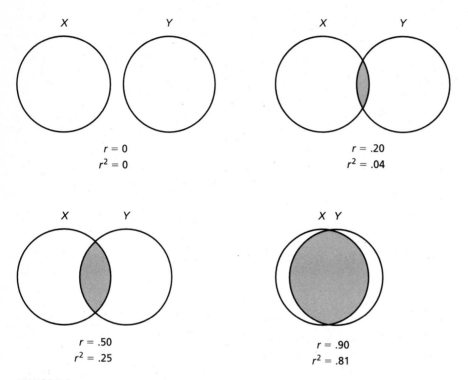

FIGURE 5.6
Illustration of the coefficient of determination (r^2) as overlapping areas representing variance

distribution of standard scores equals 1.0, we can assume the variance of both variable X and variable Y to be equal to 1.0. Thus we define the area within each circle—the variance—as equal to 1.0. The varying degrees of overlap in the circles in Figure 5.6 reflect the proportion of the variance in Y that can be associated with the variance in X.

The square of the correlation coefficient (r^2), or the coefficient of determination, indicates the proportion of the variance in one variable that can be associated with the variance in the other variable.

Computer Example: The Survey of High School Seniors Data Set

In Chapters 2 and 3, we used the FREQUENCIES and DESCRIPTIVES procedures in SPSS to generate frequency data and descriptive statistics for the variables in the survey of high school seniors data set. In this chapter, we use the CORRELATIONS procedure to compute correlation coefficients illustrating the relationship between the variables, MATHACH, VISUAL, and MOSAIC. The SPSS printout is found in Table 5.7; note that the CORRELATIONS procedure generates descriptive statistics for each of the variables. You might want to compare these descriptive statistics with those in Chapter 3.

Below the descriptive statistics in the printout is a "matrix of correlation coefficients" for these three variables. Above the correlation matrix, you see the line

```
Pearson Correlation Coefficients/Prob>|R| under Ho: Rho=0/N=500
```

This line indicates that the correlation coefficient is the top number; the number of observations is the middle number; and the "Probability" is the lower number. (We will discuss this lower number in Chapter 10 when describing the test of statistical significance for a correlation coefficient.)

Note that the correlation coefficient for any of the variables with itself is 1.0, i.e., the diagonal elements of the matrix are all 1.0. The correlation between MATH-ACH and VISUAL is 0.4385; the correlation between MATHACH and MOSAIC

TABLE 5.7
SPSS Printout-Descriptive Statistics and Correlation Coefficients for VISUAL, MOSAIC, and MATHACH

```
Variable     Cases      Mean     Std Dev
VISUAL       500       5.6970    3.8865
MOSAIC       500      26.8890    9.2882
MATHACH      500      13.0980    6.6046


Pearson Correlation Coefficients/Prob>|R| under Ho: Rho=0/N=500
             VISUAL     MOSAIC     MATHACH
VISUAL       1.0000      .2833      .4385
            (  500)    (  500)    (  500)
             P= .       P= .000    P= .000


MOSAIC        .2833     1.0000      .2918
            (  500)    (  500)    (  500)
             P= .000    P= .       P= .000


MATHACH       .4385      .2918     1.0000
            (  500)    (  500)    (  500)
             P= .000    P= .000    P= .


(Coefficient / (Cases) / 2-tailed Significance)

''.'' is printed if a coefficient cannot be computed
```

is 0.2918; and the correlation between VISUAL and MOSAIC is 0.2833. Looking specifically at the relationship between MATHACH and VISUAL and/or the relationship between MATHACH and MOSAIC, we would say that the correlations coefficients were low and positive.

Spearman rho (ρ)

There are some research studies in which the data consist of ranks, and there are other studies in which the raw scores are converted to rankings. In these situations, we can use the Spearman rho (ρ) correlation coefficient, which is a special case of the Pearson r. Because rankings are ordinal data, the Pearson r is not applicable to them.

Let us return to our example of SAT scores and final examination scores and convert the scores to ranks to find the Spearman rho. Table 5.8 contains the original lists of scores and the ranking of those scores. There are some special considerations about rankings that should be noted. First, we have given the highest SAT score the rank of 1, and the lowest score a ranking of 15. The highest final examination score, likewise, has been ranked 1, and the lowest score 15. We could have just as easily given the lowest score a rank of 1; the important thing is to use the same procedure on both columns of scores.

TABLE 5.8
Data for Computing Spearman rho

X	Y	X rank	Y rank	d	d^2
595	68	4	1	3	9
520	55	8	7	1	1
715	65	1	2	−1	1
405	42	14	12	2	4
680	64	2	3	−1	1
490	45	11	10	1	1
565	56	6.5	5.5	1	1
580	59	5	4	1	1
615	56	3	5.5	−2.5	6.25
435	42	13	12	1	1
440	38	12	14	−2	4
515	50	9	9	0	0
380	37	15	15	0	0
510	42	10	12	−2	4
565	53	6.5	8	−1.5	2.25
Σ				0	36.50

Tied Ranks

Note also that some students had the same raw scores, so they have tied ranks. Two students scoring 565 on the SAT would have occupied rank positions 6 and 7, so they received the average of those positions, which was 6.5. There were three scores of 42 on the final examination. These students would have occupied rank positions 11, 12, and 13, so they received a rank of 12, which was the average of these positions.

Calculating Spearman rho

The formula for calculating Spearman rho uses the squared differences between pairs of ranks:

$$\rho = 1 - \frac{6\Sigma d^2}{n(n^2 - 1)} \tag{5.8}$$

where

$n =$ number of paired ranks

$d =$ difference between the paired ranks

For the data in Table 5.8, the calculations are as follows:

$$\rho = 1 - \frac{6(36.50)}{15(225 - 1)}$$

$$= 1 - 0.07$$

$$= 0.93$$

Note that the calculated value of Spearman rho (.93) is close to the Pearson r that was calculated from the raw scores (.90). The difference is the result of some tied scores. When there are no tied scores, Spearman rho will equal Pearson r.

Correlation and Causality

Correlation indicates relationship or association between two variables; it does not necessarily follow that scores on one variable are caused by scores on the other variable. A third variable or a combination of other variables may be causing the two correlated variables to relate as they do.

Consider the GPA-aptitude example. It is reasonable to assume that the high correlation between GPA and aptitude reflects a causal relationship. However, this interpretation is not made strictly on the basis of the correlation coefficient. It is made only after we consider the nature of the variables involved.

Suppose a school principal measures the reading comprehension and the running speed of all of the students at his grade K–6 elementary school. He finds a strong positive correlation and concludes that training children to run faster will improve their reading comprehension. What he has really found is that sixth graders run faster and read better than kindergarteners. The third variable, age of the student, is confounding his correlation coefficient. His cause-and-effect conclusion is not warranted.

Thus, when interpreting the size of the correlation coefficient in terms of variance, the primary interpretation is of the proportion of the variance in one variable that is *associated* with the variance in the other. An interpretation in terms of causation is appropriate if and only if the variables under investigation provide a logical basis for such interpretation.

> A high value of r does not necessarily imply causation between correlated variables. Causation must be established through a consideration of the specific variables.

Summary

Correlation is one of the most widely used analytic procedures in the behavioral sciences. The correlations in this chapter were limited to cases in which the relationships between the variables were linear. This is not a problem for most of the variables studied in the behavioral sciences. Pearson r was introduced for application in situations in which both variables are measured on at least interval scales. Spearman rho was introduced for application to ranked data.

The correlation coefficient is a number (between -1.0 and $+1.0$, inclusive) that indicates the degree of relationship between two variables. The size of the correlation coefficient is affected by the homogeneity of the scores on the variables. If a relationship exists between two variables and that relationship is linear, then the more heterogeneous the scores and the greater the range of measurement, the greater the absolute value of r. The number of observations of each variable generally does not affect the absolute value of r.

The value of r that indicates a substantive relationship depends on the specific variables correlated. Causality between variables may or may not be inferred, again depending on the specifics of the situation. The square of r indicates the proportion of "shared" variance between the variables, irrespective of causality, that is, the proportion of the variance in one variable associated with the variance in the other.

We have used the correlation coefficient in a descriptive way in this chapter. Correlation coefficients are also used (1) to predict scores on one variable from scores on another variable and (2) in inferential statistics, where a correlation in a sample is used to draw a conclusion about a correlation in a larger population. Both of these applications are developed in later chapters.

Exercises

1. The following measurements were taken for five individuals on variables X and Y:

Individual	Variable X	Variable Y
A	2	6
B	6	14
C	5	12
D	4	10
E	1	4

   pearson

   The corresponding points, when plotted on a scatterplot, lie in a straight line with positive slope.

   a. Draw a scatterplot for these data to show their perfect linear relationship.

   b. Compute the Pearson correlation coefficient and show its actual value to be 1.0.

*2. A psychologist has 10 patients to whom he administers two performance tests, PTA and PTB. The following are the scores of the 10 patients on the tests:

Patient	PTA	PTB
A	15	20
B	12	15
C	10	12
D	14	18
E	10	10
F	8	13
G	6	12
H	15	10
I	16	18
J	13	15

   pearson (reliability estimate test)

   $n = 10$ $\Sigma XY = 1{,}761$
   $\Sigma X = 119$ $\Sigma Y = 143$
   $\Sigma X^2 = 1{,}515$ $\Sigma Y^2 = 2{,}155$

   a. Compute the Pearson correlation coefficient between the scores on the two tests.

   b. Find the percentage of the variance in the scores of PTB associated with the variance in PTA scores.

3. A statistical analysis has shown that 60 percent of the variance in variable X can be associated with variance in variable Y. What is the value of the Pearson correlation coefficient?

   $r = \sqrt{.60}$

4. Give an example of two variables for which the Pearson r might be expected to be a high negative value ($-.70$ to -1.0).

5. Below are the scores collected from two tests that measure verbal ability.

Individual	Test 1	Test 2
1	55	94
2	52	91
3	51	88
4	48	84
5	44	86
6	40	81
7	37	85
8	34	76
9	32	79
10	30	74

$n = 10$ $\Sigma XY = 35{,}923$
$\Sigma X = 423$ $\Sigma Y = 838$
$\Sigma X^2 = 18{,}619$ $\Sigma Y^2 = 70{,}592$

  a. Compute r for these data.

  b. Using only the first five individuals, compute r.

  c. Compare the results found in parts a and b and discuss.

  d. What percentage of the variance in test 2 is associated with the variance in test 1 when $n = 10$? $n = 5$?

6. A researcher is interested in knowing the Pearson r correlation between number of years married and level of marital satisfaction. Below are the data collected from 20 couples. Do the data exhibit a curvilinear trend? If so, is the Pearson r an appropriate measure of the relationship between these two sets of data?

Years of marriage	Level of marital satisfaction
2	55
3	52
4	48
6	41
6	45
7	42
7	37
8	30
8	28
9	27
10	27
10	30
12	34
14	33
14	37
14	36
16	38
18	42
19	44
19	45

7. A graduate student in education is interested in knowing the correlation between SAT scores for verbal ability and GPA after the first year of college. The following data were collected for 20 students randomly selected from the freshman class.

SAT score	GPA	SAT score	GPA
480	3.20	567	2.30
482	2.62	581	2.50
484	2.21	593	2.82
493	3.15	601	3.32
501	2.95	611	3.17
512	2.75	627	2.75
522	2.43	640	2.95
530	2.13	643	3.50
542	3.30	662	3.20
555	2.80	670	3.62

$n = 20$ $\Sigma XY = 32,814.94$

$\Sigma X = 11,296$ $\Sigma Y = 57.67$

$\Sigma X^2 = 6,456,890$ $\Sigma Y^2 = 169.77$

a. Find r.

b. What percentage of the variance in SAT scores is associated with the variance in GPA?

8. A researcher believes that there is a correlation between number of cigarettes smoked per day and intelligence. The following data were collected on 15 smokers. Find r and r^2. Discuss the results.

Number of cigarettes	Measure of IQ (coded)
7	10
49	6
41	15
38	5
37	12
19	4
35	19
40	11
1	3
10	3
18	22
21	17
15	12
7	9
38	13

$n = 15$ $\Sigma XY = 4,317$

$\Sigma X = 376$ $\Sigma Y = 161$

$\Sigma X^2 = 12,714$ $\Sigma Y^2 = 2,213$

9. A researcher was interested in the correlation between the number of seconds that it takes to solve a mathematics problem (X) and scores on a test of "short-term memory" (Y). He tested 60 twelve-year-olds and found:

$$\overline{X} = 22.3 \quad \overline{Y} = 56.5 \quad s_X = 5.6 \quad s_Y = 8.7 \quad s_{XY} = -28.26$$

a. Find r. $r = -.58$ $r^2 = .3364 \ (34\%)$

b. Interpret r in words.

(moderate negative relation)

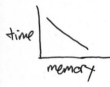

time / memory

10. A science teacher gave a botany test and a geology test to 25 high school juniors. The botany test had a mean of 88 and a standard deviation of 6.4. The geology test had a mean of 50 and a standard deviation of 4.5. The Pearson correlation coefficient between the two tests was .62. What was the covariance?

11. Ten figure skaters were rated in a competition by two judges. Below are composite scores, which include form, style, and content. Use the Spearman rho (ρ) to determine the relationship between the ratings of the two judges.

Skater	Judge X	Judge Y
1	38	36
2	24	27
3	31	33
4	27	24
5	19	22
6	44	40
7	35	32
8	20	25
9	29	26
10	37	30

12. An educational researcher wants to determine the relationship between the IQ scores of 15 elementary school students and a teacher's subjective opinion of their intelligence. The teacher is asked to rank these 15 students in terms of her perceptions of their intellectual abilities (1 = highest; 15 = lowest). This is done without the teacher's knowing the students' actual IQ scores. For the following data, compute the Spearman rho (ρ).

Student	Teacher ranking	Test score	Student	Teacher ranking	Test score
1	15	88	9	8	123
2	13	92	10	9	126
3	14	97	11	3	130
4	7	102	12	5	133
5	11	108	13	2	137
6	12	115	14	1	140
7	6	117	15	4	145
8	10	120			

13. A psychologist gathered the following scores from children who were involved in an experiment about learning and anxiety.

Anxiety score	Performance score
10	4
8	8
4	9
15	6
15	4
3	8
6	5
9	5
7	3
5	6

a. Compute the Pearson r.

b. Interpret r in words.

c. What would be the Pearson r if the psychologist added 2 points to each person's performance score?

d. Comment on whether or not this correlation coefficient represents a cause-and-effect relationship between anxiety and performance.

14. An elementary school principal correlated the reading comprehension scores and the mathematics concepts scores from a standardized achievement test (1) for all children between grades 2 and 6, (2) for all fourth graders, and (3) for fourth graders with vocabulary scores above the school median.

a. Which correlation was probably the highest? ①

b. Which correlation was probably the lowest?

6

Linear Regression: Prediction

Key Concepts

Prediction
Criterion variable
Predictor variable
Linear regression line
Slope
Intercept
Method of least squares

Regression coefficient
Regression constant
Errors in prediction
Standard error of estimate
Conditional distributions
Homoscedasticity

I n Chapter 5, we developed the concept of correlation between pairs of variables. For example, when the correlation is moderate and positive, we can expect that a score above the mean on the X variable is likely to be paired with a score that is above the mean on the Y variable. The focus of this chapter is to use the concepts of correlation and a straight line to develop the concept of linear regression. This involves predicting the score on the Y variable based on knowledge of the score on the X variable. For example, if we know that creativity (Y) and logical reasoning (X) are positively related, what is the best estimate of an individual's score on a creativity measure if we know this individual's score on a logical reasoning measure? If we have the predicted creativity scores for a group of individuals, how much error is there in the prediction process? We will deal with questions of accuracy in the prediction process in Chapter 17.

Correlation and Prediction

Prediction, in its simplest sense, is the process of estimating scores on one variable (Y), the **criterion variable**, on the basis of knowledge of scores on another variable

(X), the **predictor variable**. The concept of correlation is used in this process. Consider the example in which we want to predict an individual's score on a creativity measure (Y) based on this individual's score on a logical reasoning measure (X). Assume that a moderate to high positive relationship exists between the two variables. That is, high scores on the creativity measure are associated with high scores on the logical reasoning measure and vice versa. Hence, it would be possible to conclude that, if a person's score on a logical reasoning measure is high, chances are good that this person's creativity score will also be high.

In Chapter 5, we illustrated the concept of linear correlation graphically with the scatterplot. In the scatterplot shown in Figure 5.1, the points tend to fall along a straight line. If we could draw this straight line, it would represent, on the average, how change in one variable (X) is associated with change in another variable (Y). This line is called the **linear regression line**. If we use the variable X to predict the variable Y, the line is called the *regression of Y on X*.

> Prediction is the process of estimating scores on one variable from knowledge of scores on another variable.

Correlation and Causation

The mere fact that we use our knowledge of the relationship between two variables, X and Y, to predict values of Y from known values of X does not imply that changes in X *cause* changes in Y. For example, consider the old saying "An apple a day keeps the doctor away." A moderate negative correlation could undoubtedly be found between number of apples consumed during a year and number of visits to a physician's office. This does not imply that a person had many visits to a physician's office *because* she or he ate an insufficient number of apples. This person could have been severely injured in an automobile accident (even while eating an apple).

There are some variables for which a change cannot be attributed to the change in another variable. There may be a positive relationship between performance on a physical task and chronological age. But we cannot argue that chronological age is affected by performance on the physical task. Chronological age is affected by only one factor: the passage of time since an individual's birth.

The order of variables' occurrence may indicate whether or not a cause-and-effect relationship is possible. Suppose that the X variable consists of treatments, such as various levels of drug dosage, and that the Y variable is the performance of a rat running a maze. X cannot affect Y until the drug is administered. If the rat runs the maze before receiving the drug dose, the dose cannot have affected the rat's performance.

There are combinations of variables that are highly correlated, and in such cases, one variable is an accurate predictor of the other. But, again, accurate prediction does not necessarily imply that the predictor variable *causes* the scores on the criterion variable. How, then, does a researcher infer cause and effect? Basically, the researcher does so through understanding the variables under study and the context in which they are operating. The statistics used in correlation and regression indicate whether a relationship exists and how effective the prediction will be based on this relationship. The researcher must still interpret the statistics in the context of the situation.

The Regression Line

The process of prediction involves two steps. The first step is determining the regression line, which is a mathematical equation. The second step is using this mathematical equation to predict scores. Because we are limiting our discussion to linear regression, the mathematical equation is the equation of a straight line.

The mathematical equation of a straight line expresses a functional relationship between two variables. In predicting Y scores from X scores, we say that Y is a function of X and use the *slope-intercept form* of the equation for a straight line. The general form of this equation is

$$\hat{Y} = bX + a \tag{6.1}$$

where

$\hat{Y}$ = predicted score

b = slope of the line

$a = Y$ intercept

Figure 6.1 shows the graph of a straight line that we will use to illustrate the definition of the slope-intercept form. This line extends indefinitely in both directions but, for the sake of simplicity, we will consider only that portion of the line between $X = 0$ and $X = 5$. The plot points are

$$X = 0, \; Y = 2 \quad X = 3, \; Y = 3.5$$
$$X = 1, \; Y = 2.5 \quad X = 4, \; Y = 4$$
$$X = 2, \; Y = 3 \quad X = 5, \; Y = 4.5$$

The *slope* of a line is defined as the amount of change in Y that corresponds to a change of 1 unit in X. Notice in Figure 6.1 that, corresponding to a 1-unit increase in X from 1 to 2, there is an increase of 0.5 unit in Y from 2.5 to 3. This relationship is depicted by dashed lines in Figure 6.1. Hence the slope of this line is $+0.5$. The slope of a line can be positive or negative and can be less than or

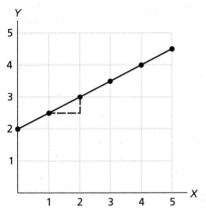

FIGURE 6.1
Graph of the straight line represented by the equation
$Y = 0.5X + 2.0$

greater than 1. A slope equal to 0 would indicate that the line is parallel to the X axis—that is, it is horizontal.

The **intercept** of the line is defined as the value of Y when X equals 0. Note that, in Figure 6.1, the intercept is the value of Y where the line crosses, or intercepts, the Y axis. For this example, the intercept is 2. Now that we know the values of the slope and the Y intercept, we can define the equation of the straight line that appears in Figure 6.1. Using the formula for a straight line (formula 6.1), we substitute the values for b and a. Thus the equation for this example is

$$Y = 0.5X + 2.0$$

The equation for a straight line used in prediction is $\hat{Y} = bX + a$, where b is the slope of the line and a is the Y intercept.

Determining the Regression Line

Now that we know the general form of the equation of a straight line, how do we fit a line to the scatterplot of points and then use this line in the prediction process? Consider again the example of predicting scores on a creativity measure (Y) from scores on a logical reasoning measure (X); hypothetical data for 20 middle-school students are found in Table 6.1, and the scatterplot is shown in Figure 6.2.[1] Notice

[1] In Chapter 3, we defined the standard deviation (s) as the standard deviation of the sample that is used to estimate the standard deviation of the population (σ). For consistency throughout the book, we continue to use this definition, so that

$$s = \sqrt{\frac{\Sigma X^2 - \Sigma(X)^2/n}{n-1}}$$

TABLE 6.1
Data for Calculating the Regression Equation for Predicting Creativity Scores (Y) from Logical Reasoning Scores (X)

Student	X	Y	X²	Y²	XY
1	15	12	225	144	180
2	10	13	100	169	130
3	7	9	49	81	63
4	18	18	324	324	324
5	5	7	25	49	35
6	10	9	100	81	90
7	7	14	49	196	98
8	17	16	289	256	272
9	15	10	225	100	150
10	9	12	81	144	108
11	8	7	64	49	56
12	15	13	225	169	195
13	11	14	121	196	154
14	17	19	289	361	323
15	8	10	64	100	80
16	11	16	121	256	176
17	12	12	144	144	144
18	13	16	169	256	208
19	18	19	324	361	342
20	7	11	49	121	77
Σ	233	257	3037	3557	3205

$$\overline{X} = 11.65 \qquad \overline{Y} = 12.85$$
$$s_X = 4.12 \qquad s_Y = 3.66$$
$$r = .74 \qquad s_{XY} = 11.16$$

that for these data there is a positive relationship between the scores on the two measures. Using formula 5.3, we find the correlation (r) between these two sets of scores to be $+.74$.

For these data, we want to determine the line that describes a trend in the data such that a change in X will be reflected in an "average" change in Y. The specific line in question is "fitted" to the data by what is called the **method of least squares**. The method of least squares fits the line in such a way that the sum of squared distances from the data points to the line is a minimum. This regression line of best fit is illustrated in Figure 6.3.

Recall that the regression line is used to predict or estimate the value of Y for a given value of X. For all values of X, the predicted values of Y (denoted $\hat{Y}$) are located on the regression line shown in Figure 6.3. Thus, another way to illustrate the least-squares method (or criterion) for fitting the regression line is to consider the errors (e) in predicting Y from X—that is, the difference between the actual value of Y and the predicted value ($\hat{Y}$). Note that this difference ($Y - \hat{Y}$)

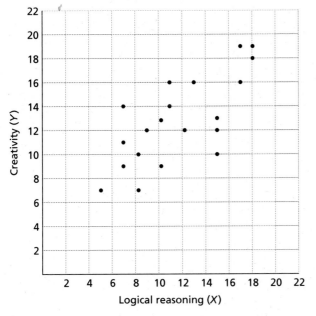

FIGURE 6.2
Scatterplot of logical reasoning (X) and creativity (Y) scores

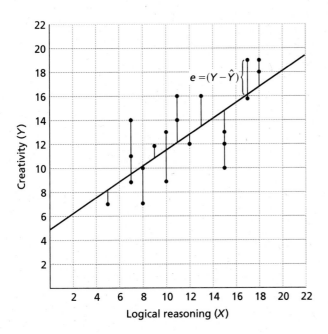

FIGURE 6.3
Regression line for predicting creativity scores (Y) from logical reasoning scores (X)

is actually the vertical distance (parallel to the Y axis) between the data point and the regression line. Thus the least-squares criterion can be expressed symbolically as minimizing $\Sigma(Y - \hat{Y})^2$.

> We fit the regression line to the data points in a scatterplot using the least-squares criterion, which requires $\Sigma(Y - \hat{Y})^2$ to be a minimum.

The formulas for the slope (b) and the Y intercept (a) are derived using calculus; this derivation is beyond the scope of this book.[2] However, the formulas for b and a are as follows:

$$b = \frac{n\Sigma XY - \Sigma X\Sigma Y}{n\Sigma X^2 - (\Sigma X)^2} \tag{6.2}$$

$$a = \frac{\Sigma Y - b\Sigma X}{n} = \overline{Y} - b\overline{X} \tag{6.3}$$

The value of b, which is the slope of the regression line, is called the **regression coefficient**. The intercept (a) is called the **regression constant**. In order to show the relationships between correlation and regression, we can use the following alternative formula for the regression coefficient:

$$b = (r)\frac{s_Y}{s_X} \tag{6.4}$$

where

$r =$ correlation between variables X and Y

$s_Y =$ standard deviation of the Y scores

$s_X =$ standard deviation of the X scores

> For the regression equation $\hat{Y} = bX + a$, b is called the regression coefficient, and a is called the regression constant.

Applying formula 6.2 to the data in Table 6.1, the regression coefficient (b) is computed as follows:

$$b = \frac{20(3205) - (233)(257)}{20(3037) - (233)^2}$$

$$= 0.65$$

[2] See B. Ostle and R. W. Mensing, *Statistics in Research* (Ames, Iowa: The Iowa State Press, 1975), pp. 167–168.

The regression constant (a) is computed using formula 6.3:

$$a = \frac{257 - (0.65)(233)}{20}$$

$$= 5.28$$

Alternatively, b can be computed using formula 6.4:

$$b = (0.74)\frac{3.66}{4.12}$$

$$= 0.65$$

A third way to compute the slope is by using the covariance. Since

$$r = \frac{s_{XY}}{s_X s_Y} \qquad \text{and} \qquad b = r\frac{s_Y}{s_X}$$

then

$$b = \frac{s_{XY}}{s_X s_Y} \cdot \frac{s_Y}{s_X} = \frac{s_{XY}}{s_X^2} \tag{6.5}$$

For the data in our example, this is:

$$b = \frac{11.16}{(4.12)^2} = 0.65$$

Thus, for the data in Table 6.1, the regression equation for predicting creativity scores from logical reasoning scores is

$$\hat{Y} = 0.65X + 5.28$$

Once the regression equation has been determined, the prediction of the creativity scores from logical reasoning scores is straightforward; we simply substitute the X value into the equation and solve it for the corresponding $\hat{Y}$ value. For example, the predicted creativity score ($\hat{Y}$) for a student with a logical reasoning score of 12 would be

$$\hat{Y} = (0.65)(12) + 5.28$$

$$= 13.08$$

The predicted creativity scores for each of the logical reasoning scores in this example have been computed in Table 6.2. We will use these predicted scores in a later section of this chapter.

To find predicted scores, substitute the value of the known variable in the regression equation and solve the equation.

TABLE 6.2
Predicted Scores and Errors in Prediction of Creativity Scores (Y) from Logical Reasoning Scores (X)

Student	X	Y	$\hat{Y}$	$e = Y - \hat{Y}$	$e^2 = (Y - \hat{Y})^2$
1	15	12	15.03	−3.03	9.18
2	10	13	11.78	1.22	1.49
3	7	9	9.83	−0.83	0.69
4	18	18	16.98	1.02	1.04
5	5	7	8.53	−1.53	2.34
6	10	9	11.78	−2.78	7.73
7	7	14	9.83	4.17	17.39
8	17	16	16.33	−0.33	0.11
9	15	10	15.03	−5.03	25.30
10	9	12	11.13	0.87	0.76
11	8	7	10.48	−3.48	12.11
12	15	13	15.03	−2.03	4.12
13	11	14	12.43	1.57	2.46
14	17	19	16.33	2.67	7.13
15	8	10	10.48	−0.48	0.23
16	11	16	12.43	3.57	12.74
17	12	12	13.08	−1.08	1.17
18	13	16	13.73	2.27	5.15
19	18	19	16.98	2.02	4.08
20	7	11	9.83	1.17	1.37
Σ				0.00*	116.59

*$\Sigma(Y - \hat{Y}) = 0.05$, and does not equal exactly zero because of rounding errors.

The Second Regression Line

So far, we have discussed the regression equation for predicting Y from X, as well as the formulas for the regression coefficient (b) and the regression constant (a). For the most part, this is the conventional way of labeling the criterion variable (Y) and the predictor variable (X) in behavioral science research. However, it is possible to develop a regression equation for predicting X from Y. In our example, we could develop an equation for logical reasoning scores from creativity scores. The slope (b) and intercept (a) for this equation will differ from the values computed above. It should be noted that the two regression lines intersect at the point ($\bar{X}, \bar{Y}$), but would coincide if there were a perfect relationship between Y and X. In the latter case, all the data points lie in a straight line.

We could have included the computational formulas for the regression coefficient and regression constant for predicting X from Y. These formulas are essentially the same as formulas 6.2, 6.3, 6.4, and 6.5, but each X must be replaced with a Y, and each Y with an X. However, rather than having two sets of formulas, we have chosen to follow standard research convention and to present only the one set for predicting Y from X.

Predicting Standard Y Scores from Standard X Scores

In Chapter 3, we defined a standard score (z) as

$$z = \frac{X - \overline{X}}{s}$$

Suppose we want to develop the regression equation for predicting standard Y scores (z_Y) from standard X scores (z_X). First, consider the regression equation for predicting Y scores from X scores (formula 6.1),

$$\hat{Y} = bX + a$$

where a is defined as $\overline{Y} - b\overline{X}$ (formula 6.3). Next, by substituting for a in the above prediction equation and algebraically manipulating the equation, we have

$$\hat{Y} = bX + (\overline{Y} - b\overline{X})$$

So,

$$\hat{Y} = \overline{Y} + b(X - \overline{X})$$

Now, by substituting the value from formula 6.4 for b, the equation becomes

$$\hat{Y} = \overline{Y} + (r)\frac{s_Y}{s_X}(X - \overline{X})$$

Finally, manipulating the equation algebraically, we find

$$\frac{\hat{Y} - \overline{Y}}{s_Y} = r\left(\frac{X - \overline{X}}{s_X}\right)$$

or

$$z_{\hat{Y}} = r z_X \tag{6.6}$$

That is, in order to determine predicted standard Y scores ($z_{\hat{Y}}$) from standard X scores (z_X), we multiply z_X by the correlation coefficient (r). For the data in Table 6.1, consider the standard X score for student 1.

$$z_X = \frac{15 - 11.65}{4.12}$$

$$= 0.81$$

Multiplying this standard score by the correlation coefficient ($r = .74$), we find the predicted standard Y score

$$z_{\hat{Y}} = (.74)(0.81)$$

$$= 0.60$$

We will use this concept of predicting standard Y scores from standard X scores again in Chapter 18.

Errors in Prediction

If there is a perfect relationship between two variables, all data points in their scatterplot lie in a straight line, and it is easy to find the equation for that line. However, when the relationship is less than perfect, we must fit the regression line to the data using the method of least squares. Recall that this process involves determining the line such that the sum of squared distances from the data points to the line is a minimum. These distances are the differences between the actual Y scores and the predicted scores ($\hat{Y}$). They are defined as the **errors in prediction**. Symbolically,

$$e = (Y - \hat{Y}) \tag{6.7}$$

Because the points in a scatterplot are both above and below the regression line, the errors are both positive and negative (see Table 6.2 and Figure 6.3).

Now consider the distribution of these errors in prediction. Recall from Chapter 3 that, in order to describe a distribution of scores adequately, we need to know (1) the shape of the distribution, (2) the central tendency, and (3) the variation of the scores. As we will discuss later in this chapter, the distribution of the errors in prediction (e) is assumed to be normal at fixed values of X. Regarding the mean of the errors ($\overline{e}$), we see in Table 6.2 that the sum of errors should equal zero; thus,

$$\overline{e} = \frac{\Sigma e}{n}$$

$$= \frac{0}{n} = 0$$

It is also possible to compute the variance and standard deviation of the distribution of error scores. Symbolically, the estimated variance of the error scores ($s_{Y \cdot X}^2$) is

$$s_{Y \cdot X}^2 = \frac{\Sigma (e - \overline{e})^2}{n - 2} \tag{6.8}$$

Since the mean of the error scores equals zero, the formula is simplified to

$$s_{Y \cdot X}^2 = \frac{\Sigma e^2}{n - 2} \tag{6.9}$$

The standard deviation is called the **standard error of estimate** ($s_{Y \cdot X}$) and is defined as follows:

$$s_{Y \cdot X} = \sqrt{\frac{\Sigma e^2}{n - 2}} \tag{6.10}$$

For the data in Table 6.2, we find

$$s_{Y \cdot X} = \sqrt{\frac{116.59}{18}}$$

$$= \sqrt{6.48}$$

$$= 2.55$$

Using formula 6.10 for the standard error of estimate requires that we compute each error first. Such a process becomes very tedious in studies involving large numbers of subjects. The following alternative formula offers a more direct method for computing $s_{Y \cdot X}$:

$$s_{Y \cdot X} = s_Y \left[\sqrt{1 - r^2} \right] \left[\sqrt{(n-1)/(n-2)} \right] \tag{6.11}$$

Using formula 6.11 for the data in Table 6.2, we find $s_{Y \cdot X}$ to be:[3]

$$3.66 \left[\sqrt{(1 - (0.74)^2)} \right] \left[\sqrt{(19/18)} \right]$$

$$= 2.53$$

Notice the term $\left[\sqrt{(n-1)/(n-2)} \right]$ in formula 6.11. For large samples, this ratio is essentially equal to 1; therefore, the formula reduces to

$$s_{Y \cdot X} = s_Y \sqrt{1 - r^2} \tag{6.12}$$

This formula appears in most statistics texts. However, for small samples, it will underestimate the standard error. Thus, for small samples, the term $\left[\sqrt{(n-1)/(n-2)} \right]$ should be included in the computation.

In formula 6.12, we can see the relationship between correlation and regression. Notice that, when the correlation between Y and X is high, the standard error is small and vice versa. In fact, when the correlation equals $+1.0$ or -1.0, the standard error is 0. Thus a high correlation between Y and X reduces the standard error of estimate and therefore enhances the accuracy of prediction.

In regression, the larger the correlation between Y and X, the smaller the standard error of estimate and the greater the accuracy of prediction.

Conditional Distributions

Another way to look at the standard error of estimate is to consider the standard deviation of the Y scores around the predicted Y for a given value of

[3] The difference in the standard error computed using formula 6.10 and that computed using formula 6.11 is due to rounding error.

X. The predicted score is on the regression line, and the actual Y scores are above and below the line at the particular value of X. The distribution of Y scores for all those with the same value of X is called the *conditional distribution*. Each of these conditional distributions is considered to be normal, and the standard deviations of each of these distributions are assumed to be equal. This latter property is referred to as *homoscedasticity*. If we assume that the conditional distributions exhibit normality and homoscedasticity, we can use the properties of the normal distribution to make probability statements about the predicted scores.

Assume that Figure 6.4 represents the scatterplot of the creativity and logical reasoning scores, with conditional distributions for $X_1 = 7$, $X_2 = 12$, and $X_3 = 17$. For the logical reasoning score $X_1 = 7$, we can find the corresponding predicted creativity score ($\hat{Y}_1$) by using the previous regression equation.

$$\hat{Y}_1 = 0.65(7) + 5.28$$
$$= 9.83$$

This predicted score is a point on the regression line. However, the predicted score is only an estimated creativity score ($\hat{Y}$) given a logical reasoning score of 7. There would actually be an entire distribution of creativity scores given the logical reasoning score of 7; in other words, the conditional distribution for $X_1 = 7$. The conditional distributions for $X_2 = 12$ and $X_3 = 17$ would be similarly defined,

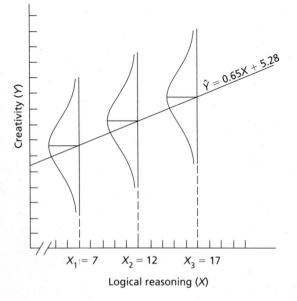

FIGURE 6.4
Conditional distributions of Y for $X_1 = 7$, $X_2 = 12$, and $X_3 = 17$.

with the two predicted scores ($\hat{Y}_2$ and $\hat{Y}_3$) computed as follows:

$$\hat{Y}_2 = 0.65(12) + 5.28$$
$$= 13.08$$
$$\hat{Y}_3 = 0.65(17) + 5.28$$
$$= 16.33$$

The standard error of estimate ($s_{Y \cdot X}$) was previously calculated and found to be 2.55. Thus, because of the assumption of homoscedasticity, each of these conditional distributions would have a standard deviation of 2.55.

Computer Example: The Survey of High School Seniors

In Chapter 5, we computed the correlation coefficents illustrating the relationship between MATHACH and VISUAL, and the relationship between MATHACH and MOSAIC using the CORRELATIONS procedure in SPSS. It is also possible to generate the regression equations for predicting MATHACH scores from MOSAIC scores. Using the computational procedures from this chapter, the two equations would be:

$$\text{MATHACH} = 0.745083 \text{ VISUAL} + 8.853289$$

$$\text{MATHACH} = 0.207464 \text{ MOSAIC} + 7.519532$$

In Chapter 17, we will use the REGRESSION procedure in SPSS to generate these two equations. In that chapter, we will be talking about not only the regression equation, but also the test of significance for the regression coefficient.

For the purposes of this chapter, it is important to note the use of these regression equations. Consider the first equation and how we would compute a MATHACH score for a VISUAL score of 10. The following calculation would be necessary.

$$\text{MATHACH} = 0.745083 \text{ VISUAL} + 8.853289$$
$$= 0.745083(10) + 8.853289$$
$$= 7.450830 + 8.853289$$
$$= 16.304119$$

For the second equation, consider the MATHACH score for a MOSAIC score of 30.

$$\text{MATHACH} = 0.207464\,\text{MOSAIC} + 7.519532$$
$$= 0.207464(30) + 7.519532$$
$$= 6.223920 + 7.519532$$
$$= 13.743452$$

Summary

Prediction is an important practical application of the concept of correlation. In this chapter, we have discussed predicting scores on one variable (Y) based on knowledge of scores on another variable (X), using linear regression. One necessary assumption is that the two variables are linearly related. That is, a straight line is an adequate fit to the set of points determined by the values of the two variables.

There are actually two regression equations; however, we have considered only the equation for predicting Y scores from X scores, or the regression of Y on X. This regression line is fitted by the method of least squares, which minimizes the variance of errors of prediction. The slope of this line (b) is called the regression coefficient, and the Y intercept (a) is called the regression constant.

Errors of prediction can be illustrated using conditional distributions, which are defined as the distributions of scores on the Y variable for a given value of X: These conditional distributions are assumed to be normally distributed and homoscedastic; this latter assumption is that the standard deviations for all of these distributions are equal. With these two assumptions, it is possible to use the normal distribution to make probability statements and develop confidence intervals for predicted scores.

The correlation coefficient quantifies the nature of the association between two variables. The goal of regression is predicting scores on a variable Y given scores on a variable X. As the absolute value of the correlation coefficient increases, the errors in prediction decrease. However, the ability to make predictions and the ability to quantify the accuracy of these predictions should not be confused with causal explanations of the relationship between two variables. In our example, performance on a creativity measure may have been correlated substantially with performance on a logical reasoning measure. This does not imply that X *caused* Y or vice versa. The statistics merely confirm the predictive relationship between the two variables. The causality of the relationship depends on the specifics of the research context.

The concepts of linear regression with a single predictor variable can be extended to linear regression with more than one predictor variable. Although the logic underlying the development and use of the regression equation remains the same, the formulas for computing the regression coefficients and the tests of significance are more complicated. This process, called *multiple linear regression*, is discussed in Chapter 18.

Exercises

*1. A group of athletic trainers use a test of physical conditioning that is very comprehensive, but has the disadvantage that its administration is long and complex. A researcher develops a shorter, simpler test of physical conditioning and administers both tests to a group of 24 young adult, male athletes. The intention is to use the shorter test as a substitute for the longer, more complex test, but that will depend on how good a predictor scores on the short test are of the scores on the long test. The following are pairs of scores for the 24 adult subjects.

Subject	Short (X)	Long (Y)	Subject	Short (X)	Long (Y)
1	3	2.8	13	5	3.9
2	4	2.8	14	4	3.3
3	2	2.6	15	3	2.5
4	5	4.1	16	8	6.1
5	2	2.7	17	9	6.0
6	6	5.8	18	6	5.4
7	8	6.3	19	4	3.5
8	4	3.5	20	3	3.1
9	7	6.5	21	8	6.8
10	3	3.7	22	2	2.4
11	9	6.6	23	7	6.5
12	3	4.0	24	3	3.4

$n = 24$ $\quad\quad$ $\Sigma XY = 591.50$
$\Sigma X = 118$ $\quad$ $\Sigma Y = 104.3$
$\Sigma X^2 = 704$ $\quad$ $\Sigma Y^2 = 510.01$

a. Plot the data in a scatterplot (Y axis—long test; X axis—short test).

b. Determine the regression equation for predicting scores on the long test (Y) from scores on the short test (X).

c. Draw the regression line on the scatterplot.

d. Find the predicted scores on Y (long test) from scores on X (short test) of 5 and 7.

e. Compute the standard error of estimate using formulas 6.11 and 6.12.

f. Comment on the adequacy of the regression.

2. A psychologist is interested in predicting levels of depression from the number of social events attended per week for a sample of high school seniors; the data

collected were

Student	Social events per week	Depression score	Student	Social events per week	Depression score
1	0	15	9	3	2
2	2	3	10	3	4
3	2	12	11	4	2
4	1	11	12	1	8
5	3	5	13	1	10
6	1	8	14	1	12
7	2	15	15	2	8
8	0	13			

$n = 15$ $\Sigma XY = 166$
$\Sigma X = 26$ $\Sigma Y = 128$
$\Sigma X^2 = 64$ $\Sigma Y^2 = 1,378$

a. Plot the data in a scatterplot. (Y axis—Depression; X axis—Social Events)

b. Determine the regression equation for predicting depression (Y) from social events (X).

c. Draw the regression line on the scatterplot.

d. Predict the depression score of a student who attended three social events.

e. Compute the standard error of estimate using formulas 6.11 ~~and 6.12.~~

f. calculate 68% confidence level for ind. who attends 3 social events per week

3. Suppose a statistics professor is interested in predicting final exam scores (Y) from SAT mathematics scores (X), using the following data:

Student	SAT score	Final score	Student	SAT score	Final score
1	440	40	8	590	68
2	465	47	9	607	44
3	482	43	10	619	57
4	521	54	11	630	54
5	535	64	12	636	62
6	552	52	13	657	61
7	572	59			

$n = 13$ $\Sigma XY = 400,647$
$\Sigma X = 7,306$ $\Sigma Y = 705$
$\Sigma X^2 = 4,164,458$ $\Sigma Y^2 = 39,145$

a. Plot the data in a scatterplot (Y axis—Final; X axis—SAT).

b. Determine the regression equation for predicting scores on the final (Y) from SAT scores (X).

c. Draw the regression line on the scatterplot.

d. From an SAT score of 550, predict the score on the final.

e. Compute the standard error of estimate using formulas 6.11 and 6.12.

4. An industrial psychologist is interested in the relationship between age (X) and an efficiency rating (Y). Data were collected from 20 employees of a small

manufacturing company. Some initial data analyses were completed; the summary data were

$$n = 20 \quad r = .64$$
$$\Sigma X = 796 \quad \Sigma Y = 1,266$$
$$s_X = 9 \quad s_Y = 12$$

a. Determine the regression equation for predicting the efficiency rating (Y) from age (X).

b. Using this regression equation, predict the efficiency rating for a person 45 years old.

c. Compute the standard error of estimate using formulas 6.11 and 6.12.

5. The director of a national research institute believes that faculty at larger, more "comprehensive" institutions have lighter teaching and advisory responsibilities and more time for research and scholarly writing. The director develops a measure of institutional comprehensiveness and a measure of scholarly productivity for psychology departments. In a pilot project, the director selects 18 colleges and universities and determines the institutional comprehensiveness and scholarly productivity of the psychology departments in these institutions. The data were as follows:

Institution	Comprehensiveness	Scholarly productivity
1	54	70
2	74	62
3	89	78
4	115	92
5	150	75
6	155	75
7	162	82
8	180	62
9	185	65
10	194	70
11	200	81
12	210	83
13	210	94
14	222	82
15	240	84
16	242	77
17	250	88
18	270	86

$$n = 18 \qquad \Sigma XY = 254,402$$
$$\Sigma X = 3,202 \qquad \Sigma Y = 1,406$$
$$\Sigma X^2 = 635,616 \qquad \Sigma Y^2 = 111,370$$

a. Plot the data in a scatterplot.

b. Determine the regression equation for predicting scholarly productivity (Y) from institutional comprehensiveness (X).

c. Draw the regression line on the scatterplot.

d. For an institution with a comprehensiveness score of 175, predict the psychology department's scholarly productivity score.

e. Compute the standard error of estimate using formulas 6.11 and 6.12.

6. For the data in Exercise 4, suppose the industrial psychologist wants to develop a regression equation for predicting standardized $\hat{Y}$ values ($z_{\hat{Y}}$) from standardized X values (z_X).

a. Determine the regression equation for predicting $z_{\hat{Y}}$ from z_X.

b. For the following z_X values, what are the corresponding $z_{\hat{Y}}$ values?

$$
\begin{array}{ll}
z_X = -1.25 & z_{\hat{Y}} = \\
z_X = -0.72 & z_{\hat{Y}} = \\
z_X = 0.0 & z_{\hat{Y}} = \\
z_X = 0.66 & z_{\hat{Y}} = \\
z_X = 1.64 & z_{\hat{Y}} =
\end{array}
$$

c. For the following X values, compute the z_X value and use the regression equation to find the corresponding $z_{\hat{Y}}$ values.

$$
\begin{array}{lll}
X = 45 & z_X = & z_{\hat{Y}} = \\
X = 33 & z_X = & z_{\hat{Y}} = \\
X = 68 & z_X = & z_{\hat{Y}} =
\end{array}
$$

d. Transform the $z_{\hat{Y}}$ values in part c back into $\hat{Y}$ values. Compare these transformed $\hat{Y}$ values to those predicted values using the equation determined in Exercise 4.

7. The dean of students at Southeastern State University is interested in the relationship between students' grades and their part-time work. Data from 25 students who hold part-time jobs were collected on number of hours worked per week (X) and last semester's grade-point average (Y).

Student	Hours worked	GPA	Student	Hours worked	GPA
1	17	2.9	14	19	3.1
2	7	3.2	15	23	2.5
3	10	2.5	16	27	2.4
4	32	1.9	17	18	3.2
5	20	3.0	18	10	3.5
6	22	2.1	19	32	2.2
7	15	2.4	20	18	3.0
8	12	3.3	21	22	2.5
9	19	2.7	22	15	3.3
10	13	3.1	23	16	3.1
11	26	2.3	24	19	2.3
12	23	2.7	25	22	2.7
13	25	3.3			

$n = 25$ $\quad\quad$ $\Sigma XY = 1,290.70$
$\Sigma X = 482$ $\quad\quad$ $\Sigma Y = 69.20$
$\Sigma X^2 = 10,276$ $\quad$ $\Sigma Y^2 = 196.22$

a. Plot the data in a scatterplot.

b. Determine the regression equation for predicting GPA (Y) from hours worked (X).

c. Draw the regression line on the scatterplot.

d. Predict the GPA for a student who worked 25 hours.

e. Compute the standard error of estimate using formulas 6.11 and 6.12.

8. A counseling psychologist is studying the relationship between marital satisfaction (X) and job satisfaction (Y). Data were collected from 35 high-level executives from local businesses and industries. Some initial data analyses have been completed; the summary data are as follows:

$$n = 35 \quad\quad r = .57$$
$$\Sigma X = 922 \quad\quad \Sigma Y = 568$$
$$s_X = 6.3 \quad\quad s_Y = 5.1$$

a. Determine the regression equation for predicting job satisfaction (Y) from marital satisfaction (X).

b. Using this regression equation, predict the job satisfaction score for an executive with a marital satisfaction score of 33.

c. Compute the standard error of estimate using formulas 6.11 and 6.12.

7

Probability, Sampling Distributions, and Sampling Procedures

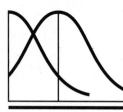

I n previous chapters, we have used statistical procedures to classify and summarize data, to describe distributions and individual scores, and to determine the relationship between variables. These procedures are called *descriptive statistics* because they are used primarily for *describing* the data collected. However, researchers need other statistical procedures, techniques that they can use to formulate and test hypotheses about the effects of experimental treatments, to generalize results to a given population, and to estimate the parameters of a population.

In order to deal with the tasks implied above, we turn to a part of statistics called **inferential statistics,** so called because we will be making inferences about a population from a sample selected from that population. The concepts and procedures of inferential statistics are based on probability. So before discussing inferential statistics procedures, which pretty well occupy the remainder of this text, it is necessary to cover basic concepts of probability and how they are used in inferential statistics.

Laws of Probability

We encounter concepts of probability every day. Weather reporters tell us the chance of rain tomorrow, and professional economists predict the likelihood of a recession or an upturn in business next year. In such situations the method of assigning probabilities to events is personal judgment, usually the judgment of "experts." The probability assigned is based on the individual's experience, and the accuracy of such probabilities depends on the expert's ability to evaluate the situation. Probabilities so determined are called subjective probabilities, and they are not the kind of probabilities that provide the basis for inferential statistics. The latter probabilities are based on mathematical concepts and theory.

Consider first a definition of probability and the ways of expressing probability. The **probability** of an event is the ratio of the number of outcomes including the event to the total number of possible outcomes. For example, a coin toss has two possible outcomes — heads or tails — and the probability of obtaining heads is 1/2. In general, if A is the event we are interested in, then the probability of A is symbolized $P(A)$ and is defined as follows:

$$P(A) = \frac{\text{Number of outcomes that include } A}{\text{Total number of outcomes}} \qquad (7.1)$$

Consider another example. Suppose we roll one die. What is the probability of rolling a 3? We might roll a 1, 2, 3, 4, 5, or 6; so there are six possible outcomes, and only one of these outcomes contains the stated event — rolling a 3. So the probability of rolling a 3 is 1/6, or .167. Sometimes probabilities are expressed as percentages. We might say that the probability of getting tails in one toss of an unbiased coin is 50%. However, technically, probabilities are expressed as decimals between 0 and 1.00, inclusive.

As probabilities are expressed in decimals, the larger the decimal, the greater the probability that the event will occur. A value of 0 indicates that there is no chance that the event will occur. A value of 1 indicates that the event is absolutely certain to occur. For example, the probability that any of us would be able to run nonstop from Chicago to Boston is 0, whereas the probability that income taxes will be paid next year is 1.

The probability of event *A*, *P(A)*, is the ratio of the number of outcomes that include event *A* to the total number of possible outcomes.

If we know the probability of event A, we can also state the probability that A will not occur, which is written $P(\overline{A})$ (read as "the probability of *not A*"). The sum of $P(A)$ and $P(\overline{A})$ always equals 1. For example, we said that if event A is obtaining heads when a coin is tossed, then $P(A) = 1/2$ or .50. The probability of obtaining tails — in other words, the probability that A will not occur — is $1 - P(A) = 1 - .50 = .50$. If event A is rolling a 3 when we roll one die, $P(A) = .167$, as we explained previously; the probability of *not* rolling a 3 therefore is $1 - .167$, or .833.

These are simple examples, but probability theory has a specified set of laws that enable us to compute the probability of an event or a combination of events in far more complicated situations. These *laws of probability* describe the behavior of events given certain conditions. We will begin by considering single events and the application of the addition rules.

Single Observations: The Addition Rules

Suppose we ask a child to select a number from 1 to 10. Picking any one of these numbers represents a *single event* in the total set of possible events. Figure 7.1 illustrates all possible events for this example in what is called an Euler diagram. The set of events represented in the diagram is called the *sample space*. If we define selecting a specified number as a single event and assume that there is an equal probability that any of the possible events will occur, then for each number, the probability that it will be selected equals 1/10 or .10.

Mutually Exclusive Compound Events   In contrast to a single event, we can define a **compound event** as one that involves two or more single events. For example, we could define the compound event A as selecting from the sample space a number greater than 8 — in other words, selecting the number 9 *or* the number 10. In Figure 7.2, compound event A is indicated by the circle drawn around the sample points that represent the numbers 9 and 10. Notice that these two events are *mutually exclusive*, which means that only one of the two can occur: the occurrence of one excludes the possibility that the other can occur. In this case, if we select 9, we cannot also select 10.

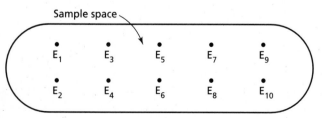

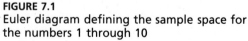

FIGURE 7.1
Euler diagram defining the sample space for the numbers 1 through 10

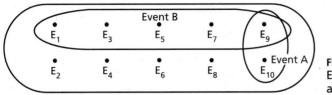

FIGURE 7.2
Euler diagram defining compound events *A* and *B* for the sample space

How do we determine the probability of a compound event's occurring? When two events are mutually exclusive, the probability that one *or* the other will occur is equal to the sum of the probabilities of the single events. This is the *addition rule for mutually exclusive events.*

The probability of the compound event *X* or *Y*, where *X* and *Y* are mutually exclusive events, equals

$$P(X \text{ or } Y) = P(X) + P(Y) \qquad (7.2)$$

Applying this rule to our example, we find that the probability of compound event *A* — selecting either the number 9 or the number 10 — is:

$$P(A) = P(9 \text{ or } 10) = P(9) + P(10)$$

$$= .10 + .10$$

$$= .20$$

Similarly, if we define event *B* as selecting an odd number from the sample space, it is a compound event, because selecting 1 or 3 or 5 or 7 or 9 satisfies the definition of the event (see Figure 7.2). Again, the single events are mutually exclusive; again, we use formula 7.2 to determine the probability of compound event *B*.

$$P(B) = P(1 \text{ or } 3 \text{ or } 5 \text{ or } 7 \text{ or } 9)$$

$$= P(1) + P(3) + P(5) + P(7) + P(9)$$

$$= .10 + .10 + .10 + .10 + .10$$

$$= .50$$

This result certainly is reasonable because $P(1 \text{ or } 3 \text{ or } 5 \text{ or } 7 \text{ or } 9)$ is the probability of selecting an odd number from ten numbers, half of which are odd and half of which are even.

Nonmutually Exclusive Events   Now consider a more complex example. Suppose we want to know the probability of selecting a number greater than 8 (event *A*) *or* selecting an odd number (event *B*). Notice that these two events are *not* mutually exclusive; if we select the number 9, we have satisfied the definition of both events (see Figure 7.2). As a result, if we simply add the probability of event *A* and the probability of event *B*, we end up counting the probability of selecting a 9 two times. Therefore, for nonmutually exclusive events, the probability that one of the events will occur is calculated by the following general addition rule.

The probability of the compound events *X* or *Y*, where *X* and *Y* are not mutually exclusive events, equals

$$P(X \text{ or } Y) = P(X) + P(Y) - P(X \text{ and } Y) \tag{7.3}$$

$P(X$ and $Y)$ refers to the probability of the outcome that satisfies the definition of *both* compound events. Note that, when events are mutually exclusive, $P(X$ and $Y) = 0$. Thus, this rule also applies to nonmutually exclusive events.

In our example, $P(A$ and $B)$ is the probability of selecting the number 9 — that is, $P(9)$, which we saw is .10. Thus, applying formula 7.3 to this example, we find that

$$P(A \text{ or } B) = P(A) + P(B) - P(A \text{ and } B)$$
$$= .20 + .50 - .10$$
$$= .60$$

Joint or Successive Events: The Multiplication Rules

So far, we have determined the probability of an event when there is only one trial or observation, but we also can consider the probability that two or more events will occur jointly or successively when more than one observation or trial is made. In other words, we want to know the probability that some event *A* and some event *B* will both occur. In general, we find the probability that two events will both occur by *multiplying* the separate probabilities of each. But to calculate the probability, we must first determine whether the events are statistically independent.

Statistically Independent Events Events are *statistically independent* if the probability that one event will occur is not affected by the occurrence of the other. For example, suppose we want to determine the probability of selecting the number 2 from the sample space and then, after replacing that number, selecting the number 4. This procedure is called *sampling with replacement*. Because the first number is returned to the sample space before the second number is selected, the first selection does not influence the outcome of the second selection; therefore, the two events are statistically independent.

When two events are statistically independent, the probability that *both* will occur equals the product of their individual probabilities. Thus, the multiplication rule for statistically independent events is as follows.

Given statistically independent events X and Y, the probability of their joint occurrence is

$$P(X \text{ and } Y) = P(X) \cdot P(Y) \tag{7.4}$$

Using the sample space defined in Figure 7.2, let us find the probability of selecting the number 2 from the sample space and then, after replacement, selecting the number 4.

$$P(2 \text{ and } 4) = P(2) \cdot P(4)$$
$$= (.10)(.10)$$
$$= .01$$

To illustrate this rule with a more complex example, consider our compound events A and B. $P(B \text{ and } A)$ is defined as the probability of selecting an odd number and then, after replacement, selecting the number 9 or the number 10. Using the multiplication rule for statistically independent events, we find that

$$P(B \text{ and } A) = P(B) \cdot P(A)$$
$$= (.50)(.20)$$
$$= .10$$

Nonindependent Events   When *sampling without replacement* is used, two events are no longer statistically independent. For example, consider again $P(2 \text{ and } 4)$. Suppose we do *not* replace the first number, 2, before selecting the second number. Then the probability of selecting a particular number, 4, on the second draw depends on the result of the first draw. In other words, the result of the first draw influences the probability of the result of the second draw. These events are said to be *nonindependent*, and the probability of the second event is called *conditional*. We indicate a conditional probability by writing $P(Y|X)$, which reads, "the probability of Y given that X has occurred." To calculate the probability that two nonindependent events will both occur, we use the following multiplication rule for statistically nonindependent events.

> Given events X and Y, which are nonindependent, the probability of both X and Y occurring jointly is the product of the probability of obtaining one of these events times the conditional probability of obtaining the other event, given that the first event has occurred.
>
> $$P(X \text{ and } Y) = P(X) \cdot P(Y|X) \text{ or } P(Y) \cdot P(X|Y) \qquad (7.5)$$

Consider again the probability of selecting a 2 and then a 4, but this time suppose the first number selected is not replaced. The probability of selecting the number 2 on the first draw is still .10. Assuming that we did select the number 2, only nine numbers would be left in the sample space, and the probability of selecting the number 4 on the second draw is 1/9 or .111. Therefore, the probability of selecting the numbers 2 and 4 is:

$$P(2 \text{ and } 4) = P(2) \cdot P(4|2)$$
$$= (.10)(.111)$$
$$= .011$$

Consider a slightly more complex example. Suppose we have an urn that contains 100 balls, 25 each of the colors red, green, blue, and yellow. The balls are well mixed in the urn, and as we draw out a ball, we cannot see into the urn. What is the probability of selecting a red ball on the first draw (event A) and a yellow ball on the second draw (event B), assuming we do not replace the first ball?

Of the 100 balls, 25 are red, so the probability of event A is $P(A) = 25/100 = .25$. Suppose we obtain a red ball on the first draw and lay it aside. Now there are 99 balls left in the urn, of which 25 are yellow. Therefore, the probability of event B, given event A has occurred, is $P(B|A) = 25/99 = .2525$. Therefore, the probability of first selecting a red ball and then a yellow ball is:

$$P(A) \cdot P(B|A) = (.25)(.2525)$$
$$= .063$$

The examples discussed thus far in this chapter are based on the concept that events are random events. The concept of randomness underlies probability and inferential statistics. It is relatively easy to define a random sample, as we do later in this chapter. But the concept of random event or randomness is difficult to define, and the idea of its being haphazard or accidental are actually misleading. Statisticians are quite systematic in their selection of random samples or the assignment of procedures or individuals at random.

Operationally, we can generate random events by having an unbiased coin or die for the coin-tossing and die-tossing examples. The balls-in-the-urn example specified that the balls were well mixed, that we did not put all the balls of one color on the bottom, etc. If we have a random selection of ten numbers, the probability of selecting any one number at random is the same for all of them, namely, .10.

In helping readers to understand the meaning of random events, we can identify the following characteristics. For any one, specific event we cannot predict the outcome with certainty, but for many, many events we can predict the pattern of outcomes. Suppose we had a single, unbiased die and threw it once. What is the probability of obtaining a 3? We saw earlier that this probability is .167, relatively low probability, and before we throw the die there is no way of knowing for certain whether or not we will obtain a 3. On the other hand, suppose we throw the die 4,000 times. How many 3s will be obtained? We can predict that we will obtain approximately $4000 \times .167 = 688$ 3s. As we increase the number of throws, the actual number of 3s obtained will come closer and closer to 16.7% (actually, 1/6) of the total number of throws.

As we discuss randomness relative to specific situations such as sampling, we will describe specific procedures. At this point, it is adequate that the reader understand that the concept of randomness is part of the conceptual foundation of probability, underlying distributions and inferential statistics, as discussed in this text. When we say that events are due to chance, we are talking about events happening at random.

Probability and Underlying Distributions

The application of the rules of probability to relatively limited situations is straightforward and illustrates the nature of probability, but for the procedures of inferential statistics, we turn to underlying distributions of an event. The same type of idea is applied in gambling casinos in terms of the long-run occurrence of events. Casino owners are not concerned about the occurrences of gamblers' winning (unless they suspect that some system other than random occurrence is being used), because they know that over the long run of events, the probabilities of winning combinations favor the casino. For example, the distribution of possible outcomes at a blackjack table is known, more of those favor the casino than the gambler, and over the long run the outcomes will follow that distribution.

An underlying distribution represents *all possible outcomes* of a particular event. From an underlying distribution, we can determine the relative frequency of an event — that is, the proportion of the number of outcomes including an event to the total number of possible outcomes. The area of a distribution (the area under the curve of the distribution) can be designated as 1.00, for example, as was done earlier with the standard normal distribution. Thus, we can interpret proportions of area in underlying distributions as probabilities of specified outcomes. The proportion of area is equivalent to the proportion of the number of specified outcomes including the desired event to the total number of possible outcomes. Underlying distributions are for the most part theoretical distributions or curves (such as the normal), but knowing that all possible outcomes of an event follow a known distribution enables us to determine the probabilities of specified outcomes.

The concept of underlying distribution is introduced with a limited example: what is the underlying distribution of all possible sums of the roll of two dice, one green and one red? There are 11 different possible sums, ranging from 2 (when two 1s are rolled) to 12 (when two 6s are rolled), inclusive. But most of these sums can occur in more than one way. For example, there are two ways in which the sum could equal 3: if the green die reads 1 and the red die 2; or if the green die reads 2 and the red die 1. Similarly, there are three ways in which the sum can equal 4: the green die could read 3 and the red die 1; the green die could read 1 and the red die 3; or both could read 2. In all, there are 36 different possible outcomes. Figure 7.3 is a frequency distribution of these possibilities, showing the number of ways in which each sum could be obtained. In other words, it shows the *underlying distribution* of possible outcomes when we roll two dice.

From Figure 7.3, we can see the probability of each possible outcome. For example, what is the probability of obtaining a sum equal to 7 on a single roll of two dice? Figure 7.3 shows that this outcome can be obtained in six ways, that there are six possible outcomes that include this event, out of the total number of possible outcomes, 36. Thus, the probability of obtaining a sum of 7 is 6/36, or 0.167.

Once we know the underlying distribution for an event, we can use that distribution to determine the probability of any combination of outcomes. Suppose we want to know the probability of rolling a sum of 2, 3, 10, 11, or 12. Using Figure 7.3, we can determine the probability of obtaining each of these sums.

Sum	*Frequency/Total Outcomes = Probability*	
2	1/36	.028
3	2/36	.056
10	3/36	.083
11	2/36	.056
12	1/36	.028
	9/36	.251[1]

Then, since these are mutually exclusive events, we use the addition rule for mutually exclusive events (formula 7.2) to find the probability of obtaining either a 3 or less, or a 10 or more.

$$P(2 \ or \ 3 \ or \ 10 \ or \ 11 \ or \ 12) = P(2) + P(3) + P(10) + P(11) + P(12)$$

$$= .028 + .056 + .083 + .056 + .028$$

$$= .251$$

Few variables in the behavioral sciences have the underlying distribution of the sum of a single roll of two dice; we have introduced it merely to illustrate the concept of an underlying distribution. In practice, statisticians do not spend

[1]Note that the sum of these probabilities really equals .250, not .251. The difference is due to rounding error. If we enumerate the outcomes that produce these sums, there will be nine, which is .25 or 25% of the total 36 possible outcomes.

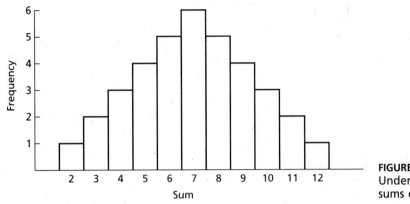

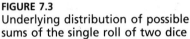

FIGURE 7.3
Underlying distribution of possible sums of the single roll of two dice

time generating or plotting underlying distributions such as the distribution shown in Figure 7.3. Instead of generating an underlying distribution from observations, they use *theoretical distributions* that have been developed through mathematical statistics. A commonly used theoretical distribution, although by no means the only one, is the normal distribution. However, next we discuss the binomial distribution.

An underlying distribution is the distribution of all possible outcomes of a particular event.

A Second Example of a Constructed Underlying Distribution

To further illustrate the concept of an underlying distribution, we use a situation more complex than the dice-throwing example, but before doing so, it is necessary to consider two mathematical concepts, factorials and combinations. The **factorial** of a number is defined as the product of all integers of 1 through that number, although the integers are usually ordered in reverse. The symbol for a factorial is the number followed by an exclamation point. Thus,

$$3! = (3)(2)(1) = 6$$
$$6! = (6)(5)(4)(3)(2)(1) = 720$$
$$N! = N(N - 1)(N - 2) \cdots (2)(1)$$

In general, in order to keep the system consistent, the zero factorial is defined as 1, that is, $0! = 1$. We can see from the examples above that as N gets large, $N!$ increases rapidly.

Combinations are what the name implies and address the question, "In how many ways can we select a *different* subset of things or individuals from a population of such things or individuals?" Suppose we have seven basketball players, all able to play the different team positions. How many different teams of five players can be formed? Any time one or more players are changed, we have a different team. The answer is: the combinations of seven players taken five at a time. This is symbolized by $_7C_5$ and

$$_7C_5 = \frac{7!}{5!(7-5)!} = \frac{7!}{5! \ 2!}$$

If we write out the factorials, we have

$$\frac{7!}{5! \ 2!} = \frac{(7)(6)(5)(4)(3)(2)(1)}{(5)(4)(3)(2)(1)(2)(1)} = 21$$

Therefore, there are 21 different teams of five players that can be formed from the seven players.[2] In general, if we have N things taken n at a time, with $N \leq n$, the number of possible combinations is:

$$_NC_n = \frac{N!}{n!(N-n)!} \qquad (7.6)$$

There are a couple of characteristics of factorials, because they are the products of integers, that simplify working with them. For one thing, when writing out a factorial, it is not necessary to put in the factor (1) because any number multiplied by 1 is that number. When dividing factorials, suppose we have $\frac{8!}{6!}$. This can be written

$$\frac{(8)(7)(6!)}{6!}$$

which equals (8)(7), or 56, because the 6!s cancel. In general

$$\frac{N!}{n!} = \frac{N(N-1)\cdots(n+1)\cdot n!}{n!} = N(N-1)\cdots(n+1) \qquad (7.7)$$

Consider the hypothetical situation of a university department having ten assistant professors, five men and five women, who are eligible for tenure. Six receive tenure and four do not; of the six receiving tenure, two are men and four are women. Is there any evidence of gender bias in this decision? If six individuals of the ten had been selected randomly for tenure, what is the probability that a sample of four women and two men, or a sample even less favorable to men — five women and one man — would have been selected?[3]

---

[2]This result can be verified by taking seven things or individuals and determining how many different groups of five can be formed.

[3]In sex-discrimination and age-discrimination litigation, any number of factors may be brought to bear in a specific case, but the federal courts have allowed evidence about possible outcomes had random samples been selected.

If six individuals are selected randomly from a group of five men and five women, the total number of possible samples is:

$$_{10}C_6 = \frac{10!}{6!\ 4!} = 210$$

and the possible gender combinations are:

5 men, 1 woman

4 men, 2 women

3 men, 3 women

2 men, 4 women

1 man, 5 women

In order to determine the number of possible samples for each of these gender combinations, consider the first listed, five men and one woman. Of the five men, all five are required. Thus $_5C_5$ (read as the combination of five individuals taken five at a time) gives the number of combinations for the men.

$$_5C_5 = \frac{5!}{5!\ 0!} = 1$$

For the women, there are five women and one is to be selected; thus we have

$$_5C_1 = \frac{5!}{4!\ 1!} = 5$$

The selections of men and women are independent events, so the number of different possible samples of five men and one woman is (the first subscript on the n is the number of men, the second subscript the number of women):

$$n_{5,1} = {_5C_5} \cdot {_5C_1} = (1)(5) = 5$$

Likewise, the numbers of possible samples of the remaining gender combinations are:

$$n_{4,2} = {_5C_4} \cdot {_5C_2} = 50$$
$$n_{3,3} = {_5C_3} \cdot {_5C_3} = 100$$
$$n_{2,4} = {_5C_2} \cdot {_5C_4} = 50$$
$$n_{1,5} = {_5C_1} \cdot {_5C_5} = 5$$

Note that the sum of the possible samples of the five gender combinations is 210, the total number of possible samples, because these five gender combinations exhaust all the possibilities.

The underlying distribution for this example is given in Figure 7.4. It is a symmetrical distribution but rather steep in the middle. The shaded area represents the proportion of the total area that, through random sampling, would give a sample of two men and four women or a sample even less favorable for men. Actually, we

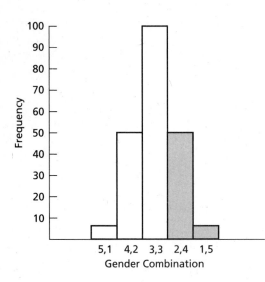

FIGURE 7.4
Underlying distribution of possible samples for the gender combinations example

do not need the figure to obtain this proportion or probability. We know that there are 50 possible samples with two men and four women, and five possible samples with one man and five women, so the probability is:

$$P(2,4 \text{ or } 1,5) = \frac{50}{210} + \frac{5}{210} = \frac{55}{210} = .262$$

Thus, the probability is .262, or slightly greater than one in four, that with random sampling a sample of two men and four women, or one even less favorable for men, would have been selected.

The Binomial Distribution as an Underlying Distribution

The binomial distribution illustrates situations in which two outcomes are possible on each attempt or trial and the trials are independent of each other. The outcomes of flipping coins (heads or tails being the two possible outcomes) follow the binomial. Just as there are many possible normal distributions (conceptually, an infinite number), there is a family of binomial distributions, the specific distribution dependent on the conditions of the situation, namely, the number of trials.

Thus, the binomial distribution can be used to determine the probability of statistically independent events. Consider the toss of two unbiased coins. There are, of course, four possible outcomes: (1) heads and heads (HH); (2) heads and tails (HT); (3) tails and heads (TH); and (4) tails and tails (TT). The probability of obtaining HH is determined by the multiplication rule for statistically independent events (formula 7.4).

$$P(HH) = P(heads) \cdot P(heads)$$

$$= (.50) \cdot (.50)$$

$$= .25$$

We can determine the probabilities of the other three outcomes with similar calculations. The resulting probabilities are:

Outcome	First coin	Second coin	Probability
1	H	H	1/4 or .25
2	H	T	1/4 or .25
3	T	H	1/4 or .25
4	T	T	1/4 or .25

Notice that, although there are four possible outcomes, two of them are, in effect, the same. That is, there are two ways to obtain heads and tails: HT and TH. Thus, there are only three distinct possible outcomes:

Outcome		Probability
H	T	
2	0	.25
1	1	.50
0	2	.25

We can obtain these same results mathematically by squaring a binomial. Symbolically,

$$(X + Y)^2 = X^2 + 2XY + Y^2$$

Therefore, if X equals heads (H) and Y equals tails (T), then

$$(H + T)^2 = HH + 2HT + TT$$

Now if we let H equal the probability of obtaining heads, which is .50, and T equal the probability of obtaining tails (.50), then by substituting these probabilities into the expanded binomial, we obtain

$$(.50 + .50)^2 = (.50)(.50) + 2(.50)(.50) + (.50)(.50)$$

$$= .25 + .50 + .25$$

These are the respective probabilities of obtaining two heads (.25), heads and tails (.50), and two tails (.25).

Now consider the simultaneous toss of three coins. There are three statistically independent events (the toss of each coin), so the probability of each possible outcome is determined by the multiplication rule for statistically independent events

TABLE 7.1
Outcomes of the Toss of Three Coins

	Possible Outcomes			
Outcome	First Coin	Second Coin	Third Coin	Probability
1	H	H	H	$1/8 = .125$
2	H	H	T	$1/8 = .125$
3	H	T	H	$1/8 = .125$
4	H	T	T	$1/8 = .125$
5	T	H	H	$1/8 = .125$
6	T	H	T	$1/8 = .125$
7	T	T	H	$1/8 = .125$
8	T	T	T	$1/8 = .125$

Distinct Outcomes		
H	T	Probability
3	0	$1/8 = .125$
2	1	$3/8 = .375$
1	2	$3/8 = .375$
0	3	$1/8 = .125$

(formula 7.4). For example,

$$P(HHH) = P(heads) \cdot P(heads) \cdot P(heads)$$

$$= (.50)(.50)(.50)$$

$$= .125$$

Table 7.1 shows the eight possible outcomes of tossing three coins and their probabilities. But, as in the previous example, some of these outcomes are in effect the same, so there are only four distinct possible outcomes. Table 7.1 also shows these outcomes and their probabilities.

We can obtain the same results as those shown in Table 7.1 simply by taking the binomial $(X + Y)$ to the third power; that is,

$$(X + Y)^3 = X^3 + 3X^2Y + 3XY^2 + Y^3$$

If we let X equal the probability of obtaining heads and Y the probability of obtaining tails, we can find the probability of each outcome.

$$(X + Y)^3 = X^3 + 3X^2Y + 3XY^2 + Y^3$$

$$(H + T)^3 = H^3 + 3H^2T + 3HT^2 + T^3$$

$$(.50 + .50)^3 = (.50)^3 + 3(.50)^2(.50) + 3(.50)(.50)^2 + (.50)^3$$

$$= .125 + .375 + .375 + .125$$

That is, .125 is the probability of obtaining three heads; .375 is the probability of obtaining two heads and one tail; .375 is the probability of obtaining one head and two tails; and .125 is the probability of obtaining three tails.

Using these probabilities, we can answer certain questions about the probability that a specific event will occur. For example, according to the addition rule in formula 7.2, the probability of obtaining at least one head in the toss of three coins equals the probability of obtaining three heads (.125), plus the probability of obtaining two heads (.375), plus the probability of obtaining one head (.375)—a probability of .875.

Though our examples have been limited to coin-tossing ones, the binomial distribution applies to many situations, those in which individual events have two possible outcomes. Any binomial expression such as $(X + Y)$ can be raised to any power, but expanding the binomial to higher powers is quite arduous. There is a formula, somewhat complex, called the binomial formula, which can be used to generate the expansion of a binomial to any desired power. Each term of that formula has a coefficient, which can be calculated using factorials. The specific factorials depend upon the power to which the binomial is raised. However, it is not necessary to compute binomial coefficients because they can be found in tables of binomial coefficients.

The preceding discussion of probability and underlying distributions was included for the purpose of enhancing the reader's understanding of the inferential-statistics concepts to follow. In practice we seldom construct underlying distributions unless we have an unusual, limited situation for which there is no known distribution. Practically, we use known distributions, which apply, of course, to the specific situations and which are contained in statistical tables. It was mentioned earlier that the binomial distribution, although an important conceptual underlying distribution that applies to many situations, is not used frequently. The reason is that as the number of events or trials increases (a binomial expression is raised to higher powers), the specific binomial distribution becomes more and more like the normal distribution. Thus, for most practical situations in inferential statistics for which the binomial distribution applies, the sample size is large enough that the normal distribution approximates the binomial distribution.

Normal Distribution as an Underlying Distribution

One of the more commonly used underlying distributions in behavioral science research is the normal distribution. As we noted in Chapter 4, the frequency distributions of many variables—such as measures of intelligence, aptitude, and achievement—approximate a normal distribution. Also, as will be seen later, many statistics generated through inferential statistics procedures are normally distributed, or are so close to being normally distributed that the normal distribution can be used as the underlying distribution.

At this point, we will illustrate the use of the normal distribution as an underlying distribution by considering a variable that is normally distributed and by using the normal distribution to determine the probability of randomly selecting particular scores from that distribution.

Suppose for 15,000 third- and fourth-grade students, we have mathematics achievement test scores recorded on slips of paper and placed in a large container. Suppose further that we know that the scores from the mathematics test are normally distributed, with a mean (μ) equal to 60 and a standard deviation (σ) equal to 10, as shown in Figure 7.5. What is the probability of drawing a slip of paper with a number between 50 and 60?

From Chapter 4 we know that the proportion of scores that falls within a certain interval equals the percentage of the area under the normal curve that covers this interval; in other words, the proportion of the area tells us the relative frequency of a range of scores. The relative frequency of an event, we have seen, is its probability. Therefore, when the underlying distribution for an event is normal, we can use the normal distribution table (Table C.1) to determine the probability of various outcomes. In this example, since 50 is one standard deviation from the mean, 60, we see from Table C.1 that 34.13 percent of possible scores fall between 50 and 60; thus, the probability of drawing a number between 50 and 60 is .3413.

Now consider the probability of drawing a number that is greater than 80 or less than 40. The first step is to convert the raw scores 80 and 40 to standard scores.

$$z_{80} = \frac{80 - 60}{10} = +2.00$$

$$z_{40} = \frac{40 - 60}{10} = -2.00$$

We use Table C.1, as discussed in Chapter 4, and find that 0.0228, or 2.28 percent, of the area is above a z score of 2.00 and that the same amount is below a z score of -2.00. Adding these two areas, we find a probability of $.0228 + .0228 = .0456$ that a score selected at random will be greater than 80

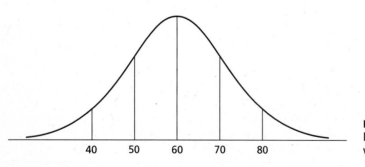

FIGURE 7.5
Normal distribution of 15,000 scores
with $\mu = 60$ and $\sigma = 10$

or less than 40. In other words, it is highly unlikely (there is a low probability) that a number less than 40 or greater than 80 will be drawn from the container.

In the same manner, we can determine the probability of selecting at random a score between any two scores, or beyond or below a given score. We use the procedure of determining area in the standard normal distribution, as described in Chapter 4. Consider the probability of selecting a score between 50 and 70. These scores convert to standard scores of -1.00 and $+1.00$, respectively. By using the standard normal distribution, we find that the actual probability is .6826; in other words, it is very likely (highly probable) that a number between 50 and 70 will be drawn. Again, the sum of the probabilities of all possible outcomes is 1.00, the total area under the curve.

> The normal distribution is one of the more commonly used underlying distributions in inferential statistics.

Chain of Reasoning in Inferential Statistics

Up to this point, the discussion in this chapter has laid the groundwork for the reasoning and procedures of inferential statistics and for the concepts of probability, random sample, and underlying distribution. When using inferential statistics, we measure a sample to generate data, and these data and their resulting measures are called statistics. Statistics are the data we have, and they are the only data we have. Generally, we do two things in inferential statistics: (1) test hypotheses about parameters, which are measures of the population sampled, and (2) estimate parameters. However, it is important to realize that the parameters are not known. If they were known, it would not be necessary to make inferences about them and we would not need inferential statistics.

We begin with an overview of the chain of reasoning in inferential statistics, using an example involving the three concepts mentioned above and the sample mean. In Chapter 1, we noted that the mean of a sample $(\overline{X})$ is a *statistic* (a descriptive measure of a sample), whereas the mean of a population (μ) is a *parameter* (a descriptive measure of a population). This distinction takes on added importance in inferential statistics because the statistics are the facts on which we draw conclusions about parameters. There are many situations in which it is not feasible, necessary, or desirable to collect data on, or to measure, an entire population. Therefore, we select a random sample and draw conclusions about the population on the basis of the sample results.

Suppose we have a distribution of SAT scores from a sample of 144 freshman psychology students randomly selected from the population of all freshman psychology students enrolled at a major state university. We find that the mean of this sample is 535. What does this sample mean represent? Since it is based on a small subset of the population, we cannot assume that it tells us the mean of the population. If we select another sample from the population and compute the sample mean, we would not expect this second sample mean to be exactly 535. It could be larger or smaller. However, since both means were selected from the same population, both reflect the population mean (μ). In this regard, the sample mean is only an *estimate* of the population mean.

To infer the characteristics of a population from the characteristics of a sample, we need to know how the sample we have taken compares with other samples that might be taken from the population. Moreover, we also need to know the likelihood that those other samples would behave in a similar way, exhibiting characteristics similar to those in the sample we have selected.

This chain of reasoning is outlined in the following steps and is illustrated in Figure 7.6.

Step 1. To draw inferences about the parameter on the basis of an estimate from a sample, the sample must be selected *randomly*.

Step 2. The estimate from this sample must be compared to an *underlying distribution* of estimates from all other samples of the same size that might be selected from this population.

Step 3. Based on this comparison and the probability associated with outcomes when random sampling is used, we can make inferences about parameters.

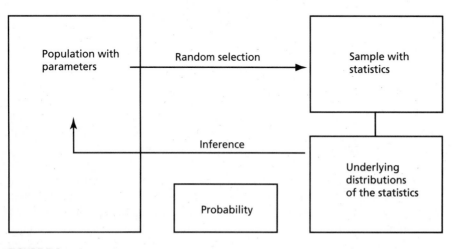

FIGURE 7.6
Chain of reasoning for inferential statistics

Sampling Distribution of the Mean

Continuing the SAT score example, we have a mean of 535 based on a sample of 144 freshman psychology students. We know that this mean reflects not only the population mean, but also fluctuation due to random sampling, which is called sampling error. If we took other samples, we would obtain other sample means. Therefore, in order to make inferences about the population, we need first to interpret the statistics in light of their underlying distribution and probability.

Earlier we used the normal distribution as the underlying distribution for determining the probability of selecting a particular score from the population. Now we are interested in sample means (not scores), and we will use the normal distribution differently as an underlying distribution. For the above example of SAT scores, we will consider the distribution of the sample means ($\overline{X}$) for all possible samples of a given size that might be obtained from the population.

Suppose we could record on a slip of paper the quantitative SAT score for every freshman psychology student at a major state university and place these slips of paper in a container. This would constitute a distribution of *all* quantitative SAT scores in the *population*. Further suppose we randomly select 144 slips of paper from this container. This would constitute a distribution of quantitative SAT scores for a sample of size 144 — that is, a distribution of scores for a *sample* from the population. Under the assumption that the scores were selected randomly, the distribution of scores in the sample should be representative of the distribution of scores in the original population.

Now suppose that all possible random samples of size 144 are selected (with replacement) from the population, that the mean of each sample ($\overline{X}$) is computed and recorded on a slip of paper, and that these slips of paper are placed in a second container. The slips of paper in the second container constitute the distribution of sample means for samples of size 144. This distribution is called the **sampling distribution of the mean**. Assuming that *all* possible samples are selected and all sample means are included in the container, this sampling distribution is the underlying distribution of all possible outcomes, the outcomes being the sample means. It should be noted that if the population of freshman psychology students is relatively large — say, any number 500 or greater — the second container would contain a very, very large number of slips. The number would be the combination of 500 (or more) scores taken 144 at a time. This is a conceptual distribution; it is not a distribution that would be generated in practice.

The reader should not become confused about the terms *sampling distribution* and *underlying distribution*. In inferential statistics, the sampling distribution is the underlying distribution of a statistic, and as a component in the chain of reasoning, it takes on utmost importance. Underlying distributions, for the most part, are theoretical distributions that are known through mathematical statistics. They are *not* the distributions from which the statistics are computed. A sampling distribution is the distribution of all values of the statistic under consideration, from all possible random samples of a given size.

In inferential statistics, three distributions are considered:
1. The distribution of scores in the population.
2. The distribution of scores in a sample randomly selected from the population.
3. The distribution of sample statistics (in this case, means) for all samples of a given size randomly selected from the population.

This last distribution is the sampling distribution of the mean.[4]

Properties of the Sampling Distribution of the Mean

Researchers and statisticians do not spend their time putting slips of paper into containers, drawing samples, and developing sampling distributions. Rather, they use **theoretical sampling distributions**, which are defined by mathematical theorems. The mathematical theorems provide the necessary information for describing (1) the shape, (2) the central tendency, and (3) the variability of these distributions.

The shape of the theoretical sampling distribution of the mean is defined by the **central limit theorem**, which states that

as the sample size (n) increases, the sampling distribution of the mean for simple random samples of n cases, taken from a population with a mean equal to μ and a finite variance equal to σ^2, approximates a normal distribution.

When the sample size is greater than 30, the approximation of the sampling distribution to a normal distribution is usually quite close even if the population is not normally distributed. Even with a sample size smaller than 30, the approximation to a normal distribution is generally reasonable, as we will illustrate later.

For determining the other characteristics of the sampling distribution of the mean (central tendency and variability), we use mathematical theorems dealing with unbiased estimators and their expected values.[5] Here we have a new term, unbiased estimator. Statistics are estimators because they estimate parameters. In general, a statistic is an unbiased estimator if the mean of its sampling distribution equals the parameter being estimated. In other words, in the long run or over repeated sampling, the expected value of the statistic is the value of the parameter.

[4]We will consider the sampling distributions for other statistics (e.g., proportions, correlation coefficients, etc.) in Chapters 10 to 12.

[5]W. L. Hays, *Statistics*, 4th ed. (New York: Holt, Rinehart, and Winston, 1988), pp. 197–201.

Considering the sampling distribution of the mean and its central tendency, it can be shown that the expected value of the sample mean $(\overline{X})$ is the mean of the population (μ).

We use the same theorems to define the variance and standard deviation of the sampling distribution of the mean. Again, the expected value of the variance of the sampling distribution, denoted $\sigma_{\overline{X}}^2$, is defined as

$$\sigma_{\overline{X}}^2 = \frac{\sigma^2}{n} \tag{7.8}$$

where

$\sigma^2 = $ variance in the population

$n = $ size of sample

Correspondingly, the standard deviation of the sampling distribution, denoted $\sigma_{\overline{X}}$, is defined as

$$\sigma_{\overline{X}} = \sqrt{\sigma_{\overline{X}}^2} \tag{7.9}$$

$$= \sqrt{\frac{\sigma^2}{n}}$$

$$= \frac{\sigma}{\sqrt{n}}$$

The standard deviation of the sampling distribution of the mean is called the **standard error of the mean**.

As the sample size (n) increases, the sampling distribution of the mean has the following properties:

1. The distribution is normal.
2. The mean of the distribution equals μ.
3. The standard deviation of the distribution is called the standard error of the mean and equals $\sigma/\sqrt{n}$.

Using the Normal Distribution as the Sampling Distribution of the Mean

For our example of SAT scores, we have only *one* of all possible samples of size 144 that could be drawn from the population of freshman psychology students, and the mean of this sample $(\overline{X} = 535)$ is thus only *one* of all possible sample means. The theoretical sampling distribution of all possible sample means can be defined by the mathematical theorems mentioned above. According to these theorems, this distribution (1) is normally distributed, (2) has a mean equal to μ, and (3) has a standard error equal to $\sigma/\sqrt{n}$. Assuming that the quantitative SAT test has a mean (μ) of 455 and a standard deviation (σ) of 100, the theoretical sampling distribution for this example would be defined as follows:

1. The distribution is normal.

2. The mean of the distribution equals 455 (the assumed value for μ).

3. The standard error of the mean is $100/\sqrt{144} = 8.33$ (where the assumed value for σ is 100).

This distribution, which is an underlying distribution of all possible outcomes, is illustrated in Figure 7.7.

Using the properties of the normal distribution, we can determine the probability of randomly selecting a particular sample mean ($\overline{X}$) from this underlying distribution. For example, consider the probability of selecting a sample mean between 455 and 463.33. The point 463.33 is 1.0 standard error *above* the mean ($\mu = 455$); and from Table C.1, we know the probability is .3413. Similarly, we could find the probability of selecting a sample mean that is between 446.67 and 455. Since the point 446.67 is 1.0 standard error *below* the mean, the probability is also .3413. Summing these two probabilities, we can say that the probability of selecting from this sampling distribution a sample mean that is between 446.67 and 463.33 is .6826. In the same way, we can find the probability of selecting a sample mean that is between 438.33 and 471.67. Notice that these two points are -2.0 and $+2.0$ standard errors below and above the mean ($\mu = 455$), respectively. Thus, using Table C.1, we find the probability to be $.4772 + .4772 = .9544$.

Now suppose we want to determine the probability of selecting from this distribution a sample mean that is *less* than 450 or *greater* than 465. In order to do so, we must use the concept of the standard score discussed in Chapter 4 — as we did earlier in this chapter. Recall that, in these earlier discussions, the standard score (z) indicated the number of standard deviations a corresponding raw score ($\overline{X}$) was above or below the mean. When we use z scores with the theoretical

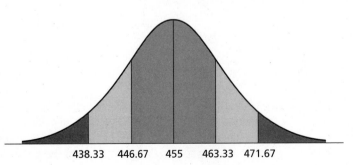

438.33   446.67   455   463.33   471.67

FIGURE 7.7
Sampling distribution of the mean for $\mu = 455$ and $\sigma_{\overline{X}} = 8.33$

sampling distribution of the mean, the z score is the number of *standard errors* the corresponding sample mean ($\overline{X}$) is above or below μ. Thus, to find the probability of obtaining a sample mean ($\overline{X}$) less than 450, we calculate the z score as follows:

$$z = \frac{\overline{X} - \mu}{\sigma_{\overline{X}}}$$

$$= \frac{450 - 455}{8.33}$$

$$= -0.60$$

Using Table C.1, we find the associated probability to be .2743 (see Figure 7.8). Similarly, we can find the probability of obtaining a sample mean greater than 465.

$$z = \frac{465 - 455}{8.33}$$

$$= 1.20$$

Using Table C.1, we see that the associated probability is .1151 (see Figure 7.8).

Earlier we used the concept of standard score to find the probability of selecting a particular score (X) from a distribution. The formula for the z score was

$$z = \frac{X - \mu}{\sigma}$$

Notice that the denominator of this formula is the *standard deviation* of this distribution of *scores*.

In this section, we have used the same concepts to find the probability of selecting a particular sample mean ($\overline{X}$) from the theoretical sampling distribution of the mean. In this case, the formula for the z score was

$$z = \frac{\overline{X} - \mu}{\sigma_{\overline{X}}}$$

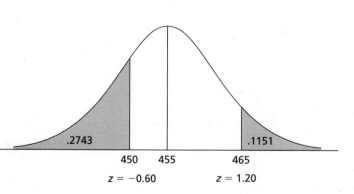

.2743

.1151

450   455   465

$z = -0.60$ $z = 1.20$

FIGURE 7.8
Sampling distribution of the mean for $\mu = 455$ and $\sigma_{\overline{X}} = 8.33$

In this formula, the denominator is the *standard error of the mean*, which was defined as the standard deviation of this distribution of sample *means*.

Two Generalizations about the Sampling Distribution of the Mean

When the sampling distribution of the mean is used in inferential statistics, it is very important to remember two generalizations. The first is:

> As the sample size (n) increases, the variability of the sampling distribution of the mean decreases (the standard error decreases).

This decrease in variability can be seen in the formula for the standard error of the mean $\sigma_{\bar{X}} = \sigma/\sqrt{n}$. Notice that as n increases, the standard error decreases; that is, as the sample size increases, the variability of the sampling distribution of the mean decreases.

This generalization is illustrated in Figure 7.9, which shows the distribution of SAT scores to be normal with $\mu = 455$ and $\sigma = 100$ (see Figure 7.9A) and the sampling distribution of the mean for sample sizes 1, 2, 4, 16, and 25. Assume that each distribution is based on 100,000 different samples. Figure 7.9B shows that the sampling distribution of the mean for $n = 1$ resembles the original distribution of scores. In fact, the standard error of the mean for a sample size of 1 is equal to the standard deviation of the original distribution of scores. However, as illustrated in Figures 7.9 C, D, E, and F, the variation of the sampling distribution of the mean decreases as the sample size increases. For example, the standard error of the mean for $n = 16$ is

$$\sigma_{\bar{X}} = \frac{\sigma}{\sqrt{n}} = \frac{100}{\sqrt{16}} = 25$$

while the standard error of the mean for $n = 25$ is

$$\sigma_{\bar{X}} = \frac{\sigma}{\sqrt{n}} = \frac{100}{\sqrt{25}} = 20$$

The second generalization based on the central limit theorem is:

> Even when the population is not normally distributed, the shape of the sampling distribution of the mean becomes more like the normal as the sample size increases.

Figure 7.10 illustrates this second generalization. It shows two population distributions that are not normal; one is rectangular and one is skewed to the right. For both, the mean (μ) is 455 and the standard deviation (σ) is 100 (see Figure 7.10A). As with Figure 7.9, Figure 7.10 illustrates the sampling distribution of the mean (assuming 100,000 different samples) for sample sizes 1, 2, 4, 16, and 25. Notice that, for both the rectangular and the skewed population distributions, the sampling distribution of the mean becomes normal even for sample sizes as small as 25.

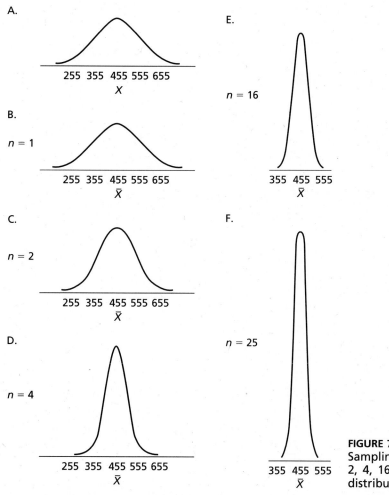

FIGURE 7.9
Sampling distributions for $n = 1$, 2, 4, 16, and 25; original score distribution—normal

As sample size increases, the sampling distribution of the mean:

1. decreases in variability—that is, its standard error decreases—and

2. becomes more like the normal distribution in shape, even when the population distribution is not normal.

In this section we have described in detail the sampling distribution of the sample mean. The sample mean is a statistic used extensively in research, so it is a logical statistic with which to introduce these concepts. However, all statistics have sam-

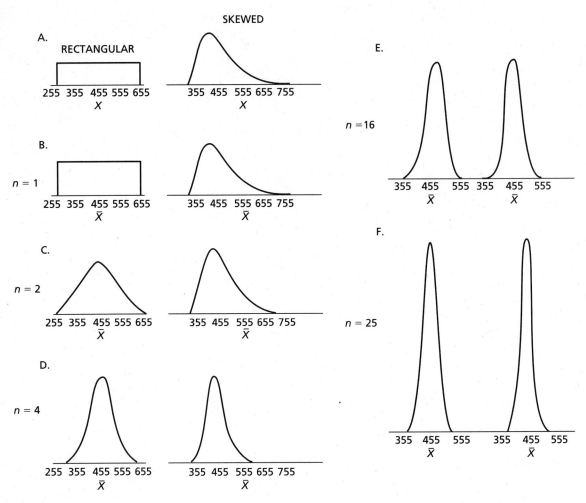

FIGURE 7.10
Sampling distributions for *n* = 1, 2, 4, 16, and 25; original score distribution—rectangular and skewed

pling distributions, and in discussions of other statistics in later chapters, the same concepts will apply to their sampling distributions. Again, these will be theoretical underlying distributions whose characteristics we know through mathematical statistics.

Sampling

We have noted that much behavioral science research is conducted by studying populations through the use of samples because it is often too costly or time-consuming to gather information on all members of a population. The sample, which is assumed to be representative of the population, provides the data on which research conclusions are based.

There are numerous sampling procedures, some of them complex, and entire books are written about sampling.[6] In the remainder of this chapter, we will describe some of the more general sampling methods widely used by researchers. All methods discussed involve some aspect of random selection. We will simply define and illustrate the methods, making no attempt to deal with them comprehensively.

Simple Random Sampling

As the name implies, simple random sampling is the least complex of the sampling procedures and probably the most widely used by behavioral science researchers. To define such a sample, we need to distinguish between sampling with replacement and sampling without replacement.

Recall that *sampling with replacement* means that, as members of the population are selected for the sample, they are "replaced" in the population and can be selected again. In sampling *with* replacement, a **simple random sample** is one in which all members of the population have the same probability of being selected.

In behavioral science research, however, the more common practice is to sample *without* replacement. In this case, once a population member is selected for the sample, that member is not replaced in the population and cannot be selected a second time. In sampling *without* replacement, a simple random sample is one in which all possible samples of a given size have the same probability of selection. This is only a slight adjustment. In both cases, all members of the sample are selected independently of one another.

Procedurally, there are different ways to select a simple random sample. In a dichotomous situation, we could flip an unbiased coin. We could record on slips of paper all names (or some identification code) of the population members, place these slips in a container, and mix the slips very well. Suppose we want a random sample of size 50. Then, 50 slips of paper would be drawn without bias from the container, either with or without replacement, and the names drawn would represent the simple random sample.

These types of procedures are cumbersome, and there are more practical methods for selecting a simple random sample (with or without replacement) from a finite population. Suppose we want a sample of 50 members from a population of 500. First, we assign each member an identification number from 1 to 500 (that is, N, the size of the population). Then we obtain 50 numbers by using a table of random numbers, a computer, or a programmable hand calculator that will generate

---

[6]One of the classic treatments of sampling is a book by Leslie A. Kish, *Survey Sampling* (New York: Wiley, 1965). The Kish book is quite mathematically oriented. For a less mathematical discussion, the reader is referred to Seymour Sudman, *Applied Sampling* (New York: Academic Press, 1976).

a list of random numbers. Members are included in the sample if their identification numbers are on this list of random numbers.

A simple random sample is one in which all population members have the same probability of being selected and the selection of each member is independent of the selection of all other members.

Samples That Are Not Random   Sometimes people make the incorrect inference that almost any available sample is a random sample. Haphazard samples, ad hoc samples, and samples of convenience are not random samples. A sample is random or not random. There is no sample that is almost random. In order for a sample to be random, it must have been selected through a randomizing procedure.

Samples of volunteers are not random samples. If in a health clinic the first 75 patients willing to cooperate in a study comprised the study sample, they would not comprise a random sample. If in an elementary school an intact class of 30 students participated in an experiment simply because the teacher was willing for them to do so, this sample would not be a random one.

In survey research, using mailed questionnaires typically means that some recipients will fail to return completed questionnaires. Even if the original sample of questionnaire recipients was selected at random, those returning the questionnaires would no longer constitute a random sample of the population. In essence, the returnees selected themselves for the sample, and it could not be argued that they did so on a random basis. Therefore, researchers must be careful not to assume a sample to be random and must recognize when the selection process has violated the random selection procedure.

Systematic Sampling

When a list of the members of a population is readily available, the procedure called systematic sampling is often used. In **systematic sampling** from a list, the investigator chooses every kth member of the list for the sample. As a result, systematic sampling provides sampling throughout the population by dividing the selections over the entire population list.

The first step in systematic sampling is to determine the sample size and the sampling fraction. The **sampling fraction** is the ratio of the size of the sample (n) to the size of the population (N). For example, suppose a researcher has a population of 15,000 members and wants to select a 2 percent sample—that is, 2 percent of the population, or 300 members. The sampling fraction is

$$\frac{n}{N} = \frac{300}{15,000} = \frac{1}{50}$$

Assuming that a list containing the names of the 15,000 population members is available and that the list is in random order, systematic sampling involves three steps.

Step 1. Randomly select a number between 1 and the denominator of the sampling fraction, inclusive, using the methods described above. This is the first sample member selected. For our example, this means randomly selecting a number between 1 and 50. If we select 37, then the first sample member is the 37th member on the list.

Step 2. To identify subsequent members, add multiples of the denominator of the sampling fraction to the first number selected. In our example, then, the second member is number $(1)(50) + 37 = 87$; the third member is number 137; and so on.

Step 3. Continue this procedure until the list is exhausted, at which point the sample members will have been selected.

Systematic sampling is more convenient than simple random sampling when a list of the population members is available, but it may not be as easy as this example suggests. Notice that our example is simple because the denominator of the sampling fraction is the whole number 50. The systematic sampling procedure becomes slightly more complex when the denominator is a decimal fraction.[7]

> A systematic sample is a probability sample in which every *k*th member of the population is selected and in which 1/*k* is the sampling fraction. The first member of the sample is determined by randomly selecting an integer between 1 and *k*.

When systematic samples are used, it is usually because the selection process is easier procedurally than the selection process for a simple random sample. But systematic sampling is not quite the same as simple random sampling; for one thing, there are fewer possible, different samples. If an alphabetical list were used, for example, people whose names were close together (such as twins with the same last name) could get into the same sample. If the list were put into random order before being used, the extent of work is such that a simple random sample might as well have been selected.

With systematic sampling it is important that no periodic factor is operating within the list that may bias the sample. Alphabetical lists usually can be assumed to be free of such factors. It is possible that if ethnic background were a factor, certain last names might group together, but even this situation can be circumvented because the entire list is used for sample selection. Therefore, although systematic

[7] Sudman, pp. 54–55.

samples are not quite the same as simple random samples, in most situations they can be used as simple random samples. They are used when their selection is more convenient than selecting a simple random sample.

Cluster Sampling

Sometimes members of a population occur in naturally formed groups called clusters. For example, classrooms of children, or even schools, constitute clusters. When clusters like these exist and it is impractical to select individual members for a sample, we can use the procedure called cluster sampling.

In **cluster sampling**, clusters — not individual members — are randomly selected from the population of clusters. Each member of the population must belong to only one cluster, but the clusters may have different numbers of members. Once a cluster is selected, *all* members of that cluster are included in the sample.

Suppose the research director in a large school system is asked to survey parents about the counseling services available in the elementary schools. It would be logistically difficult to select either a simple random sample or a systematic sample, but it would be relatively easy to randomly select third-, fourth-, and fifth-grade classes and then mail the survey to the parents of students in those classes. In this case, the class is the cluster. Suppose the classes average 26 students and the research director decides that the total number in the sample should be about 750. Then the research director will randomly select $750/26 = 28.85$, or 29, classes from all the third-, fourth-, and fifth-grade classes in the school system. After the 29 classes are selected, the survey is mailed to the parents of *all* the students in the selected classes.

> Cluster sampling involves the random selection of clusters (groups of population members) rather than individual population members. When a cluster is selected for the sample, all members of that cluster are included in the sample.

The above example is quite straightforward, and the sample is selected in one step. However, in practice, cluster sampling can be used effectively if the sample is selected in more than one step, or what is called multistage sampling. If two steps are used, sampling is called two-stage sampling. For example, if a survey were conducted in a large geographical area, precincts might be randomly selected in the first stage. Then blocks of residents might be selected, again at random, from these precincts. However, the blocks would be selected as clusters, and all residents within selected blocks would be surveyed.

Stratified Random Sampling

In simple random sampling, systematic sampling, and cluster sampling, the population is assumed to be generally homogeneous. However, a population may instead be heterogeneous and consist of several subpopulations, which are called *strata*. For example, suppose we are studying the political party affiliations of all registered voters in a small midwestern town. The population might consist of three subpopulations, or strata: those registered as Republicans, those registered as Democrats, and those registered as independent voters. To study these voters, we would use **stratified random sampling**; we would first define the strata and then take random samples of members of each stratum.

When stratified random sampling is used, the researcher not only defines the strata but also determines how many members of each stratum to include in the sample. There are different ways of allocating the sample; for example, equal numbers could be selected from the strata regardless of their sizes. However, the procedure most commonly used in determining the number of members from each stratum is *proportional allocation*. For example, if 30 percent of the registered voters in the town are registered as Republicans, then 30 percent of our sample will be Republicans. With proportional allocation, then, each stratum contributes to the sample a number of members proportional to its size.

Intuitively, stratified random sampling with proportional allocation is attractive because it ensures that all strata are represented proportional to their sizes in the population. However, this usually is not an important reason for using stratified random sampling because simple random samples tend to distribute themselves proportionally. Stratified random sampling does ensure that members from each stratum are included in the sample and that no stratum is missed. More importantly, if strata are heterogeneous, stratified random sampling will enhance statistical precision. This means that the variability of the sampling distributions of whatever statistics are involved in the research will be decreased. As we will see in later chapters, enhancing statistical precision is a desirable outcome of the selection process.

> In stratified random sampling, the population is divided into subpopulations called strata. All strata are represented in the sample, often through proportional allocation.

Summary

Statistical inference involves studying a population by using a sample drawn from the population. When a sample is selected and a statistic such as the sample mean $(\overline{X})$ is computed, the sample mean reflects not only the parameter of the population

(μ) but also sampling error. As a result, in order to make inferences about the characteristics of a population from measures of a sample, we must use the key concepts of probability, underlying distributions, and sampling.

The probability of an event, as we have seen, is the ratio of the number of outcomes including the event to the total number of possible outcomes; the underlying distribution represents the distribution of all possible outcomes. In this chapter, we constructed some relatively simple underlying distributions and also discussed the binomial distribution and normal distribution as underlying distributions. When used for statistics, underlying distributions are called sampling distributions. There are many different sampling procedures that can be used in research studies, some of which are very complex. Regardless of the procedure used, the process of statistical inference involves making inferences about population parameters based on the observed sample statistics drawn from the population.

The theoretical sampling distribution of the mean is the distribution of means computed for all possible samples of a given size. This distribution is normally distributed, with a mean equal to μ and standard error equal to $\sigma_{\overline{X}} = \sigma/\sqrt{n}$. Thus, it is possible to use the properties of the normal distribution to determine the probability of obtaining a given sample mean ($\overline{X}$) when μ and σ are known. In the next chapter, we will use the concept of the sampling distribution and the associated probability of a given event to test whether a hypothesized value for μ is tenable, given the corresponding value of $\overline{X}$.

The concepts of this chapter provide the basis for the reasoning and procedures of inferential statistics. These are based on probability, and although in applied statistics not much coin-flipping and dice-throwing are done, the concepts of probability apply throughout inferential statistics procedures. Probability is represented by area in sampling distributions, which are known through mathematical statistics; these are applied in a quite straightforward manner with the aid of statistical tables.

Exercises

1. Consider the Euler diagram shown in Figure 7.2. Use the addition rules to determine the following probabilities:
   a. P(selecting the number 6).
   b. P(selecting the number 6 or 4).
   c. P(selecting an odd number).
   d. P(selecting a number that is a multiple of 3).

2. Consider the Euler diagram shown in Figure 7.2. Define the following compound events:

   Event A = selecting an odd number
   Event B = selecting an even number
   Event C = selecting a number that is a multiple of 3

Use the addition rules to determine the following probabilities:

a. $P(A \text{ or } B)$

b. $P(A \text{ or } C)$

c. $P(B \text{ or } C)$

3. Consider the compound events described in Exercise 2. Use the multiplication rules for statistically independent events to determine the following probabilities:

a. $P(A \text{ and } B)$

b. $P(A \text{ and } C)$

c. $P(B \text{ and } C)$

4. Consider the compound events described in Exercise 2. Use the multiplication rules for statistically dependent events to determine $P(A \text{ and } B)$.

5. Given the numbers from 1 to 5, determine the underlying distribution of all possible outcomes that are the sum of three of the numbers. Now consider the numbers from 1 to 10. Determine the underlying distribution of all possible outcomes that are the sum of two of the numbers.

6. Consider a deck of cards as the sample space (that is, the sample space contains 52 sample points). Use the rules of probability to determine the following:

a. P(drawing the ace of spades).

b. P(drawing an ace).

c. P(drawing a heart).

d. P(drawing the ace of spades *and* the ace of hearts) with and without replacement.

e. P(drawing three aces) with and without replacement.

f. P(drawing ace, king, and queen) with and without replacement.

7. An urn contains 20 red balls, 20 green balls, and 20 yellow balls, thoroughly mixed. By making three random selections from the urn, what is the probability of selecting:

a. three red balls, sampling with replacement.

b. three green balls, sampling without replacement.

c. three balls of any one color, sampling without replacement.

d. three balls, each of a different color, sampling with replacement.

*__8.__ The superintendent of a school system with seven elementary schools selects a committee of school principals to develop a common policy on student involvement in extracurricular activities. Four of the principals are men and three are women; the committee is to have three members. If a random sample of principals is selected for the committee, what is the probability that the committee will consist of two or more women?

9. Consider only the red balls in the urn of Exercise 7; how many different samples of size 18, all red balls, are possible?

10. Assume that the probability of a girl being born is .5; this is also the probability of a boy being born. If we have 1,000 families with three children each, what would be the underlying distribution of the gender of the children? What would be the probability of randomly selecting a family with two girls and one boy?

11. Given a distribution of 3,000 numbers that are assumed to be normally distributed with $\mu = 120$ and $\sigma = 25$, use the properties of the normal distribution to determine the following probabilities:

   a. P(selecting a number less than 100).

   b. P(selecting a number greater than 150).

   c. P(selecting a number between 110 and 125).

   d. P(selecting a number less than 75 or greater than 160).

12. Over the past ten years, a high school Spanish teacher has been giving the same comprehensive final examination to all students in first-year Spanish. The mean on this examination is 78.4, and the standard deviation is 14.8. To pass the course, a student must score at least 55. To be placed in the honors section, a student must score at least 99. If a student is selected at random, use the properties of the normal distribution to determine the following probabilities:

   a. P(student will pass the course).

   b. P(student will be placed in the honors section).

   c. P(student will score higher than 80).

   d. P(student will score lower than 85).

13. Consider the distribution of scores given in Exercise 11. Use the central limit theorem to describe the sampling distribution of the mean for samples of size 144 and 400.

14. A university department has a five-person tenure committee consisting of professors and associate professors. If there are three professors and seven associate professors in the department, how many different committees are possible if members of the committee are selected randomly? What is the probability that a randomly selected committee would consist entirely of associate professors? Would include all three professors?

15. Using the sampling distribution of the mean for the sample of size 144 in Exercise 13 and the properties of the normal distribution, determine the following probabilities:

   a. $P(\bar{X} > 121.4)$ b. $P(\bar{X} < 118.2)$
   c. $P(\bar{X} < 120.8)$ d. $P(\bar{X} > 119.4)$

16. Using the sampling distribution of the mean for the sample of size 400 in Exercise 13 and the properties of the normal distribution, determine the following probabilities:

   a. $P(\bar{X} > 121.4)$ b. $P(\bar{X} < 118.2)$
   c. $P(\bar{X} < 120.8)$ d. $P(\bar{X} > 119.4)$

17. Suppose that, in the latest census for a large region, the average income for a family unit was $\mu = \$35,285$ and the standard deviation was $\sigma = \$7,200$. Suppose

a sample of size 400 is randomly selected. Develop and describe the sampling distribution of the mean, and then use the properties of the normal distribution to determine the following probabilities:

a. $P(\overline{X} < \$35,000)$ b. $P(\overline{X} > \$35,700)$
c. $P(\$34,900 < \overline{X} < \$35,500)$ d. $P(\overline{X} > \$36,000)$

18. For the census information in Exercise 17, what proportion of the families in this region has incomes exceeding \$40,000? incomes below \$25,000?

19. Describe how random selection is involved in each of the following sampling procedures; simple random sampling, cluster sampling, stratified random sampling, and systematic sampling.

20. In the chain of reasoning for inferential statistics, what is the connection between the underlying distribution of the statistic and probability?

8

Hypothesis Testing: One-Sample Case for the Mean

I n this chapter, we apply the concepts of probability and theoretical sampling distributions to hypothesis testing. The logic for testing hypotheses follows the chain of reasoning in inferential statistics illustrated in Figure 7.6 of the previous chapter. Hypothesis testing involves making inferences about the nature of the population on the basis of observations of a sample drawn from the population.

For example, we can test the hypothesis that the population of freshman psychology students has a mean quantitative SAT score (μ) of 455. Even though this example is somewhat contrived, we can use it in this chapter to illustrate the process of testing hypotheses. As we will see, this process involves determining the difference between the hypothesized value for the population mean (μ) and the mean of the sample selected from the population ($\overline{X}$). If this difference is very large, we reject the hypothesis. But if the difference is very small, we do not.

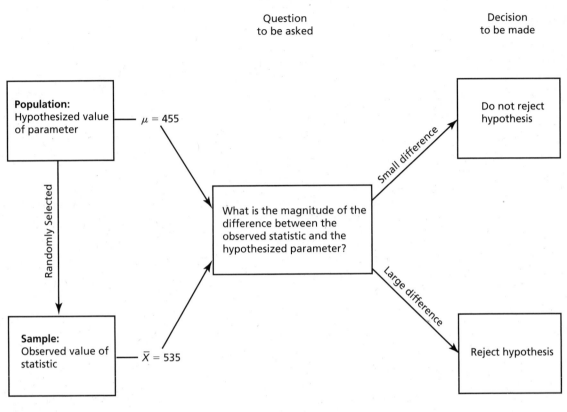

FIGURE 8.1
Logic of hypothesis testing

In other words, hypothesis testing involves determining the magnitude of the difference between an observed value of the statistic ($\overline{X}$) and the hypothesized value of the parameter (μ) and then deciding whether the magnitude of this difference justifies rejection of the hypothesis. Figure 8.1 provides an overview of this process. In this chapter we will test one type of hypothesis: a hypothesis about the value of the population mean. In the following sections we will discuss the four steps of this process. These four steps are very important because they will be used here and in later chapters for testing hypotheses about other parameters.

Hypothesis testing in inferential statistics involves making inferences about the nature of the population on the basis of observations of a sample drawn from the population.

Step 1: State the Hypothesis

In general terms, a hypothesis is simply a conjecture about some phenomenon or set of facts. When conducting research, a hypothesis provides the general framework for the investigation and delineates the problem and the variables under study. The researcher then collects data from a sample and uses those data to determine whether the hypothesis is tenable. Testing the specific hypothesis does *not* prove or disprove the conjecture. Rather, the outcome of hypothesis testing either supports or refutes the tenability of the hypothesis.

In inferential statistics, the term **hypothesis** has a very specific meaning: conjecture about one or more population parameters. The hypothesis to be tested is called the **null hypothesis** and is given the symbol H_0. The null hypothesis is the hypothesis of no relationship or no difference. Ordinarily in behavioral science research, the null hypothesis is stated in terms of *no* relationship between variables or *no* difference between treatment groups. (We will consider hypotheses like these in Chapters 10 and 12). For our example, we will use a simple null hypothesis that the mean quantitative SAT score of the population of freshman psychology students is 455. Thus, our null hypothesis, written in symbols, is

$$H_0: \mu = 455 \qquad \text{or} \qquad H_0: \mu - 455 = 0$$

where

H_0 = symbol for the null hypothesis
μ = population mean
455 = hypothesized value to be tested

A hypothesis is a conjecture about one or more population parameters.

We test the null hypothesis (H_0) against an **alternative hypothesis** (symbolized H_a), which includes the possible outcomes not covered by the null hypothesis. For the above example, we will use the alternative hypothesis that the mean SAT score of the population of freshman psychology students is *not* 455. In symbols, then, our alternative hypothesis is

$$H_a: \mu \neq 455$$

In general, although the null hypothesis specifies no relationship or no difference, the behavioral science researcher is actually trying to establish that a relationship exists between variables or that a difference exists between groups. The process of establishing such a relationship or difference involves rejecting H_0 in favor of H_a.

The logic of this hypothesis-testing procedure has been established in scientific practice. The strategy is similar to the method of indirect proof used by mathematicians. In indirect proof, a proposition is eliminated when it is found to be or to lead to a contradiction of known fact. As we will see, a null hypothesis is rejected when an observed statistic (for example, a sample mean) is shown to be highly unlikely if the null hypothesis is true. In this way, the value specified for the parameter in the hypothesis (in the example, $\mu = 455$) is shown to be untenable.

When conducting research, the alternative hypothesis is often considered the research hypothesis. That is, the researcher is looking for a relationship or a difference. Once the alternative hypothesis has been established, the null hypothesis is determined. The data are then collected and analyzed, and the researcher attempts to reject the null hypothesis in favor of the alternative (research) hypothesis. In this sense, the null hypothesis is a "straw man" to be knocked down in order to support the alternative. However, the null hypothesis is by no means a trivial hypothesis. It is the basis for the procedures of hypothesis testing. Also, in practice, there are many research studies in which null hypotheses cannot be rejected, regardless of the desired or anticipated result.

The alternative hypothesis, often considered the research hypothesis, can be supported only by rejecting the null hypothesis.

To summarize the first step in the process of hypothesis testing for this example: we test the null hypothesis that the mean quantitative SAT score for the freshman psychology students is 455 against the alternative hypothesis that the mean SAT score is not 455. In symbols,

$$H_0: \mu = 455$$

$$H_a: \mu \neq 455$$

Step 2: Set the Criterion for Rejecting H_0

After stating the hypotheses, the next step in hypothesis testing is determining how different the sample statistic $(\overline{X})$ must be from the hypothesized population parameter (μ) before the null hypothesis can be rejected. For our example, suppose we randomly select 144 freshman psychology students from the population and find the sample mean $(\overline{X})$ to be 535. Is this sample mean $(\overline{X} = 535)$ sufficiently different from what we hypothesized for the population mean $(\mu = 455)$ to warrant rejecting the null hypothesis? Before answering this question, we need to consider three concepts: (1) errors in hypothesis testing, (2) level of significance, and (3) region of rejection.

Errors in Hypothesis Testing

When we decide to reject or not reject the null hypothesis, there are four possible situations:

1. A true hypothesis is rejected.

2. A true hypothesis is not rejected.

3. A false hypothesis is not rejected.

4. A false hypothesis is rejected.

If the hypothesis is true and we do not reject it, we have made the proper decision; and if the hypothesis is false and we reject it, we have made the proper decision. But if we reject a true hypothesis or do not reject a false hypothesis, our decision is in error. Thus, in a specific situation, we may make one of two types of errors, as Figure 8.2 shows:

Type I error is when we reject a true null hypothesis.

Type II error is when we do not reject a false null hypothesis.

Although we cannot eliminate the possibility of making an error in hypothesis testing, we can control the criterion for rejecting the null hypothesis, as we will see below. In general, under constant conditions, we increase the likelihood of one type of error as we decrease the likelihood of the other. Therefore, a judgment must be made depending on the consequences of each type of error.

Which of the two errors is more serious? Suppose that patients with a serious disease are randomly assigned to one of two treatment groups. One group is given a newly developed, expensive drug to treat the disease, and a second group receives the traditional drug treatment. The hypothesis tested in this investigation is that there is no difference between the two treatments' effects.

Now consider the consequences of making each type of error. If the new drug is not more effective than the existing drug but the hypothesis is rejected anyway (Type I error), the new drug will begin to be used to treat the disease even though

	State of nature	
	Null hypothesis is true	Null hypothesis is false
Decision made		
Reject null hypothesis	Type I error	Correct decision
Do not reject null hypothesis	Correct decision	Type II error

FIGURE 8.2
The four possible outcomes in hypothesis testing

it is considerably more expensive and no more helpful. On the other hand, if the hypothesis is not rejected even though it is false (Type II error), the new drug will not be used. The consequence of the Type I error is that patients will incur additional cost for the new drug even though it is not more effective. The consequence of the Type II error is that the patients will not incur the additional expense of the new drug but will not receive its additional benefits.

This example illustrates that a value judgment is required to determine which of the errors is more serious. The general approach to hypothesis testing is to argue that the alternative hypothesis is true by rejecting the null hypothesis. This approach focuses on the Type I error, rejecting the null hypothesis when in fact it is true. This is not to say that the Type I error is more serious. The specific conditions of the research situation determine which type is more serious. Ideally, both types of errors should be minimized in any research investigation.

> The two types of errors possible in hypothesis testing are Type I (rejecting a null hypothesis when it is true) and Type II (not rejecting a null hypothesis when it is false).

Level of Significance

To choose the criterion for rejecting H_0, the researcher must first select what is called the level of significance. The **level of significance** or **alpha (α) level** is defined as the probability of making a Type I error when testing a null hypothesis. Rather than computing the actual probability of making a Type I error, researchers usually establish the level of significance before collecting any data.

The most frequently used levels of significance are .05 and .01. In deciding to reject the hypothesis at one of these levels, the researcher knows that the decision to reject the hypothesis may be incorrect 5 percent or 1 percent of the time, respectively. The researcher is willing to accept this risk. When the null hypothesis is rejected at the .05 level of significance, we say that the result, or the difference between the observed statistic and the hypothesized value of the parameter, is "statistically significant at the .05 level."

> The level of significance is the probability of making a Type I error: rejecting H_0 when it is true.

The probability of making a Type II error—of not rejecting a false null hypothesis—is not as easy to determine.[1] This probability is called beta (β) and is related to the level of significance. If other factors are constant, raising the level of significance (that is, raising the probability of making a Type I error) from .05 to .10 decreases the probability of making a Type II error, and lowering the level of significance from .05 to .01 increases the probability of making a Type II error.

When selecting the level of significance, the researcher should carefully consider the consequences of making either type of error in a specific situation. For example, if major expenditures or changes will result from rejecting the null hypothesis, the researcher will want to reduce the probability of making a Type I error. The researcher in such a situation will probably reduce the level of significance to .01 or below—a choice known as "going in a more conservative direction"—using a very conservative level of significance, such as .005 or .001. In other situations, indications of the direction of a trend might be important, and a less substantial departure from the null hypothesis might provide legitimate evidence of the trend. In these cases, we might use a less conservative level of signicance, such as .10 or .20.

Region of Rejection

The level of significance is a probability, and will be used in connection with a sampling distribution. Remember from Chapter 7 that when probability was represented as area in an underlying (or sampling) distribution, the probability of an event was equal to the proportion of the area occupied by the event in the distribution. So too the level of significance represents a proportion of area in a sampling distribution that equals the probability of rejecting the null hypothesis if it is true. This area in the sampling distribution is called the **region of rejection.** Before we identify the region of rejection in the sampling distribution for the SAT example, we will identify the sampling distribution.

Assume for our SAT example that the level of significance is set at .05. Now we must decide whether the difference between the observed sample mean ($\overline{X}$) and the hypothesized value for the mean in the population (μ) is sufficient to reject the null hypothesis. To do this, we must use the sampling distribution of the mean, a theoretical distribution, which was introduced in Chapter 7. Recall from Chapter 7 that our sample of 144 freshman psychology students is only one of many samples of size 144 that could have been selected from the population. Correspondingly, the mean quantitative SAT score for this sample ($\overline{X} = 535$) is only one of many possible sample means in this theoretical sampling distribution.

---

[1] To determine the probability of making a Type II error (β), the value of the alternative hypothesis must be a specific hypothesized value for the parameter. We will discuss this concept in greater detail in Chapter 13.

This sampling distribution is defined as follows by the central limit theorem (see Figure 8.3). As the sample size (n) increases,

1. The distribution of sample means is normal.

2. The mean of the distribution is μ. Under the null hypothesis stated in step 1, we assume that $\mu = 455$.

3. The standard error of the mean ($\sigma_{\overline{x}}$) equals $\sigma/\sqrt{n} = 100/\sqrt{144} = 8.33$ (assuming, as we did in Chapter 7, that $\sigma = 100$).

The reader is reminded that the standard error of the mean is the standard deviation of this sampling distribution of means.

From our earlier discussions of the normal distribution, recall that we use the normal distribution table (Table C.1 of the Appendix) to find area. Using this table, we find that the proportion of area under the curve between one standard error (standard deviation) below the mean (446.67) and one standard error *above* the mean (463.33) equals $0.3413 + 0.3413 = 0.6826$. In the same way, we find the proportion of the area between two standard errors *below* the mean (438.33) and two standard errors *above* the mean (471.67) to be 0.9544. For this example, Figure 8.3 shows these points and locations; if the null hypothesis is *true*, then 95.44 percent of all possible sample means from samples of size 144 will be between 438.33 and 471.67. Correspondingly, 4.56 percent of all possible sample means will be in the tails of the distribution, either less than 438.33 or greater than 471.67.

But our level of significance is .05, not .0456, so now consider Figure 8.4, which illustrates the sampling distribution of the mean for this example in a slightly different way. Notice that the shaded areas under the curve begin at 1.96 standard errors below *and* above the mean.

$$\mu - 1.96\sigma_{\overline{x}} = 455 - (1.96)(8.33) = 455 - 16.33 = 438.67$$

$$\mu + 1.96\sigma_{\overline{x}} = 455 + (1.96)(8.33) = 455 + 16.33 = 471.33$$

These shaded areas represent the extreme 5 percent of the area under the normal curve, with 2.5 percent in each tail of the distribution. When testing the null

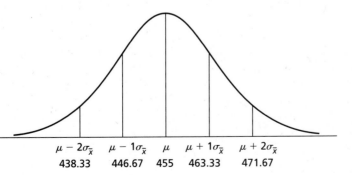

FIGURE 8.3
Sampling distribution of the mean for the hypothesis H_0: $\mu = 455$ and $\sigma_{\overline{x}} = 8.33$

$\mu - 2\sigma_{\overline{x}}$	$\mu - 1\sigma_{\overline{x}}$	μ	$\mu + 1\sigma_{\overline{x}}$	$\mu + 2\sigma_{\overline{x}}$
438.33	446.67	455	463.33	471.67

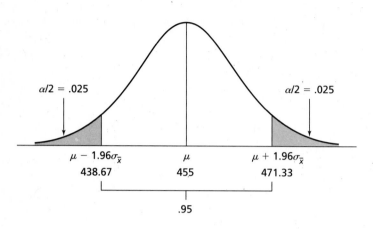

$\alpha/2 = .025$ $\alpha/2 = .025$

$\mu - 1.96\sigma_{\bar{x}}$ μ $\mu + 1.96\sigma_{\bar{x}}$

438.67   455   471.33

.95

FIGURE 8.4
Region of rejection for sampling distribution of the mean for null hypothesis $H_0 : \mu = 455$ and $\sigma_{\bar{X}} = 8.33$

hypothesis ($H_0 : \mu = 455$) against the alternative hypothesis ($H_a : \mu \neq 455$), these areas represent values for sample means that are highly *unlikely* if the null hypothesis is true. That is, the probability of obtaining a sample mean less than 438.67 *or* greater than 471.33 is .05. The level of significance, .05, equals the proportion of the total area in the distribution that is shaded. Because the level of significance is a probability, this is consistent with having probability represented by area in a distribution, as was discussed previously.

The shaded areas in Figure 8.4 are referred to as the **region of rejection**: the area of the sampling distribution that represents values for the sample mean that are improbable if the null hypothesis is true. The unshaded area under the curve, above 438.67 and below 471.33, is called the *region of nonrejection* because it represents values of the sample mean that are more probable if the null hypothesis is true.

> The region of rejection is the area of the sampling distribution that represents those values of the sample mean that are improbable if the null hypothesis is true.

If we assume that the null hypothesis ($H_0 : \mu = 455$) is true, then the region of rejection reflects values of the sample mean that are less than or equal to 438.67 and greater than or equal to 471.33. These values, 438.67 and 471.33, are called the **critical values** used for testing $H_0 : \mu = 455$. How do these critical values relate to the level of significance? Recall that, since we were willing to take a 5 percent chance of making a Type I error — rejecting the null hypothesis when it is true — the level of significance was set at .05. Therefore, if the observed sample mean ($\overline{X}$) is less than or equal to 438.67 *or* greater than or equal to 471.33, we reject the null hypothesis because such values are improbable if H_0 is true. In other words, they are possible but highly unlikely. By rejecting the null hypothesis, we are saying that the difference between the observed sample mean $(\overline{X})$ and the hypothesized population mean (μ) is too great to be attributed only to chance fluctuations in sampling. However, we must realize that there is a small probability that the difference is due to chance fluctuation and, by rejecting the null hypothesis, a Type I error may have been made. Since the probability of such an occurrence is .05 (the level of significance), we decide that a 1-in-20 chance is worth the risk and stand by the decision. The result, then, is that the sample mean is *significantly* different from the hypothesized value at the .05 level of significance.

> **The critical values of the test statistic are those values in the sampling distribution that represent the beginning of the region of rejection.**

Notice that the region of rejection in Figure 8.4 is equally divided between the two tails of the distribution. The reason is that the null hypothesis in this case is to be tested against a *nondirectional alternative hypothesis* — an alternative hypothesis that states merely that the parameter is *not* equal to a hypothesized value, such as $H_a: \mu \neq 455$. Thus, we place half of the area associated with the significance level $(\alpha/2)$ in each of the tails. When the null hypothesis is tested against a nondirectional alternative hypothesis, the procedure is called a **two-tailed**, or **nondirectional, test** of the null hypothesis.

> **When the alternative hypothesis is nondirectional, the region of rejection is located in both tails of the sampling distribution. The test of the null hypothesis against this nondirectional alternative is called a two-tailed test.**

Just a reminder about probability: in inferential statistics, probability relates to the statistic, a measure from the sample. Suppose we reject the null hypothesis. It is *not correct* to conclude that the probability is .95 that H_0 is false. If we fail to reject the null hypothesis, it is *not correct* to conclude that the probability is .95 that H_0 is true. In this chapter and subsequent chapters, we will be presenting numerous probability statements. The reader is encouraged to note these statements carefully because the role of probability in inferential statistics, when incorrectly conceptualized, leads to confusing and incorrect probability statements, and thus to incorrect conclusions.

Step 3: Compute the Test Statistic

After stating the hypotheses and setting the criterion for rejecting H_0, the next step in hypothesis testing is to analyze the sample data. For our example, the data are the SAT scores for the random sample of 144 freshman psychology students; the mean SAT score for this sample is 535. For the purpose of this example, we will use the population standard deviation[2] ($\sigma = 100$). Thus, the data for this example are as follows:

$\mu = 455$, the hypothesized value for the parameter
$n = 144$, the size of the sample
$\overline{X} = 535$, the observed value for the sample statistic
$\sigma = 100$, the value of the standard deviation in the population

Recall that, even though we have a random sample from the population, we do not expect the sample mean to be exactly equal to the hypothesized value of the population mean. In this case, whereas the hypothesized value for μ is 455, the sample mean is 535. So the question becomes: how different can the observed sample mean ($\overline{X}$) be from the hypothesized population mean (μ) before the null hypothesis is rejected? In other words, if the null hypothesis is true ($\mu = 455$), is it likely that we will obtain such an observed sample mean ($\overline{X} = 535$)?

To answer this question, we must consider the observed sample mean in terms of the sampling distribution of the mean; that is, the distribution of *all* possible means from samples of size 144. This distribution is illustrated in Figures 8.3 and 8.4. Using this distribution and the properties of the normal distribution, we will determine the probability of obtaining a sample mean such as 535 when the null hypothesis is true.

First, using the concept of z scores, we determine how different $\overline{X}$ is from μ, or the number of standard errors (standard deviation units) the observed sample value is from the hypothesized value. In symbols,

$$z = \frac{\overline{X} - \mu}{\sigma_{\overline{X}}} \tag{8.1}$$

For this example,

$$z = \frac{535 - 455}{8.33}$$

$$= 9.60$$

Calculating the z score using formula 8.1 is called *computing the test statistic*. This **test statistic** tells us that the observed sample mean ($\overline{X} = 535$) is 9.60 standard errors above the hypothesized value for the population parameter ($\mu = 455$). Referring to Figure 8.4, we see that, if the null hypothesis is true, 95 percent of

---

[2] In a later section, we will consider an example in which the standard deviation of the sample (s) is used.

all possible sample means will fall between 438.67 and 471.33, which is ± 1.96 standard errors from the hypothesized value of μ. Thus, the probability of obtaining a sample mean equal to 535 if the mean of the population is 455 is very, very small, actually much less than .05. In other words, such an occurrence is *not* likely if the null hypothesis is true.

> **The test statistic for testing the null hypothesis about the population mean is a standard score indicating the difference between the observed sample mean and the hypothesized value of the population mean.**

Step 4: Decide About H_0

We have now completed the first three steps in hypothesis testing and can take the final step: deciding about H_0. Recall that, in our SAT example, the null hypothesis was that the mean performance of the population of freshman psychology students was equal to 455 ($\mu = 455$). For a random sample of 144 students, we found the mean SAT to be 535 ($\overline{X} = 535$). We then used the sampling distribution of the mean and the properties of the normal distribution to determine that the test statistic was 9.60. In other words, the observed sample mean was 9.60 standard errors above the hypothesized value of μ in the sampling distribution of the mean.

In making the decision whether or not to reject H_0, we computed the test statistic, using the hypothesized value of the mean ($\mu = 455$) and the standard error ($\sigma_{\overline{X}}$) of the sampling distribution to transform the observed sample mean ($\overline{X} = 535$) into a standard score ($z = 9.60$). In the same way, we can transform all the sample means in the sampling distribution into z scores; this transformed distribution will have a mean equal to 0 and standard error equal to 1 (see Chapter 4). This transformed distribution is illustrated in Figure 8.5. Notice that the critical values — the

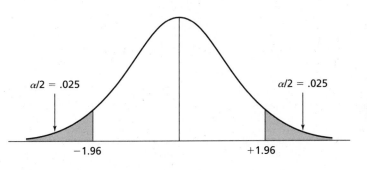

$\alpha/2 = .025$ $\alpha/2 = .025$

-1.96 $+1.96$

FIGURE 8.5
Region of rejection for theoretical sampling distribution with generalized critical values of the test statistic for $\alpha = .05$

values that begin the region of rejection—are now -1.96 and $+1.96$. That means it is not likely to observe a sample mean more than ± 1.96 standard errors from the hypothesized value of μ if the null hypothesis is true. *Since the observed value of the test statistic ($+9.60$) exceeds the critical value (± 1.96), the probability*[3] *is less than .05 that the observed sample mean ($\overline{X} = 535$) will have occurred by chance if the null hypothesis is true ($\mu = 455$)*; in symbols, $p < .05$. The result, then, is that the sample mean is considered significantly different from the hypothesized value at the .05 level of significance. Therefore, we reject the null hypothesis and conclude that the mean SAT level for the population of freshman psychology students is not equal to 455.

By rejecting the null hypothesis, we are saying that the difference between the observed sample mean and the hypothesized population mean is too great to be attributed to chance fluctuation in sampling. However, we realize that there is a small probability that the difference is due to chance fluctuation and that rejecting this null hypothesis may thus be a Type I error. The probability of this is .05 (the level of significance). When the significance level was set at .05, it was decided, in essence, that a 1-in-20 chance of making a Type I error was worth the risk, so we reject the null hypothesis.

The steps in testing a null hypothesis are:
1. State the hypothesis.
2. Set the criterion for rejecting H_0.
3. Compute the test statistic.
4. Decide whether to reject H_0.

Suppose we had found that the sample mean ($\overline{X}$) for 144 students was not 535, but 465. Our hypotheses, sampling distribution, and critical values ($+1.96$ and -1.96) remain the same, but now the test statistic is:

$$z = \frac{\overline{X} - \mu}{\sigma_{\overline{X}}} = \frac{465 - 455}{8.33} = 1.20$$

In other words, the observed sample mean ($\overline{X} = 465$) is 1.20 standard errors above the hypothesized value of the population mean. In this case, as Figure 8.6 shows,

---

[3]Because the sample mean is only a point located on the line (see Figure 8.6), and a point in a distribution has no area, this probability statement has a slight conceptual flaw. The probability is less than .05 that an observed sample mean this extreme (or more extreme) would have occurred by chance if the null hypothesis were true. For all practical purposes, we are concerned with the observed sample mean, so we will continue to use the less cumbersome probability statement.

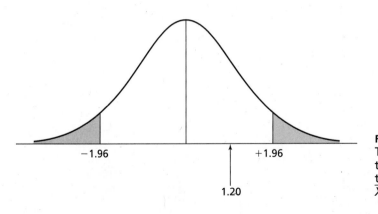

FIGURE 8.6
Theoretical sampling distribution for the hypothesis $H_0: \mu = 455$, illustrating the values of the test statistic when $\overline{X} = 465$

the test statistic does *not* exceed the critical value; it does *not* fall into the region of rejection; and we should not reject the null hypothesis. This decision is based on the properties of the sampling distribution and the fact that this observed sample mean is not sufficiently different from the hypothesized value of the population to warrant rejection of the null hypothesis. We are therefore willing to attribute this *nonsignificant* difference to sampling error.

For this example with a sample mean of 465, the probability statement would be: the probability that a sample mean of 465 would appear by chance, if the population mean is 455, is greater than .05. Note that in these probability statements for testing the null hypothesis we are *not saying* that the probability is .05, only whether it is less than or greater than .05.

Region of Rejection: Directional Alternative Hypothesis

In the SAT example, we tested the null hypothesis against a nondirectional alternative:

$$H_0: \mu = 455$$

$$H_a: \mu \neq 455$$

This test is called two-tailed or nondirectional because the region of rejection was located in both tails of the sampling distribution of the mean.

Suppose that the researcher has information about the variable under investigation so that a direction of the results is anticipated. Or suppose that the researcher is interested only in one direction of the results; for example, a new approach to teaching science would be adopted only if it improved student achievement relative to the traditional approach. In such situations, it may be more appropriate to test the null hypothesis against a *directional* alternative. A *directional alternative hypothesis* states that a parameter is either greater or less than the hypothesized value. For instance, in the SAT example we might use the alternative hypothesis that the mean SAT level of our population is *greater than* 455. In symbols,[4]

$$H_0: \mu = 455$$

$$H_a: \mu > 455$$

> An alternative hypothesis can be either nondirectional or directional. A directional alternative hypothesis states that the parameter is greater than or less than the hypothesized value. A nondirectional alternative hypothesis merely states that the parameter is different from (not equal to) the hypothesized value.

The test of the null hypothesis against a directional alternative is called a **one-tailed**, or **directional, test**. The procedures for a one-tailed test are essentially the same as for a two-tailed test: the null hypothesis is used to locate the mean of the sampling distribution; and the test statistic is calculated the same way. However, the critical values differ. There are two critical values for the test statistic when the alternative hypothesis is nondirectional, but only one critical value when the alternative hypothesis is directional. This is because we put the entire region of rejection in just one tail of the theoretical sampling distribution.

For example, suppose our alternative hypothesis is that the mean SAT level for the population of freshman psychology students is greater than 455. In this case, we are interested only in the right-hand tail of the theoretical sampling distribution. Figure 8.7 shows the sampling distribution and region of rejection for the .05 level of significance. We want to place the entire 5% of the area making up the rejection region in the right tail of the distribution. Referring to Table C.1 of the Appendix, we find the value of +1.645 marks off 5 percent of the area in the tail. The value corresponds to the beginning of the region of rejection at this level of significance and thus is the critical value of the test statistic for the null hypothesis. Recall that, for this example, we computed a sample mean of 535 and a test statistic of +9.60. This test statistic is greater than the value of +1.645, so it falls in the region of rejection.

A word of caution about what hypothesis is being rejected or not being rejected. Regardless of whether a test is directional or nondirectional (one-tailed or two-tailed), the null hypothesis is used to locate the sampling distribution of the statistic. A hypothesis such as $H_a: \mu > 455$ cannot be used to locate a sampling distribution because an infinite number of values fit this hypothesis. Therefore, because the null hypothesis is used to locate the sampling distribution, it is the hypothesis tested

---

[4] Some authors write the null hypothesis for this example as $H_0: \mu \leq 455$, using the rationale that the null and alternative hypotheses must include all possibilities. However, the specific null hypothesis tested is $H_0: \mu = 455$ since the point 455 is used as the mean of the sampling distribution. For a more thorough discussion, see R. E. Kirk, *Experimental Design: Procedures for the Behaviorial Sciences,* 2d ed. (Monterey, Calif.: Brooks/Cole, 1982), pp. 29–30.

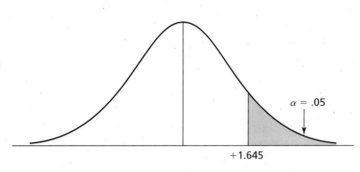

FIGURE 8.7
Theoretical sampling distribution
for the hypothesis $H_0 : \mu = 455$
and the directional alternative
$H_a : \mu > 455$; $\alpha = .05$

directly by the statistical test. In the above example, the test statistic value of $+9.60$ was statistically significant, so the null hypothesis was rejected. If the null hypothesis, $H_0 : \mu = 455$, is rejected, the alternative hypothesis, $H_a : \mu > 455$, is supported.

Note that the proportion of the area (.05) in the region of rejection of the sampling distribution of Figure 8.7 equals the probability of rejecting the null hypothesis if it is true. The proportion of the area is the level of significance. Because the null hypothesis is tested directly by the statistical test, the probability statement also relates to this hypothesis. For this situation, the probability statement is: the probability that a sample mean of 535 would appear by chance if $H_0 : \mu = 455$ is true is less than .05.

The test of the null hypothesis against a directional alternative is called a one-tailed test. In a one-tailed test, the region of rejection is located in one of the two tails of the sampling distribution. The specific tail of the distribution is determined by the direction of the alternative hypothesis.

Now suppose the alternative hypothesis stated that the mean SAT was less than 455. In symbols, the hypotheses are

$$H_0 : \mu = 455$$

$$H_a : \mu < 455$$

In this case, the region of rejection is located in the left-hand tail. We reject the null hypothesis and retain H_a only if the sample mean is substantially less than 455. Again, using the observed sample mean of 535, the computed test statistic is $+9.60$. From the normal distribution table, we find that the critical value (in the left-hand tail) is -1.645, as in Figure 8.8. In other words, the observed sample mean would have to be more than 1.645 standard errors *below* the value 455 for the researcher to reject the null hypothesis at the .05 level of significance. The computed test statistic is $+9.60$, so in this case, the test statistic falls in the region of nonrejection for the null hypothesis. Actually, in this situation, once we knew

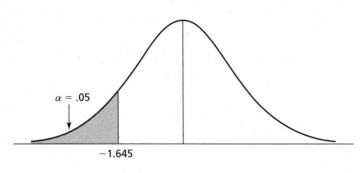

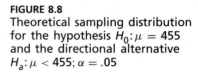

FIGURE 8.8
Theoretical sampling distribution
for the hypothesis $H_0: \mu = 455$
and the directional alternative
$H_a: \mu < 455; \alpha = .05$

that the test statistic was $+9.60$, there was no need to do more calculations or look up critical values. An inspection of Figure 8.8 shows no rejection region for $H_0: \mu = 455$ in the right tail of the sampling distribution. The right tail contains all the positive values of the test statistic, so any positive value of the test statistic would result in failing to reject $H_0: \mu = 455$.

We must be careful in locating the rejection region and in identifying the correct critical value of the test statistic, especially when dealing with directional alternatives. An easy way to remember the tail in which to establish the region of rejection and the corresponding critical value is to make an arrow out of the inequality sign ($<$ or $>$) that appears in the alternative hypothesis. This arrow will point to the appropriate tail. For example:

$$H_0: \mu = 455$$

$$H_a: \mu > 455 \text{ becomes } H_a: \mu \rightarrow 455$$

The arrow that we substitute for $>$ points to the right. This tells us that the critical value for testing this hypothesis, $+1.645$, is in the right-hand tail.

If the significance level is changed, the critical values of the test statistic also change. For example, if the researcher establishes the level of significance at .01 for a two-tailed test, the critical values are -2.576 *and* $+2.576$. For a one-tailed test at the .01 level, the critical value is either -2.326 *or* $+2.326$, depending on the direction of the alternative hypothesis. The critical values for both two-tailed and one-tailed tests using the most common levels of significance are presented in Table 8.1.

Hypothesis Testing When σ^2 Is Unknown

In the previous example, we used the normal distribution as the sampling distribution for testing the null hypothesis $H_0: \mu = 455$. We could use the normal distribution because it is known that the population standard deviation of SAT scores is 100. (If the standard deviation is known, the variance is known). When σ^2 is known, the sampling distribution of the mean is normally distributed, with

TABLE 8.1
Critical Values of the Test Statistic, Using the Normal Distribution as the Sampling Distribution

Level of Significance, Two-Tailed Test	Level of Significance, One-Tailed Test	Critical Value of Test Statistic
.20	.10	1.282
.10	.05	1.645
.05	.025	1.960
.02	.01	2.326
.01	.005	2.576
.001	.0005	3.291

μ assumed to be equal to the hypothesized value, and the standard deviation (the standard error of the mean) equal to $\sigma/\sqrt{n}$.

Usually, however, σ is not known. In these cases, the researcher estimates the standard deviation of the population (σ) by using the standard deviation of the sample (s). Correspondingly, the sample standard deviation is used to estimate the standard error of the mean (standard deviation of the sampling distribution of the mean). The estimated standard error of the mean is given by:

$$s_{\overline{X}} = s/\sqrt{n} \tag{8.2}$$

For testing the hypothesis about a population mean when σ is not known, we estimate the standard deviation of the population (σ) by using the standard deviation of the sample (s). The estimated standard error of the sampling distribution of sample means ($s_{\overline{X}}$) is then given by $s/\sqrt{n}$.

Student's t Distributions

Does this adjustment of using s to estimate σ have an effect on the statistical test? Actually, it does, especially for small samples. The effect is that the normal distribution is inappropriate as the sampling distribution of the mean. This was discovered shortly after the beginning of the twentieth century by William S. Gosset, a young chemist who worked on quality control at a brewery in Dublin, Ireland. Gosset found that, for small samples, sampling distributions departed substantially from the normal distribution and that, as sample sizes changed, the distributions changed. This gave rise to not one distribution but a family of distributions. Gosset also noted that, as the sample size increased, the distributions increasingly approximated the normal distribution. He published the results in 1908 under the pen name "Student"; thus these sampling distributions are called Student's t distributions. Later, with the assistance of R. A. Fisher, he developed a general formula for these distributions, and this work was published in 1926.

The *t* **distributions** are a family of distributions that, like the normal distribution, are symmetrical and bell-shaped, centered on the mean. As we have indicated, the distribution changes as the sample size changes. Thus, there is a specific *t* distribution for every sample of a given size. In order to determine the appropriate *t* distribution to use when testing a hypothesis, we first need to understand the concept of degrees of freedom.

> **The *t* distributions are a family of symmetrical, bell-shaped distributions that change as the sample size changes.**

Degrees of Freedom   The concept of degrees of freedom is fundamentally mathematical. The number of degrees of freedom of a statistic depends on the number of sample observations (*n*). For our discussion, it is sufficient to say that the number of **degrees of freedom** (abbreviated df) is the number of observations less the number of restrictions placed on them.

For example, if the mean of two numbers is *known* to be 50, then once the first number is known, so is the second. In this case, the sample consists of two numbers; there is one restriction — the known mean — and only one of the numbers is free to vary. Thus, $n - 1 = 2 - 1 = 1$ df.

Or suppose that the mean of a population is being estimated from a sample of five and that the five sample values are 21, 23, 24, 27, and 30. Recall that the sum of the deviations from a mean is equal to 0; this fact constitutes the restriction in this case. Because there are five scores, there are five deviations. But once the mean (in this case, 25) is specified, the fifth value is not free to vary because the last deviation score must be such that the sum of the deviations equals 0. Thus, only four of the five deviations are free to vary. In this example, then, there are $n - 1$, or 4, degrees of freedom.

Degrees of freedom also may be viewed as representing an adjusted sample size that improves estimates of the unknown population variance and standard deviation. When computing the standard deviation of a sample, $n - 1$, instead of n (or N), is used in the denominator. The value $n - 1$ is the degrees of freedom in this case.

Each *t* distribution is associated with a unique number of degrees of freedom. For example, Figure 8.9 shows *t* distributions for degrees of freedom equal to 5, 15, and infinity (∞). Notice that, as the number of degrees of freedom increases, the difference between the *t* distribution and the normal distribution decreases. The *t* distribution for infinite degrees of freedom is identical to the normal distribution.

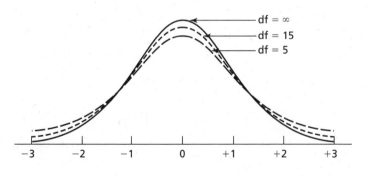

FIGURE 8.9
Student's *t* distribution for 5, 15, and ∞ degrees of freedom

The number of degrees of freedom is a mathematical concept defined as the number of observations less the number of restrictions placed on them.

Hypothesis Testing and the t *Distributions*   We use the *t* distributions in hypothesis testing in the same way we used the normal distribution when σ was known. Table C.3 in the Appendix is the table of *t* distributions, which have a mean equal to 0 and a standard deviation equal to 1. Notice that each row of this table, unlike the rows in the normal distribution table, represents a different *t* distribution; and each distribution is associated with a unique number of degrees of freedom. The column headings in the table (.10, .05, and so on) represent the portion of the area remaining in the tails of the distribution; the numbers in the column are *t* scores, and we use them just as we used *z* scores.

Consider the *t* distribution with 15 degrees of freedom in Figure 8.10. When using Table C.3, the first concern is to be in the correct *t* distribution, that is, on the correct row of the table, because each row represents a different *t* distribution. When we find the row with 15 df, we see that 2.5 percent of the area is in the right-hand tail beyond the point that is 2.131 standard deviations above the mean. Thus, 2.131 is actually the standard score that marks off 2.5 percent of the area

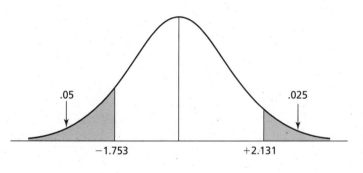

FIGURE 8.10
Areas under the curve of the Student's *t* distribution for 15 degrees of freedom

in the right-hand tail. The values in the t distribution table, with the exception of the df column, are standard scores in t distributions. They are standard scores that mark off designated areas in the tails, as indicated by the column headings. For the heading "Level of significance for one-tailed test," the area is concentrated entirely in one tail; for the heading "Level of significance for two-tailed test," the area is equally divided between the two tails.

Note that, like the normal distribution table, all standard score values in the table are positive. Yet, if the mean of these distributions is zero, all standard scores below the mean are negative. Because the t distributions are symmetrical (as is the normal distribution), those corresponding scores in the left-hand tail of a distribution are the same as those in the right-hand tail, except that they are negative. Thus, 5 percent of the area is in the left-hand tail below a point corresponding to 1.753 standard deviations below the mean, and -1.753 is the standard score that marks off 5 percent of the area in the left-hand tail. Notice in Table C.3 that, for degrees of freedom equal to ∞, the points marking off these same areas would be $+1.960$ and -1.645 — the same points that we saw before for the normal distribution. Also, in Table C.3, after 30 df, t distributions are given for 40, 60, 120, and ∞ degrees of freedom. The intervals increase because, as df increases, the t distributions become increasingly similar and more like the normal distribution. Although theoretically only the t distribution with an infinite number of df coincides with the normal distribution, for practical purposes the normal distribution is an adequate approximation of the t distribution when df exceeds 120.

As sample size increases, the difference between the normal distribution and the corresponding t distribution decreases. From a practical standpoint, the normal distribution is an adequate approximation of the t distribution when df exceeds 120.

Example Using the t Distribution

Suppose the athletic director at a large university is asked to investigate the claim that student athletes are not doing as well academically as the average student. Data supplied by the Office of Institutional Research reveal that the overall GPA for the university is 2.50 on a 4-point scale. To investigate the claim, the athletic director randomly selects 20 student athletes and obtains the current GPA for each. These data are given in Table 8.2. The sample mean for these 20 student athletes is 2.45, and the sample standard deviation is 0.54. In completing the investigation, the athletic director follows the four steps for hypothesis testing.

TABLE 8.2
GPA Data for the 20 Student Athletes

1.8	2.0	1.2	3.0
3.1	3.2	2.5	2.9
2.6	2.8	2.3	2.0
2.4	2.7	2.0	1.9
2.2	3.3	2.8	2.2

$\Sigma X = 48.9$

$\Sigma X^2 = 125.15$

$$\overline{X} = \frac{\Sigma X}{n} \qquad s^2 = \frac{\Sigma X^2 - \frac{(\Sigma X)^2}{n}}{n-1} \qquad s = \sqrt{0.29}$$

$$= \frac{48.9}{20} \qquad = \frac{125.15 - \frac{(48.9)^2}{20}}{19} \qquad = 0.54$$

$$= 2.45$$

$$= 0.29$$

Step 1. State the Hypotheses

The claim made in this example is that the mean GPA of student athletes is less than the mean of the population of students at the university. Thus, the null hypothesis is tested against the directional alternative; that is, a one-tailed test of the hypothesis should be conducted. In symbols, the hypotheses are

$$H_0: \mu = 2.50$$

$$H_a: \mu < 2.50$$

Step 2. Set the Criterion for Rejecting H_0

After stating the hypotheses, the researcher must determine whether the difference between the sample statistic $(\overline{X})$ and the hypothesized value for the population parameter is sufficient for rejecting the null hypothesis. As before, the theoretical sampling distribution of the mean must be considered. For this example, since we will use the standard deviation of the sample (s) as an estimate of the standard deviation in the population, this sampling distribution is the t distribution with $n - 1 = 20 - 1 = 19$ degrees of freedom, rather than the normal distribution. The critical value for testing this hypothesis depends on

1. The level of significance (α).

2. The nature of the alternative hypothesis (nondirectional or directional).

3. The nature of the sampling distribution (normal distribution or t distribution).

For this example, we will use the .05 level of significance ($\alpha = .05$) for testing the null hypothesis ($H_0 : \mu - 2.50 = 0$) against a directional alternative hypothesis ($H_a : \mu - 2.50 < 0$). For the t distribution with 19 degrees of freedom, we consult Table C.3 and find the critical value (t_{cv}) to be -1.73. In symbols,

$$t_{cv} = -1.73$$

Notice that this critical value is in standard score form. Therefore, to reject H_0, the observed sample mean ($\overline{X}$) must be 1.73 standard errors below the hypothesized value for μ.

Step 3. Compute the Test Statistic

When the variance in the population is known and the normal distribution is used as the sampling distribution, the test statistic is defined as z (see formula 8.1). However, when the variance of the sample is used as an estimate of the variance in the population, the test statistic is defined as t.

$$t = \frac{\overline{X} - \mu}{s_{\overline{X}}} \tag{8.3}$$

For this example, we compute the estimated standard error using formula 8.2:

$$s_{\overline{X}} = s/\sqrt{n}$$
$$= 0.54/\sqrt{20}$$
$$= 0.12$$

The test statistic is computed as

$$t = \frac{2.45 - 2.50}{0.12}$$
$$= -0.42$$

That is, the observed sample mean ($\overline{X} = 2.45$) is 0.42 standard errors below the hypothesized value for μ (2.50).

Step 4. Decide About H_0

In deciding whether to reject the null hypothesis, we compare the observed value of the test statistic (t value) to the critical value of the test statistic (t_{cv}). Because $t = -0.42$ does *not* exceed $t = -1.73$ (in absolute value), the null hypothesis is not rejected. The associated probability statement is: *the probability is greater than .05 that the observed sample mean ($\overline{X} = 2.45$) would have occurred by chance if the null hypothesis were true ($\mu - 2.50 = 0$)*. That is, the difference between the sample mean and the hypothesized value of the population mean is not sufficient to attribute it to anything other than sampling error. Thus, the athletic director concludes that the mean academic performance of student athletes is *not* different from the mean academic performance of the student body.

A couple of comments about symbols are in order. Note in the above example that $\alpha = .05$. α is a probability and determines the criterion for rejecting H_0. When we have completed the statistical test, we can use the symbol $p > .05$, which indicates that the probability is *greater than* .05 (because the statistical test was not significant) that a sample mean of 2.45 would appear by chance if $\mu = 2.50$. Alpha is always specified as equal to some value. After the statistical test, the probability (p) can be designated as greater than ($p >$) or less than ($p <$) the significance level, depending on whether or not the null hypothesis was rejected. As we will see in later chapters, for some computer printouts, the exact probability is given for the statistical test.

Statistical Significance versus Practical Importance

In the SAT example, the statistical test was significant at the .05 level, so we concluded that the null hypothesis, $H_0: \mu = 455$, was not tenable and that the population mean was not 455. But what is meant by statistical significance? Technically, the difference between the hypothesized population parameter and the corresponding sample statistic is said to be *statistically significant* when the probability that the difference occurred by chance is less than the significance level (α level). But how large must the difference be? And is it necessarily of practical importance just because it is statistically significant?

One consideration in answering these questions is the level of significance established for testing the hypothesis. The selection of the level of significance in behavioral science research does not often receive the attention it merits. The .05 and .01 levels are commonly used, but sometimes there seems to be no rationale for their use other than the fact that R. A. Fisher used them in his agricultural experiments several decades ago. In previous sections, we indicated that the criterion for selecting the appropriate level of significance for a specific research study is the importance of the consequences of making either a Type I error (rejecting a true hypothesis) or a Type II error (retaining a false hypothesis). There has been a tendency in behavioral science research to guard against the Type I error, but some critics assert that this tendency has resulted in setting α levels that are too conservative. Sometimes this results in retaining the null hypothesis even though the probability that the statistic would occur if the hypothesis were true is relatively small, say, .07 or .08. Thus, the area of the research and the characteristics of the specific study must be considered and given careful attention by the researcher when the level of significance is chosen.

A second consideration in answering these questions is related to the concept of statistical precision. In testing hypotheses, we define **statistical precision** as the inverse of the standard error of whatever statistic is being tested. In other words, the smaller the standard error, the greater the statistical precision. (Statistical precision is discussed further in Chapter 9.) Consider formula 8.1 for z. A smaller value of the standard error of the mean will result in a larger absolute value of the test statistic z and thus a greater likelihood that we will reject the null hypothesis.

But how can the researcher increase statistical precision? One way is to increase the sample size. Increasing the sample size yields a more precise (smaller) estimate of the standard error of the statistic because the statistic is based on more observations. Consider the formula for the standard error of the mean (formula 8.2). Note that the standard deviation (s) is divided by $\sqrt{n}$. Thus, as n increases, $s_{\overline{X}}$ decreases, and the statistical precision is increased.

What statistical precision is adequate for research in the behavioral sciences? The observant reader may have noted that a sufficiently large sample size leads to the rejection of any null hypothesis based on a *fixed* difference between the hypothesized parameter and the observed sample statistic. On the other hand, a researcher might *not* reject a null hypothesis, even though there is a seemingly large difference between the parameter and the corresponding statistic, because the sample size is simply too small. Critics of inferential statistics have charged that this makes statistically based research a "numbers game." This charge is unfounded if the researcher has been thorough in designing and conducting the investigation and if the results are interpreted in light of practical considerations, as well as theoretical, statistical implications.

Finally, note that statistical tests do not establish the practical importance of statistical significance. As we have discussed, increasing the statistical precision can render almost any difference between the observed statistic and the hypothesized parameter statistically significant. But what does statistical significance mean in the practical context of the research situation? Further, when does a difference become large enough to warrant corrective action? In the student athletes example, does the difference of .05 points in the athletes' mean GPA indicate that the institution should establish a remedial academic program for athletes? These kinds of questions are not fully answered by inferential statistics. They can be answered only on the basis of a thorough knowledge and understanding of the research area and the variables under consideration. Inferential statistics are only tools for analyzing data. They are not substitutes for knowledgeable interpretation of those data.

Summary

Hypothesis testing involves determining whether a hypothesized value of a parameter (for example, μ) is tenable. The researcher establishes this value on the basis of the research literature or previous research experience. A sample is drawn, and the appropriate sample statistic (for example, $\overline{X}$) is computed. The decision to reject the null hypothesis is based on the magnitude of the difference between the observed statistic ($\overline{X}$) and the hypothesized parameter (μ).

The first step in hypothesis testing is to state the null and alternative hypotheses. Next, the criterion for rejecting H_0 is established by determining the critical values; that is, the beginning values for the region of rejection in the sampling distribution. The critical values are found by using the tables for either the normal distribution or Student's t distributions when testing hypotheses about a single mean. These critical values depend on the directional nature of the alternative hypothesis and the level of significance.

The third step is to compute the test statistic, which is a standard score—either z or t, in the case of a mean—indicating the number of standard errors the observed sample statistic ($\overline{X}$) is from the hypothesized parameter (μ). The general formula for the test statistic is

$$\text{Test statistic} = \frac{\text{statistic} - \text{parameter}}{\text{standard error of the statistic}}$$

This test statistic is then compared to the critical value. If the test statistic exceeds the critical value in absolute value, then the null hypothesis is rejected.

In subsequent chapters, we apply the same basic logic of hypothesis testing to testing hypotheses about other population parameters. The general formula for the test statistic will remain the same, but additional test statistics will be introduced.

Exercises

*1. In a psychological study of abnormally hyperactive children, there is an interest in determining whether such children differ from "normal" children in creativity. A creativity inventory has been standardized on a large population of children. The mean score on the inventory for this population is 150 and the standard deviation of scores is 16. A random sample of size 50 is selected from a population of hyperactive children. The sample is administered the creativity inventory; the sample mean is 152.1 and the standard deviation is 15.1. The research question is, "Do abnormally hyperactive children differ in creativity (as measured by this creativity inventory) from normal children? Test the appropriate null hypothesis against the nondirectional alternative hypothesis.

2. The dean of a large college is concerned that the students' grade-point averages have changed dramatically in recent years. The graduating seniors' mean GPA over the past five years is 2.75. The dean randomly samples 256 seniors from the last graduating class and finds that their mean ($\overline{X}$) GPA is 2.85, with a standard deviation (s) of 0.65. Test the null hypothesis that the mean GPA for graduating seniors is 2.75 ($H_0: \mu = 2.75$) against the nondirectional alternative ($H_a: \mu \neq 2.75$). Use alpha (α) = .05.

3. Suppose that the dean in Exercise 2 samples only 30 students. Using the appropriate t distribution, test the same null hypothesis ($H_0: \mu = 2.75$) against the nondirectional alternative ($H_a: \mu \neq 2.75$).

4. The scores on a physical-performance test for boys of junior high school age have been standardized with a mean of 175 and a standard deviation of 12 for the general population. In a large city school system, a random sample of 225 junior high school boys is tested. The sample mean is 173.6.

   a. Find the standard error of the mean. A standard error is a standard deviation. Describe the distribution of which this standard error is the standard deviation.

   b. Test the hypothesis that the mean performance test score for the population of junior high school boys of this school system is 175 against the alternative hypothesis that it is not 175. Use $\alpha = .05$. Identify the four steps in testing the hypothesis, $H_0: \mu = 175$.

   c. After completing the test in part b, give the conclusion and the probability statement.

   d. Suppose school officials are interested in testing the hypothesis that the population of junior high school boys in this school system performs lower on the test than the general population does. Test this hypothesis, again using $\alpha = .05$.

   e. Is there an inconsistency between the results of parts b and d? Explain.

   f. Suppose that, instead of a sample size of 225, one of size 40 had been selected. Would a t distribution now be the appropriate sampling distribution for the mean? Why or why not?

   g. If a sample of size 40 had been selected, would attaining statistical significance require a difference between $\overline{X}$ and the hypothesized value of μ larger or smaller than if a sample of size 225 had been selected? Explain.

5. It is believed that the average weight of a population of males is 162 pounds. A random sample of 200 males is selected from the population. The mean weight $\overline{X}$ is 160 pounds, with a standard deviation (s) of 20 pounds. Test the hypothesis that $H_0: \mu = 162$ against the nondirectional alternative ($H_a: \mu \neq 162$). Use $\alpha = .10$. Give the probability statement after the statistical test is done.

6. Indicate if the region of rejection is in the right- or the left-hand tail of the sampling distribution for the following directional hypotheses:

   a. The mean IQ of all college students is greater than 110.

   b. The mean annual income for high school graduates is more than $21,000.

   c. The mean SAT verbal scores of 1993 high school graduates are less than 520.

   d. The mean weight of females born in 1963 is less than 167 pounds.

7. If a null hypothesis is rejected at the .05 level of significance, what can be said about its rejection at the .01 level? On the other hand, if a null hypothesis is rejected at the .01 level of significance, what can be said about its rejection at the .05 level?

8. In your own words, explain why the null hypothesis is rejected when the computed test statistic is in the region of rejection.

9. Find the first, fifth, 95th, 98th, and 99th percentiles for the t distribution with 18 df.

10. In the t distribution with 9 df, what proportion of the area lies between the scores of -1.383 and $+3.250$?

11. One would expect a student to be able to guess the answers to 10 items on a 20-item true-false test. The scores for 14 students on a 20-item true-false test follow. Test the null hypothesis ($H_0: \mu = 10$) against the directional alternative ($H_a: \mu > 10$). Use $\alpha = .01$.

12	10	9	13	13
8	11	7	14	11
15	17	11	12	

12. A researcher tests the null hypothesis $H_0: \mu = 80$. The sample mean is 85, and the test is statistically significant at $\alpha = .05$. The following probability statements are incorrect. Identify what is wrong with each. Finally, provide the correct probability statement for this result.

   a. The probability is .95 that H_0 is true.

   b. The probability that the population mean is 80 if the sample mean is 85 is less than .05.

   c. The probability is .05 that a false hypothesis has been rejected.

   d. The probability is .05 that a Type II error (failing to reject a false hypothesis) has been made.

13. Which of the following are more likely to lead to the rejection of the null hypothesis?

   a. a one-tailed or two-tailed test.

   b. .05 or .01 level of significance.

   c. $n = 144$ or $n = 444$.

14. A study on the reaction time of children with cerebral palsy reports a mean of 1.6 seconds on a particular task. A researcher believes that the reaction time can be reduced by using a motivating set of directions. An equivalent set of children is located, and they complete the same task with the motivating set of directions. The reaction times for the 12 children follow. Test the null hypothesis ($H_0: \mu = 1.6$) against the directional alternative ($H_a: \mu < 1.6$). Use $\alpha = .01$.

Child	Reaction Time	Child	Reaction Time
A	1.4	G	1.5
B	1.8	H	2.0
C	1.1	I	1.4
D	1.3	J	1.9
E	1.6	K	1.8
F	0.8	L	1.3

15. Statistical tables are very helpful when testing hypotheses, but they must be used correctly. Find the critical values for rejecting H_0 in the following t distributions under the given conditions.

    a. df $= 4$, $\alpha = .01$, nondirectional test.

    b. df $= 9$, $\alpha = .05$, directional test.

    c. df $= 17$, $\alpha = .01$, directional test.

    d. df $= 22$, $\alpha = .10$, nondirectional test.

    e. df $= 28$, $\alpha = .005$, directional test.

    f. df $= 50$, $\alpha = .05$, directional test.

16. Given any t distribution, why are the critical values for rejecting H_0 the same for $\alpha = .05$, directional test, as they are for $\alpha = .10$, nondirectional test?

17. For the t distribution with df $= 24$, find the values between which lies 90 percent of the area closest to the mean.

18. In your field of interest, find two research articles that use a level of significance other than .01 or .05. Explain why the different α level might have been used.

9

Estimation: One-Sample Case for the Mean

Key Concepts   Statistical estimation   Confidence interval
                     Point estimate   Level of confidence
                     Interval estimation   Statistical precision

I n Chapter 8, we tested hypotheses based on the assumption that the researcher can logically establish a hypothesized value for a population parameter. But instead of testing the tenability of a hypothesized value for, say, the population mean, we might want to estimate its value. For example, rather than testing the hypotheses in Chapter 8, we might try to answer these questions: what *is* the mean SAT level for freshman psychology students? or what *is* the mean level of academic performance for student athletes? or, because we probably would not settle on a single value as an estimate, what are tenable or reasonable values for the population mean? Because we never know the population mean for sure unless we measure every member of the population, we are still *inferring* from the sample statistic to the population parameter. This process is called **statistical estimation**: we are making inferences about the population on the basis of what we observe in the sample.

In this chapter, we discuss how to estimate the mean of a population. The reasoning we will use is applicable to the estimation of other parameters as well.

Statistical estimation is the process of estimating a parameter of a population from a corresponding sample statistic.

Point Estimates and Interval Estimates

There are two approaches to the statistical estimation of a parameter. The first is called point estimation. A **point estimate** is a single value that represents the best estimate of the population value. If we are estimating the mean of a population (μ), then the sample mean ($\overline{X}$) is the best point estimate.

If the point estimate is the "best" estimate of a parameter, why do we need another estimating approach? Recall that the sample mean reflects not only the population mean but also sampling error. As a result, we do not expect the sample mean to equal the population mean exactly. Therefore, if estimating a population parameter is the main purpose of an investigation, we can use a second approach to estimation: interval estimation. **Interval estimation** builds on point estimation to arrive at a *range* of values that are tenable for the parameter and that define an interval that we are confident contains the parameter.

For example, suppose a team of research psychologists is interested in studying the level of emotional stress among recently divorced women. Even though the psychologists plan to use a standardized stress inventory, they have no good reason to hypothesize a specific mean for this population, and for the moment at least, there is no specific null hypothesis to test. Their primary interest is in *estimating* the population mean. For this example, the process involves randomly selecting a sample of, say, 400 recently divorced women in a major metropolitan area, giving them the stress inventory, and then computing the sample mean ($\overline{X}$). The sample mean provides one point estimate of the population mean. If the psychologists want to report only a single value as the best estimate, then the mean is the value to report. But since the psychologists know that this sample mean represents the population mean only within the constraints of sampling error, they might prefer to determine a range of values that has a certain degree of confidence, say, 95-percent, of containing the population mean. This range of values is called a *confidence interval*.

Confidence Intervals

To understand what a confidence interval is, suppose for a moment that the mean score for the population of recently divorced women taking the stress inventory is 43 ($\mu = 43$) and the standard deviation is 10 ($\sigma = 10$). Now suppose we can select from this population all possible samples of size 400 and compute the sample mean for each; this theoretical sampling distribution of the mean is illustrated in Figure 9.1. As in Chapter 8, we can use the properties of the normal distribution to determine that 95 percent of all possible means for samples of size 400 fall between $+1.96$ and -1.96 standard errors of μ.

$$\mu \pm (1.96)(\sigma_{\overline{X}})$$

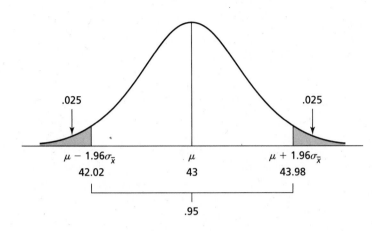

FIGURE 9.1
Sampling distribution of the mean under the assumption that $\mu = 43$ and $\sigma_{\overline{X}} = .50$

For this example,

$$\mu \pm (1.96)(10/\sqrt{400})$$

$$\mu \pm (1.96)(0.50)$$

$$\mu \pm 0.98$$

Since we are assuming for the moment that $\mu = 43$, we can say that 95 percent of all sample means will fall between $43 - 0.98 = 42.02$ and $43 + 0.98 = 43.98$ (see Figure 9.1).

When we do *not* know the value of the parameter, such as μ, we can estimate it by constructing an interval centered on an *observed statistic*, such as $\overline{X}$. The result is called a **confidence interval** (CI), a range of values that we are confident (but not certain) contains the population parameter. The general formula for constructing this interval is

$$\text{CI} = \text{statistic} \pm (\text{critical value})(\text{standard error of the statistic}) \qquad (9.1)$$

If we are estimating the population mean (μ), with the sample mean ($\overline{X}$) as the point estimate and with the population variance (σ^2) *known*, the sampling distribution of the mean is normally distributed (see Figure 9.1). Therefore, the critical value for the confidence interval is found in a table of areas under the normal curve (Table C.1). The standard error of the statistic is $\sigma_{\overline{X}} = \sigma/\sqrt{n}$. Thus, for estimating the population mean when the variance is *known*, the formula for the confidence interval becomes

$$\text{CI} = \overline{X} \pm (z_{\text{cv}})(\sigma_{\overline{X}}) \qquad (9.2)$$

where

$\overline{X}$ = sample mean

z_{cv} = critical value using the normal distribution

$\sigma_{\overline{X}}$ = standard error of the mean

The confidence interval is a range of values that we are confident contains the population parameter.

In order to determine the critical value to be used in formula 9.2, we must first choose the **level of confidence**, which is defined as the degree of confidence we have that the computed interval contains the parameter being estimated. Typically, in a specific research situation, the level of confidence is the complement of the level of significance $(1-\alpha)$. For example, suppose an appropriate hypothesis can be specified and tested using the .05 level of significance (α). The corresponding level of confidence for constructing the confidence interval is $1 - .05 = .95$. The specific critical value is the same as the critical value for the test statistic (see Chapter 8) for a two-tailed test at $\alpha = .05$. Therefore, the formula for the 95-percent confidence interval is

$$\text{CI}_{95} = \overline{X} \pm (1.96)(\sigma_{\overline{X}})$$

For our example, suppose the team of research psychologists finds the mean score on the emotional stress inventory for the sample of 400 women to be 44.6. They compute the 95-percent confidence interval as follows:

$$\text{CI}_{95} = \overline{X} \pm (1.96)(\sigma_{\overline{X}})$$

$$= 44.6 \pm (1.96)(10/\sqrt{400})$$

$$= 44.6 \pm (1.96)(0.50)$$

$$= 44.6 \pm 0.98$$

$$= (43.62,\ 45.58)$$

The psychologists can conclude that they are 95 percent confident that the interval from 43.62 to 45.58 contains μ.

Confidence Intervals When σ^2 Is Unknown

In the stress inventory example, we assumed that the population variance (σ^2) of the scores on the stress inventory was *known*. As we indicated in Chapter 8, knowing the population variance is the exception rather than the rule. Generally, we must use the variance of the sample (s^2) to estimate σ^2. This being the case, we use the critical values from Student's t distributions, rather than the normal distribution, when computing the confidence interval.

In addition, we must use the estimated standard error of the mean ($s_{\overline{X}}$ rather than $\sigma_{\overline{X}}$). Thus, the formula for the confidence interval becomes

$$\text{CI} = \overline{X} \pm (t_{\text{cv}})(s_{\overline{X}}) \tag{9.3}$$

where

$\overline{X}$ = sample mean

t_{cv} = critical value using the appropriate t distribution

$s_{\overline{X}}$ = estimated standard error of the mean

To illustrate the use of formula 9.3, consider the example from Chapter 8 of student athletes. In that example, we found the sample mean ($\overline{X}$) for the 20 student athletes to be 2.45 and the estimated standard error ($s_{\overline{X}}$) to be 0.12. The critical value for the confidence interval is found in the same way we found the critical value for the test statistic. Suppose we want to compute the 95-percent confidence interval; that is, the level of confidence is .95. Since the variance is unknown, we use the t distribution for $n - 1 = 20 - 1 = 19$ degrees of freedom. From Table C.3, we find that the critical value is 2.09. Thus, applying formula 9.3,

$$CI_{95} = \overline{X} \pm (t_{cv})(s_{\overline{X}})$$

$$= 2.45 \pm (2.09)(0.12)$$

$$= 2.45 \pm 0.25$$

$$= (2.20, 2.70)$$

We conclude that we are 95 percent confident that this interval contains μ, the mean GPA for the population of student athletes at the university. Thus, for this sample, the 95 percent confidence interval has the lower limit 2.20 and the upper limit 2.70.

When computing confidence intervals for which σ^2 is unknown, find the critical value by using the t distributions rather than the normal distribution.

Interpreting Confidence Intervals

What does this result mean? If the CI_{95} is 2.20 through 2.70, then we can conclude with 95 percent confidence that the interval contains μ, the mean GPA of the student athletes. If we take another sample from the same population and compute a different sample mean, we will obtain a different confidence interval. The confidence interval would differ in its location—that is, it would have a mean different from that of the first interval, and it most likely would have a different standard deviation. These statistics from the second sample would be within random-sampling fluctuation of those from the first sample. Thus, we would expect the standard error of the mean ($s_{\overline{X}}$) computed using this second sample to be slightly different from the 0.12 value for $s_{\overline{X}}$ computed using the first sample.

Theoretically, suppose we computed the sample means of all possible samples of size 20 and constructed the 95-percent confidence intervals for the population mean using all of these sample means. Then, 95 percent of these intervals would contain μ and 5 percent would *not*. Note that we *cannot* say that the probability is .95 that the interval from 2.20 to 2.70 contains μ. Either the interval contains μ or it does not. Therefore, we call the intervals confidence intervals, and we talk of our confidence that the interval contains the mean.

When we talk about the probability related to confidence intervals, we are talking about the probability that the confidence intervals constructed *from all possible samples of a given size* for a specific population will include μ. This point is somewhat difficult to illustrate graphically because both $\overline{X}$ and s will vary due to sampling fluctuation as we select repeated samples. Therefore, as mentioned above, not only would the intervals vary in location due to different means, but the interval widths would vary slightly due to small differences in the $s_{\overline{X}}$'s. For the sake of illustration, assume that $s_{\overline{X}}$ for our example remains the same across samples and is 0.12. Table 9.1 illustrates the confidence intervals computed using 30 means from samples of size 20, given that μ is in fact 2.50. Notice that two of the 30 confidence intervals do not contain the point 2.50. These are the 95-percent confidence intervals for $\overline{X} = 2.21$ and $\overline{X} = 2.84$. If we had the confidence intervals from all possible samples of size 20 (a large number of samples), we could expect that only 95 percent of these intervals constructed around the sample mean would contain μ. Note that the discussion above about all of these intervals reflects the theoretical basis for interval estimation. In practice, we construct only one interval for an interval estimate of a parameter, and base our conclusions on that interval.

Relationship Between Hypothesis Testing and Estimation

Statistical estimation is an extremely valuable tool in research investigations because it is not limited to situations in which a hypothesized value for the population parameter can be logically established. For example, in Chapter 8, we rejected the null hypothesis (H_0: $\mu = 455$) for SAT scores when the sample mean ($\overline{X} = 535$) was found to be significantly different from the hypothesized value of the population mean, and we concluded that the population mean was probably different from 455. The conclusion did not specify an estimate for μ. To do so, we would have had to develop an appropriate confidence interval. In a more general sense, rejecting

TABLE 9.1
95-Percent Confidence Intervals from 30 Sample Means for H_0: $\mu = 2.50$, with $s_{\overline{X}} = 0.12$ and $n = 20$

$\overline{X}$	CI_{95}	2.00	2.25	2.50	2.75	3.00
2.48	2.23, 2.73					
2.64	2.39, 2.89					
2.21	1.96, 2.46					
2.32	2.07, 2.57					
2.39	2.14, 2.64					
2.70	2.45, 2.95					
2.57	2.32, 2.82					
2.35	2.10, 2.60					
2.43	2.18, 2.68					
2.26	2.01, 2.51					
2.60	2.35, 2.85					
2.52	2.27, 2.77					
2.84	2.59, 3.09					
2.31	2.06, 2.56					
2.27	2.02, 2.52					
2.47	2.22, 2.72					
2.54	2.29, 2.79					
2.68	2.43, 2.93					
2.38	2.13, 2.63					
2.73	2.48, 2.98					
2.40	2.15, 2.65					
2.62	2.37, 2.87					
2.29	2.04, 2.54					
2.65	2.40, 2.90					
2.45	2.20, 2.70					
2.34	2.09, 2.59					
2.51	2.26, 2.76					
2.59	2.34, 2.84					
2.42	2.17, 2.67					
2.55	2.30, 2.80					

the null hypothesis yields little information about the variable under investigation. Therefore, statistical estimation is especially important in research studies in which it is important to make inferences about the magnitude of parameters, and to know the tenable values on the scale of measurement.

In both hypothesis testing and estimation, the computation involves the sample mean ($\overline{X}$) and the standard error of the mean (either $\sigma_{\overline{X}}$ or $s_{\overline{X}}$). In addition, the critical value (z_{cv} or t_{cv}, respectively) is used both in interpreting the test statistic and in computing the confidence interval. Hence, it is not surprising that the two procedures are related.

To illustrate this relationship, consider the SAT example. Recall that, using a two-tailed test, we tested and rejected the following null hypothesis at the .05 level of significance:

$$H_0: \mu = 455$$

$$H_a: \mu \neq 455$$

Using the values from this example,

$$\overline{X} = 535 \qquad \text{critical values} = \pm 1.96 \qquad \sigma_{\overline{X}} = 8.33$$

we can compute the 95-percent confidence interval as follows:

$$\text{CI}_{95} = 535 \pm (1.96)(8.33)$$

$$= 535 \pm 16.33$$

$$= (518.67, 551.33)$$

We conclude that we can be 95 percent confident that the interval (518.67, 551.33) contains the mean SAT score for the freshman psychology students.

Notice that the hypothesized population value ($\mu = 455$) is *not* contained in the interval. It should not be in the interval because the interval contains only tenable values for μ based on the observed sample mean ($\overline{X} = 535$). In other words, rejecting the null hypothesis means that we do not consider $\mu = 455$ a tenable value and thus would *not* expect it to be contained in the interval.

Constructing a confidence interval can be a means of testing a large number of hypotheses simultaneously. For example, the hypotheses

$$H_0: \mu = 535 \qquad H_0: \mu = 528 \qquad H_0: \mu = 522$$

would *not* be rejected because each of these values for μ is within the confidence interval. In contrast, the hypotheses

$$H_0: \mu = 455 \qquad H_0: \mu = 500 \qquad H_0: \mu = 560$$

would be rejected because each of these values is outside the confidence interval. In summary, any values between 518.67 and 551.33 are tenable values for the population mean, and all other values are not.

Developing a confidence interval can be viewed as a means of testing many hypotheses simultaneously. Any value within the interval is a tenable value for the population parameter. All values outside the interval are not tenable.

Statistical Precision in Estimation

If a study is to determine estimates of population parameters, much of the preliminary planning is concerned with enhancing the accuracy of the estimates. The accuracy with which a confidence interval can be used to estimate a population parameter can be thought of as the **statistical precision** of the estimate. Generally speaking, the *narrower the width of the confidence interval, the more precise the estimate*.

In planning studies that use confidence intervals to estimate parameters, the major concerns are the level of confidence to be used and the appropriate sample size. Both are important considerations, but the sample size is more directly related to statistical precision.

Consider the example of the GPA for student athletes. On the basis of the sample size of 20, we found that the 95-percent confidence interval was

$$CI_{95} = 2.45 \pm (2.09)(0.12)$$

$$= (2.20,\ 2.70)$$

The width of this interval is 0.50. We can enhance statistical precision by increasing the sample size. Suppose that the sample size is 121 rather than 20, and assume that both the sample mean and the standard deviation are the same ($\overline{X} = 2.45$, $s = 0.54$). How much will this affect the CI_{95}? First, the standard error will be smaller.

$$s_{\overline{X}} = s/\sqrt{n}$$

$$= 0.54/\sqrt{121}$$

$$= 0.049$$

Second, the t_{cv} will be based on 120 degrees of freedom rather than 19, so $t_{cv} = 1.98$. The CI_{95} will then be

$$CI_{95} = 2.45 \pm (1.98)(0.049)$$

$$= (2.35, 2.55)$$

This confidence interval has a width of 0.20. Thus, by increasing the sample size, we make the standard error of the mean smaller and the width of the confidence interval narrower. In other words, the width of the confidence interval, which we have defined as the accuracy of the estimation, is related to the size of the sample in this way: as we increase sample size, we reduce the width of the confidence interval, thus increasing statistical precision.

In interval estimation, as the width of the interval decreases, statistical precision increases. The width of the interval can be decreased by increasing the sample size.

The researcher can also affect the width of the confidence interval by manipulating the level of confidence. To see the effect of the level of confidence on the width of the confidence interval, consider again the student athlete example with the sample size 20. We found CI_{95} to be

$$CI_{95} = X \pm (t_{cv})(s_{\overline{X}})$$

$$= 2.45 \pm (2.09)(0.12)$$

$$= (2.20, 2.70)$$

Now consider the CI_{99} for the same example. The only change will be in t_{cv}. From Table C.3, we find $t_{cv} = 2.861$. Thus, CI_{99} is

$$CI_{99} = 2.45 \pm (2.861)(0.12)$$

$$= (2.11, 2.79)$$

Notice that the width of the confidence interval has increased from 0.50 to 0.68. This will always be true. That is, the greater the level of confidence, the wider the interval.

> As the level of confidence increases, say, from .95 to .99, the width of the interval also increases if other conditions are kept constant.

Summary

This relatively brief chapter has introduced the concept of estimating parameters from sample statistics. Estimation of parameters is an important and useful procedure of inferential statistics, especially for research studies in which it is important to know tenable values for the parameters. Estimation is not used as extensively in behavioral science research as hypothesis testing, and it effectively could be used more, especially by providing additional information in studies that emphasize hypothesis testing.

The first step in estimating a population parameter, such as the population mean (μ), is to determine the best point estimate. In estimating μ, the sample mean ($\overline{X}$) is the best point estimate. However, although the sample statistic is the best single estimate, it is seldom considered an adequate estimate because it is subject to random-sampling fluctuation and it would be extremely unlikely that the statistic would equal the parameter being estimated. Therefore, an interval estimate is developed that contains tenable values for the parameter. This interval, called the confidence interval, is defined as a range of values that has a specified degree of confidence of containing the parameter. If μ is being estimated, the development of this interval is based on the sample mean ($\overline{X}$), the standard error of the mean ($s_{\overline{X}}$), and the sampling distribution of the mean. In general, the formula for the confidence interval in the estimation of any population parameter is

$$CI = \text{statistic} \pm (\text{critical value})(\text{standard error of the statistic})$$

The width of the confidence interval depends on the sample size and the level of confidence. As the interval becomes narrower, statistical precision increases. In other words, the narrower the confidence interval, the more accurate the estimate of the parameter.

Hypothesis testing and estimation are the two basic procedures of inferential statistics. In our examples, we have tested hypotheses about means and estimated means. But the basic reasoning underlying hypothesis testing and estimation is the same regardless of the parameters and statistics involved. It is the chain of reasoning in inferential statistics. As you proceed with the following chapters, be sure to keep this underlying logic in mind.

Exercises

1. Compute the 90- and 95-percent confidence intervals for the data in Exercise 8.2.

2. Compute the 95- and 99-percent confidence intervals for the data in Exercise 8.3.

*3. The physical education teachers in a large school district administer a physical performance test to a random sample of seventh grade boys, $n = 64$. The sample mean is 84.20 and the standard deviation is 8.48. Construct the 95-percent confidence interval for the mean of the population of seventh grade boys in this district.

4. A social scientist believes that working people today want to retire at a later age than workers did previously. She reviews the available literature and concludes that the desired retirement age five years ago was 65. To investigate this conjecture, the social scientist randomly selects 400 workers between the ages of 58 and 62 and interviews them. Based on these interviews, she finds the mean desired retirement age to be 65.6 and the standard deviation to be 4.0.

   a. Test $H_0: \mu = 65$ against $H_a: \mu > 65$. Use $\alpha = .05$.

   b. Construct CI_{95}. (*Hint:* The critical value for the confidence interval will *not* be the same as for the test statistic.)

5. The 95-percent confidence interval for a population mean is (4.12, 10.88). Without any further information, find the 99-percent and 90-percent confidence intervals. The sample size is greater than 120. (*Hint:* Find $\overline{X}$, the critical value of t or z, and the standard deviation of the sampling distribution of the mean—in other words, the standard error of the mean.)

6. Given that $\overline{X} = 22.6$ and $s = 2.34$, compute and compare the 95-percent confidence intervals for sample sizes $n = 150$ and $n = 10$.

7. The 95-percent confidence interval for the mean of a population is (21.84, 40.56), based on a sample of size 22. Find CI_{90} and CI_{99}.

8. Given a desired level of confidence of 95 percent and $\sigma = 15$, what sample size do we need to estimate the mean IQ in a given population to within ± 3 points?

9. A 95-percent confidence interval is constructed for a population mean, and the interval is found to be (70, 80). What is conceptually incorrect about the following probability statements?

   a. The probability is .95 that μ falls in the interval.

   b. The probability is .05 that μ lies outside the interval.

   c. The probability is .025 that μ is greater than 80.

   d. The probability is .025 that μ is less than 70.

10. Suppose the interval in Exercise 9 was based on a sample of size 225. What were the mean and standard deviation of the sample, and the standard error of the mean?

11. A researcher constructs a 99-percent confidence interval for a population mean based on a sample of size 50. The sample mean is $\overline{X} = 82.5$, and the interval is given as 80.15 to 84.95. What is wrong with the construction of this interval, and why is it wrong?

12. A population mean is being estimated based on a sample of size 49. The sample mean is 95, and the sample standard deviation is 21.

   a. Construct CI_{95}.

   b. Give the confidence statement for the interval in part a.

   c. Suppose CI_{99} is constructed. How does this interval compare to the interval in part a?

   d. Assume that another sample has a standard deviation of 21 (or very close to 21). What sample size is required to have a CI_{95} that does not exceed seven points in width? (*Hint:* Estimate the critical value needed based on the information from the solution of part a.)

13. Describe what is meant by statistical precision when constructing confidence intervals. Why is statistical precision enhanced as sample size is increased?

14. A researcher tests $H_0: \mu = 95$ against the alternative hypothesis, $H_a: \mu \neq 95$. He selects a sample of size 200 and computes the statistical test with the data. The sample mean is 96.5 and he rejects H_0, with $\alpha = .05$. Then a 95-percent confidence interval is constructed going from 93.5 to 97.5. Identify an inconsistency in the results and an error in the computation.

15. A 95-percent confidence interval for the mean, based on a sample of size 169, goes from 80.57 to 87.43.

   a. What are the mean and standard deviation of the sample?

   b. Assume that the standard deviation of the sample remains the same, how large a sample would be necessary to obtain a 95-percent confidence interval one-half the length of the one above?

10

Hypothesis Testing: One-Sample Case for Other Statistics

I n Chapters 8 and 9, we introduced hypothesis testing, which involves determining whether a pre-established value of a parameter is tenable, and interval estimation, which involves obtaining an interval with a designated degree of confidence of containing the parameter. We saw that, in both hypothesis testing and interval estimation, the researcher bases decisions about the nature of a parameter on knowledge of the corresponding sample statistic. In this chapter, we extend the concepts of hypothesis testing and interval estimation to population values of correlation (ρ) and proportions (P).[1] We will test each of the following hypotheses, which have a as the hypothesized value of the population parameter.

$$H_0: \rho = a$$

$$H_0: P = a$$

where

$\rho = $ population correlation coefficient

$P = $ population proportion

---

[1] In keeping with the notation convention for parameters, the Greek letter rho (ρ) represents the population correlation coefficient. However, a capital P is used for the population proportion because the Greek letter pi (π) has been reserved for the mathematical constant, 3.1416 (see Chapter 4).

For each of these hypotheses, we modify and extend the steps for hypothesis testing in Chapter 8 to include a step for constructing confidence intervals. The steps are:

1. State the hypotheses.

2. Set the criterion for rejecting H_0.

3. Compute the test statistic.

4. Construct the confidence interval.

5. Interpret the results.

These steps provide a logical, consistent procedure for testing hypotheses and developing confidence intervals. It is important to follow them in order to understand the chain of reasoning for inferential statistics.

For most of the hypotheses in this chapter, the general formula for the test statistic given in Chapter 8 can be used in step 3:

$$\text{Test statistic} = \frac{\text{statistic} - \text{parameter}}{\text{standard error of the statistic}} \tag{10.1}$$

This formula indicates the magnitude of the difference between the observed sample statistic and the hypothesized population parameter in standardized form.

Similarly, for step 4, we can use the general formula for the confidence interval that was given in Chapter 9.

$$\text{CI} = \text{statistic} \pm (\text{critical value})(\text{standard error of the statistic}) \tag{10.2}$$

Testing H_0: $\rho = a$

Suppose a team of sociologists wants to study the relationship between family income and child rearing attitudes. As we saw in Chapter 5, the *correlation coefficient* provides an index of the linear relationship between two variables. Therefore, this research team would be interested in testing hypotheses about the value of the correlation coefficient (ρ) or constructing confidence intervals to estimate its value. To do so, they would follow the same steps that we followed to test hypotheses about μ and to estimate its value. But in this case, they must consider the sampling distribution of the correlation coefficient.

Fisher's z transformation

The sampling distribution of the correlation coefficient is conceptually similar to the sampling distribution of the mean. Recall that the sampling distribution of the mean is the distribution of all possible outcomes—the outcome being the sample mean. When we test hypotheses about the correlation coefficient, the sampling distribution is the distribution of all possible correlation coefficients from samples of a given size.

In previous chapters, we saw that it is possible to define the sampling distribution of the mean using mathematical theorems. However, we *cannot* use these theorems directly to define the sampling distribution of the correlation coefficient because correlation coefficients are restricted to values between -1.0 and $+1.0$. As a result, the sampling distribution of the correlation coefficient changes its shape as a function of both the magnitude and the sign of the coefficient. As the magnitude increases, the sampling distribution becomes increasingly skewed to the left for positive correlations and increasingly skewed to the right for negative correlations (see Figure 10.1). However, when the population value (ρ) is near zero, the sampling distribution is symmetrical and approximately normal. The approximation of normality improves as the sample size increases.

Because the theoretical sampling distribution becomes increasingly skewed as the magnitude of the population correlation coefficient increases, it is inappropriate to use the normal distribution as the sampling distribution for testing hypotheses about population values for ρ that are not zero. Faced with this problem, R. A. Fisher[2] developed a statistic called the **Fisher z transformation** (z_r). The transformation is

$$z_r = \left(\frac{1}{2}\right) \log_e \left(\frac{1+r}{1-r}\right) \tag{10.3}[3]$$

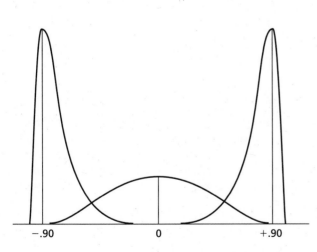

-.90   0   +.90

FIGURE 10.1
Sampling distribution for the correlation coefficient when $\rho = -.90$, 0, and $+.90$

[2]R. A. Fisher, "Frequency Distributions of Values of the Correlation Coefficient in Samples of an Infinitely Large Population," *Biometrika* 10 (1915): 507–521.

[3]Log is the natural log function.

Fisher showed that, for large samples, the sampling distribution of the transformed correlation coefficient (z_r) is approximately normal for any value of the population correlation, positive or negative. He also showed that the estimated standard deviation of the sampling distribution — the **standard error of z_r** — is

$$s_{z_r} = \sqrt{\frac{1}{n-3}} \qquad (10.4)$$

where

$n =$ sample size

The sampling distributions of transformed correlation coefficients are illustrated in Figure 10.2. It is important to emphasize that, regardless of the value of ρ in the population, we can use the normal distribution as the sampling distribution when testing hypotheses, if we have transformed r by using the Fisher z transformation.

The process of transforming r to z_r using formula 10.3 is somewhat complicated, so Table C.6 in the Appendix lists the transformed values. For any given value of r, this table indicates the corresponding value of z_r. We will illustrate the use of this table by testing a hypothesis.

> The sampling distribution of r is not normal and changes as the value of ρ changes; therefore, r is transformed to z_r. The sampling distribution of z_r is approximately normally distributed, with an estimated standard error (s_{z_r}) of $\sqrt{1/(n-3)}$.

Example

Suppose a team of sociologists is conducting a study that involves the relationship in a suburban community between family income and fathers' attitudes about democratic child rearing practices. Adequate measures are available for both variables. From a review of the literature and knowledge of the variables, it is expected that the relationship between the variables will be rather high and in the positive

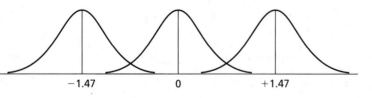

-1.47 0 $+1.47$

FIGURE 10.2
Sampling distribution for the transformed correlation coefficient (z_r) for $\rho = -.90, 0, +.90 \Rightarrow$ $z_\rho = -1.472, 0, +1.472$

direction. The research literature implies that the correlation between these variables is around .65, so the sociologists decide to focus on this value. Suppose, for this investigation, that the team randomly selects 30 families from the population and collects and records data on total family income and fathers' attitude scores. The Pearson product-moment correlation coefficient (r) is computed and found to be .61.

Step 1. State the hypothesis The null hypothesis that the population correlation coefficient is .65 is to be tested against a nondirectional alternative. Symbolically:

$$H_0: \rho = .65$$

$$H_a: \rho \neq .65$$

The level of significance is set at .10 ($\alpha = .10$).

Step 2. Set the criterion for rejecting H_0 As we noted, the sampling distribution of the transformed correlation coefficient is approximately normally distributed, with the estimated standard error determined by

$$s_{z_r} = \sqrt{\frac{1}{n-3}}$$

$$= \sqrt{\frac{1}{27}}$$

$$= 0.192$$

This distribution is illustrated in Figure 10.3.

Since the null hypothesis is tested against a nondirectional alternative (two-tailed test) and the α level is set at .10, the critical values for the test statistic are ± 1.645. That is, if the transformed value of the *sample* correlation coefficient is greater than or less than 1.645 standard errors from the transformed value of the *hypothesized* correlation coefficient, then the null hypothesis is rejected.

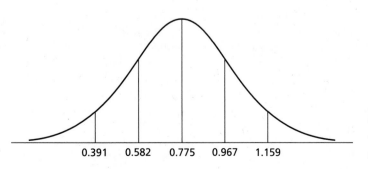

0.391   0.582   0.775   0.967   1.159

FIGURE 10.3
Theoretical sampling distribution for hypothesis $H_0: \rho = .65$ with $z_\rho = .775$, $n = 30$

Step 3. Compute the test statistic To test the hypothesis, both the hypothesized value of ρ (.65) and the observed value of r (.61) must be transformed using the Fisher z transformation. From Table C.6, we find the transformed value for ρ by reading the column labeled r until we reach .65; the corresponding value in the column labeled z_r is .775. This value would be the transformed value, z_ρ. Similarly, the z_r value for $r = .61$ is 0.709. Therefore, the parameter and the statistic that we use in the computation are 0.775 and 0.709, respectively.

$$\rho = .65 \Rightarrow z_\rho = 0.775$$

$$r = .61 \Rightarrow z_r = 0.709$$

Now we can apply the general formula for the test statistic (formula 10.1) to testing the hypothesis,[4] using (1) the observed value of the sample statistic (z_r); (2) the hypothesized value of the population parameter (z_ρ); and (3) the standard error of the statistic (s_{z_r}).

$$z = \frac{z_r - z_\rho}{s_{z_r}} \tag{10.5}$$

$$= \frac{0.709 - 0.775}{0.192}$$

$$= -0.344$$

Because the value of the test statistic ($z = -0.344$) does *not* exceed the critical value (± 1.645), the sociologists would *not* reject the hypothesis that the correlation between family income and fathers' attitudes toward democratic child rearing is .65.

Step 4. Construct the confidence interval Whether or not the null hypothesis is rejected, it can be informative to develop a confidence interval for estimating the parameter. For this example, because the α level is set at .10 for testing the null hypothesis, the corresponding level of confidence is $1 - \alpha = .90$. Therefore, a 90-percent confidence interval will be constructed.

Because of the Fisher z transformation, building a confidence interval for a correlation coefficient is slightly more complicated than, for example, building a confidence interval for a mean. First, it is necessary to build a confidence interval around z_r and then convert the end points of this interval, which are in z_r values, back to r values. The sampling distribution for the Fisher z transformation (z_r) is the normal distribution, and because $\alpha = .10$, the critical value for CI_{90} is 1.645.

---

[4] Notice that we use the symbol z as the standard score for testing the null hypothesis $H_0: \rho = a$ because the normal distribution, not the t distribution, is the sampling distribution.

The confidence interval is constructed using general formula 10.2 as follows:

$$CI_{90} = z_r + 1.645(s_{z_r}) \qquad (10.6)$$

$$= 0.709 \pm 1.645(0.192)$$

$$= 0.709 \pm 0.316$$

$$= (0.393, \ 1.025)$$

To convert these end points back to r values, we use Table C.6 in reverse. Going down the column labeled z_r, we find the values 0.393 and 1.025. The corresponding r values are .374 and .773, respectively. Therefore, the CI_{90} for r is as follows:

$$CI_{90} = (0.374, \ 0.773)$$

Step 5. Interpret the results In this example, the sociologists initially test the null hypothesis $\rho = .65$ and then develop the 90-percent confidence interval. In testing the hypothesis, the observed value of the test statistic ($z = -0.344$) does not exceed the critical values (± 1.645); thus, the hypothesized value for ρ (.65) is not rejected as a tenable value. That is, the null hypothesis is *not* rejected at the .10 level of significance. The conclusion is: *the probability that the observed statistic ($r = .61$) would have occurred by chance, if in fact the null hypothesis ($\rho = .65$) is true, is greater than* .10. In this case, the difference between the observed sample correlation coefficient and the value hypothesized for ρ is not sufficient to discount sampling error as the contributing factor. Since the level of significance is set at .10, the sociologists are unwilling to take a greater risk of making a Type I error. Therefore, the null hypothesis is not rejected.

The null hypothesis that $\rho = .65$ cannot be rejected, but the population correlation coefficient may be different from .65. The best point estimate would be the sample correlation coefficient $r = .61$. However, an interval estimate is more informative, and the CI_{90} for the population correlation coefficient is found to be (0.374, 0.773). The sociologists are 90 percent confident that this interval contains the population correlation coefficient. Notice that the hypothesized value of .65 is contained in the interval, which is consistent with failing to reject H_0: $\rho = .65$.

Testing H_0: $\rho = 0$

The professional literature in the behavioral sciences abounds with research studies that attempt to determine whether a relationship exists between two variables. Exploratory studies often involve the search for relationships regardless of the magnitude of the relationships. This gives rise to the null hypothesis of zero correlation in the population (H_0: $\rho = 0$).

In these studies, there is an alternative procedure for testing the null hypothesis. Earlier we indicated that the sampling distribution of the correlation coefficient (r) is symmetrical and approximately normal when the population correlation (ρ) is zero. Actually, the sampling distribution when $\rho = 0$ is the t distribution with $n - 2$ degrees of freedom.[5] Therefore, to test $H_0: \rho = 0$, we can use the theoretical sampling distribution of r rather than z_r. The formula for the test statistic is

$$t = r\sqrt{\frac{n - 2}{1 - r^2}} \tag{10.7}$$

For testing $H_0: \rho = 0$, the Student's t distribution for $n - 2$ degrees of freedom can be used as the sampling distribution for the test statistic.

Consider an example. A psychologist wants to know if there is a relationship between teenagers' scores on an attitude measure and an anxiety scale. The attitude measure is scored so that the higher the score, the more positive the attitude. The research question might be phrased: is there a tendency for high attitude scores to be associated with either high or low anxiety scores? The psychologist tests the null hypothesis ($H_0: \rho = 0$) against the nondirectional alternative ($H_a: \rho \neq 0$) at the .05 level of significance.

A sample of 32 teenagers is selected randomly and measured on both variables, and the sample correlation coefficient is found to be $-.375$. Computing the test statistic using formula 10.7, we get:

$$t = -0.375\sqrt{\frac{32 - 2}{1 - (-0.375)^2}} = -2.215$$

The critical values of t for 30 degrees of freedom at the .05 level are ± 2.042. Since the absolute value of the test statistic exceeds the critical value, the null hypothesis is rejected in favor of the nondirectional alternative $H_a: \rho \neq 0$. The conclusion is that, if in fact $\rho = 0$, then the probability that the researcher would have observed by chance a sample correlation coefficient of $r = -0.375$ is less than .05. The psychologist concludes that there is a relationship in the population between attitude and anxiety.

When the null hypothesis ($H_0: \rho = 0$) is rejected, the usual statement made in the professional literature is that the relationship between attitude and anxiety is "statistically significant." In fact, it is the statistic r that is (statistically) significantly different from zero. The conclusion is that there is a relationship between the variables in the population, but this conclusion should not be construed to imply

[5]Recall the definition and discussion of degrees of freedom from Chapter 8. Two degrees of freedom are lost in this case, since we are estimating the variance for both variable X and variable Y, losing one degree of freedom for each variable.

anything about the magnitude of the relationship — only that the relationship is different from zero.

To explore the magnitude of the relationship, we construct and interpret the appropriate confidence interval. We recommend transforming the observed sample correlation coefficient using the Fisher z transformation and the procedures outlined in the section on the previous example. The CI_{95} will be constructed to be consistent with using $\alpha = .05$ for testing the null hypothesis above.

For this example, we transform $r = -.375$ to $z_r = -0.394$ and then compute the confidence interval using the general formula for the confidence interval.

$$\text{CI}_{95} = z_r \pm 1.96(s_{z_r})$$

$$= -0.394 \pm 1.96 \left(\sqrt{1/(32 - 3)} \right)$$

$$= -0.394 \pm 0.364$$

$$= (-0.758, -0.030)$$

Transforming these z_r values back to r values, we get

$$\text{CI}_{95} = (-0.640, -0.029)$$

The conclusion for this confidence interval is that the psychologist can be 95 percent confident that the interval from -0.640 to -0.029 contains ρ; that is, we are 95 percent confident that the correlation could be as low as $-.029$ or as high as $-.640$. It is this conclusion about the confidence interval that should serve as the basis for interpreting the magnitude of the relationship between attitude and anxiety in the population. Note that zero is not contained in this interval, which is consistent with the fact that H_0: $\rho = 0$ was rejected.

Actually, it is not necessary to do the computation using formula 10.7 to test H_0: $\rho = 0$. An inspection of that formula shows that it contains three variables: n, r, and t. The degrees of freedom are $n - 2$, and these determine the specific t distribution. Given a t distribution, the critical values for a specified level of significance are known. Therefore, the values of r are determined uniquely if n and the critical values of t are known, as they would be in a practical situation. Table C.7 contains values for r required for significance at various levels when testing H_0: $\rho = 0$ and when using either a one-tailed or two-tailed test. Notice that, when the degrees of freedom $(n - 2)$ are small, a large value of r is required. For example, to reject the null hypothesis at the .05 level, an r value greater than or equal to .576 for df $= 10$ is required. However, when df $= 60$, an r value greater than or equal to .250 is needed.

Note that, because under the hypothesis H_0: $\rho = 0$ the sampling distribution of r is symmetrical, the r values of Table C.7 can be considered as absolute values. That is, in order to reach statistical significance, the absolute value of a computed r (whether the r is positive or negative) must equal or exceed the value in the table. For example, suppose a computed $r = -.428$ based on a sample of size 62. This correlation coefficient is statistically significant at $\alpha = .05$, using a two-tailed test, because its absolute value is greater than .250, the value in the table required for significance.

$H_0: \rho = 0$ can be tested directly, without using the Fisher z transformation, with tabled values. For a given df value, if the absolute value of r is equal to or exceeds the tabled value for a given α level, then the null hypothesis ($H_0: \rho = 0$) is rejected.

Statistical Precision and Correlation Coefficients

Statistical precision is related to the standard error of the statistic, and the statistic is z_r, the Fisher z transformation, when testing a null hypothesis such as $H_0: \rho = a$ when $a \neq 0$. As formula 10.4 shows, the standard error of s_{z_r} will decrease as n increases. Because s_{z_r} is in the denominator of the formula for the test statistic, the smaller its value, the smaller the difference between r and the hypothesized value of the population correlation coefficient required to reach statistical significance. Because the width of the confidence interval depends on the size of the standard error of z_r, as s_{z_r} decreases, the confidence interval for a specified level of confidence will become narrower and thus more precise.

When testing the null hypothesis $H_0: \rho = 0$, consider the critical values of the correlation coefficient in Table C.7. As sample size increases, degrees of freedom increase. For a specified level of significance, the values required for statistical significance become smaller as degrees of freedom — and hence, sample size — increase. As we saw before, statistical precision is enhanced as sample size increases.

When correlation coefficients are analyzed using inferential statistics, statistical precision is enhanced as sample size is increased.

Testing $H_0: P = a$

There are many times in behavioral science research when we are studying a dichotomous variable in a population, a variable that has two mutually exclusive classes. That is, we study the proportion of the population — the fraction of the population — that belongs to each class. Proportions are very visible in election years, when polling agencies estimate the proportions of voters that will vote for a specific candidate, typically calling these proportions "percentages."

Consider an example of a dichotomous population. We can designate a specified work force as a dichotomous population, and the mutually exclusive classes are those who are employed and those who are unemployed. The proportion that is unemployed is then the ratio of the number who are unemployed to the total population. If the total population consists of 10,000 people and 540 are unemployed, the proportion that is unemployed is 540/10,000, or 0.054. We use the symbol P to designate the population proportion, and the symbol p to indicate a sample proportion.

How do we test hypotheses about proportions? We use the same chain of reasoning for inferential statistics and the same steps for hypothesis testing and estimation introduced earlier. Now, however, we have the sample proportion as the statistic and therefore must use its sampling distribution.

Consider an example. A researcher wants to test a hypothesis about the proportion of a population of community college students who either transfer to a four-year college or terminate their education with an associate's degree. For this example, suppose the researcher hypothesizes that this proportion in the population equals 0.60:

$$H_0: P = 0.60$$

Further, suppose the researcher tests this null hypothesis against the nondirectional alternative hypothesis:

$$H_a: P \neq 0.60$$

Because it is impractical to measure every member of the population, the researcher hypothesizes a value for the population proportion (P), randomly selects a sample, and then uses the sample proportion (p) to test the tenability of the stated hypothesis. The sampling distribution of this sample proportion is used in the same way the sampling distributions of the mean ($\overline{X}$) and the Fisher z transformation (z_r) are used.

What is the sampling distribution of p? Actually, it is the binomial distribution (described in Chapter 7). But if the sample size is reasonably large, the normal distribution is an adequate approximation of the binomial distribution. As was the case for testing hypotheses about means and correlation coefficients, the mean of the sampling distribution of p is defined by the null hypothesis, i.e., $H_0: P = 0.60$. The standard deviation of this sampling distribution,[6] called the **standard error of the proportion**, is defined as follows:

$$s_P = \sqrt{\frac{PQ}{n}} \tag{10.8}$$

[6]The values P and Q are used in formula 10.8 since they are the expected values from the null hypothesis.

where

$P =$ hypothesized value of the proportion possessing the characteristic

$Q =$ proportion not possessing the characteristic

$n =$ sample size

For our community college example, consider the sampling distribution of the proportion when the null hypothesis is H_0: $P = 0.60$. The standard error of the proportion is found by using formula 10.8.

$$s_p = \sqrt{\frac{(0.60)(0.40)}{400}} = 0.0245$$

Thus, the sampling distribution of p for this example has a mean equal to the hypothesized value of P, 0.60, and a standard deviation of 0.0245. This sampling distribution is illustrated in Figure 10.4.

Like the correlation coefficient, P has a restricted range of values. It can take on values from 0 to 1.0, inclusive. If P approaches either of these extreme values, say, less than 0.25 or greater than 0.75, and if the sample size is small, the normal distribution may not be an adequate approximation of the binomial distribution. A rule of thumb for determining the size of the sample necessary to use the normal curve as the sampling distribution is that nP or nQ, whichever is smaller, must be greater than 5, so the normal distribution can be used as the sampling distribution.

The sampling distribution of the proportion is the binomial distribution. With a large sample size, the normal distribution is an adequate approximation of the binomial. The standard error of the proportion is given by

$$s_p = \sqrt{\left(\frac{PQ}{n}\right)}$$

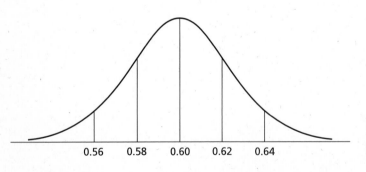

0.56   0.58   0.60   0.62   0.64

FIGURE 10.4
Theoretical sampling distribution for hypothesis H_0: $P = .60$ and $n = 400$

First Example

Suppose a state senator cannot decide how to vote on an environmental protection bill. Various opinion polls disagree on the proportion of the state's registered voters who support the bill. The senator decides to request his own survey. If the proportion of registered voters supporting the amendment exceeds 0.60, the senator will vote for it. A random sample of 750 voters is selected, and 495 are found to support the bill.

Step 1. State the hypotheses For this example, the null hypothesis will be tested against the directional alternative. Symbolically:

$$H_0 : P = 0.60$$

$$H_a : P > 0.60$$

Step 2. Set the criterion for rejecting H_0 Since the alternative hypothesis is directional, the region of rejection for H_0 is in only one tail — the right tail of the sampling distribution. It is decided to use a conservative level of significance, so α is set at .01. Using the properties of the normal distribution, we find that for $\alpha = .01$ the critical value for the test statistic is +2.33.

$$z_{cv} = +2.33$$

Step 3. Compute the test statistic As indicated above, 495 of 750 voters surveyed support the bill, so the sample proportion is $495/750 = 0.66$. Thus, the hypothesized value of the population proportion (P) and the observed value of the sample proportion (p) are

$$P = 0.60$$

$$p = 495/750 = 0.66$$

Notice that, if the sample proportion (p) is equal to or less than 0.60, the computational process is terminated because there is no rejection region for H_0 in the left tail of the distribution. There is no way of rejecting the null hypothesis if the sample proportion is equal to or less than 0.60. However, since $p = 0.66$, we can still pose the question: what is the probability that $p = 0.66$ will appear due to sampling error if $P = 0.60$?

Formula 10.1 for the test statistic can now be applied.

$$z = \frac{p - P}{s_p} \tag{10.9}$$

For this example, the standard error of the proportion is

$$s_p = \sqrt{\frac{PQ}{n}} = \sqrt{\frac{(0.60)(0.40)}{750}} = 0.0179$$

Thus,

$$z = \frac{0.66 - 0.60}{0.0179}$$

$$= 3.35$$

Since the observed value of the test statistic (3.35) exceeds the critical value (2.33), the null hypothesis (H_0: $P = 0.60$) is rejected.

Step 4. *Construct the confidence interval*[7] In this case, because the senator has decided to vote for the amendment if the proportion of voters supporting it exceeds 0.60, there would be no point in constructing a confidence interval. But the senator is still uneasy about voting for the bill and wants an estimate of the proportion of voters in the population who are likely to vote for the bill. Now building a confidence interval is appropriate.

An adjustment in the critical value is necessary because the hypothesis tested above was directional and the confidence interval requires two end points. To correspond with the α level of .01, we will construct a 99-percent confidence interval. Under the normal curve, 99 percent of the area is between $z = -2.58$ and $z = +2.58$. The confidence interval for the population proportion is constructed symmetrically around the sample proportion. The CI_{99} for this example is:

$$\text{CI}_{99} = p \pm 2.58(s_p) \tag{10.10}$$

$$= 0.66 \pm (2.58)(0.0245)$$

$$= 0.66 \pm 0.046$$

$$= (0.614, \ 0.706)$$

---

[7]If the researcher was only interested in developing a confidence interval for the sample proportion, without first testing the null hypothesis, the standard error of the proportion would be calculated using the sample value $p = 0.66$ and correspondingly $q = 1 - p = 1 - 0.66 = 0.34$, i.e.,

$$s_p = \sqrt{\frac{pq}{n}} = \sqrt{\frac{(0.66)(0.34)}{750}} = 0.0173$$

then the CI_{99} would be computed as follows:

$$\text{CI}_{99} = 0.66 \pm (2.58)(0.0173)$$

$$= 0.66 \pm 0.045$$

$$= (0.615, 0.705)$$

Step 5. Interpret the results It is concluded that the population proportion is greater than 0.60. The probability is less than 0.01 that a sample proportion of 0.66 would appear due to sampling error if in fact the population proportion is equal to 0.60. With respect to the interval estimate of P, it is concluded with 99 percent confidence that the interval from 0.61 to 0.71 contains the population proportion (P).

With regard to the senator's vote on the environmental bill, it is pretty safe to conclude that more than 60 percent of the voters favor the bill. The confidence interval shows that perhaps as many as 70 percent of the voters favor the bill, but it is very unlikely that less than 61 percent favor the bill. Therefore, regardless of other consequences of voting for the bill, voting for it would be the course of action supported by most voters.

When a one-tailed test is used, the critical values for the test and for constructing the confidence interval are *not* the same even though the level of confidence equals $1 - \alpha$. This is because the region of rejection for H_0 is in only one tail of the distribution in a one-tailed test, whereas the confidence interval is constructed symmetrically around the observed sample statistic. In this example, $\alpha = .01$ and the region of rejection is the 1 percent of the area in the right-hand tail beyond the critical value $z = +2.33$. However, for CI_{99}, the critical value for the confidence interval is 2.58.[8]

Second Example

A family services agency serving an urban area of approximately 120,000 households is conducting a survey to determine the level of need for its services. In the survey, several needs are addressed, but we will focus on one: the need for day care for children under six years of age.

A random sample of households is selected, and a telephone interview is conducted. Data are collected on a sample of 400 households. Staff at the agency hypothesize that 30 percent or .30 of the households in the population need day care for children under six years old. In addition, agency staff want to estimate the number of households in the population with this need so they can plan accordingly to provide services. The sample results show that 88 households indicated a need for day care for children under six years old.

Step 1. State the hypotheses The null hypothesis that the population proportion is .30 will be tested against the nondirectional alternative that the population proportion is not .30. Symbolically:

$$H_0: P = .30$$

$$H_a: P \neq .30$$

---

[8] It is possible to construct one-tailed intervals, but they are rarely used and the discussion of such intervals is not included in this text.

Step 2. *Set the criterion for rejecting H_0* The alternative hypothesis is nondirectional, so the rejection region for H_0 is split between the two tails of the sampling distribution for p. The .05 level of significance is used ($\alpha = .05$). The normal distribution is the appropriate sampling distribution, and for $\alpha = .05$, the critical values of the test statistic are ± 1.96.

$$z_{cv} = \pm 1.96$$

Step 3. *Compute the test statistic* The sample results showed 88 of 400 households indicated a need for day care. Thus, the sample proportion is:

$$p = 88/400 = 0.22$$

Next, we determine the standard error of p by using formula 10.8.

$$s_p = \sqrt{\frac{PQ}{n}} = \sqrt{\frac{(0.30)(0.70)}{400}} = 0.0229$$

The hypothesized value for P is .30, so we can now apply the general formula for the test statistic.

$$z = \frac{p - P}{s_p}$$
$$= \frac{0.22 - 0.30}{0.0229}$$
$$= \frac{-0.08}{0.0229} = -3.49$$

The absolute value of the test statistic exceeds the critical value of 1.96, and the null hypothesis ($H_0: P = .30$) is rejected.

Step 4. *Construct the confidence interval*[9] The staff at the family services agency wants an estimate of the population proportion of households in need

[9]Like the previous example, if the researcher was only interested in developing a confidence interval, without first testing the null hypothesis, the standard error of the proportion would be calculated using the sample value $p = 0.22$, i.e.,

$$s_p = \sqrt{\frac{pq}{n}} = \sqrt{\frac{(0.22)(0.78)}{400}} = 0.0207$$

Then the CI_{95} would be computed as follows

$$CI_{95} = 0.22 \pm (1.96)(0.0207)$$
$$= 0.22 \pm 0.041$$
$$= (0.179, 0.261)$$

of day care for children under six years of age. Because $\alpha = .05$ for testing the null hypothesis above, the level of confidence is set at .95 ($1 - \alpha = 1 - .05 = .95$). The CI_{95} is constructed symmetrically around the sample proportion; in this case, $p = .22$. In the normal curve, 95 percent of the area lies between $z = -1.96$ and $z = +1.96$. Therefore, the CI_{95} for the population proportion is constructed by

$$\text{CI}_{95} = p \pm 1.96(s_p)$$
$$= 0.22 \pm (1.96)(0.0229)$$
$$= 0.22 \pm 0.045$$
$$= (0.175, \, 0.265)$$

Step 5. Interpret the results The null hypothesis that the population proportion is .30 is rejected, so it is concluded that $P \neq .30$. In fact, an inspection of the sample results leads to the conclusion that in the population less than 30 percent of the households need day care for children under six years of age. The probability that a sample proportion of $p = .22$ would appear by chance, if the population proportion were .30 (that is, if H_0: $P = .30$ were true), is less than .05.

The results of constructing the confidence interval show that we are 95 percent confident that the interval .175 to .265 contains the population proportion. The agency staff can use this information to estimate the actual need for day care in this urban area. There are approximately 120,000 households, so the low estimate is 120,000 (0.175) = 21,000 families. The high estimate is 120,000 (0.265) = 31,800 families. The family services agency staff can now plan accordingly.

Statistical Precision and Proportions

As with other statistics, statistical precision is directly reflected in the standard error of the statistic, the standard error of p in this case. As the standard error of p is decreased, statistical precision will increase.

Consider formula 10.8, the formula for the standard error of p. As n is increased, the denominator of the formula will increase and the value of s_p will decrease. The result of decreasing s_p is that a smaller difference between p and the hypothesized value of P will be required to reach statistical significance when testing hypotheses. When constructing a confidence interval, the interval becomes smaller for a specified level of confidence as s_p becomes smaller. Thus, the interval will be more "precise" for estimating the population proportion. For example, assume that a sample size of 1,000 had been used in the immediately preceding example.

$$s_p = \sqrt{\frac{(0.30)(0.70)}{1,000}} = 0.0145$$

and the corresponding CI_{95} would be

$$CI_{95} = 0.22 \pm 1.96\,(0.0145)$$

$$= 0.22 \pm 0.028$$

$$= (0.192,\ 0.248)$$

This interval is narrower and a more precise estimate than the interval constructed with a sample size of 400.

> **When applying inferential statistics to proportions, statistical precision is enhanced as sample size is increased.**

Summary

In this chapter, we extended the concepts of hypothesis testing and interval estimation to include hypotheses about the population values of correlation coefficients (ρ) and proportions (P). When testing hypotheses and developing confidence intervals, we used the same general formulas presented in Chapters 8 and 9. The reasoning underlying each of the examples discussed in this chapter is the chain of reasoning for inferential statistics presented earlier, in Figure 7.6. Random samples are selected, and the statistics (sample measures) reflect the corresponding parameters and sampling errors. When a hypothesis is posed, the question is raised: what is the probability that the statistic would have occurred by chance if the hypothesis were true? If the probability is less than the significance level, the hypothesis is rejected; if greater, the hypothesis is not rejected. In building confidence intervals, we use the sampling distribution for establishing an interval of tenable values for the parameter being estimated.

Statistical precision is enhanced when sample size is increased, other conditions remaining constant. This is because the standard error of the statistic is decreased with increased n. Operationally, enhancing statistical precision means that a smaller difference between the statistic and the hypothesized value of the parameter is required to reach statistical significance, and the confidence interval is narrower for a specified level of confidence.

TABLE 10.1
Information on Testing Hypotheses: One-Sample Case

Hypothesis Tested	Parameter	Statistic	Standard Error of the Statistic	Test Statistic	Degrees of Freedom	Sampling Distribution
$H_0: \mu = a$	μ	$\overline{X}$	$s_{\overline{X}} = s/\sqrt{n}$	$t = \dfrac{\overline{X} - \mu}{s_{\overline{X}}}$	$n - 1$	Student's t distributions
$H_0: \rho = a$	z_ρ	z_r	$s_{z_r} = \sqrt{\dfrac{1}{n-3}}$	$z = \dfrac{z_r - z_\rho}{s_{z_r}}$	—	Normal distribution
$H_0: \rho = 0$	ρ	r	—	$t = r\sqrt{\dfrac{n-2}{1-r^2}}$	$n - 2$	Student's t distributions
$H_0: P = a$	P	p	$s_p = \sqrt{\dfrac{PQ}{n}}$	$z = \dfrac{p - P}{s_p}$	—	Normal distribution

TABLE 10.2
Interval Estimation of Parameters: One-Sample Case

Parameter Estimated	Statistic	Confidence Interval
μ	$\overline{X}$	$\overline{X} \pm (t_{cv})(s/\sqrt{n})$ or $\overline{X} \pm (t_{cv})s_{\overline{X}}$
z	z_r	$z_r \pm (z_{cv}) \left(\sqrt{\dfrac{1}{n-3}} \right)$ or $z_r \pm (z_{cv})s_{z_r}$ (Converted back to values of r)
P	p	$p \pm (z_{cv}) \left(\sqrt{\dfrac{PQ}{n}} \right)$ or $p \pm (z_{cv})s_p$

Table 10.1 summarizes information needed to test hypotheses. It includes the parameter, the statistic, the standard error of the statistic, the test statistic, degrees of freedom (when applicable), and the sampling distribution associated with the test statistic. Table 10.2 summarizes information for constructing confidence intervals.

Exercises

For Exercises 1 to 7, use the following format:

1. State the hypotheses.
2. Set the criterion for rejecting H_0.
3. Compute the test statistic.
4. Construct the confidence interval.
5. Interpret the results.

*1. The leadership of a large labor union has proposed a set of salary and benefits demands for its next round of negotiations with management. However, the demands will be proposed only if the leadership has support from significantly more than 65% of the membership. A random sample of 200 union members is surveyed and 134 are favorable toward the demands, and 66 are unfavorable. Test the hypothesis that, in the population, at least 65 percent of the membership favors the demands.

2. A high school teacher is interested in testing the relationship between high school achievement scores and scores on the SAT verbal section. A literature review of similar studies indicates the relationship to be .35. Suppose the teacher selects a random sample of 50 students and finds $r = +.38$.

   a. Test $H_0: \rho = 0$ against $H_a: \rho \neq 0$. Use formula 10.5 and $\alpha = .05$.

   b. Test $H_0: \rho = .35$ against $H_a: \rho \neq .35$. Use formula 10.5 and $\alpha = .05$.

   c. Construct the 95-percent confidence interval for ρ using formula 10.6.

3. A graduate student in psychology wants to investigate the relationship between self-esteem and weekly intake of alcoholic beverages. The student randomly selects 30 undergraduate students and asks them to complete a scale measuring self-esteem and to record the weekly amount of alcohol they consume. Computing the Pearson correlation coefficient, the student finds $r = +.48$.

   a. Test $H_0: \rho = 0$ against $H_a: \rho \neq 0$. Use *both* formulas 10.5 and 10.7 and $\alpha = .05$.

   b. Construct CI_{95} for ρ using formula 10.6.

4. The head of the housing office at a large state university wants to know if there will be enough rooms on campus for returning students. Only 70 percent of students who are already on campus can be housed. Out of a random sample of 250 students, 150 say that they will be returning and 100 say they will not.

   a. Test $H_0: P = .70$ against $H_a: P \neq .70$ to determine whether there will be enough housing for all the returning students. Use formula 10.9 and $\alpha = .01$.

   b. Construct CI_{99} for P using formula 10.10.

   c. Although part a of this exercise specifies a nondirectional test, why might a directional test be more appropriate? What would be the alternative hypothesis for a directional test?

5. A political science graduate student wants to know if gun owners are more likely to be Republican than Democrat. The student randomly selects 225 members of the National Rifle Association and finds that 135 are Republican and 90 Democrat.

   a. Test $H_0: P = .50$ against $H_a: P > .50$. Use $\alpha = .10$.

   b. Construct CI_{90} for P. Remember that the critical value for the confidence interval will be different from that for the test of the hypothesis.

6. A sociologist is interested in the relationship between education (measured in years of formal schooling) and income (measured in thousands of dollars). Data

are collected from nine individuals as follows:

Individual	Education	Income
1	4	6
2	6	12
3	8	14
4	11	10
5	12	17
6	14	16
7	16	13
8	17	16
9	20	19

$\Sigma X = 108$ $\qquad$ $\Sigma Y = 123$

$\Sigma X^2 = 1,522$ $\qquad$ $\Sigma Y^2 = 1,807$

$\overline{X} = 12$ $\qquad$ $\overline{Y} = 13.67$

$\Sigma XY = 1,606$

a. Compute the Pearson r for these data.

b. Using these data, test the null hypothesis $H_0: \rho = 0$ against $H_a: \rho > 0$. Use $\alpha = .05$.

c. Construct CI_{95} for ρ. Remember that the critical value for the confidence interval will be different from that for the test of the hypothesis.

7. A congressman decides to vote for a bill that he personally opposes if more than 60 percent of his constituents favor the bill. In a survey of 200 randomly selected voters in the congressman's district, 140 indicate that they favor the bill.

a. Test $H_0: P = .60$ against $H_a: P > .60$ Use $\alpha = .05$.

b. Construct CI_{95} for P. Remember that the critical value for the confidence interval will be different from that for the test of the hypothesis.

c. Give the probability statement after completing the statistical test in part a.

d. Give the confidence statement after constructing the CI_{95} in part b.

8. Use Table C.7 to answer the following questions.

a. Given $n = 37$, $\alpha = .05$, and a two-tailed test, is a correlation coefficient of .30 "statistically significant"?

b. For part a, if $n = 72$, would $r = .30$ be statistically significant?

c. If $n = 42$, how large must r be in order to reject the null hypothesis $H_0: \rho = 0$, given $\alpha = .01$ and a two-tailed test?

d. If for part c a one-tailed (directional) test were used, how large must r be to reject $H_0: \rho = 0$?

e. A computed r is found to be .43, based on a sample of size 37. Test $H_0: \rho = 0$, using $\alpha = .01$ and a two-tailed test. What is the conclusion? Give the probability statement after completing the test.

f. With a sample size of 23, $r = .40$. Is this sample correlation coefficient statistically significant given $H_0: \rho = 0$, $\alpha = .10$, and a two-tailed test? Given $\alpha = .05$ and a two-tailed test, is it statistically significant? Relative to the sampling distribution of r, why does the conclusion change when the alpha level changes?

9. Explain why, for any given sample size and a specified alpha level (such as .05), the critical values in Table C.7 are less when using a one-tailed test than when using a two-tailed test. (*Hint*: Consider the sampling distribution of the statistic.)

10. An exercise physiologist is interested in the correlation between two physical-performance variables. An entire population cannot be measured, so a random sample of 150 individuals is selected and measured on both variables. The correlation coefficient is .55.

   a. Construct the 95-percent confidence interval for the population correlation coefficient. Give the confidence statement after constructing the interval.

   b. Suppose someone had constructed the CI_{95} for part a and found it to be (.486, .535). Why should it be concluded that there is an error in the computation?

   c. The exercise physiologist decides that the interval in part a is not precise enough. How large a sample size is necessary to have an interval no larger than .150?

   d. If additional population members were selected to meet the condition in part c, would it make any difference whether or not r remained at .55? Why or why not?

   e. Suppose the exercise physiologist cannot add to the sample but wants a narrower confidence interval. What could be done to obtain a narrower interval?

11. A large university is planning to provide a new financial aid counseling service for its student body. In order to estimate the required personnel resources (counselors and support staff) for the service, an estimate is needed of the number of students desiring the service. A random sample of 300 students is surveyed, and 35 percent of them indicate that they will use the service.

   a. Construct the 99-percent confidence interval for the population proportion of those students who will use the service.

   b. The university enrolls approximately 30,500 students. Based on the CI_{99} in part a, provide an interval estimate of the number of students who will use the service.

   c. A university official argues that the interval estimate is too wide and that a 90-percent confidence interval is adequate for the estimation. Do additional students need to be selected in order to construct CI_{90}? Why or why not? Construct CI_{90} and compare the results with those of parts a and b above. What argument must the university official use to justify going from CI_{99} to CI_{90}?

11

Hypothesis Testing: Two-Sample Case for the Mean

I n Chapter 10, we extended the concepts and procedures of hypothesis testing and interval estimation to correlation coefficients and proportions. However, throughout that chapter and those dealing with testing hypotheses and estimation for the mean, only one sample was involved. In this chapter, we apply the principles of hypothesis testing and parameter estimation to the *two-sample* case. There are many situations in behavioral sciences research that involve two samples. Suppose, for example, we want to test a hypothesis about the difference between the means of men and women on a test of physical endurance and to estimate the difference between the means. Or we may have one group receiving an experimental treatment and another receiving a control or traditional treatment, and we want to test for a difference between the mean performances of these groups. In both cases, we would analyze statistics from two samples, and the hypothesis and confidence interval would deal with the difference between two population means ($\mu_1 - \mu_2$).

The logic and procedures for testing hypotheses and constructing confidence intervals that were developed in the previous chapters also apply to the two-sample case. In other words, hypothesis testing and confidence intervals for both one-sample and two-sample cases are based on sampling and probability theory and specifically on (1) the sampling distribution of the statistic, (2) the errors inherent in hypothesis testing and estimation, (3) the level of significance and the level of confidence, and (4) the directional nature of the alternative hypothesis.

To illustrate the general strategy for testing hypotheses and developing the corresponding confidence intervals in the two-sample case, we follow the same general procedure introduced earlier with the one-sample case. First, we test the hypothesis about the difference between two population means and then develop the corresponding confidence interval. In Chapter 12, we will extend these procedures to deal with other parameters. We will follow the same steps used in Chapter 10:

1. State the hypotheses.

2. Set the criterion for rejecting H_0.

3. Compute the test statistic.

4. Construct the confidence interval.

5. Interpret the results.

> The basic reasoning for inferential statistics is the same whether we are testing hypotheses about a single mean or about the difference between two means.

Two-Sample Case: Testing H_0: $\mu_1 = \mu_2$

When testing a hypothesis for the one-sample case for the mean, the statistic was the sample mean ($\overline{X}$), which we compared with the hypothesized value of the population mean (μ). Using the sampling distribution of the mean, we determined the magnitude of the difference between the observed value of $\overline{X}$ and the hypothesized value of μ and then decided whether this magnitude was sufficient to reject the null hypothesis.

The strategy for testing hypotheses for the two-sample case for the mean is the same as for the one-sample case. For the two-sample case, the statistic is the *difference* between the two sample means ($\overline{X}_1 - \overline{X}_2$), and the parameter is the hypothesized difference between μ_1 and μ_2. This difference between the means is hypothesized to be zero—that is, the null hypothesis in the two-sample case for the mean is

$$H_0: \mu_1 = \mu_2$$

or

$$H_0: \mu_1 - \mu_2 = 0$$

To test this null hypothesis, we determine the magnitude of the difference between the statistic $(\overline{X}_1 - \overline{X}_2)$ and the hypothesized value for the parameter (H_0: $\mu_1 - \mu_2 = 0$) and then decide whether the magnitude is sufficient for rejecting the null hypothesis.

To make this decision, we use the same general formula for the test statistic as we used in the one-sample case, but the sampling distribution is different. Instead of using the sampling distribution of the mean, we are interested in the sampling distribution of all possible *differences* between the means of the two samples. Remember, the statistic of concern is now the difference between the two sample means.

How are these differences distributed? When we were testing the null hypothesis H_0: $\mu = a$, the central limit theorem was applied and the sampling distribution of the mean was the normal distribution if σ^2 were known. But when σ^2 is unknown and s^2 is used as the unbiased estimate, the sampling distribution is the t distribution with the appropriate degrees of freedom. Similarly, for the two-sample case, if σ^2 is known, the sampling distribution of differences is normally distributed. But if σ^2 is unknown and is estimated by s^2, the sampling distribution is the t distribution for the appropriate degrees of freedom. In order to define the standard error for the sampling distribution of differences, however, we must consider two assumptions that underlie H_0: $\mu_1 - \mu_2 = 0$: (1) the two samples drawn from the respective populations are independent, and (2) the variances of the two populations are equal.[1]

Assumption of Independent Samples

We have stated that the logic of inferential statistics involves selecting a random sample from the population, taking measurements on that sample, and then, using probability theory, drawing inferences about the population. For the one-sample case, the assumption that the sample was random provided the basis for making inferences about the population. When we are dealing with two samples, an extension of the concept of randomness is necessary in order to ensure that we have what are called *independent samples*.

The assumption of independent samples for the two-sample case essentially means that the scores (data) of one sample in no way influence the scores of the other sample. In this way, the scores of the individuals in one sample are unrelated to the scores of the individuals in the other sample. If the samples are selected at random, or if subjects are assigned randomly to experimental treatments, the two samples will initially vary only in terms of random fluctuation.

The traditional approach to ensuring this independence is to first select a random sample from the population, and then to randomly assign half of the subjects to an experimental treatment condition (called an *experimental group*) and half to a nonexperimental condition (called a *control group*). It is assumed that the

---

[1] For a more extensive discussion of these assumptions, see F. N. Kerlinger, *Foundations of Behavioral Research* (New York: Holt, Rinehart and Winston, 1986), pp. 267–269.

random assignment of subjects to the two conditions ensures that the two groups are equivalent at the beginning of the experiment. That is, any differences that exist among subjects are randomly distributed between the groups. Equivalency of the groups before the experiment begins is necessary to ensure that the treatment effect is not confounded by initial differences in the groups. If both random selection and random assignment are employed in the investigation, then statistical inference can be made to the entire population.

Another common research situation involves testing the difference between two fixed populations. In this case, random samples are drawn from each of the two populations, and appropriate measurements are taken. The difference between the two samples is determined and the results are generalized back to the respective populations. Because both samples are randomly selected from the respective populations, we can make inferences about each population from the samples.

Assumption of Homogeneity of Variance

When a sample of subjects is selected from a single population and then randomly assigned to two treatment groups, σ^2 is estimated by s^2 and is computed using the sample data from both treatment groups. This result is called the *pooled estimate of* σ^2. We also estimate σ^2 from s^2 when samples are randomly drawn from two distinct populations. Again we use data from both samples in the estimation. However, in order to use the pooled estimate, we must assume that the variance for population 1 is equal to the variance for population 2, i.e., $\sigma_1^2 = \sigma_2^2$. This assumption is referred to as *homogeneity of variance*.

In testing the null hypothesis H_0: $\mu_1 = \mu_2$, we must first consider the assumption of homogeneity of variance. It is important to note that this assumption concerns population variances, not sample variances. Because of random sampling fluctuation, we would not expect the sample variances to be exactly equal, even if the population variances were equal.

The effect of violating this assumption depends primarily upon the sizes of the two samples. If the two samples are of equal size, i.e., $n_1 = n_2$, then the effect of the violation of the homogeneity assumption is not serious. However, if $n_1 \neq n_2$ and $\sigma_1^2 \neq \sigma_2^2$, an alternative procedure is used in testing the null hypothesis H_0: $\mu_1 = \mu_2$; this procedure is discussed later in the chapter.

Testing H_0: $\mu_1 = \mu_2$ for Independent Samples When $\sigma_1^2 = \sigma_2^2$

Suppose we are interested in the difference in job satisfaction between middle-aged men who graduated from college and those who ended their formal education after receiving a high school diploma. We call those who graduated from college population 1, and those who received only a high school diploma population 2.

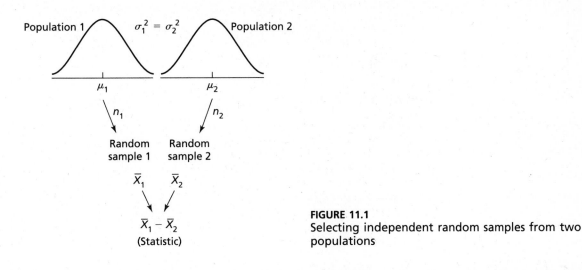

FIGURE 11.1
Selecting independent random samples from two populations

Under the assumption of homogeneity of variance, we assume that the variance of population 1 (σ_1^2) is equal to the variance of population 2 (σ_2^2) (see Figure 11.1). Further, under the assumption of independence, we assume that selecting a subject from population 1 is independent of selecting any subject from population 2. This does not mean, however, that the distribution of scores for these populations cannot overlap. (In Figure 11.1, the distributions are separated simply to set them off.) Indeed, if H_0 is true (that is, if $\mu_1 = \mu_2$), there will be considerable overlap.

Suppose we select a sample of size n_1 from the first population and a sample of size n_2 from the second population. (It is not necessary that $n_1 = n_2$.) Let's assume that one sample has been randomly selected from each of the two populations, as illustrated in Figure 11.1. For both samples, we can compute the mean and then determine the difference between the two sample means ($\overline{X}_1 - \overline{X}_2$). We can repeat this procedure for a second pair of means, and then a third pair, and so on, using samples of size n_1 and n_2, respectively. If we continue this process indefinitely, we will generate a distribution of differences between sample means. This distribution is the *sampling distribution of differences between two sample means.*

Sampling Distribution of Differences

The sampling distribution of the differences is analogous to the sampling distribution of the mean discussed in Chapters 7 and 8; it, too, can be defined by using mathematical theorems. For the two-sample case, the mathematical theorems give us the following:

As the sample sizes for populations 1 and 2 (n_1 and n_2, respectively) increase, the sampling distribution of the differences between the sample means

$(\overline{X}_1 - \overline{X}_2)$ of random samples of size n_1 and n_2—taken from populations with means equal to μ_1 and μ_2, respectively, and with finite variances equal to σ^2—has the following properties:

1. *Shape*: The distribution of differences between sample means approaches a normal distribution.

2. *Central tendency*: The mean of the distribution of differences of sample means equals $\mu_1 - \mu_2$.

3. *Variability*: The standard deviation of the distribution of differences of sample means—called the **standard error of the difference between means**—equals:

$$\sigma_{\overline{X}_1 - \overline{X}_2} = \sqrt{\sigma^2 \left(\frac{1}{n_1} + \frac{1}{n_2} \right)}$$

Notice that the standard error of differences for this sampling distribution is based on the assumption of homogeneity of variance $(\sigma_1^2 = \sigma_2^2 = \sigma^2)$. If there is a substantial difference between σ_1^2 and σ_2^2, procedural adjustments are required; these will be discussed later in this chapter.

Usually, because we do not know σ^2, we must use an estimate of the population variance. Earlier in Chapter 8, for the one-sample case in which σ^2 was unknown and s^2 was used as the unbiased estimate, the standard error of the mean, $s_{\overline{X}}$ was estimated by $s_{\overline{X}} = s/\sqrt{n}$. For the two-sample case, s_1^2 is the unbiased estimate of σ_1^2, and s_2^2 is the unbiased estimate of σ_2^2. Under the assumption of homogeneity of variance, σ^2 is estimated by s^2; correspondingly, the formula for the estimated standard error of the difference between two means is as follows:

$$s_{\overline{X}_1 - \overline{X}_2} = \sqrt{s^2 \left(\frac{1}{n_1} + \frac{1}{n_2} \right)} \tag{11.1}$$

In testing the null hypothesis $H_0: \mu_1 = \mu_2$, the standard error of the difference between sample means is defined as the standard deviation of the sampling distribution of the difference. The formula for the estimated standard error of the difference is:

$$s_{\overline{X}_1 - \overline{X}_2} = \sqrt{s^2 \left(\frac{1}{n_1} + \frac{1}{n_2} \right)}$$

Pooled estimate of variance　　In formula 11.1, s^2 is called the **pooled estimate of the population variance**. A pooled estimate of variance is obtained by finding the sums of squared deviations around the respective sample means, adding these sums, and then dividing the total by the sum of the degrees of freedom in the two samples. In deviation form, the pooled estimate of population variance is given by the formula:

$$s^2 = \frac{\Sigma(X_{i1} - \overline{X}_1)^2 + \Sigma(X_{i2} - \overline{X}_2)^2}{n_1 + n_2 - 2} \tag{11.2}$$

For the two-sample case, it is necessary to use notation with double subscripts. For example, X_{31} refers to the score for the third subject in the first group, and X_{72} refers to the score for the seventh subject in the second group. The summation notation ΣX_{i1} refers to the summing of all the scores for the subjects in the first group.

Formula 11.2 reflects the fact that the sum of squared deviations around the individual group means (SS_1 and SS_2) are added together and divided by the degrees of freedom associated with this estimate of the population variance. For the one-sample case discussed earlier, s^2 was the estimate of σ^2 and there were $n - 1$ degrees of freedom associated with the estimate. For the two-sample case, the two samples have $(n_1 - 1)$ and $(n_2 - 1)$ degrees of freedom, respectively, so the pooled estimate of population variance has $(n_1 - 1) + (n_2 - 1) = n_1 + n_2 - 2$ degrees of freedom associated with it. Thus, formula 11.2 could be expressed as:

$$s^2 = \frac{SS_1 + SS_2}{n_1 + n_2 - 2} \tag{11.3}$$

Formula 11.2 is analogous to the deviation score formula for the variance (see Chapter 3); it is the deviation score formula for the pooled estimate of the population variance. As we indicated in Chapter 3, when the mean is not an integer and the number of cases is large, the deviation score formula is difficult to apply. The preferred alternative is the corresponding raw score formula. This formula can be derived from the deviation score formula by substitution in a manner similar to that used in Chapter 3; the raw score formula for s^2 is

$$s^2 = \frac{\left[\Sigma X_{i1}^2 - (\Sigma X_{i1})^2/n_1\right] + \left[\Sigma X_{i2}^2 - (\Sigma X_{i2})^2/n_2\right]}{n_1 + n_2 - 2} \tag{11.4}$$

Notice that the numerator in formula 11.4 is just the sums of squares for the two groups added together; the denominator is the sum of the degrees of freedom:

$$s^2 = \frac{SS_1 + SS_2}{n_1 + n_2 - 2}$$

TABLE 11.1
Data for Illustrating the Equivalence of Formulas 11.2, 11.4, and 11.5 for Computing the Pooled Estimate of the Population Variance (s^2)

Sample 1	Sample 2
7	12
8	14
10	18
4	13
6	11
	10
$n_1 = 5$	$n_2 = 6$
$\Sigma X_1 = 35$	$\Sigma X_2 = 78$
$\overline{X}_1 = 7.0$	$\overline{X}_2 = 13.0$
$\Sigma X_1^2 = 265$	$\Sigma X_2^2 = 1{,}054$
$\Sigma(X_{i1} - \overline{X}_1)^2 = 20.0$	$\Sigma(X_{i2} - \overline{X}_2)^2 = 40.0$
$s_1^2 = 5.0$	$s_2^2 = 8.0$
$s_1 = 2.236$	$s_2 = 2.828$

To show the equivalence of formulas 11.2 and 11.4, consider the data in Table 11.1. Using formula 11.2 with these data, we find

$$s^2 = \frac{20.0 + 40.0}{6 + 5 - 2}$$

$$= \frac{60.0}{9}$$

$$= 6.67$$

Using formula 11.4 for the same data, we find

$$s^2 = \frac{[265 - (35)^2/5] + [1{,}054 - (78)^2/6]}{6 + 5 - 2}$$

$$= \frac{20.0 + 40.0}{6 + 5 - 2}$$

$$= \frac{60.0}{9}$$

$$= 6.67$$

When the data to be analyzed are given as raw scores, formula 11.4 should be used to calculate s^2. However, when only the values of the variances of the two samples are given (s_1^2 and s_2^2), formula 11.4 reduces to the following:

$$s^2 = \frac{(n_1 - 1)s_1^2 + (n_2 - 1)s_2^2}{n_1 + n_2 - 2} \tag{11.5}$$

Using formula 11.5 with the data from Table 11.1, we can show the equivalence of the three formulas for the pooled estimate of the population variance (s^2):

$$s^2 = \frac{(5 - 1)5.0 + (6 - 1)8.0}{6 + 5 - 2}$$

$$= \frac{20.0 + 40.0}{6 + 5 - 2}$$

$$= \frac{60.0}{9}$$

$$= 6.67$$

The pooled estimate of the population variance is defined as

$$s^2 = \frac{SS_1 + SS_2}{n_1 + n_2 - 2}$$

Three formulas can be used for computing s^2; they are:

1. Formula 11.2 for deviation scores.
2. Formula 11.4 for raw scores.
3. Formula 11.5 when using individual group variances.

Example

Suppose an educator is interested in the difference in mathematics achievement of third-grade students who are taught under two conditions. The first condition is a highly structured environment in which a fixed time is given for the teaching and learning of mathematical concepts. The second condition is a less structured environment in which the same concepts are presented but with no set time for completion of the teaching-learning process. We will call the group assigned to the highly structured environment population 1 and the group assigned to the less structured environment population 2. At the end of six weeks, a random sample of each group of students takes an achievement test. The sample for the structured, fixed-time condition consists of nine students, and the other sample of eight students. The focus of interest is whether or not the structured, fixed-time condition produces lower mathematics-achievement scores.

Step 1: State the Hypotheses

The focus of interest implies a direction, so the null hypothesis[2] is tested against a directional alternative.

$$H_0: \mu_1 - \mu_2 = 0$$

$$H_a: \mu_1 - \mu_2 < 0$$

The level of significance for testing this hypothesis is set at .05.

Step 2: Set the Criterion for Rejecting H_0

For this example, we have two independent, randomly selected samples from the populations, as illustrated in Figure 11.1. The sample mean ($\overline{X}_1$) from population 1 reflects the mean of population 1 (μ_1) and sampling error, whereas the sample mean ($\overline{X}_2$) from population 2 reflects the mean of population 2 (μ_2) and sampling error. We also assume homogeneity of variance, that is, $\sigma_1^2 = \sigma_2^2$.

The sampling distribution for the difference between two means is the normal distribution when σ_1^2 and σ_2^2 are known. But when σ_1^2 and σ_2^2 are estimated using s_1^2 and s_2^2—and the common variance of the populations, σ^2, is estimated using the pooled estimate s^2—the appropriate distribution is Student's t distribution. The specific t distribution to be used depends on the degrees of freedom associated with the estimate of the variance. For the two-sample case, the degrees of freedom are $(n_1 - 1) + (n_2 - 1) = n_1 + n_2 - 2$. Thus, when testing $H_0: \mu_1 = \mu_2$, the t distribution with $n_1 + n_2 - 2$ degrees of freedom is the appropriate sampling distribution.

In this example, the null hypothesis is tested against the directional alternative at the .05 level of significance. The critical value of the test statistic is found in Table C.3. Using the t distribution with $n_1 + n_2 - 2 = 9 + 8 - 2 = 15$ degrees of freedom, we see that the critical value (t_{cv}) is -1.753.

---

[2]As we indicated in a footnote in Chapter 8, some authors write the null hypothesis for this example as $H_0: \mu_1 - \mu_2 \geq 0$, using the rationale that the null and alternative hypotheses must include all possibilities. However, the specific null hypothesis tested is $H_0: \mu_1 - \mu_2 = 0$ since the point 0 is used as the mean of the sampling distribution. For a more thorough discussion, see R. E. Kirk, *Experimental Design: Procedures for the Behavioral Sciences*, 2d. ed. (Monterey, Calif: Brooks/Cole, 1982), pp. 29–30.

Step 3: Compute the Test Statistic

The data on the achievement test taken by a random sample of students in each of the two groups are as follows:

Group 1 (Structured—Fixed time)	Group 2 (Less structured—Flexible time)
35	52
51	87
66	76
42	62
37	81
46	71
60	55
55	67
53	

$n_1 = 9$ $n_2 = 8$

$\Sigma X_{i1} = 445$ $\Sigma X_{i2} = 551$

$\overline{X}_1 = 49.44$ $\overline{X}_2 = 68.88$

$\Sigma X_{i1}^2 = 22{,}865$ $\Sigma X_{i2}^2 = 39{,}009$

$s_1^2 = 107.78$ $s_2^2 = 151.27$

$s_1 = 10.38$ $s_2 = 12.30$

For this example, the hypothesized value of the population parameter is the difference between the two population means $(\mu_1 - \mu_2) = 0$. The corresponding statistic is the difference between the two sample means $(\overline{X}_1 - \overline{X}_2) = (49.44 - 68.88) = -19.44$. Notice that the order of subtraction is the same for both; $\overline{X}_2$ is subtracted from $\overline{X}_1$, since μ_2 is subtracted from μ_1. This order is especially important when the null hypothesis is tested against a directional alternative.

The standard error of the statistic was defined as the standard deviation of the distribution of differences. For this example, since the variance in the population is unknown, we need to use the estimated standard error of the difference (formula 11.1). The first step, however, is to compute s^2, the pooled estimate of σ^2.

Using formula 11.4, we compute s^2 as follows:

$$s^2 = \frac{\left[\Sigma X_{i1}^2 - (\Sigma X_{i1})^2/n_1\right] + \left[\Sigma X_{i2}^2 - (\Sigma X_{i2})^2/n_2\right]}{n_1 + n_2 - 2}$$

$$= \frac{[22{,}865 - (445)^2/9] + [39{,}009 - (551)^2/8]}{9 + 8 - 2}$$

$$= \frac{862.22 + 1{,}058.88}{9 + 8 - 2}$$

$$= 128.07$$

Now we can calculate the estimated standard error of the difference of the sample means using formula 11.1.

$$s_{\overline{X}_1 - \overline{X}_2} = \sqrt{s^2 \left(\frac{1}{n_1} + \frac{1}{n_2} \right)}$$

$$= \sqrt{128.07 \left(\frac{1}{9} + \frac{1}{8} \right)}$$

$$= 5.50$$

As in the one-sample case, with knowledge of the hypothesized parameter $(\mu_1 - \mu_2)$, the corresponding statistic $(\overline{X}_1 - \overline{X}_2)$, the sampling distribution of the difference, and the standard error of the difference, we now apply the general formula for the test statistic.

$$\text{Test statistic} = \frac{\text{Statistic} - \text{Parameter}}{\text{Standard Error of the Statistic}}$$

For the null hypothesis $(H_0: \mu_1 = \mu_2)$, the test statistic is

$$t = \frac{(\overline{X}_1 - \overline{X}_2) - (\mu_1 - \mu_2)}{s_{\overline{X}_1 - \overline{X}_2}} \tag{11.6}$$

Using the data from the mathematics-achievement example,

$$t = \frac{-19.44 - 0}{5.50}$$

$$= -3.53$$

Thus, because the observed difference of -19.44 is 3.53 standard errors *below* the hypothesized value of the parameter $(\mu_1 - \mu_2 = 0)$ and because -3.53 exceeds -1.753 (in absolute value), the null hypothesis is rejected.

Step 4: Construct the Confidence Interval

Since the null hypothesis is rejected, the educator might want to develop a *confidence interval* around the observed difference in the sample means. A confidence interval for the two-sample case is developed in exactly the same way as for the one-sample case. The general formula for the confidence interval is

$$\text{CI} = \text{statistic} \pm (\text{critical value})(\text{standard error of the statistic})$$

For this example, the statistic is the difference between the two sample means $(\overline{X}_1 - \overline{X}_2)$, and the standard error of the statistic $(s_{\overline{X}_1 - \overline{X}_2})$ is defined by formula 11.1. Thus, the formula for computing the confidence interval for the difference between two means is

$$CI_{95} = (\overline{X}_1 - \overline{X}_2) \pm \text{cv}(s_{\overline{X}_1 - \overline{X}_2}) \tag{11.7}$$

Now consider the critical value for the confidence interval. Because the level of significance (α) was set at .05, the corresponding level of confidence is $1 - \alpha$, or .95. However, when a one-tailed test of a hypothesis is conducted, the critical values for hypothesis testing and constructing the confidence interval are *not* the same. This is so because the confidence interval is constructed *symmetrically* around the statistic, whereas the region of rejection for the one-tailed test is located in only one tail of the sampling distribution. Thus, for this example, in which the sampling distribution is the t distribution for 15 degrees of freedom, we see from Table C.3 that the critical value for constructing CI_{95} is 2.131. Using formula 11.7,

$$CI_{95} = -19.44 \pm (2.131)(5.50)$$

$$= -19.44 \pm 11.72$$

$$= (-31.16, -7.72)$$

Thus, we can say with 95 percent confidence that the difference between the two means could be as small as 7.72 points or as large as 31.16 points.

Notice that this interval does *not* contain the point zero, which was the hypothesized difference between the two population means ($\mu_1 - \mu_2 = 0$). This result illustrates the relationship between testing hypotheses and constructing confidence intervals. That is, if the null hypothesis is rejected at a specific α level, the corresponding confidence interval will *not* contain the hypothesized value of the parameter.[3]

Step 5: Interpret the Results

In this example, the educator tests the null hypothesis (H_0: $\mu_1 - \mu_2 = 0$) and then constructs the 95-percent confidence interval. The null hypothesis is tested against the alternative hypothesis (H_a: $\mu_1 - \mu_2 < 0$). Since the observed value of $t(-3.53)$ exceeds the critical value (-1.753) in absolute value in this one-tailed test, the null hypothesis is rejected in favor of the directional alternative hypothesis. The educator concludes that the structured, fixed-time condition is less effective for teaching third-grade mathematics than the less structured, flexible-time condition.

The associated probability statement for the statistical test is tied to the null hypothesis. The probability that the observed difference between the sample means ($\overline{X}_1 - \overline{X}_2 = -19.44$) would have occurred by chance, if in fact the null hypothesis

---

[3] *A note of caution:* We have used different critical values for testing the null hypothesis and developing the confidence interval. If the observed t in the test of the hypothesis is between -2.131 and -1.753, the confidence interval will contain 0. Thus, we must be cautious in developing confidence intervals using the general formula when we have used a one-tailed test of H_0.

$(H_0: \mu_1 - \mu_2 = 0)$ were true, is less than .05.[4] In other words, the discrepancy between the difference between the sample means and the hypothesized value of no difference is too great to attribute to sampling error. Since the probability of making a Type I error is less than .05, the educator is willing to accept the consequences of making such an error and rejects the null hypothesis.

The 95-percent confidence interval for this example is $(-31.16, -7.72)$; in other words, the educator can be 95 percent confident that this interval contains the difference between μ_1 and μ_2. (Notice that zero is not in the interval.) Thus, we now have an estimate of the magnitude of the difference between the population means. The size of this difference should have implications regarding the relative effectiveness of the two methods.

For testing $H_0: \mu_1 - \mu_2$ when σ^2 is unknown, the general formulas for the test statistic and for developing the confidence interval are

$$t = \frac{(\overline{X}_1 - \overline{X}_2) - (\mu_1 - \mu_2)}{s_{\overline{X}_1 - \overline{X}_2}}$$

$$\text{CI} = (\overline{X}_1 - \overline{X}_2) \pm \text{cv}(s_{\overline{X}_1 - \overline{X}_2})$$

The t distribution with $n_1 + n_2 - 2$ degrees of freedom is the appropriate sampling distribution for determining the critical values of the test statistic and for constructing the confidence interval.

Testing $\mu_1 = \mu_2$ for Independent Samples When $\sigma_1^2 \neq \sigma_2^2$

In the previous examples, we used the central limit theorem to develop the sampling distribution of the difference between sample means and the procedure for estimating the standard error of the difference; this sampling distribution was based upon the assumption of homogeneity of variance. As we mentioned earlier in the chapter, the effect of violating this assumption in testing the null hypothesis $H_0: \mu_1 = \mu_2$ can be quite serious. When $n_1 = n_2$, the violation of the assumption has been shown to be unimportant. However, if $n_1 \neq n_2$, the violation of the assumption can seriously affect the result of the statistical test. In such a case, adjustments must be made to the procedures for estimating the standard error of the difference

---

[4] As we indicated in Chapter 8, we use this wording in an effort to shorten the statement of probability and to focus on the specific value of the sample statistic. The probability is actually less than .05 that a difference of -19.44 or greater (in absolute value) would occur by chance if in fact the null hypothesis were true. For a more detailed discussion of the probability statement, see Paul A. Games and George R. Klare, *Elementary Statistics: Data Analysis for the Behavioral Sciences* (New York: McGraw-Hill, 1967), pp. 289–295.

and to the degrees of freedom used in the statistical test. Before discussing these procedures we will illustrate the test of the assumption of homogeneity of variance.

Test for the Homogeneity of Variance Assumption

The test for the homogeneity of variance assumption tests the null hypothesis $H_0: \sigma_1^2 = \sigma_2^2$, i.e., that the variances in the populations from which the samples were selected are equal. When this hypothesis is *not* rejected, homogeneity of variance is confirmed. Rejecting this null hypothesis means that the homogeneity assumption is not tenable and adjustments are necessary for testing the null hypothesis $H_0: \mu_1 = \mu_2$.

If we hypothesize that the two population variances are equal, it is equivalent to hypothesizing that the ratio of the variances equals 1.00. Therefore, the null and alternative hypotheses for testing the assumption of homogeneity of variance can be expressed as follows:

$$H_0: \sigma_1^2 = \sigma_2^2 \quad \text{or} \quad H_0: \frac{\sigma_1^2}{\sigma_2^2} = 1$$

$$H_a: \sigma_1^2 \neq \sigma_2^2 \quad \text{or} \quad H_a: \frac{\sigma_1^2}{\sigma_2^2} \neq 1$$

The corresponding test statistic, called the **F ratio**, is the ratio of the two sample variances, i.e.,

$$F = \frac{s_1^2}{s_2^2} \tag{11.8}$$

The sampling distribution for this test statistic cannot be approximated by the normal distribution or Student's t distributions. Therefore, a different sampling distribution must be used. It is called the **F distribution**[5], named for R. A. Fisher. Like the t distribution, the F distribution is a family of distributions. The specific F distribution for testing $H_0: \sigma_1^2 = \sigma_2^2$ is determined by two degrees of freedom values, one associated with each of the variance estimates for the two populations. Unlike the normal and t distributions, the F distributions are not symmetrical, and go from zero to $+\infty$ rather than from $-\infty$ to $+\infty$. In addition, the shapes of the F distributions vary drastically, especially when the degrees of freedom values are small. These characteristics make the determination of critical values for the F distribution (F_{cv}) more complicated than for the normal and t distributions.

The critical values for the test statistic using the F distribution (F_{cv}) are found in Table C.5 in the Appendix. The table is organized as follows. To identify the specific F distribution, it is necessary to determine the degrees of freedom associated with the sample variance in the numerator of the F ratio (n_1) and the degrees of freedom associated with the sample variance in the denominator (n_2). For the

---

[5]The F distribution is also used in Chapter 14 in the analysis of variance (ANOVA).

purpose of testing the assumption of homogeneity of variance, we recommend that the larger sample variance be placed in the numerator and the smaller variance in the denominator. By doing so, the F ratio will always be greater than 1.00; this will simplify the determination of the critical value (F_{cv}) from Table C.5. As we can see from Table C.5, only four percentile points are given for each F distribution: those beyond which lie 25, 10, 5 and 1 percent of the area in the right-hand tail. It is important to note that critical values for these percentage points are in the *right-hand tail only*—thus the name of the table, "Upper Percentage Points."

Consider an example. Suppose we are testing the null hypothesis H_0: $\mu_1 = \mu_2$. Before we conduct our test, we want to test the assumption of homogeneity of variance; that is, we want to test the null hypothesis

$$H_0:\ \sigma_1^2 = \sigma_2^2 \qquad \text{or} \qquad H_0:\ \frac{\sigma_1^2}{\sigma_2^2} = 1$$

against the nondirectional alternative hypothesis

$$H_a:\ \sigma_1^2 \neq \sigma_2^2 \qquad \text{or} \qquad H_a:\ \frac{\sigma_1^2}{\sigma_2^2} \neq 1$$

The data for this example are

$$n_1 = 31 \qquad\qquad n_2 = 41$$
$$s_1^2 = 105.96 \qquad s_2^2 = 36.42$$

For this test, we will set the level of significance (α) at .02. The test statistic (F) for these data is computed as follows, using formula 11.8.

$$F = \frac{s_1^2}{s_2^2}$$
$$= \frac{105.96}{36.42}$$
$$= 2.91$$

The critical value for this test statistic is found in Table C.5. The degrees of freedom associated with the F ratio are $n_1 - 1 = 30$ and $n_2 - 1 = 40$. Note that the null hypothesis is tested against a nondirectional alternative hypothesis. Therefore, we will locate the region of rejection in both tails of the sampling distribution. For this example, we would have 1 percent of the area in each of the tails of the F distribution. Consulting Table C.5, we look across the table until we come to the column for 30 degrees of freedom (df for the numerator); then we go down the column until we come to the row for 40 degrees of freedom (df for the denominator). The entry in the table for .01 is 2.20, which is the critical value (F_{cv}). Since the observed value of the test statistic ($F = 2.91$) exceeds the critical value ($F_{cv} = 2.20$), we reject the null hypothesis. We conclude that the assumption of homogeneity of variance is *not* tenable and that we need to use alternative procedures for determining the standard error of the difference in means for testing the null hypothesis H_0: $\mu_1 = \mu_2$. The procedures are outlined in the examples that follow this discussion of the homogeneity of variance assumption.

Note that we specified that the larger sample variance should go into the numerator of the F ratio. As we indicated, this strategy will insure that the F ratio exceeds 1.00. However, it is possible to have the larger sample variance in the denominator and thus to have an F ratio less than 1.00. In this case we would use the region of rejection in the left-hand tail of the F distribution. Consider the previous example and suppose that we placed the smaller variance in the numerator, i.e.,

$$F = \frac{36.42}{105.96}$$

$$= 0.34$$

The critical value from Table C.5 would be determined by looking across to the column for 40 degrees of freedom (df for the numerator) and down to the column for 30 degrees of freedom (df for the denominator). The value from the table is 2.30. The actual critical value is the reciprocal of this value, $1/2.30 = 0.44$. Note that the observed value of the test statistic ($F = 0.34$) is less than (more extreme than) the critical value ($F_{cv} = 0.44$). Therefore, as above, the null hypothesis would be rejected and the homogeneity of variance assumption would not be met.

The latter procedure for determining the critical value from the F distribution is more cumbersome than the former. Thus, we recommend that the test of the homogeneity of variance assumption be conducted by computing the F ratio with the larger sample variance in the numerator. However, caution is recommended when establishing the region of rejection and the level of significance when conducting a two-tailed test of this null hypothesis.

The Standard Error of the Difference Between the Means When $\sigma_1^2 \neq \sigma_2^2$

When the assumption of homogeneity of variance is not tenable, an alternative procedure for testing H_0: $\mu_1 = \mu_2$ is used in which (1) the standard error of difference is estimated differently and (2) degrees of freedom used to test the hypothesis are adjusted. Rather than using the pooled estimate of the population variance in the formula for the estimated standard error, Cochran and Cox[6] suggest the following formula:

$$s_{\bar{X}_1 - \bar{X}_2} = \sqrt{\frac{s_1^2}{n_1} + \frac{s_2^2}{n_2}} \tag{11.9}$$

$$= \sqrt{s_{\bar{X}_1}^2 + s_{\bar{X}_2}^2}$$

[6]W. G. Cochran and G. M. Cox, *Experimental Designs* (New York: John Wiley & Sons, 1957), pp. 79–80.

The degrees of freedom are computed using the following formula, suggested by Satterthwaite.[7]

$$df = \frac{(s_1^2/n_1 + s_2^2/n_2)^2}{(s_1^2/n_1)^2/(n_1 - 1) + (s_2^2/n_2)^2/(n_2 - 1)}$$

$$= \frac{(s_{\overline{X}_1}^2 + s_{\overline{X}_2}^2)^2}{(s_{\overline{X}_1}^2)^2/(n_1 - 1) + (s_{\overline{X}_2}^2)^2/(n_2 - 1)} \tag{11.10}$$

An Example

Consider the following data for testing the null hypothesis $H_0: \mu_1 = \mu_2$ against the nondirectional alternative $H_a: \mu_1 \neq \mu_2$ when $\sigma_1^2 \neq \sigma_2^2$:

	Sample 1	*Sample 2*
$\overline{X}$	27.20	19.60
n	16	31
s^2	145.90	62.60
s	12.07	7.09

For these data, the first step would be to test for the assumption of homogeneity of variance, i.e., we would test the null hypothesis $H_0: \sigma_1^2 = \sigma_2^2$ using the F ratio as the test statistic. For testing both the hypothesis about the difference between the means and the hypothesis about the homogeneity of variance assumption, we will use $\alpha = .10$.

$$F = \frac{s_1^2}{s_2^2}$$

$$= \frac{145.90}{62.60}$$

$$= 2.33$$

The degrees of freedom associated with this test statistic are $n_1 - 1 = 16 - 1 = 15$ for the variance in the numerator and $n_2 - 1 = 31 - 1 = 30$ for the denominator. Referring to Table C.5 for 15 and 30 degrees of freedom, we find the critical value of the test statistic to be 2.01. Since the observed value of the test statistic ($F = 2.33$) exceeds the critical value ($F_{cv} = 2.01$), we conclude that the assumption of homogeneity of variance is *not* tenable and that we need to use alternative procedures for determining the standard error of the difference in means and the degrees of freedom associated with the test of $H_0: \mu_1 = \mu_2$ against $H_a: \mu_1 \neq \mu_2$.

---

[7]F. W. Satterthwaite, "An Approximate Distribution of Estimates of Variance Components," *Biometrics Bulletin 2* (1946): 110–114.

Using formula 11.9 for computing the standard error of the difference, we find

$$s_{\overline{X}_1 - \overline{X}_2} = \sqrt{\frac{145.90}{16} + \frac{62.60}{31}}$$

$$= \sqrt{9.12 + 2.02}$$

$$= 3.34$$

Using formula 11.10 for computing the degrees of freedom associated with the critical values of the test statistic, we find

$$df = \frac{(9.12 + 2.02)^2}{\dfrac{(9.12)^2}{15} + \dfrac{(2.02)^2}{30}}$$

$$= \frac{124.10}{5.54 + 0.14}$$

$$= 21.85$$

The value of df will seldom be a whole number; therefore, we must round the value to the nearest whole number in order to find the critical values for the test statistic. For a two-tailed test at $\alpha = .10$, the critical values for 22 degrees of freedom (see Table C.3) are ± 1.717.

Computing the test statistic using the general formula and the standard error of the difference computed above, we find

$$t = \frac{(\overline{X}_1 - \overline{X}_2) - (\mu_1 - \mu_2)}{s_{\overline{X}_1 - \overline{X}_2}}$$

$$= \frac{(27.2 - 19.6) - (0)}{3.34}$$

$$= 2.28$$

Therefore, by using the Cochran and Cox/Satterthwaite procedures, we find that the observed value of the test statistic ($t = 2.28$) exceeds the critical value ($t_{cv} = 1.717$). Thus we would reject the null hypothesis.

When testing the null hypothesis H_0: $\mu_1 = \mu_2$ when $n_1 \neq n_2$ and $\sigma_1^2 \neq \sigma_2^2$, the standard error of the difference is defined as follows:

$$s_{\overline{X}_1 - \overline{X}_2} = \sqrt{\frac{s_1^2}{n_1} + \frac{s_2^2}{n_2}}$$

In addition to using this standard error of the difference in the denominator of the general formula for the test statistic, an adjustment should be made to the degrees of freedom.

TABLE 11.2
**Computer Solution for the Satisfaction-of-Marital-Status Example, Testing the
Hypothesis H_0: $\mu_1 - \mu_2 = 0$**

```
t-tests for independent samples of MARITAL

GROUP 1 - MARITAL EQ 1: SINGLE
GROUP 2 - MARITAL EQ 2: MARRIED
```

Variable	Number of Cases	Mean	Standard Deviation	Standard Error	F Value	2-tail Prob.
GROUP 1	61	2.6557	.602	.077		
					1.70	.009
GROUP 2	161	2.7516	.461	.036		

Pooled Variance Estimate			Separate Variance Estimate		
t Value	Degrees of Freedom	2-tail Prob.	t Value	Degrees of Freedom	2-tail Prob.
-1.27	220	.207	-1.12	88.01	.264

A Computer Example

Let us consider a study that focused on whether there was a difference between
single and married parents in terms of satisfaction with their marital status. Random
samples of 61 single parents and 161 married parents were selected to participate
in the study and given a marital status satisfaction scale. On this scale, the higher
the score, the greater the satisfaction was. The data for this study are found in
the SPSS printout in Table 11.2. Note that this table contains the means, standard
deviations, and standard errors for the two samples, along with other pertinent
statistics.

Step 1: State the Hypotheses

There is no reason to assume that one population is more satisfied than the other;
therefore, the null hypothesis is tested against the nondirectional alternative hypothesis.

$$H_0: \mu_1 = \mu_2$$
$$H_a: \mu_1 \neq \mu_2$$

The level of significance for testing this hypothesis is set, a priori, at $\alpha = .05$.

Before testing the above hypothesis, we must first test for the assumption of homogeneity of variance by testing the following null hypothesis against the nondirectional alternative.

$$H_0: \sigma_1^2 = \sigma_2^2 \quad \text{or} \quad H_0: \frac{\sigma_1^2}{\sigma_2^2} = 1$$

$$H_a: \sigma_1^2 \neq \sigma_2^2 \quad \text{or} \quad H_a: \frac{\sigma_1^2}{\sigma_2^2} \neq 1$$

The test statistic for this hypothesis is the F ratio. In Table 11.2, the printout contains entries for "F value" and "2-tail prob" (two-tailed probability). The F value is simply the F ratio of the two sample variances; the two sample variances are computed by squaring the two "Standard Deviation" values found in Table 11.2.

$$F = \frac{(.602)^2}{(.461)^2}$$

$$= 1.70$$

The two-tailed probability ($p = .009$) is the actual probability of finding an F ratio of 1.70 by chance if the null hypothesis were true. Since this probability is less than the probability set, a priori, as the level of significance (α), the null hypothesis would be rejected and we would conclude that the assumption of homogeneity of variance was *not* met.

Step 2: Set the Criterion for Rejecting H_0

Note that the computer printout has two t values, one for "pooled variance estimates" and one for "separate variance estimates." The t value for the pooled variance estimate is used when the assumption of homogeneity of variance is met. For this t value, the standard error of the difference in means would be computed using formula 11.1, where s^2 is the pooled estimate of the population variance (see formulas 11.2, 11.3, 11.4 and 11.5). The degrees of freedom associated with this pooled estimate are $n_1 + n_2 - 2 = 61 + 161 - 2 = 220$. Thus, the critical value (t_{cv}) for testing $H_0: \mu_1 = \mu_2$ against $H_a: \mu_1 \neq \mu_2$ at $\alpha = .05$ under the assumption of homogeneity of variance (see Table C.3) is ± 1.96.

The t value for the separate variance estimates is used when the assumption of homogeneity of variance is *not* met. For this case, the standard error of the difference in means would be computed using the Cochran and Cox procedure (formula 11.9) and the degrees of freedom would be computed using the Satterthwaite procedure (formula 11.10). Note that the computer printout shows 88.01 degrees of freedom for this test. Thus, from Table C.3, the t_{cv} is ± 1.99 for the two-tailed test at $\alpha = .05$.

Step 3: Compute the Test Statistic

The test statistic for testing the null hypothesis is computed using the general formula

$$\text{Test statistic} = \frac{\text{Statistic} - \text{Parameter}}{\text{Standard Error of the Statistic}}$$

For this example, the statistic is the difference between the two sample means, $\overline{X}_1 - \overline{X}_2 = 2.6557 - 2.7516 = -.0959$. The parameter is the hypothesized difference between the population means, $\mu_1 - \mu_2 = 0$. Finally, the standard error of the statistic could be computed using either the pooled variance estimate or the separate variance estimate. However, since the homogeneity of variance assumption was *not* met, only the standard error for the separate variance estimate would be used. For the data in Table 11.2, the standard error of the difference using the Cochran and Cox procedure would be computed as follows:

$$s_{\overline{X}_1 - \overline{X}_2} = \sqrt{(.077)^2 + (.036)^2}$$

$$= \sqrt{.0059 + .0013}$$

$$= \sqrt{.0072}$$

$$= .085$$

Therefore, the test statistic is computed by

$$t = \frac{-.0959 - 0}{.085}$$

$$= -1.12$$

Since the observed value of $t(-1.12)$ does not exceed the critical value (-1.99) in absolute value, the null hypothesis would not be rejected. Note that the "2-tail prob" in Table 11.2 associated with this t value is .264. Just as we compared observed values of t with critical values, we can also compare the actual probability associated with the probability set, a priori, as the level of significance. Since the actual value is larger than α, the null hypothesis would not be rejected.

Step 4: Construct the Confidence Interval

The computer printout does not contain the confidence interval for the difference between means. However, it is possible to construct the interval using the general formula and the data from the above steps.

$$\text{CI} = \text{statistic} \pm (\text{critical value})(\text{standard error})$$

For this example, suppose we wanted to compute the 95 percent confidence interval, i.e., CI_{95}.

$$CI_{95} = -.0959 \pm (1.99)(.085)$$
$$= -.0959 \pm .1692$$
$$= (-.2651, .0733)$$

Note that this confidence interval spans zero, which is consistent with the earlier result of failing to reject the hypothesis of no difference between the population means.

Step 5: Interpret the Results

For this example, the null hypothesis was not rejected. Therefore, it can be concluded that, since the data do not contradict the null hypothesis, the means of the two populations do not differ. Satisfaction with marital status is the same for single parents and married parents. The specific probability statement would be, "the probability that a difference in sample means of $-.0959$ would have occurred by chance if $\mu_1 - \mu_2 = 0$ is greater than .05." From the computer printout, the actual probability of that occurrence is .264.

In practice, because the null hypothesis was not rejected, it would be unlikely that a confidence interval would be constructed. For this example, we constructed this interval to illustrate how the data from the computer printout would be used in the calculations. The confidence statement regarding the interval constructed would be that we are 95 percent confident that the interval from $-.2651$ to $.0733$ spans the difference between the two population means.

Computer Exercise: The Survey of High School Students Data Set

For the survey of high school students data set, consider the differences between female and male students on two of the interval scale measures, VISUAL and MOSAIC. The T-TEST procedure from SPSS was used to generate the computer printouts for these two analyses (see Table 11.3).

Step 1: State the Hypotheses

For each of the analyses, the null hypothesis would be that there is no difference between female and male students on the respective measures. For the alternative hypothesis, there is no reason to assume directionality, i.e., that no one gender

TABLE 11.3
SPSS Printout: *t*-tests for VISUAL and MOSAIC

```
t-tests for Independent Samples of GENDER      gender


                            Number
Variable                    of Cases     Mean        SD    SE of Mean
-----------------------------------------------------------------------
VISUAL   visualization score

female                        280       5.1589     3.450      .206
male                          220       6.3818     4.291      .289
-----------------------------------------------------------------------


        Mean Difference = -1.2229



        t-test for Equality of Means                            95%
Variances   t-value     df    2-Tail Sig    SE of Diff      CI for Diff
-----------------------------------------------------------------------
Equal        -3.53     498       .000          .346      (-1.903, -.543)
Unequal      -3.44   414.06      .001          .355      (-1.921, -.525)
-----------------------------------------------------------------------



                            Number
Variable                    of Cases     Mean        SD    SE of Mean
-----------------------------------------------------------------------
MOSAIC   pattern test score

female                        280      27.2446     9.375      .560
male                          220      26.4364     9.178      .619
-----------------------------------------------------------------------


        Mean Difference = .8083



        t-test for Equality of Means
Variances   t-value     df    2-Tail Sig    SE of Diff      CI for Diff
-----------------------------------------------------------------------
Equal         .97      498       .335          .837      (-.836, 2.453)
Unequal       .97    474.79      .333          .835      (-.832, 2.449)
-----------------------------------------------------------------------
```

would score higher than the other on the measures. Thus, the null and alternative hypotheses are:

$$H_0: \mu_F = \mu_M$$

$$H_a: \mu_F \neq \mu_M$$

The level of significance set a priori for each of these analyses is $\alpha = .05$.

As was the case in the previous examples, the first step in these analyses is to test the assumption of homogeneity of variance. For the variable, VISUAL, the F value from the printout is computed by squaring the two "Standard Deviation" values to obtain the variances for the two samples and then by determining the ratio of these two sample variances.

$$F = \frac{(4.2914)^2}{(3.4497)^2}$$

$$= 1.55$$

Since the two-tail probability equals .0006 and is less than .05, we reject the hypothesis of equal variances for the two samples and thus reject the assumption of homogeneity of variance. Therefore, we must use the Cochran and Cox procedure for computing the estimated standard error (formula 11.9) and the Satterthwaite procedure for determining the degrees of freedom for the t-test (formula 11.10).

For the variable MOSAIC, the F ratio for testing the assumption of homogeneity of variance was $F = 1.04$. Since the probability was .7438 and is greater than .05, the null hypothesis of equal variances was not rejected and the assumption of homogeneity of variance was met. Therefore, the t-test analysis for the variable MOSAIC will use the estimated standard error computed using the pooled estimate of the population variance (formula 11.4).

Step 2: Set the Criterion for Rejecting H_0

Note that the SPSS printouts contain two t values; one for the "Pooled Variance Estimate" and one for "Separate Variance Estimate." When the assumption of homogeneity of variance is met, the t value for the pooled variance estimate must be used; when the assumption is *not* met, the t value for the separate variance estimate must be used. For the variable, VISUAL, the t value for the separate variance estimate must be used and the degrees of freedom (414.1) were computed using formula 11.10. Thus from Table C.3 in the Appendix, the t_{cv} for a two-tailed test with 414.1 degrees of freedom at $\alpha = .05$ are ± 1.96.

For the variable, MOSAIC, the t value for the pooled variance estimate must be used and the degrees of freedom are $n_F + n_M - 2 = 280 + 220 - 2 = 498$. Thus from Table C.3, the t_{cv} for a two-tailed test with 498 degrees at $\alpha = .05$ is ± 1.96.

Step 3: Compute the Test Statistic

For the variable, VISUAL, the test statistic is computed using formula 11.6.

$$t = \frac{(\overline{X}_1 - \overline{X}_2) - (\mu_1 - \mu_2)}{s_{\overline{X}_1 - \overline{X}_2}}$$

Recall that the estimated standard error used in the denominator of the above formula would be computed using the Cochran and Cox procedure. Using the data from the computer printout, the estimated standard error is computed as follows:

$$s_{\overline{X}_1 - \overline{X}_2} = \sqrt{s_{\overline{X}_1}^2 + s_{\overline{X}_2}^2}$$
$$= \sqrt{(.206)^2 + (.289)^2}$$
$$= \sqrt{0.0424 + 0.0835}$$
$$= \sqrt{0.1259}$$
$$= 0.355$$

The test statistic is

$$t = \frac{(5.1589 - 6.3818) - (0)}{0.355}$$
$$= \frac{-1.2229 - 0}{0.355}$$
$$= -3.44$$

Since the observed t value (-3.44) exceeds the critical t value (± 1.96), the null hypotheses would be rejected. Note that the "2-tail Prob" in the computer printout associated with this t value is .0006. Since this value is less than the level of significance set a priori, this also indicates that the null hypothesis is rejected.

For the variable, MOSAIC, the test statistic is also computed using formula 11.6; however, the estimated standard error would be computed using the formula for the pooled estimate of the population variance.

$$s_{\overline{X}_1 - \overline{X}_2} = \sqrt{s^2 \left(\frac{1}{n_1} + \frac{1}{n_2} \right)}$$

where

$$s^2 = \frac{\left[\Sigma X_{i1}^2 - (\Sigma X_{i1})^2 / n_1 \right] + \left[\Sigma X_{i2}^2 - (\Sigma X_{i2})^2 / n_2 \right]}{n_1 + n_2 - 2}$$

Computing the pooled estimate using the data in the printout,

$$s^2 = \frac{(280-1)(9.375)^2 + (220-1)(9.178)^2}{280 + 220 - 2}$$

$$= \frac{(279)(87.891) + (219)(84.235)}{498}$$

$$= \frac{24521.589 + 18447.465}{498}$$

$$= \frac{42969.054}{498}$$

$$= 86.283$$

Thus the estimated standard error is

$$s_{\bar{X}_1 - \bar{X}_2} = \sqrt{86.283(1/280 + 1/220)} = 0.837$$

And the test statistic is

$$t = \frac{(27.2446 - 26.4363) - (0)}{0.837}$$

$$= \frac{0.8083 - 0}{0.837}$$

$$= 0.97$$

Since the observed t value (0.97) does not exceed the critical t value (± 1.96), the null hypotheses is not rejected. Note that the "2-tail Prob" in the computer printout associated with this t value is .3346. Since this value is greater than the level of significance set a priori, this also indicates that the null hypothesis is not rejected.

Step 4: Construct the Confidence Interval

The computer printout also contains the confidence interval for the difference between means; however, it was computed using the general formula and the data from the above steps.[8]

$$CI = \text{statistic} \pm (\text{critical value})(\text{standard error})$$

For the variable, VISUAL, the CI_{95} is

$$CI_{95} = -1.2229 \pm (1.96)(0.355)$$

$$= -1.2229 \pm 0.696$$

$$= (-1.919, -.527)$$

---

[8] Note very slight differences in the confidence intervals on the printout and those calculated below; the differences are attributed to rounding.

Note that this interval does not contain zero, which is consistent with the previous result that the null hypothesis was rejected, i.e., there is a difference between the population means.

For the variable, MOSAIC, the CI_{95} is

$$CI_{95} = .8083 \pm (1.96)(0.837)$$

$$= .8083 \pm 1.641$$

$$= (-0.833, 2.45)$$

Note that this interval does contain zero, which is consistent with the previous result that the null hypothesis was not rejected, i.e., there is no difference between the population means.

Step 5: Interpret the Results

For the variable, VISUAL, the null hypothesis was rejected; we conclude that, based upon these data, the population means for females and males differ. The specific probability statement would be, "the probability that a difference in sample means of -1.2229 would have occurred by chance if $\mu_F - \mu_M = 0$ is less than .05." The actual probability of this occurrence, from the printout, is less than .001.

For the variable, MOSAIC, the null hypothesis was not rejected; we conclude that, based upon these data, the population means for females and males do not differ. The specific probability statement would be, "the probability that a difference in sample means of .8083 would have occurred by chance if $\mu_F - \mu_M = 0$ is greater than .05." The actual probability of this occurrence, from the printout, is .335.

Dependent Samples

In the previous tests of $H_0: \mu_1 = \mu_2$, we had two independent samples. The subjects were randomly selected from two populations, and measurements taken; in the case of an experiment, they were assigned randomly to two treatment conditions. Because of this random assignment, the two groups were assumed to be equivalent at the beginning of the experiment.

An alternative situation is one in which we have dependent or related samples. Two samples of data are dependent when each score in one sample is paired with a specific score in the other sample. Dependent samples can occur in two ways. A group may be measured twice, such as in a pretest–posttest situation. The two scores from the same individual are then the dependent scores. In this situation, sometimes we say that a subject acts as his or her own control.

The second situation is that of matched samples, in which each subject in one sample is matched on some relevant variable with a subject in the other sample. For example, suppose we have an instructional experiment in which the dependent variable is the score on a measure of academic performance, such as a mathematics test. There are two experimental treatments, and 50 subjects have been selected to participate in the experiment. IQ test scores are available on the subjects. In constructing the matched samples, we would take the two subjects with the highest IQ test scores and randomly assign them to the two experimental treatments. Then we would take the pair with the next two highest scores and randomly assign them. We would continue this process until all 25 pairs were assigned randomly to the two experimental treatments. The data from this experiment then would be from dependent samples.

Such data are correlated data. For example, in the case of measuring the same subject at pretest and posttest, a subject scoring high on one test will tend to score high on the other test, and vice versa. In the case of matched samples, we have a corresponding situation: a subject in one sample and his or her matched subject in the other sample will tend to have similar scores. These pairs of scores are dependent, or related, to each other. Thus we have **dependent samples** for testing the hypothesis H_0: $\mu_1 = \mu_2$.

In order to distinguish between tests of hypotheses for independent samples and those for dependent samples, we will use different symbols for the hypothesized population parameter and the corresponding sample statistic for the dependent-sample case. The hypothesized population parameter, defined by the null hypothesis, will be $\delta = 0$ where δ (delta) is defined as the mean of the difference scores across the two measurements. That is,

$$H_0: \delta = \mu_1 - \mu_2 = 0$$

The corresponding sample statistic is $\overline{d}$, which is the mean of the difference scores across the two measurements for the sample. That is,

$$\overline{d} = \overline{X}_1 - \overline{X}_2$$

Step 1: State the Hypotheses

Suppose a psychologist wants to investigate the reaction time necessary to brake an automobile when subjects are under the influence of three ounces of alcohol. The psychologist designs an experiment to compare the subjects' reaction times before any alcohol has been ingested with their reaction times after three ounces have been ingested.

In the experiment, the psychologist tests the null hypothesis that the mean difference in reaction time under the two conditions is zero. This hypothesis is tested against the directional alternative hypothesis that the mean difference in reaction time is greater than zero, that is, that the reaction time under the influence of alcohol is greater than the reaction time when free of such influence. If we let the subscript 1 represent the condition of three ounces of alcohol, and subscript 2 no alcohol, the hypotheses are

$$H_0: \delta = \mu_1 - \mu_2 = 0$$

$$H_a: \delta = \mu_1 - \mu_2 > 0$$

The psychologist sets the level of significance at .01.

Step 2: Set the Criterion for Rejecting H_0

Suppose 28 subjects are randomly selected and tested under the two conditions. The data for this experiment are found in Table 11.4. The psychologist hypothesizes that the mean difference in the reaction times under the two conditions is zero: $\delta = 0$. From Table 11.4, we find the corresponding statistic is $\overline{d} = 11.57$. As in the case for independent samples, the distribution of differences between the means of dependent samples is known, and this sampling distribution is defined as follows:

> As the sample size for the paired observations (n) increases, the sampling distribution of the means of differences of simple random samples of size n taken from any population with mean equal to $\delta = 0$ and finite variance equal to σ_d^2 has the following properties:

1. *Shape*: The distribution of differences approaches a normal distribution.

2. *Central tendency*: The mean of the distribution of differences is

$$\delta = 0$$

3. *Variability*: The standard deviation of the distribution of differences—called the *standard error of the difference*—is

$$\sigma_{\overline{d}} = \sigma_d / \sqrt{n}$$

Since the population variance (σ_d^2) is unknown, it is estimated by

$$s_d^2 = \frac{\Sigma(d - \overline{d})^2}{n - 1} \tag{11.11}$$

The corresponding raw score formula is

$$s_d^2 = \frac{\Sigma d^2 - (\Sigma d)^2 / n}{n - 1} \tag{11.12}$$

The estimated standard deviation is thus

$$s_d = \sqrt{\frac{\Sigma(d - \overline{d})^2}{n - 1}} \tag{11.13}$$

$$= \sqrt{\frac{\Sigma d^2 - (\Sigma d)^2 / n}{n - 1}} \tag{11.14}$$

TABLE 11.4
Data for Computing the Dependent *t* test (reaction time in hundredths of seconds—decimal not included)

Subject	Alcohol	No Alcohol	d = difference	d^2
1	46	33	13	169
2	51	41	10	100
3	41	29	12	144
4	32	18	14	196
5	37	26	11	121
6	48	40	8	64
7	37	23	14	196
8	36	25	11	121
9	38	28	10	100
10	30	21	9	81
11	42	31	11	121
12	43	36	7	49
13	53	39	14	196
14	38	27	11	121
15	47	38	9	81
16	36	19	17	289
17	33	22	11	121
18	36	24	12	144
19	42	33	9	81
20	54	42	12	144
21	36	27	9	81
22	48	35	13	169
23	46	34	12	144
24	33	22	11	121
25	48	36	12	144
26	54	39	15	225
27	43	29	14	196
28	50	37	13	169
Sum	1178	854	324	3888
Mean	$\bar{X}_1 = 42.07$	$\bar{X}_2 = 30.50$	$\bar{d} = 11.57$	
Standard deviation	$s_1 = 7.08$	$s_2 = 7.17$	$s_d = 2.27$	
Standard error of the difference			$s_{\bar{d}} = 0.43$	

Correspondingly, the standard error of the difference is estimated by the following formula:

$$s_{\bar{d}} = \frac{s_d}{\sqrt{n}}$$

(11.15)

Using the data in Table 11.4 and formula 11.14, we see that s_d for this example is

$$s_d = \sqrt{\frac{3888 - (324)^2/28}{27}}$$

$$= 2.27$$

Therefore, the estimated standard error of the difference (using formula 11.15) is

$$s_{\overline{d}} = \frac{2.27}{\sqrt{28}}$$

$$= 0.43$$

In this example, there are observations on 28 individuals under both conditions, or 28 paired observations. Therefore, there are $n - 1$, or 27, degrees of freedom associated with estimating the population variance (σ_d^2). Because the alternative hypothesis is directional, the critical value of t is $+2.473$ (for $\alpha = .01$).

Step 3: Compute the Test Statistic

The psychologist in this example hypothesizes that the mean difference in the reaction times under the two conditions is zero $(\delta = 0)$. The corresponding sample statistic computed from the data is $\overline{d} = 11.57$, and the standard error of the statistic, computed using formula 11.14, is 0.43. Using the general formula for the test statistic,

$$t = \frac{\overline{d} - \delta}{s_{\overline{d}}}$$

$$= \frac{11.57 - 0}{0.43}$$

$$= 26.91$$

Since the observed value of the test statistic $(t = 26.91)$ exceeds the critical value $(t_{cv} = +2.473)$, the null hypothesis is rejected.

Step 4: Construct the Confidence Interval

The null hypothesis is rejected, and the psychologist might want to construct the 99-percent confidence interval. That confidence interval (as defined with the general formula) is developed symmetrically around the sample statistic, and the critical value for the confidence interval is different from the critical value of the test statistic when a one-tailed test is conducted. Thus, for this example, using Student's t distribution with 27 degrees of freedom, the critical value for the confidence interval is 2.771, and CI_{99} is computed as follows:

$$CI_{99} = \overline{d} \pm t_{cv}(s_{\overline{d}}) \tag{11.16}$$

$$= 11.57 \pm (2.771)(0.43)$$

$$= 11.57 \pm 1.19$$

$$= (10.38, 12.76)$$

Step 5: Interpret the Results

The psychologist tests the null hypothesis (H_0: $\delta = 0$) against the directional alternative hypothesis (H_a: $\delta > 0$). The different symbols reflect the fact that dependent samples rather than independent samples are used to test the hypothesis. Because the observed value of t exceeds the critical value of t, the null hypothesis is rejected at the .01 level of significance. The conclusion is that reaction time is longer under the influence of three ounces of alcohol. The associated probability statement is: the probability that the observed difference ($\overline{d} = 11.57$) would have occurred by chance if the null hypothesis ($\delta = 0$) were true is *less than* .01. Actually, the probability may be much less than .01, but since the level of significance was established a priori, the null hypothesis is rejected at that predetermined level.

In the example, the null hypothesis of no difference is rejected, but it is also useful to have an estimate of the population difference. Thus, a 99-percent confidence interval is constructed and is found to range from 10.38 to 12.76. That is, we are 99 percent confident the interval 10.38 to 12.76 spans the difference in the population.

When testing the null hypothesis H_0: $\delta = 0$ using sample means from dependent samples—or when estimating δ, the difference between the population means—the estimated standard error of the difference is defined as

$$s_{\overline{d}} = \frac{s_d}{\sqrt{n}}$$

This value is used in the general formula for the test statistic and the general formula for the confidence interval.

Statistical Significance versus Practical Importance: A Return to Reality

In earlier chapters, we discussed the concept of statistical precision. We asked whether a statistically significant difference between the observed sample statistic and the hypothesized population parameter is also practically important. Inferential statistics has been criticized for being only a "numbers game." On one hand, a sufficiently large sample size (n) will lead to the rejection of any null hypothesis, given a fixed difference between the hypothesized parameter and the observed statistic. On the other hand, research findings based on large samples are more reliable. When samples are inappropriately small, a difference that could be of practical importance might be found statistically nonsignificant.

The selection of less conservative α levels was discussed in Chapter 8. Using conservative α levels (.05 or .01) protects against making a Type I error (rejecting a true hypothesis), but at the expense of increasing the probability of making a Type II error. In the reaction-time example in this chapter, the α level was set at .01. The rationale for this conservative level could be that, if the null hypothesis were rejected, an expensive educational program for drunk drivers might be implemented. Thus, because of the consequence of making the Type I error—that is, implementing the expensive educational program when in reality there is no difference in the mean reaction times—the Type I error might be considered the more serious. But a rationale for using a less conservative α level could be developed. For example, we might argue that, if there is any possibility that alcohol slows reaction time, we should document the fact. In this case, the Type II error of failing to reject the null hypothesis when it is false would be the more serious error. Thus a less conservative level, say .10, would have been more appropriate. However, the statistical test was significant, the null hypothesis was rejected at the .01 level, and there was no chance of making a Type II error.

We must again emphasize that practical importance cannot be measured by the inferential statistical test. Rejecting or not rejecting the null hypothesis is a function of the probability that a difference between the observed statistic and the hypothesized parameter is due to chance, given the fixed α level. The statistical test of significance cannot by itself answer the question about whether the magnitude of the difference dictates a corrective course of action. Inferential statistics are tools for analyzing and intepreting data; they are not a substitute for the knowledgeable interpretation of results.

Summary

The purpose of this chapter was to extend the principles of hypothesis testing and construction of confidence intervals from the one-sample case to the two-sample case. The logic and procedures for testing H_0: $\mu_1 - \mu_2 = 0$ and for constructing a confidence interval around $\overline{X}_1 - \overline{X}_2$ for the two-sample case are similar to those for the one-sample case. First, it was necessary to describe the sampling distribution of the statistic, which in this case was the difference between the sample means. Second, the probability associated with this sampling distribution was used to determine whether the difference between the observed statistic $(\overline{X}_1 - \overline{X}_2)$ and the hypothesized population parameter $(\mu_1 - \mu_2 = 0)$ was sufficient to reject the null hypothesis. Finally, this probability was also used to construct the confidence interval around $\overline{X}_1 - \overline{X}_2$.

In this chapter, we introduced two assumptions associated with testing H_0: $\mu_1 - \mu_2 = 0$: homogeneity of variance and independence. A statistical test of the homogeneity of variance assumption—the hypothesis that the populations from which samples are selected have equal variance—was described. To test this hypothesis, we used the ratio of two sample variances as the test statistic. A new distribution, the F distribution, was introduced as the sampling distribution of the ratio of two sample variances. There is a family of F distributions, which requires two degrees-of-freedom values to identify a specific distribution. If the homogeneity of variance assumption is not met, the Cochran and Cox/Satterthwaite procedure is recommended for testing H_0: $\mu_1 - \mu_2 = 0$ for independent samples.

When we have two samples of data on the same individuals or we have matched samples, the samples are called dependent, or related. We then use the t test for the difference between the means of dependent samples.

In the next chapter, we will extend the logic and procedures for the two-sample case for the mean to the two-sample case for other statistics.

Exercises

For Exercises 1 to 7, use the following five-step process:

1. State the hypotheses.

2. Set the criterion for rejecting H_0.

3. Compute the test statistic.

4. Construct the confidence interval.

5. Interpret the results.

*1. A study is conducted on the effects of two drugs (simply called Drug A and Drug B) upon hyperactivity in laboratory rats. Two random samples of rats are used for the study; one sample receiving Drug A, the other sample receiving Drug B. After two weeks, the rats are measured on hyperactivity with the following results:

	Drug A	Drug B
X	75.6	72.8
n	18	24
s^2	12.25	10.24
s	3.5	3.2

a. Test the null hypothesis H_0: $\mu_A = \mu_B$ against H_a: $\mu_A \neq \mu_B$. Use $\alpha = .05$

b. Construct CI_{95}.

Note: Assume that the two population variances are homogeneous, which is a valid assumption given the above data.

2. The instructor of an introductory psychology course is interested in knowing if there is a difference in the mean grades on the final exam between the fall and winter classes. Summary data for the two samples from these classes follow:

	Fall class	Winter class
$\overline{X}$	82.4	84.2
n	150	150
s^2	11.56	11.44

   a. Test the null hypothesis $H_0: \mu_1 = \mu_2$ against $H_a: \mu_1 \neq \mu_2$. Use $\alpha = .10$.

   b. Construct CI_{90}.

3. A researcher wants to determine whether children will learn concepts better with positive examples alone or with both positive and negative ones. Ten children are randomly assigned to each of the two experimental conditions; their scores on the concept-formation task are given below. Decide whether there is a difference between the two methods. Use $\alpha = .01$.

Positive (Group 1)	Positive + negative (Group 2)	Positive (Group 1)	Positive + negative (Group 2)
8	14	9	6
10	8	10	15
7	7	11	11
12	10	6	9
6	12	13	8

   a. Test the null hypothesis $H_0: \mu_1 = \mu_2$ against $H_a: \mu_1 \neq \mu_2$.

   b. Construct CI_{99} for the difference between the population means.

4. An industrial psychologist is interested in the differences between high-performing and low-performing salespeople on several psychological factors. Random samples are selected from two groups of salespeople, and a battery of standardized tests is given to each sample. The results of one of the tests are as follows:

Low-performing group	High-performing group
8	23
6	11
4	17
12	16
16	6
17	14
12	15
10	19
11	121
13	
109	

   a. Test the null hypothesis $H_0: \mu_1 = \mu_2$ against $H_a: \mu_1 \neq \mu_2$. Carry out the test of this hypothesis, computing the standard error of the difference (formulas 11.1 and 11.3) with both the pooled estimate of the population variance and the Cochran and Cox/Satterthwaite procedure (formulas 11.9 and 11.10). Use $\alpha = .10$.

b. Construct CI_{90} for the difference between population means.

5. A survey is conducted on attitudes toward smoking. A random sample of eight married couples is selected, and the husbands and wives respond to an attitude-toward-smoking scale. The scores are as follows:

Husbands	Wives
16	15
20	18
10	13
15	10
8	12
19	16
14	11
15	12

a. Test the null hypothesis $H_0: \delta = 0$ against $H_a: \delta \neq 0$. Use $\alpha = .05$.

b. Construct CI_{95} for δ.

c. Give the probability statements for the results of parts a and b.

6. An automobile manufacturer wants to know if its new model has better gasoline performance than the old model. A random sample of 30 new-model cars is selected and the gasoline performance determined. These data are then compared with data on 25 randomly selected older-model automobiles. Using the data below, determine whether gasoline performance is better for the new cars.

	Old model	New model
$\overline{X}$	52	56
n	25	30
s^2	220	210

a. Test the homogeneity of variance assumption. What is the conclusion about this assumption? Use $\alpha = .10$.

b. Test the null hypothesis $H_0: \mu_1 = \mu_2$ against $H_a: \mu_1 < \mu_2$. Use $\alpha = .05$.

c. Construct CI_{95}. Remember that the critical value for the confidence interval will be different from that for the hypothesis test.

7. A psychologist interested in the relationship between stress and short-term memory tests ten subjects prior to and after their exposure to a stressful situation. Test the hypothesis of no difference between stressful and unstressful situations, using the data below. Use $\alpha = .10$.

Prestress	Poststress
12	11
14	14
10	8
14	15
14	11
17	14
16	16
11	9
12	11
16	13

a. Test the null hypothesis H_0: $\delta = 0$ against H_a: $\delta \neq 0$.

b. Construct CI_{90} for δ.

c. Give the probability statement for the test in part a.

8. Given the following pairs of sample variances from independent samples, test the homogeneity of variance assumption for each pair. Use $\alpha = .10$.

a. $s_1^2 = 80.6$, $s_2^2 = 23.0$, $n_1 = 21$, $n_2 = 10$

b. $s_1^2 = 50$, $s_2^2 = 100$, $n_1 = 19$, $n_2 = 43$

c. $s_1^2 = 90.6$, $s_2^2 = 25.3$, $n_1 = 61$, $n_2 = 16$

d. $s_1^2 = 52.0$, $s_2^2 = 81.2$, $n_1 = 40$, $n_2 = 121$

9. For parts c and d of Exercise 8, give the probability statements after completing the test for the assumption of homogeneity of variance.

10. Given two independent samples of sizes 101 and 31, how large must the ratio of the sample variances be to violate the homogeneity of variance assumption? Use $\alpha = .10$. How large would the ratio need to be to violate the assumption if the significance level is set at .02?

11. Two random, independent samples of high school seniors are selected and administered a mathematics exam. The information about the samples and the test scores is as follows:

$$n_1 = 41, \ n_2 = 50$$

$$\overline{X}_1 = 70.0, \ \overline{X}_2 = 62.1$$

$$s_1^2 = 200, \ s_2^2 = 220$$

a. Verify that the homogeneity of variance assumption is tenable. Use $\alpha = .10$.

b. Compute the pooled estimate of the population variance.

c. Test the null hypothesis, H_0: $\mu_1 - \mu_2 = 0$, against the alternative hypothesis, H_a: $\mu_1 - \mu_2 \neq 0$. Compute the standard error of the difference between sample means using the pooled estimate (formula 11.1). Use $\alpha = .10$.

d. Construct CI_{95} for the difference between the population means.

e. Summarize the conclusions from parts c and d.

12. Random samples are selected from two populations of women, one aged 16 to 20, and the other aged 26 to 30. The women are then measured individually in a physical-exercise laboratory on a physical-performance test. The size of the sample of younger women is 16, and that of the older women 121. The results for the two samples are as follows:

$$\overline{X}_1 = 88.6, \ \overline{X}_2 = 94.8$$

$$s_1^2 = 100, \ s_2^2 = 40$$

   a. Explain why these samples are independent.

   b. Test the assumption of homogeneity of variance in the populations from which these samples were selected. Use $\alpha = .10$.

   c. Based on the result of part b, test the null hypothesis, H_0: $\mu_1 - \mu_2 = 0$, against the alternative hypothesis, H_a: $\mu_1 - \mu_2 \neq 0$, applying the appropriate procedure. Use $\alpha = .05$.

   d. Construct the CI_{95} for the difference between the population means.

   e. Summarize the conclusions from parts b, c, and d.

12

Hypothesis Testing: Two-Sample Case for Other Statistics

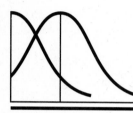

Key Concepts
Standard error of the difference between correlation coefficients
Standard error of the difference between independent proportions

Standard error of the difference between dependent proportions

W e presented hypothesis testing and the construction of confidence intervals for the two-sample case for the mean in Chapter 11. The logic and procedures for this case are identical to those discussed for the one-sample case. That is, hypothesis testing involves determining whether the difference between the observed sample statistic and the hypothesized population parameter is statistically significant; interval estimation involves determining a range of values with a designated confidence of containing the parameter being estimated. Whether testing a hypothesis or constructing a confidence interval, the researcher bases decisions about the nature of the parameter on knowledge of the corresponding sample statistic. Thus, the inference is made from the statistic to the parameter.

In this chapter, we extend the principles and procedures of hypothesis testing and interval estimation to deal with differences in correlation coefficients ($\rho_1 - \rho_2$) and differences in proportions ($P_1 - P_2$). We will use the five steps followed in previous chapters for testing hypotheses and constructing confidence intervals in order to establish a logical and consistent procedure.

Testing $H_0: \rho_1 = \rho_2$ for Independent Samples

In this section, we will test the null hypothesis $H_0: \rho_1 = \rho_2$, the hypothesis of no difference between two population correlation coefficients. This hypothesis will be tested using two sample correlation coefficients computed from data from two independent samples.[1] The test of this hypothesis is analogous to the test of the difference between two means for independent samples.

Unlike the sampling distribution of the mean, which retains its symmetrical shape (the normal distribution or the appropriate t distribution) regardless of the value of the population mean, the sampling distribution of the correlation coefficient becomes increasingly skewed as the absolute value of the population correlation coefficient increases. Thus, as we saw in Chapter 10, the normal distribution cannot be used as the sampling distribution for this test statistic. In order to overcome the problem, we apply the Fisher z transformation, which produces a statistic whose sampling distribution is nearly normal for any value of the statistic. The Fisher z transformation will also be used in this section for testing the null hypothesis concerning the difference between correlation coefficients for independent samples.

Suppose a large university typically admits the freshman class in two groups or populations: those admitted early in the year, between January 5 and March 31; and those admitted later, between May 1 and June 30. An admissions officer is interested in whether there is a difference in the relationship between SAT scores and scores on the College Placement English Examinations for freshmen from these two populations. Relationship implies correlation, and if there is a difference in the correlation coefficients, this difference may have implications for the placement of freshmen in the English classes. It is decided to select two random samples of freshmen, 73 from the population of those admitted early and 95 from the population of those admitted later. These are independent samples because any individual belongs to only one population, and the prospective freshmen independently take the examinations. The correlation coefficient for sample 1 is .562 and that for sample 2 is .479. We now turn to our five-step process.

Step 1: State the Hypotheses

The null hypothesis is that there is no difference between the correlation coefficients for these two populations. There is no reason to hypothesize that one coefficient

---

[1] There is a test for $H_0: \rho_1 = \rho_2$ for dependent samples, but it is used infrequently and is not discussed in this text. For a discussion of the test for dependent samples, the reader is referred to G. V Glass and K. D. Hopkins, *Statistical Methods in Education and Psychology*, 3rd ed. (Needham Heights, MA: Allyn and Bacon, 1996), pp. 362–363.

would be any larger than the other, so the null hypothesis will be tested against the nondirectional alternative hypothesis. In symbols,

$$H_0: \rho_1 = \rho_2 \qquad \text{or} \qquad H_0: \rho_1 - \rho_2 = 0$$

$$H_a: \rho_1 \neq \rho_2 \qquad \text{or} \qquad H_a: \rho_1 - \rho_2 \neq 0$$

The parameter here is $\rho_1 - \rho_2$, so the corresponding statistic is $r_1 - r_2$. The .10 level of significance will be used for testing the null hypothesis.

Step 2: Set the Criterion for Rejecting H_0

Given the sample correlation coefficients above, the necessary Fisher z transformations for the parameters and the statistic are:

parameter: $\rho_1 - \rho_2 = 0$ $z_{\rho_1} - z_{\rho_2} = 0$

statistic: $r_1 - r_2 = .562 - .479$ $z_{r_1} - z_{r_2} = 0.636 - 0.522$

$$= 0.114$$

Like the sampling distribution of the correlation coefficient, the sampling distribution of the difference changes shape when values of the correlation coefficients increase. However, if the two correlation coefficients from independent samples are transformed using the Fisher z transformation, which was introduced in Chapter 10, the sampling distribution of the difference between the z_r values is defined. It is approximately normally distributed, with a mean of $z_{\rho_1} - z_{\rho_2}$ or, for the null hypothesis, a mean of zero. The estimated standard deviation of the sampling distribution, called the *estimated standard error of the difference between independent transformed correlation coefficients*, is given by the following formula:

$$s_{z_{r_1} - z_{r_2}} = \sqrt{\frac{1}{n_1 - 3} + \frac{1}{n_2 - 3}} \qquad (12.1)$$

For the above data the standard error of the difference between the transformed correlation coefficients is thus

$$s_{z_{r_1} - z_{r_2}} = \sqrt{\frac{1}{73 - 3} + \frac{1}{95 - 3}}$$

$$= 0.159$$

For testing the null hypothesis of no difference between two population correlation coefficients from independent samples, the sampling distribution of the test statistic is the normal distribution and thus the test statistic is z.

Because the normal distribution is the appropriate sampling distribution for testing the null hypothesis of no difference between two population correlation coefficients, the critical values are found using Table C.1. For this example, the null hypothesis is tested against the nondirectional alternative hypothesis, and the level of significance is set at .10. Thus, the critical values for the test statistic (z_{cv}) are ± 1.645.

Step 3: Compute the Test Statistic

For this example, the null hypothesis is that there is no difference between the correlation coefficients for these two populations. The corresponding sample statistic is found to be $0.636 - 0.522 = 0.114$; the standard error of the difference $(s_{z_{r_1} - z_{r_2}})$, defined by formula 12.1, is found to be 0.159. With these values we can now apply the general formula for the test statistic:

$$z = \frac{(z_{r_1} - z_{r_2}) - (z_{\rho_1} - z_{\rho_2})}{s_{z_{r_1} - z_{r_2}}} \tag{12.2}$$

$$= \frac{(0.636 - 0.522) - 0}{0.159}$$

$$= 0.717$$

Since the observed value of the test statistic $(z = 0.717)$ is less than the critical value $(z_{cv} = \pm 1.645)$, the null hypothesis is not rejected.

Step 4: Construct the Confidence Interval

In a study like this, it is not likely that constructing a confidence interval for the difference between the correlation coefficients will contribute much additional information. Since the null hypothesis was not rejected, we know that CI_{90} will span zero (the hypothesized population parameter). However, we shall illustrate the procedure by constructing the 90-percent confidence interval for the difference.

The first step is to construct a confidence interval around the difference between the two transformed correlation coefficients $(z_{r_1} - z_{r_2} = 0.114)$. Then the limits for this interval will be transformed back to limits for $r_1 - r_2$, using Table C.6. These steps are outlined as follows:

$$CI_{90} = 0.114 \pm (1.645)(0.159)$$

$$= 0.114 \pm 0.262$$

$$= (-0.148, \ 0.376)$$

So far, we have the limits for $z_{r_1} - z_{r_2}$. Transforming these limits, we obtain:

$$CI_{90} = (-0.147, \ 0.359)$$

Thus, the interval -0.147 to 0.359 is the 90-percent confidence interval for the difference between the two population correlation coefficients.

Step 5: Interpret the Results

Because the observed value of the test statistic (z) does not exceed the critical value (z_{cv}), the null hypothesis is not rejected. Thus, there is no reason to infer that there is any difference between the correlation coefficients for the two populations. The probability statement is: the probability that the observed difference between the two sample correlation coefficients would have occurred by chance if the null hypothesis were true is greater than .10. Thus, the admissions officer may decide to combine the data from the two samples and recompute the correlation coefficient for the combined group.

The 90-percent confidence interval for the difference spans zero, which is consistent with the hypothesis-testing results. We are 90 percent confident that the interval -0.147 to 0.359 spans the difference between the correlation coefficients of the populations.

For testing the null hypothesis $H_0: \rho_1 = \rho_2$ using independent samples, the estimated standard error of the difference of the transformed correlation coefficients is defined as

$$s_{z_{r_1} - z_{r_2}} = \sqrt{\frac{1}{n_1 - 3} + \frac{1}{n_2 - 3}}$$

and is used in the general formulas for the test statistic and the confidence interval.

Testing $H_0: P_1 = P_2$ for Independent Samples

Next, we turn to testing hypotheses about differences in proportions. The term proportion was defined in Chapter 10 as the fractional part of a dichotomous population that has a specified characteristic of interest. Proportions are used extensively in survey research, especially by polling agencies attempting to predict election outcomes. In this section, we discuss the procedure for testing the hypothesis of no difference between two population proportions, P_1 and P_2, using data from two independent samples.

Suppose the American Automobile Association (AAA) is interested in knowing the proportion of uninsured motorists in two metropolitan areas of different sizes. Since automobile insurance rates tend to be higher in more populated areas, it is possible that this is due, at least in part, to greater proportions of uninsured motorists in the more populated areas. The AAA decides to conduct an investigation in a state that has the uninsured-motorist fee and selects both a large metropolitan area and a small metropolitan area in that state. A random sample of 10 percent of all registered motorists is selected from both metropolitan areas. In the large metropolitan area, 9,260 of the sample of 43,146 (0.215) registered motorists paid the uninsured-motorist fee, whereas only 583 of the sample of 6,230 (0.094) registered motorists in the smaller metropolitan area paid the uninsured-motorist fee.

Step 1: State the Hypotheses

In this example, the null hypothesis is that the proportion of uninsured motorists in the large metropolitan area (P_1) is equal to the proportion of uninsured motorists in the smaller metropolitan area (P_2). This hypothesis is tested against the directional alternative that the larger metropolitan area will have a larger proportion of uninsured motorists. The hypothesized parameter is that there is no difference between the two population proportions ($P_1 - P_2 = 0$). In symbol form:

$$H_0: P_1 = P_2 \quad \text{or} \quad H_0: P_1 - P_2 = 0$$
$$H_a: P_1 > P_2 \quad \text{or} \quad H_a: P_1 - P_2 > 0$$

Note that the order of the P's is important in the alternative hypothesis because it is a directional hypothesis. The level of significance is set at .01.

Step 2: Set the Criterion for Rejecting H_0

The statistic for testing the hypothesis is the difference between the two observed sample proportions, p_1 and p_2:

$$p_1 - p_2 = 0.215 - 0.094 = 0.121$$

In order to test the hypothesis, we also need to know the sampling distribution of the statistic. As stated earlier, to know a distribution is to know its shape, location, and variability. When the sample sizes n_1 and n_2 are large, and $n_1 p_1$ (and $n_1 q_1$) and $n_2 p_2$ (and $n_2 q_2$) are greater[2] than 5, the sampling distribution of the difference between two independent sample proportions is approximately normal, and the mean is equal to the hypothesized difference between the population proportions ($H_0: P_1 = P_2$ or $H_0: P_1 - P_2 = 0$). For this example, $n_1 p_2 = (0.215)(43,146) = 9,260$ and $n_2 p_2 = (0.094)(6,230) = 583$. Thus, we can use the normal distribution to describe the sampling distribution of differences between proportions.

[2] When this condition is not met, nonparametric procedures may be used. See L. A. Marascuilo and M. McSweeney, *Nonparametric and Distribution-free Methods for the Social Sciences* (Monterey, Calif.: Brooks/Cole, 1977).

The estimated standard deviation of this sampling distribution—called the **standard error of the difference between independent proportions**—is estimated by the following formula:

$$s_{p1-p2} = \sqrt{pq\left(\frac{1}{n_1} + \frac{1}{n_2}\right)} \qquad (12.3)$$

where

$$p = \frac{f_1 + f_2}{n_1 + n_2}$$

$$q = 1 - p$$

f_1 = frequency of occurrence in the first sample

f_2 = frequency of occurrence in the second sample

In this case, f_1 is 9,260, f_2 is 583, n_1 is 43,146, and n_2 is 6,230. Therefore, for these data, [3]

$$p = \frac{9,260 + 583}{43,146 + 6,230} = 0.199$$

$$q = 1 - p = 0.801$$

By applying formula 12.3, we find that the standard error in this case is

$$s_{p1-p2} = \sqrt{(0.199)(0.801)\left(\frac{1}{43,146} + \frac{1}{6,230}\right)}$$

$$= 0.005$$

Since the null hypothesis is tested against the directional alternative hypothesis, $H_a: P_1 > P_2$, the region of rejection is located in the right-hand tail of the sampling distribution. In this example, the sampling distribution is approximated by the normal distribution. Since the level of significance was established at .01, the critical value of the test statistic, consulting Table C.1, is +2.33.

Step 3: Compute the Test Statistic

The null hypothesis for this example is that there is no difference between the proportions of uninsured motorists in a large metropolitan area and those in a small metropolitan area: $H_0: P_1 - P_2 = 0$. The corresponding sample statistic is found to be $p_1 - p_2 = 0.215 - 0.094 = 0.121$; the standard error of the difference

---

[3] All numbers have been rounded to three decimal places.

$(s_{p_1-p_2})$ is found to be 0.005. With these values, we can now compute the test statistic. Applying the general formula for the test statistic to this example,

$$z = \frac{(p_1 - p_2) - (P_1 - P_2)}{s_{p_1-p_2}}$$

$$= \frac{(0.215 - 0.094) - 0}{0.005}$$

$$= \frac{0.121 - 0}{0.005}$$

$$= 24.200$$

Because the observed value of the test statistic ($z = 24.200$) exceeds the critical value ($z_{cv} = +2.33$), the null hypothesis is rejected.

Step 4: Construct the Confidence Interval

The development of the confidence interval for the difference between two population proportions uses the general formula for developing any confidence interval, which is

$$CI = \text{statistic} \pm (\text{critical value}) \times (\text{standard error of the statistic})$$

For this example, the statistic is the difference between the two sample proportions ($p_1 - p_2$), and the standard error of the statistic ($s_{p_1-p_2}$) is defined by formula 12.3. Because the sampling distribution of the test statistic is the normal distribution, we refer to Table C.1 and find that the critical value for CI_{99} is 2.58. Note that this critical value is different from the critical value for testing the hypotheses because the alternative hypothesis was directional. Using the general formula for constructing CI_{99} for these data, we find

$$CI_{99} = (p_1 - p_2) \pm (2.58)(s_{p_1-p_2})$$

$$= 0.121 \pm (2.58)(.005)$$

$$= 0.121 \pm (0.013)$$

$$= (0.108, 0.134)$$

Step 5: Interpret the Results

As in the one-sample case, the conclusion is based on the test of the null hypothesis ($H_0: P_1 = P_2$) and on the 99-percent confidence interval developed around $p_1 - p_2$. The null hypothesis was rejected in favor of the alternative. In other words, the probability that the observed difference in the sample proportions ($p_1 - p_2 = 0.121$)

would have occurred by chance if the null hypothesis (H_0: $P_1 = P_2$) were true is less than .01. The discrepancy between the observed difference of the sample proportions and the hypothesized value of no difference is too great to attribute to chance. Thus, the AAA can conclude that the larger metropolitan area has a greater proportion of uninsured motorists than the smaller metropolitan area. There is still a remote chance that a Type I error could have been made, but since the level of significance was set at .01, the AAA can be confident of the conclusion.

CI_{99} was developed and found to be 0.108 to 0.134. Thus, we are 99 percent confident that this interval contains the difference between the population proportions, P_1 and P_2. This information may be useful to the AAA in understanding differences in insurance premiums between the two areas due to the proportions of uninsured motorists.

For testing the null hypothesis H_0: $P_1 = P_2$ using independent samples, as well as for estimating $P_1 - P_2$, the estimated standard error of the difference is defined as

$$s_{p_1 - p_2} = \sqrt{pq \left(\frac{1}{n_1} + \frac{1}{n_2} \right)}$$

and is used in the general formulas for both the test statistic and the confidence interval.

Testing H_0: $P_1 = P_2$ for Dependent Samples

The null hypothesis that there is no difference between two population proportions may also be tested for dependent samples. As with the difference between the means of dependent samples, the same group may be measured twice, or matched samples may be measured. Typical research situations in which this occurs are as follows:

1. Testing the difference in the proportions of individuals who pass each of two similar items on an achievement subtest.

2. Testing the difference in the proportions of individuals who support a critical issue before and after discussion of the pertinent points underlying the issue.

3. Comparing the proportions of husbands and wives who have favorable attitudes toward an issue.

The logic underlying the test of the hypothesis of no difference between two population proportions using dependent samples is analogous to the test for two population means using dependent samples. We again have paired observations.

The two observations of a pair are not independent; they are correlated, and that correlation must be taken into account when testing hypotheses about the difference between proportions.

Consider this example. A researcher for a large corporation is interested in employees' opinions about various employment factors. The researcher wants to determine whether there is a statistically significant difference between the proportion of employees that supports the corporation's vacation policy and the proportion that supports the hospitalization policy. Rather than survey the entire population of employees, the researcher selects a random sample of 96 employees, who respond to a questionnaire. The questionnaire includes, among other items, a question about the vacation policy and a question about the hospitalization policy. The data are tabulated in a four-cell table called a 2×2 contingency table, which is shown in Table 12.1. Notice the order of the data entries in this table. Cells A and D must be those indicating disagreement or dissimilarity in responses.

TABLE 12.1
Data for Testing the Hypothesis $H_0: P_1 = P_2$ for Dependent Samples: Example of Employee Support of Vacation and Hospitalization Policies

A. Actual Frequencies

		Hospitalization-Policy Support		
		No	Yes	
Vacation-Policy Support	Yes	11 (A)	(B) 56	67
	No	23 (C)	(D) 6	29
		34	62	96

B. Percentages of Total*

		Hospitalization-Policy Support		
		No	Yes	
Vacation-Policy Support	Yes	0.115 (a)	(b) 0.583	0.698
	No	0.240 (c)	(d) 0.063	0.302
		0.354	0.646	1.000

*Addition may not work out exactly, due to rounding error.

Step 1: State the Hypotheses

In this example, the researcher is interested in testing the null hypothesis that there is no difference between the population proportions against the nondirectional alternative hypothesis. As stated in Chapter 11, the Greek letter delta is used to designate a population difference, so in symbol form the hypotheses are

$$H_0: \delta_p = P_1 - P_2 = 0$$

$$H_a: \delta_p = P_1 - P_2 \neq 0$$

The level of significance is set at .05.

Step 2: Set the Criterion for Rejecting H_0

For these data, 11/96, or 0.115, of the respondents support the vacation policy but not the hospitalization policy, while 6/96, or 0.063, of the respondents support the hospitalization policy but not the vacation policy. Note also that 67/96, or 0.698, of the employees support the vacation policy, and 62/96, or 0.646, support the hospitalization policy; 56/96, or 0.583, support both, while 23/96, or 0.240, support neither.

In this example, P_1 is the proportion of employees in the population supporting the vacation policy, and P_2 the proportion supporting the hospitalization policy. The corresponding sample proportions are p_1 and p_2. The statistic is the difference between these sample proportions, i.e.,

$$p_1 - p_2 = 0.698 - 0.646 = 0.052$$

The sampling distribution of the differences between two dependent sample proportions is known. This sampling distribution is approximately normal, with a mean equal to the hypothesized difference between the two population proportions, when the sum of the frequencies in either of the diagonal cell combinations is greater than 10.[4] For the data in Table 12.1, this restriction means that either A + D *or* B + C must be greater than 10.

The standard deviation of this sampling distribution, called the **standard error of the difference between dependent proportions**, is estimated by the following formula:

$$s_{p1-p2} = \sqrt{\frac{a+d}{n}} \tag{12.4}$$

---

[4]When this condition is not met, nonparametric procedures may be used. (See Marascuilo and McSweeney.)

where

> a = proportion of frequencies in the upper left cell in the contingency table
>
> d = proportion of frequencies in the lower right cell of the contingency table
>
> n = total number of paired observations

For the data in this example, the standard error of the difference is

$$s_{p_1 - p_2} = \sqrt{\frac{0.115 + 0.063}{96}}$$

$$= 0.043$$

Because the sampling distribution of the differences between two dependent proportions is normally distributed, the critical values are found in Table C.1. The null hypothesis is being tested against the nondirectional alternative at $\alpha = .05$, so the critical values of the test statistic (z_{cv}) are ± 1.96.

Step 3: Compute the Test Statistic

For this example, the null hypothesis is that there is no difference between the proportion of persons supporting the vacation policy and that supporting the hospitalization policy: $H_0: \delta_p = P_1 - P_2 = 0$. The corresponding sample statistic is found to be $p_1 - p_2 = 0.698 - 0.646 = 0.052$. Using formula 12.4, the standard error of the difference ($s_{p_1 - p_2}$) is found to be 0.043. With these values, we can now compute the test statistic. Applying the general formula for the test statistic to this example,

$$z = \frac{(p_1 - p_2) - \delta_p}{s_{p_1 - p_2}}$$

$$= \frac{(0.698 - 0.646) - 0}{0.043}$$

$$= \frac{0.052}{0.043}$$

$$= 1.208$$

Since the observed value of the test statistic ($z = 1.208$) is less than the critical value ($z_{cv} = \pm 1.96$), the null hypothesis is not rejected.

Alternative computational formula For testing the null hypothesis for proportions in dependent samples, the general formula for the test statistic reduces to the following:

$$z = \frac{A - D}{\sqrt{A + D}} \tag{12.5}$$

where

A = the frequency in the upper left cell of the contingency table

D = the frequency in the lower right cell of the contingency table

This is a convenient formula because A and D are frequencies and therefore integers. For the data in this example, the following value is found:

$$z = \frac{11 - 6}{\sqrt{11 + 6}}$$

$$= \frac{5}{\sqrt{17}}$$

$$= 1.213$$

The difference between this value of z and that found by using the general formula is due to rounding carried out before the general formula is applied. The formulas are algebraically equivalent.

A *note of caution*: If the data are arranged in contingency tables in the manner illustrated in Table 12.1, the algebraic sign of z will be the same for both formulas of the test statistic. However, a different arrangement of the data in the table will result in a different algebraic sign for z. This is a concern only when the null hypothesis is tested against a directional alternative.

Step 4: Construct the Confidence Interval

As before, we develop CI_{95} symmetrically around the statistic ($p_1 - p_2 = 0.052$) using the general formula for the confidence interval. For the data in this example, the critical value is 1.96, so

$$CI_{95} = 0.052 \pm (1.96)(0.043)$$

$$= 0.052 \pm 0.084$$

$$= (-0.032, 0.136)$$

Notice that zero, the hypothesized parameter ($\delta_p = P_1 - P_2 = 0$), is contained in the interval. This result is consistent with not rejecting the null hypothesis when it was tested.

Step 5: Interpret the Results

In this example, a sample of employees was measured on two items: support of the corporation's vacation policy, and support of the hospitalization policy. The null hypothesis that the proportions of employees favoring these two policies are equal (H_0: $P_1 = P_2$) was not rejected. The probability is greater than .05 that the difference between the two sample proportions ($p_1 - p_2 = 0.052$) would have occurred by chance if the null hypothesis ($H_0: P_1 = P_2$) were true. The researcher can say that there is not enough evidence to conclude that there is a difference between the proportion of employees supporting the vacation policy and that supporting the hospitalization policy. Furthermore, based on CI_{95}, we are 95 percent confident that the interval -0.032 to 0.136 contains the difference between the two population proportions. It can be concluded that support for the two policies is consistent throughout the employee population.

It should be noted that this example concerned the difference between two sample proportions. There were no tests for proportions individually, whether they were large or small or greater than or less than specified values. However, in the computations the individual proportions became known, and if there had been concern about individual proportions, hypotheses could have been tested and confidence intervals developed using the procedures of Chapter 10.

For testing the null hypothesis $H_0: P_1 - P_2 = 0$ using the dependent samples, the standard error of the difference is defined as

$$s_{p_1-p_2} = \sqrt{\frac{a+d}{n}}$$

and is used in the general formulas for both the test statistic and the confidence interval.

Summary

This chapter and Chapter 11 contain a compendium of two-sample, statistical tests. Tests for the differences between two means are used much more frequently than tests for the differences between correlation coefficients and proportions. The test for the equality of the variances of two independent populations was discussed in Chapter 11 in connection with testing the assumption of homogeneity of variance.

In earlier chapters we commented on statistical precision. We enhance statistical precision if smaller values of the test statistics are required for statistical significance and if the confidence interval is narrowed. As with other tests, statistical precision for the tests discussed in this chapter is enhanced if sample size is increased. The standard errors of the statistics are reduced if the sample size is increased (n or n_1 and n_2 appear in the denominators of the formulas for standard errors; thus, as they increase, the standard errors decrease). These standard errors, in turn, are in the denominators of the formulas for computing the test statistic. Confidence intervals also can be narrowed by decreasing the level of confidence, going from 99 percent to 95 percent, for example.

Tables 12.2 and 12.3 summarize the concepts and components for testing hypotheses for independent and dependent samples, respectively. These tables include testing hypotheses about means, as discussed in Chapter 11, and illustrate the common concepts of inferential statistics used for testing hypotheses. The formulas and sampling distributions vary as the hypotheses tested differ; however, the underlying reasoning is consistent.

Table 12.4 contains the information for constructing confidence intervals when two samples are involved. The confidence interval for estimating the difference between two population means, discussed in Chapter 11, also is included. When testing hypotheses, confidence intervals are not always constructed, but if appropriate they can add information about parameters.

The statistical tests described in this chapter are not used as frequently as tests for the means of two samples. However, they are useful when research situations that involve correlation coefficients and proportions are encountered.

Exercises

*1. A political science study about the satisfaction of the public with city government is being conducted in two large cities that are quite diverse and geographically separated. In the study, one survey question is, "Overall, are you satisfied with the leadership of your city government?" Random samples of adults from the two cities were surveyed and the results for this question were as follows:

	City 1	City 2
n	200	150
Yes	122	84
No	78	66

a. Test the null hypothesis H_0: $P_1 = P_2$ against H_a: $P_1 \neq P_2$. Use $\alpha = .05$.

b. Construct CI_{95} for the difference between the population proportions.

2. A professor wants to investigate the relationship between test anxiety and test performance in intermediate statistics classes for older students and for younger students. The professor believes that the relationship for the older students will be higher than for the younger students. To test the assumption, the professor defines "older" as over 40 and "younger" as under 26. For random samples of 52 older students and 47 younger students, the correlation between test anxiety and test performance are .48 and .22, respectively.

a. Test the null hypothesis H_0: $\rho_1 = \rho_2$ against H_a: $\rho_1 > \rho_2$. Use $\alpha = .05$.

b. Construct CI_{95} for the difference between the population correlation coefficients. Remember that the critical value for the confidence interval will be different from that for the hypothesis test.

c. Use the methods of Chapter 10 to determine whether either of these correlation coefficients is "statistically significant."

TABLE 12.2
Summary of Concepts and Components for Testing Hypotheses—Two-Sample Case: Independent Samples

Hypothesis Tested	Parameter	Statistic	Standard Error of the Statistic	Test Statistic	Degrees of Freedom	Sampling Distribution
$H_0: \mu_1 = \mu_2$	$\mu_1 - \mu_2$	$\bar{X}_1 - \bar{X}_2$	$s_{\bar{X}_1 - \bar{X}_2} = \sqrt{s^2\left(\dfrac{1}{n_1} + \dfrac{1}{n_2}\right)}$	$t = \dfrac{(\bar{X}_1 - \bar{X}_2) - (\mu_1 - \mu_2)}{s_{\bar{X}_1 - \bar{X}_2}}$	$n_1 + n_2 - 2$	Student's t distribution
$H_0: \rho_1 = \rho_2$	$z_{\rho_1} - z_{\rho_2}$	$z_{r_1} - z_{r_2}$	$s_{z_{r_1} - z_{r_2}} = \sqrt{\dfrac{1}{n_1 - 3} + \dfrac{1}{n_2 - 3}}$	$z = \dfrac{(z_{r_1} - z_{r_2}) - (z_{\rho_1} - z_{\rho_2})}{s_{z_{r_1} - z_{r_2}}}$	—	Normal distribution
$H_0: P_1 = P_2$	$P_1 - P_2$	$p_1 - p_2$	$s_{p_1 - p_2} = \sqrt{pq\left(\dfrac{1}{n_1} + \dfrac{1}{n_2}\right)}$	$z = \dfrac{(p_1 - p_2) - (P_1 - P_2)}{s_{p_1 - p_2}}$	—	Normal distribution

TABLE 12.3
Summary of Concepts and Components for Testing Hypotheses—Two-Sample Case: Dependent Samples

Hypothesis Tested	Parameter	Statistic	Standard Error of the Statistic	Test Statistic	Degrees of Freedom	Sampling Distribution
$H_0: \delta = 0$	δ	$\bar{d}$	$s_{\bar{d}} = \dfrac{s_d}{\sqrt{n}}$	$t = \dfrac{\bar{d} - \delta}{s_{\bar{d}}}$	$n - 1$	Student's t distribution
$H_0: \delta_P = P_1 - P_2 = 0$	δ_P	$p_1 - p_2$	$s_{p_1 - p_2} = \sqrt{\dfrac{a + d}{n}}$	$z = \dfrac{(p_1 - p_2) - (P_1 - P_2)}{s_{p_1 - p_2}}$	—	Normal distribution

TABLE 12.4
Information on Constructing Confidence Intervals for a Two-Sample Case

Parameter Estimated	Statistic	Confidence Interval
Independent Samples		
$\mu_1 - \mu_2$	$\bar{X}_1 - \bar{X}_2$	$(\bar{X}_1 - \bar{X}_2) \pm t_{cv}(s_{\bar{X}_1 - \bar{X}_2})$
$\rho_1 - \rho_2$	$r_1 - r_2$	$(z_{r_1} - z_{r_2}) \pm z_{cv}(s_{z_{r_1} - z_{r_2}})$
		(Converted back to values of r)
$P_1 - P_2$	$p_1 - p_2$	$(p_1 - p_2) \pm z_{cv}(s_{p_1 - p_2})$
Dependent Samples		
δ	$\bar{d}$	$\bar{d} \pm t_{cv}(s_{\bar{d}})$
δ_p	$p_1 - p_2$	$p_1 - p_2 \pm z_{cv}(s_{p_1 - p_2})$

3. A study is being conducted on the relationship between divergent thinking and science aptitude in boys and girls aged 12 to 14 years. Random samples of size 120 each are selected from populations of boys and girls in this age range. The sample members are administered a divergent-thinking test and a science-aptitude test. The correlation coefficients between the scores on these two tests are .69 for girls and .61 for boys.

   a. Test the hypothesis $H_0: \rho_G = \rho_B$ against $H_a: \rho_G \neq \rho_B$. Use $\alpha = .10$.

   b. Construct CI_{90} for the difference between the population correlation coefficients.

   c. Give the probability statements for the results of parts a and b above.

   d. Suppose a 95-percent confidence interval had been constructed. By how much does this interval differ from CI_{90}?

4. A graduate student believes that verbal scores on the SAT correlate more highly with senior-year GPAs for females than for males. The student draws a random sample of 64 females and finds that the correlation (r) is .58. For a random sample of 82 males, the correlation is .53.

   a. Test the null hypothesis $H_0: \rho_1 = \rho_2$ against $H_a: \rho_1 > \rho_2$. Use $\alpha = .05$.

   b. Construct CI_{95} for the difference between the population correlation coefficients. Remember that the critical value for the confidence interval will be different from that for the hypothesis test.

5. A researcher wants to test the hypothesis that the proportions of college men and women who smoke cigarettes are equal. The researcher selects random samples of 10 percent of both males and females at a midwestern college and finds that 84 out of 210 males and 72 out of 240 females report that they smoke.

   a. Test the null hypothesis $H_0: P_1 = P_2$ against $H_a: P_1 \neq P_2$. Use $\alpha = .05$.

   b. Construct CI_{95} for the difference between the population proportions.

c. Using the methods of Chapter 10, construct 95-percent confidence intervals for the population proportions of men and women who smoke at this college.

6. A sociologist believes that a higher proportion of Republicans than Democrats own guns. The sociologist selects a random sample of Republicans and Democrats. Of the sample of 150 Republicans, 81 own guns. Of the sample of 200 Democrats, 92 own guns.

   a. Test the null hypothesis $H_0: P_1 = P_2$ against $H_a: P_1 > P_2$. Use $\alpha = .05$.

   b. Construct CI_{95} for the difference between the population proportions. Remember that the critical value for the confidence interval will be different from that for the hypothesis test.

7. A sociologist randomly selects 50 married couples in order to study the impact of parent effectiveness training (PET) on attitudes toward the use of behavior-modification techniques. Before and after PET classes, couples are asked whether or not they favor the use of such techniques. The results are as follows:

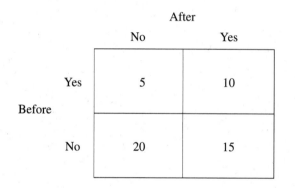

                              After

                      No   Yes

              Yes   5   10

   Before

              No   20   15

   a. Test the null hypothesis $H_0: \delta_p = 0$ against $H_a: \delta_p \neq 0$. Use $\alpha = .05$.
   b. Construct CI_{95} for δ_p.

8. A clinical psychologist, in collaboration with a psychiatrist, is interested in testing the anxiety-reducing effects of two drugs, X and Y. In a random sample of 100 persons treated with drug X, 75 report some relief of symptoms; in a second random sample, 105 of 160 persons treated with drug Y report some relief.

   a. Test the null hypothesis $H_0: P_1 = P_2$ against $H_a: P_1 \neq P_2$. Use $\alpha = .05$.
   b. Construct CI_{95} for the difference between the population proportions.
   c. What can be concluded about the relative effectiveness of the two drugs? Give the probability statements for parts a and b above.

9. A psychology instructor is completing an item analysis of the final examination given to all freshman psychology students. The instructor is interested in whether the proportion of students who answer item 5 correctly differs from the proportion who answer item 7 correctly. The data are as follows:

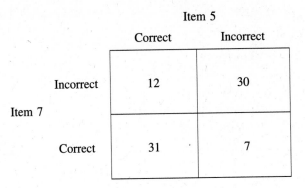

Item 5

		Correct	Incorrect
	Incorrect	12	30
Item 7			
	Correct	31	7

a. Test the null hypothesis H_0: $\delta_p = 0$ against H_a: $\delta_p \neq 0$. Use $\alpha = .05$.

b. Construct CI_{95} for δ_p.

13

Determining Power and Sample Size

U p to this point, we have discussed many different descriptive and inferential statistical procedures and applied them in various research settings. In addition to presenting formulas for computing these statistics, we have introduced certain assumptions underlying their use and concepts that are important in their interpretation. However, before discussing more advanced statistical procedures, such as analysis of variance, regression analysis, and analysis of covariance, we need to discuss two additional concepts: (1) the power of a statistical test and (2) the sample size required in designing research studies. These concepts have important implications for both research design and the interpretation of results.

Some authors deal with power exclusively as a post hoc procedure, that is, determining the power of a statistical test (such as a *t* test) *after* conducting the test. Other authors, however, consider the power of a statistical test *before* conducting the test. In the latter approach, the desired power of the statistical test is considered along with other factors in determining an appropriate sample size for the research study. This is the approach we will take in this chapter and in Chapters 14 and 16. If it is not feasible to use this approach in determining an appropriate sample size before conducting a research study, post hoc power analysis becomes important. However, discussion of power analysis is beyond the scope of this book; the interested reader is referred to the Cohen (1977) reference in the Bibliography.

How Large Should the Sample Be?

A question that researchers frequently ask is: how large should the sample be? Given the frequency and relevance of this question, one might assume that there is a simple answer. However, because so many factors are involved, no definite answer exists.[1] Nevertheless, the answer to the sample-size question is crucial because of the relationship between sample size and the statistical significance of the inferential test applied to the data. As mentioned in Chapter 8, critics of inferential tests have charged that statistically based research is often nothing more than a "numbers game" and that professional journals are concerned with publishing only "statistically significant" results. However, Hays notes that "Virtually any study can be made to show significant results if one uses enough subjects, regardless of how nonsensical the content may be."[2]

In both Chapters 8 and 11, we emphasized that there is no substitute for inferential statistical methods as a protection against interpreting differences between sample statistics and hypothesized population parameters as real differences rather than chance fluctuations due to random sampling. Within this frame of reference, there are three relevant assumptions:

1. It is sensible to view research findings based on large samples as more reliable than findings based on smaller samples, all other things being equal.

2. However, inferential statistical methods will not result in rejecting the null hypothesis if, in the design of the study, an inappropriately small sample is selected.

3. In a well-planned research study, in which the variance of the criterion variable is likely to be quite large and the treatment effects rather small, large samples are appropriate and justifiable.[3]

> In well-planned research investigations, the question of an appropriate sample size is crucial.

Procedures have been proposed for determining an appropriate sample size such that, given certain factors, the rejection of the null hypothesis is more than just a

[1] D. E. Hinkle and J. D. Oliver, "How Large Should the Sample Be? A Question with No Simple Answer? Or . . . ," *Educational and Psychological Measurement* 43 (1983): 1051–1060, and D. E. Hinkle, J. D. Oliver, and C. A. Hinkle, "How Large Should the Sample Be? Part II — The One-sample Case for Survey Research," *Educational and Psychological Measurement* 45 (1985): 271–280.

[2] W. L. Hays, *Statistics for the Social Sciences* (New York: Holt, Rinehart and Winston, 1973), p. 415.

[3] L. S. Feldt, "What Size Samples for Methods/Materials Experiments?," *Journal of Educational Measurement* 10 (1973): 221–226.

matter of "statistical significance"; it is also a matter of practical importance. In determining the appropriate sample size, the following factors[4] must be considered:

1. The level of significance (α).

2. The power of the test ($1 - \beta$).

3. The population error variance (σ^2).

4. The effect size (ES).

The level of significance (α) was defined in Chapter 8 as the probability of making a Type I error (rejecting the null hypothesis when it is true). In that chapter, we also defined the Type II error (not rejecting the null hypothesis when it is false) and indicated in a footnote that determining the probability of making a Type II error (β) requires specifying a value for the alternative hypothesis (H_a). We will discuss the strategy for determining β and subsequently the power of the test ($1 - \beta$) in the next section. The last two factors involved in determining the sample size, the population error variance (σ^2) and the effect size (ES), will be discussed later in the chapter.

Definition of Power

Recall from Chapter 8 that there are two errors inherent in hypothesis testing: the **Type I error** was defined as rejecting a true hypothesis, and the **Type II error** was defined as not rejecting a false hypothesis. In discussing these two errors, we indicated that the strategy involved in testing the null hypothesis focuses on the Type I error. This strategy involves establishing the null hypothesis as the "straw man" to be rejected so that the alternative hypothesis can be shown to be tenable. In this strategy, the researcher establishes, a priori, the acceptable risk of making a Type I error and carries out the statistical test accordingly. If the null hypothesis is rejected, the researcher concludes that the alternative hypothesis is tenable, remaining aware of the possibility that a Type I error might have been made.

But what if the null hypothesis is actually false? If the researcher rejects it, a proper decision is made; if not, a Type II error is made. Although it might seem that determining the probability of making a Type II error (β) is as straightforward as determining α, it is not: β can be determined only when values have been specified for both the alternative hypothesis (H_a) and the null hypothesis (H_0). Once the value for H_a is specified, it is possible to determine the probability of rejecting the

---

[4]Another factor that could be considered is the size of the population (N). Since knowing the size of the population is the exception rather than the rule, it is possible to determine the sample size using only the four factors listed above. For a more thorough discussion, see Hinkle and Oliver (1983) and Hinkle, Oliver, and Hinkle (1985).

null hypothesis when it is false $(1 - \beta)$. Because rejecting the false null hypothesis is precisely what the researcher wants to accomplish, the quantity $1 - \beta$ is defined as the **power** of the statistical test.

Power is defined as the probability of rejecting the null hypothesis when it is false $(1 - \beta)$.

One goal of any researcher is to minimize both Type I and Type II errors. By stating the null hypothesis (H_0) in hopes of rejecting it, the researcher must also be concerned with maximizing the power of the test $(1 - \beta)$. Obviously, by minimizing β, the power $(1 - \beta)$ is maximized. However, as we will see, there is an inverse relationship between α and β; as α increases, β decreases, and vice versa. Thus, certain tradeoffs must be made in attempting to minimize the probability of making both types of errors while maximizing the power of the test.

Whereas α is under the direct control of the researcher, β is not. The researcher can establish an α level, a priori, based on the severity of the consequences of making a Type I error. While the standard convention has been to set α at .05 or .01, lower α levels (.005 or .001) are sometimes established when the consequences are extremely serious. Higher levels (.20 or .10) are established when the consequences are not very serious. For example, in medical research, it is reasonable to set the α level very low (for example, .001) due to the potential harm to patients if a controversial drug is said to be more effective than a standard drug when it is in fact not (rejecting the null hypothesis when it is true).

Behavioral science researchers can often use higher α levels (.10) because of their concern about making Type II errors. For example, consider a researcher in education who is investigating two different teaching methods/environments and their effects on students' academic achievement. If the factors of teacher time and effort as well as the cost of the two programs are the same, the consequence of making a Type I error (implementing a program that is not better) is minimal. However, the consequences of making a Type II error (not implementing a program that is in fact superior) are more serious. The way the researcher can attempt to minimize the Type II error is to use higher α levels.

Due to the inverse relationship between α and β, the a priori selection of the α level should not be based solely on statistical convention; consideration must also be given to the consequences of making both Type I and Type II errors.

Determining the Power of the Test

We can begin discussing how to determine the power of the test by using the one-sample case for the mean from Chapter 8. In that example, we tested the null hypothesis that the mean Quantitative SAT score was equal to 455 (H_0: $\mu = 455$). Now consider testing this hypothesis against a directional alternative hypothesis:

$$H_0: \mu = 455$$

$$H_a: \mu > 455$$

We will use the standard error of the mean ($\sigma_{\overline{X}}$) from the example in Chapter 8. In that example, we used $\sigma = 100$ (the standard deviation in the population) and a sample of 144 freshman psychology students. Thus, the standard error of the mean is

$$\sigma_{\overline{X}} = \frac{\sigma}{\sqrt{n}}$$

$$= \frac{100}{\sqrt{144}}$$

$$= 8.33$$

The sampling distribution for this example is illustrated in Figure 13.1. Using the .05 level of significance (α), the critical value of the test statistic (z_{cv}) necessary for rejecting the null hypothesis is +1.645. That is, in order to reject the null hypothesis, the observed value of the sample mean ($\overline{X}$) must be greater than 1.645 standard errors above the hypothesized value of μ (455).

$$\mu + 1.645\sigma_{\overline{X}} = 455 + (1.645)(8.33)$$

$$= 455 + 13.70$$

$$= 468.70$$

In this example, the observed sample mean is 535, which is 9.60 standard errors above 455. Thus, the null hypothesis is rejected.

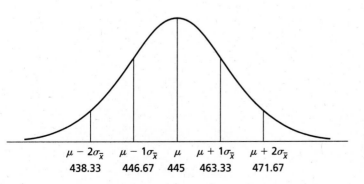

FIGURE 13.1
Sampling distribution of the mean for the hypothesis H_0: $\mu = 455$ and $\sigma_{\overline{X}} = 8.33$

$\mu - 2\sigma_{\overline{x}}$ $\mu - 1\sigma_{\overline{x}}$ μ $\mu + 1\sigma_{\overline{x}}$ $\mu + 2\sigma_{\overline{x}}$

438.33   446.67   445   463.33   471.67

Now, in order to determine the power of the statistical test of this null hypothesis, there must be a specified value for the alternative hypothesis (H_a). This value is specified in much the same way as is the value for the null hypothesis. Suppose the head of the psychology department is concerned about making a Type II error only if the value of the Quantitative SAT score is less than 10 points higher than 455. By making this assumption, an exact value for H_a can now be specified; namely, $H_a: \mu = 465$. We now have a specific value for both H_0 and H_a:

$$H_0: \mu = 455$$

$$H_a: \mu = 465$$

By specifying the value of H_a as 465, we have selected one of many possible values for H_a. In determining the power of the test of $H_0: \mu = 455$, we must assume that H_a is true and determine whether we would correctly reject the null hypothesis. In other words, we want to determine the power of our statistical test for detecting this 10-point difference.

The process is illustrated in Figure 13.2, which shows both the sampling distribution associated with H_0 and the sampling distribution associated with H_a. The

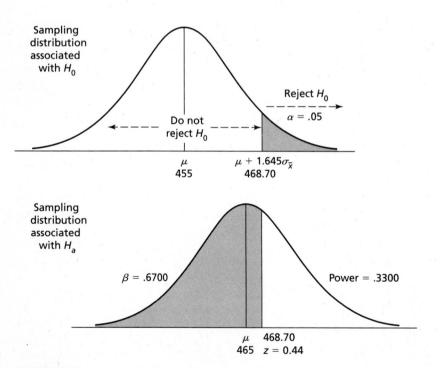

FIGURE 13.2
Power of test for $H_0: \mu = 455$ against $H_a: \mu = 465$ —
One-tailed test at $\alpha = .05$

only difference in the two distributions is the mean: 455 for H_0 and 465 for H_a. Notice that the critical value of the test statistic (z_{cv}), established in the sampling distribution associated with H_0, is not affected by the value established for μ in H_a. However, consider this critical value (468.70) when it is projected on the sampling distribution associated with H_a. The shaded area of the sampling distribution associated with H_a reflects the probability of not rejecting H_0 when it should be rejected (β). The unshaded area of the sampling distribution of H_a reflects the probability that H_0 would be correctly rejected. (It would be rejected, since, for this example, we assumed H_a: $\mu = 465$ to be true.) Notice that Figure 13.2 illustrates that the probability of correctly rejecting H_0: $\mu = 455$ when H_a is true is less than .50, since the unshaded area is less than half the area under the curve. In other words, the power of our test for detecting a 10-point difference is less than .50.

Calculating the exact value for the power of this test requires determining the unshaded area of the distribution associated with H_a, by using the methods outlined in Chapter 8. We first find the standard score of the critical value $[455 + (1.645)(8.33) = 468.70]$ when it is projected on the sampling distribution associated with H_a.

$$z = \frac{\overline{X} - \mu}{\sigma_{\overline{X}}}$$

$$= \frac{468.70 - 465}{8.33}$$

$$= \frac{3.70}{8.33}$$

$$= 0.44$$

Using Table C.1, we find that the shaded area is .6700 and the unshaded, .3300. Thus, the probability of making a Type II error (β) is .6700, and correspondingly, the power of the test ($1 - \beta$) is .3300. That is, for this example, since we assumed H_a: $\mu = 465$ to be true, there is only a .3300 probability of rejecting the null hypothesis (H_0: $\mu = 455$). In other words, this one-tailed test of the null hypothesis H_0: $\mu = 455$ at the .05 level of significance is not very powerful and has only a .3300 probability of detecting a 10-point difference.

Factors Affecting Power

The concept of power may initially seem difficult to learn and understand. However, several additional examples illustrating the factors that affect power should enhance understanding. The specific factors are:

1. The directional nature of H_a (one-tailed versus two-tailed test).

2. The level of significance (α).

3. The sample size (n).

4. The effect size (ES).

The Directional Nature of H_a   In the above example, we considered a one-tailed test of the null hypothesis H_0: $\mu = 455$ against the alternative hypothesis H_a: $\mu = 465$. Now consider the two-tailed test, again using $\alpha = .05$; this process is illustrated in Figure 13.3. Since we are conducting a two-tailed test at the .05 level of significance, the critical values of the test statistic are ± 1.96. That is, in order to reject the null hypothesis, the observed sample mean ($\overline{X}$) must be more than 1.96 standard errors below *or* 1.96 standard errors above the hypothesized value of μ.

$$\mu - 1.96\sigma_{\overline{X}} = 455 - (1.96)(8.33)$$
$$= 455 - 16.33$$
$$= 438.67$$

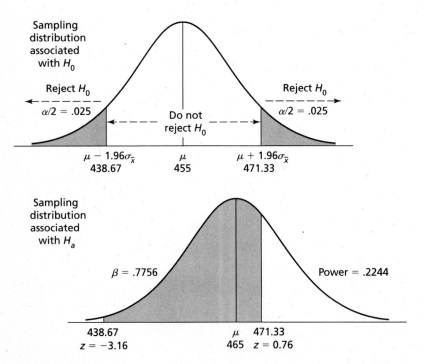

FIGURE 13.3
Power of test for H_0: $\mu = 455$ against H_a: $\mu = 465$ —
Two-tailed test at $\alpha = .05$

$$\mu + 1.96\sigma_{\overline{X}} = 455 + (1.96)(8.33)$$
$$= 455 + 16.33$$
$$= 471.33$$

Now consider the critical values [$455 - (1.96)(8.33) = 438.67$ *and* $455 + (1.96)(8.33) = 471.33$] when they are projected on the sampling distribution associated with H_a (see Figure 13.3). As before, the unshaded area of this sampling distribution of H_a reflects the power of the test (the probability of correctly rejecting H_0, since $\mu = 465$), whereas the shaded area reflects β (the probability of not rejecting H_0 when it should be rejected). Calculating the standard score of both critical values when projected on the sampling distribution associated with H_a, we have

$$z = \frac{438.67 - 465}{8.33}$$
$$= \frac{-26.33}{8.33}$$
$$= -3.16$$

for the left-hand tail and

$$z = \frac{471.33 - 465}{8.33}$$
$$= \frac{6.33}{8.33}$$
$$= 0.76$$

for the right-hand tail.

Using Table C.1, the shaded area to the left of $\mu = 465$ is .4992 and .2764 for the shaded area to the right of $\mu = 465$. Thus, the total shaded area is .4992 + .2764 = .7756, and the unshaded area is .2244. That is, the probability of making a Type II error (β) is .7756, and the power of this test ($1 - \beta$) is only .2244. Comparing this result with the power of the one-tailed test, we can see that the one-tailed test is *more* powerful than the two-tailed test (when the true population mean is in the direction hypothesized). Note, however, that neither the one-tailed test nor the two-tailed test was very powerful; the power for the one-tailed test was .3300, and the power for the two-tailed test was .2244.

With all other factors held constant, one-tailed tests are *more* powerful than two-tailed tests.

The Level of Significance   The second factor that affects power is the level of significance. As we indicated earlier in this chapter, as well as in Chapter 8, there is a relationship between α and β. With all other things held constant, if α is increased (increasing the probability of rejecting a true hypothesis) from, say, .05 to .10, then β decreases (decreasing the probability of retaining a false hypothesis); thus, the power $(1 - \beta)$ increases. Conversely, if α is decreased from .05 to .01, β increases, and thus the power $(1 - \beta)$ decreases.

Using the preceding example[5] to illustrate the relationship between α and β, recall that we found the power of the one-tailed test of H_0: $\mu = 455$, when H_a: $\mu = 465$ and using $\alpha = .05$, to be .3300. Now consider the power of a one-tailed test at the .10 level of significance; the process is illustrated in Figure 13.4. The critical value of the test statistic (z_{cv}) for testing the null hypothesis using $\alpha = .10$ is $+1.282$; that is, the observed sample mean $(\overline{X})$ must be 1.282 standard errors above the hypothesized value of μ in order to reject H_0.

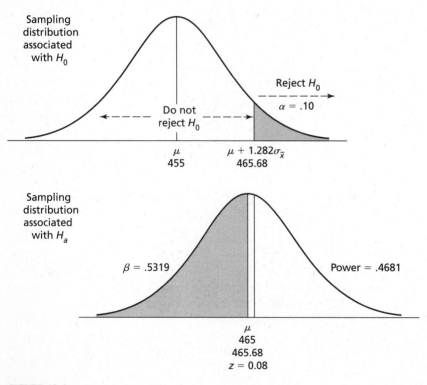

FIGURE 13.4
Power of test for H_0: $\mu = 455$ against H_a: $\mu = 465$ —
One-tailed test at $\alpha = .10$

[5] For consistency, we will use one-tailed tests in the determination of power.

$$\mu + 1.282\sigma_{\bar{x}} = 455 + (1.282)(8.33)$$

$$= 455 + 10.68$$

$$= 465.68$$

Using the methods outlined above for determining the shaded and unshaded areas of the sampling distribution associated with H_a, the standard score of the critical value (465.68) in the sampling distribution associated with H_a is found to be

$$z = \frac{465.68 - 465}{8.33}$$

$$= \frac{0.68}{8.33}$$

$$= 0.08$$

Using Table C.1, we find β to be equal to .5319 and the power $(1-\beta)$ to be .4681. Note that, for the one-tailed test when $\alpha = .05$, the power was .3300, and when $\alpha = .10$, the power was .4681. Thus, by increasing α from .05 to .10, the power of the test is increased.

> **With all other factors held constant, increasing the α level will result in a more powerful test.**

Now consider the power of the one-tailed test for this example, using $\alpha = .01$; the process is illustrated in Figure 13.5. The critical value (z_{cv}) for testing the null hypothesis at the .01 level of significance is +2.326. As before,

$$\mu + 2.326\sigma_{\bar{x}} = 455 + (2.326)(8.33)$$

$$= 455 + 19.38$$

$$= 474.38$$

Again using the methods outlined above for determining the shaded and unshaded areas of the sampling distribution associated with H_a, we can calculate the standard score for this critical value:

$$z = \frac{474.38 - 465}{8.33}$$

$$= \frac{9.38}{8.33}$$

$$= 1.13$$

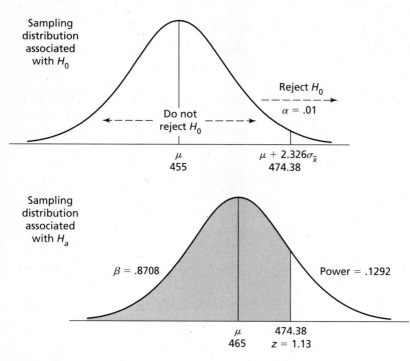

FIGURE 13.5
Power of test for H_0: $\mu = 455$ against H_a: $\mu = 465$ —
One-tailed test at $\alpha = .01$

Using Table C.1, we find β to be .8708 and the power $(1 - \beta)$ to be .1292. Thus, by decreasing α from .05 to .01, we decrease the power of the test from .3300 to .1292.

With all other factors held constant, decreasing the α level will result in a less powerful test.

The Sample Size   A third factor affecting the power of the test is the size of the sample. In Chapter 8, we discussed the relationship of the sample size to the concept of precision. We indicated that using larger sample sizes results in more precise estimates of parameters; that is, as the sample size increases, the standard error $(\sigma_{\bar{x}})$ decreases. In addition, a larger sample size results in a more powerful test of the null hypothesis if all other factors are held constant. Consider the same example. Up to this point, we have used the standard error of the mean $(\sigma_{\bar{x}})$ based on the standard deviation of the Quantitative SAT scores of the 144 freshman psychology students ($\sigma = 100$):

$$\sigma_{\overline{X}} = \frac{\sigma}{\sqrt{n}}$$

$$= \frac{100}{\sqrt{144}}$$

$$= 8.33$$

Suppose, however, we have 576 students in the example and the standard deviation is still 100. Then the standard error would be 4.17:

$$\sigma_{\overline{X}} = \frac{\sigma}{\sqrt{n}}$$

$$= \frac{100}{\sqrt{576}}$$

$$= 4.17$$

Now consider determining the power of the test of the null hypothesis H_0: $\mu = 455$ using $\alpha = .05$ for this larger sample; the process is illustrated in Figure 13.6. Notice that the shapes of the sampling distributions associated with both H_0 and H_a are different from those in earlier figures. This reflects the effect of the larger sample size. As illustrated, the critical value of the test statistic (z_{cv}) is still $+1.645$, but the standard error is smaller. In order to reject the null hypothesis for this one-tailed test at the .05 level of significance, the observed value of the sample mean ($\overline{X}$) must be greater than 1.645 standard errors above the hypothesized value of μ (455):

$$\mu + 1.645\sigma_{\overline{X}} = 455 + (1.645)(4.17)$$

$$= 455 + 6.86$$

$$= 461.86$$

Finding the standard score for 461.86 in the sampling distribution associated with H_a,

$$z = \frac{461.86 - 465}{4.17}$$

$$= \frac{-3.14}{4.17}$$

$$= -0.75$$

Using Table C.1, we find β to be .2266 and power $(1 - \beta)$ to be .7734. Thus by increasing the sample size, the power of the test is increased.

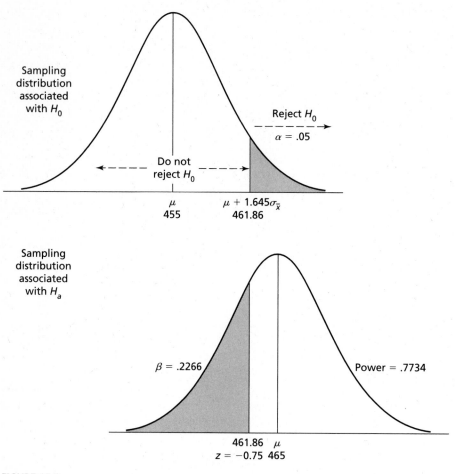

Sampling distribution associated with H_0

Reject H_0
$\alpha = .05$

Do not reject H_0

μ
455

$\mu + 1.645\sigma_{\bar{x}}$
461.86

Sampling distribution associated with H_a

$\beta = .2266$

Power $= .7734$

461.86 μ
$z = -0.75$ 465

FIGURE 13.6
Power of test for H_0: $\mu = 455$ against H_a: $\mu = 465$ —
One-tailed test at $\alpha = .05$ with $n = 576$ and $\sigma_{\bar{x}} = 4.17$

By increasing the sample size, the standard error is decreased and the power of the test is increased.

The Effect Size   The fourth factor affecting the power of the test is the effect size. Cohen defines effect size as the "degree to which a phenomenon exists." [6] For our purposes, we will define **effect size** as the difference between the value specified

[6] J. Cohen, *Statistical Power Analysis for the Behavioral Sciences*, 2d ed. (New York: Academic Press, 1977), p. 9.

in H_0 and the value specified in H_a, or the desired difference to be detected.[7] In the preceding example, we concerned ourselves with determining the power of the test of the null hypothesis for detecting a 10-point difference. As we have seen, the statistical test was not very powerful for any of the conditions that were varied, except when the larger sample size was used.

Now suppose the head of the psychology department is concerned about making a Type II error only if the population mean for the Quantitative SAT score is less than 20 points higher than 455; that is, $\mu < 475$. Given H_0: $\mu = 455$, the alternative hypothesis is H_a: $\mu = 475$ and the effect size (ES) is 20. The problem then becomes one of determining the power of the statistical test for detecting this 20-point difference. To determine the power of the test, we will use the one-tailed test of the null hypothesis H_0: $\mu = 455$ against the alternative hypothesis H_a: $\mu = 475$ at $\alpha = .05$ for the 144 freshman psychology students; the process is illustrated in Figure 13.7.

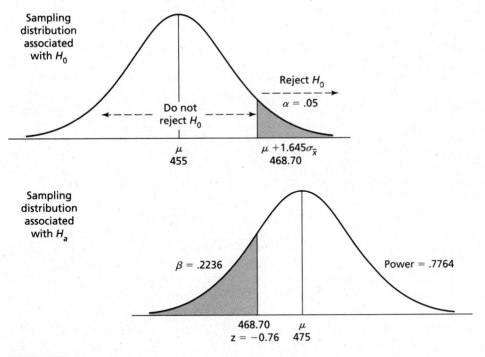

FIGURE 13.7
Power of test for H_0: $\mu = 455$ against H_a: $\mu = 475$—
One-tailed test at $\alpha = .05$ with ES $= 20$

[7] In a later section, we will define a standardized effect size as $d = \text{ES}/\sigma$.

As before, the critical value of the test statistic is not affected by the value established for H_a; the critical value (z_{cv}) is +1.645.

$$\mu + 1.645\sigma_{\overline{X}} = 455 + (1.645)(8.33)$$
$$= 455 + 13.70$$
$$= 468.70$$

Now consider the projection of this critical value on the sampling distribution associated with H_a. In determining the power, we first find the standard score of 468.70 in the sampling distribution associated with H_a:

$$z = \frac{468.70 - 475}{8.33}$$
$$= \frac{-6.30}{8.33}$$
$$= -0.76$$

Using Table C.1, we find β to be .2236 and $(1 - \beta)$, or power, to be .7764. Thus, whereas the power of the one-tailed test at $\alpha = .05$ to detect a 10-point difference is .3300, the power of this test to detect a 20-point difference is .7764.

> **With all other factors held constant, a test is more powerful when the effect size is larger.**

The Two-Sample Case

The process of determining the power of the statistical test for a two-sample case is identical to that described above for a one-sample case. To illustrate this process, we shall use the example from Chapter 11 in which a researcher studies the effects of two different teaching methods/environments on the mathematics achievement of third-grade students. In order to maintain consistency in presenting the process of determining power, we will use the one-tailed test of the null hypothesis, with the directional alternative being that the structured, fixed-time environment (group 1) will result in lower mathematics achievement scores than the less structured, flexible-time environment (group 2).

$$H_0: \mu_1 - \mu_2 = 0$$
$$H_a: \mu_1 - \mu_2 < 0$$

By using this example, two changes must be made in the process illustrated earlier. First, the critical value for this one-tailed test is in the *left-hand* tail of the sampling distribution of H_0. Second, the standard deviation of the sampling distribution for the two-sample case is the standard error of the difference $(s_{\overline{X}_1 - \overline{X}_2})$, defined by formula 11.1:

$$s_{\overline{X}_1 - \overline{X}_2} = \sqrt{s^2 \left(\frac{1}{n_1} + \frac{1}{n_2} \right)}$$

For this example, we will use the same pooled estimate for the population variance $(s^2 = 128.07)$, but we will assume that $n_1 = n_2 = 81$. Thus, the sampling distribution is the t distribution with $n_1 + n_2 - 2 = 81 + 81 - 2 = 160$ degrees of freedom (see Chapter 11).[8] Recall from Chapters 8 and 11 that, since the degrees of freedom are greater than 120, the normal distribution can be used as an adequate approximation for this t distribution. Using formula 11.1, the standard error of the difference is

$$s_{\overline{X}_1 - \overline{X}_2} = \sqrt{s^2 \left(\frac{1}{n_1} + \frac{1}{n_2} \right)}$$

$$= \sqrt{128.07 \left(\frac{1}{81} + \frac{1}{81} \right)}$$

$$= 1.78$$

Now, in order to determine the power of a statistical test, a specific value for the alternative hypothesis must be stated. Suppose the researcher in this example is concerned about making a Type II error only if the difference between μ_1 and μ_2 is less than 6 points. Therefore, assuming that group 1 will have the lower mean achievement at the end of the study, H_0 and H_a would be

$$H_0: \mu_1 - \mu_2 = 0$$
$$H_a: \mu_1 - \mu_2 = -6$$

As before, we assume that the alternative hypothesis is *true* and then determine the power of the statistical test for detecting a 6-point difference. This process is illustrated in Figure 13.8. The critical value of the test statistic is -1.645. That is, in order to reject H_0, the observed difference between the sample means $(\overline{X}_1 - \overline{X}_2)$ must be more than 1.645 standard errors (in absolute value) below the hypothesized value of the difference $(\mu_1 - \mu_2 = 0)$:

$$(\mu_1 - \mu_2) - (1.645)(s_{\overline{X}_1 - \overline{X}_2}) = 0 - (1.645)(1.78)$$

$$= -2.93$$

Notice in Figure 13.8 the projection of the critical value of the test statistic under

---

[8] For this example, since we are using the pooled estimate for the population variance, we will be using the symbol s^2 rather than σ^2, and the t distributions rather than the normal distributions for the sampling distribution.

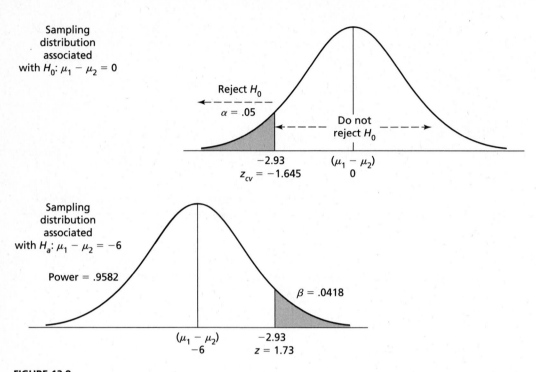

FIGURE 13.8
Power of test for $H_0: \mu_1 - \mu_2 = 0$ against $H_a: \mu_1 - \mu_2 = -6$ —
One-tailed test at $\alpha = .05$

$H_0(-2.93)$ on the sampling distribution associated with H_a.[9] Again, the unshaded area of this sampling distribution reflects the probability that H_0 would be rejected if H_a were true (it would be rejected, since, for this example, we are assuming that $\mu_1 - \mu_2 = -6$), and the shaded area reflects the probability of not rejecting H_0 when in fact H_0 is false (β).

In calculating the power of this test, we find the standard score of -2.93 in the sampling distribution associated with H_a.

$$z = \frac{-2.93 - (-6.0)}{1.78}$$

$$= \frac{3.07}{1.78}$$

$$= 1.73$$

[9]For degrees of freedom less than 50, the sampling distribution associated with H_a is the noncentral t distribution. The discussion of this distribution is beyond the scope of this book. See G. J. Resnikoff and G. J. Lieberman, *Tables of the Non-central t Distributions* (Stanford, Calif.: Stanford University Press, 1957).

Using Table C.1, we find that the shaded area is .0418 and the unshaded is .9582. Thus, the probability of making a Type II error (β) is .0418, and correspondingly, the power of the test $(1 - \beta)$ is .9582. That is, for this example, since we assumed H_a: $\mu_1 - \mu_2 = -6$ to be true, there is a .9582 probability of rejecting the null hypothesis H_0: $\mu_1 - \mu_2 = 0$. In other words, this one-tailed test of the null hypothesis at the .05 level of significance is very powerful and has a .9582 probability of detecting a 6-point difference.

> **The process of determining the power of the test for the two-sample case is identical to that for the one-sample case. The sampling distribution is the sampling distribution of the difference between means.**

It is important to note that the factors affecting the power of a statistical test for the two-sample case are exactly the same as for the one-sample case; namely,

1. The directional nature of H_a.

2. The level of significance (α).

3. The sample size (n).

4. The effect size (ES).

Examples illustrating how these factors affect the power for the two-sample case are left as exercises at the end of this chapter.

Power: An A Priori Consideration for Determining Sample Size

Now that we have defined the concept of power and illustrated how power can be determined, we return to the main question: how large should the sample be? Recall that power was only one of four factors to be considered in determining the sample size. The four factors were

1. The level of significance (α).

2. The power of the test $(1 - \beta)$.

3. The population error variance (σ^2).

4. The effect size (ES).

Thus, rather than determining the power of the test *after* collecting the data, it is more important to consider the desired power *before* collecting the data. Although the process of determining the power after the fact (a posteriori), illustrated in the previous sections, is crucial to understanding the concept of power, the question of an acceptable level of power is actually an a priori consideration.

> **The four factors to be considered in determining the sample size are:**
> 1. Level of significance (α).
> 2. Power of the test ($1 - \beta$).
> 3. Population error variance (σ^2).
> 4. Effect size (ES).

In determining, a priori, an acceptable level of power for a statistical test, the relationship between α and β must be considered. Several authors[10] have indicated that, for most behavioral science studies, Type I errors are generally more serious than Type II errors, and they have suggested a 4:1 ratio of β to α. That is, if the level of significance (α) is established, a priori, at .05, then the corresponding power is $1 - 4(.05) = .80$. For all the examples in this chapter, we will use the 4:1 ratio of β to α.

> In determining an appropriate sample size before starting a research investigation, an acceptable level of power should also be established a priori.

Methods for Determining Sample Size

Up to this point, we have discussed two of the factors that must be considered in determining, a priori, the size of the sample: (1) the level of significance and (2) the power. The other factors — (3) the population error variance and (4) the effect size — will be discussed as we illustrate the process of determining the appropriate sample size, given a priori values for α and power. Consider again the example of testing the null hypothesis H_0: $\mu = 455$. Recall that the head of the psychology department was concerned with making a Type II error only if the Quantitative SAT score were less than 10 points higher than 455. The question now becomes: how large must the sample be in order to conclude that this 10-point difference is

---

[10]Q. McNemar, *Psychological Statistics*, 4th ed. (New York: John Wiley & Sons, 1960); J. Cohen, "Some statistical issues in psychological research," in *Handbook of Clinical Psychology*, ed. B. B. Wolman (1965): 95–121; and L. J. Chase and R. K. Tucker, "Statistical power: Derivation, development and data-analytic implications," *The Psychological Record*, 26 (1976): 473–486.

statistically significant, resulting in the rejection of the H_0? For this example, the values for the four factors that affect the sample size are as follows:

1. *The level of significance.* We will use $\alpha = .05$ and a one-tailed test of H_0. Thus, the critical value of the test statistic is +1.645.

2. *The power of the test.* With $\alpha = .05$ and a 4:1 ratio of β to α, we establish power to be $1 - 4(.05) = .80$.

3. *The population error variance (σ^2).* Before the data are collected, we do not have an estimate of the population error variance. However, for this example, the SAT has been normed to have a standard deviation of 100 (variance of 10,000), so we will use this value. In the next section, we will illustrate the procedure for determining the sample size without a specific value for the population variance.

4. *The effect size.* The effect size was set at 10 by the head of the psychology department.

What we need now is the sample size that will result in a standard error of the mean ($\sigma_{\bar{X}}$) that will position the sampling distributions associated with both H_0 and H_a as illustrated in Figure 13.9. Notice that, because $\alpha = .05$, the critical value of the test statistic is +1.645. This value, projected onto the sampling distribution associated with H_a, must intercept at the z value of -0.842 in order for β to be equal to .20 and power to be equal to .80 (see Figure 13.9).

The formula[11] for determining the appropriate sample size (n) is as follows:

$$n = \frac{\sigma^2(z_\beta - z_\alpha)^2}{(\text{ES})^2} \tag{13.1}$$

where

$\sigma^2 = $ population error variance

$z_\beta = $ standard score in the sampling distribution associated with H_a corresponding to z_α for a given power

$z_\alpha = $ critical value of the test statistic in the sampling distribution associated with H_0 for a one-tailed test at a given α

$\text{ES} = $ effect size

For the above example,

$$n = \frac{(100)^2(-.842 - 1.645)^2}{(10)^2}$$

$$= \frac{(10,000)(-2.487)^2}{100}$$

---

[11]For the derivation of the formula, see J. P. Guilford and B. Fruchter, *Fundamental Statistics in Psychology and Education*, 6th ed. (New York: McGraw-Hill, 1978), p. 185.

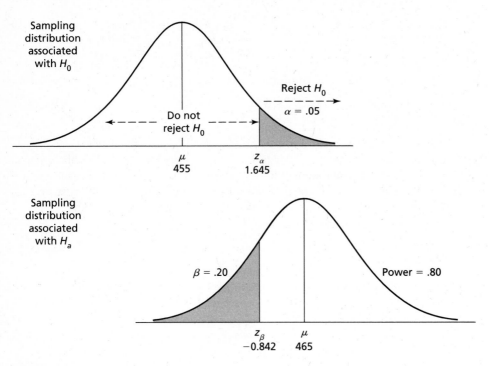

FIGURE 13.9
Determining the sample size for $H_0: \mu = 455$ against $H_a: \mu = 465$ —
One-tailed test with $\alpha = .05$ and ES $= 10$

$$= \frac{(10,000)(6.1852)}{100}$$

$$= 618.52, \text{ or } 619$$

Thus, the sample size necessary for this one-tailed test of the null hypothesis H_0: $\mu = 455$ at $\alpha = .05$, with .80 power to detect a 10-point difference, is 619. Note that, in determining the appropriate sample size, we must always round up to the next integer.

Now consider the sample size needed for a two-tailed test of this hypothesis, as illustrated in Figure 13.10. Notice that, for the two-tailed test, the critical values are in both tails of the sampling distribution associated with H_0 (± 1.96). However, when the left-hand critical value (-1.96) is projected onto the sampling distribution associated with H_a, it is in the extreme left-hand tail. Therefore, even though this is a two-tailed test, we will be concerned with only the right-hand critical value ($+1.96$) as projected on the sampling distribution associated with H_a.

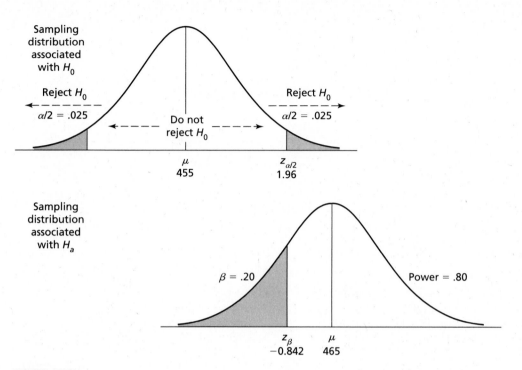

FIGURE 13.10
Determining the sample size for $H_0: \mu = 455$ against $H_a: \mu = 465$ —
Two-tailed test with $\alpha = .05$ and ES = 10

The formula[13] for determining the sample size for a two-tailed test is as follows:

$$n = \frac{\sigma^2 (z_\beta - z_{\alpha/2})^2}{(ES)^2} \tag{13.2}$$

where

$z_{\alpha/2}$ = critical value of the test statistic in the sampling distribution associated
with H_0 for a two-tailed test at a given α

z_β = standard score in the sampling distribution associated with H_a,
corresponding to $z_{\alpha/2}$ for a given power

σ^2 = population error variance

ES = effect size

---

[13] For the derivation of the formula, see Guilford and Fruchter.

For this example,

$$n = \frac{(100)^2(-.842 - 1.96)^2}{(10)^2}$$

$$= \frac{(10,000)(-2.802)^2}{100}$$

$$= \frac{(10,000)(7.8512)}{100}$$

$$= 785.12, \text{ or } 786$$

Thus, the sample size necessary for the two-tailed test of the null hypothesis H_0: $\mu = 455$ at $\alpha = .05$, with power to detect a 10-point difference, is 786. Notice that the sample size required for the two-tailed test is larger than for the one-tailed test.

> For preestablished values of α and power, a larger sample size is required to reject H_0 for a two-tailed test than for a one-tailed test for a given effect size.

Because we are considering a two-tailed test, we could have established the effect size as -10. In other words, we could have tested H_0: $\mu = 455$ against H_a: $\mu = 445$ (see Figure 13.11). In this case, the critical value of interest would be in the left-hand tail of the sampling distribution of H_0 ($z_{cv} = -1.96$). The projection of this critical value onto the sampling distribution associated with H_a would be at the point $z = +0.842$ (see Figure 13.11). This underscores the fact that, in using the process described here, it is important to sketch the distributions associated with H_0 and H_a before proceeding to determine the sample size.

Standardized Effect Size: The One-Sample Case

In behavioral science research, knowledge of the population error variance (σ^2) and thus the population standard deviation (σ), before beginning the research, is the exception rather than the rule. However, it is acceptable to use a variance cited in the research literature to determine an appropriate sample size. For example, when we considered the Quantitative SAT scores for the 144 psychology students, the standard deviation was assumed to be 100 since the SAT is a normed test with a fixed standard deviation. But knowledge of the exact value of the variance or standard deviation is *not* a prerequisite for determining the sample size, if the researcher is willing to express the effect size (ES) in terms of standard deviation units. In the SAT example, the ES was 10, and the standard deviation was 100. In

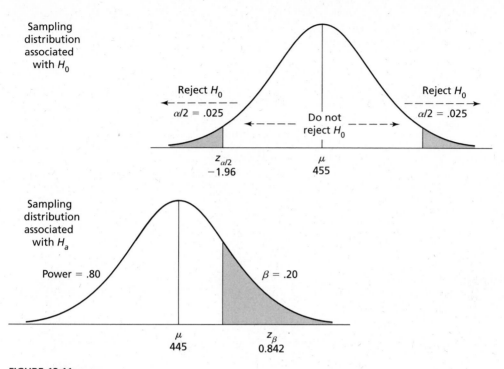

FIGURE 13.11
Determining the sample size for H_0: $\mu = 455$ against H_a: $\mu = 445$—
Two-tailed test with $\alpha = .05$ and ES $= -10$

other words, the ES was one-tenth (.10) standard deviation, and the standardized effect size would be .10. With this approach, we will define d as the effect size in standard deviation units.[13]

$$d = \frac{\text{ES}}{\sigma} \qquad (13.3)$$

When σ^2 is unknown and we must use s^2 as the estimate, the **standardized effect size** would be defined as

$$d = \frac{\text{ES}}{s}$$

With this definition, formulas 13.1 and 13.2 reduce to the following for one-tailed tests:

$$n = \frac{(z_\beta - z_\alpha)^2}{d^2} \qquad (13.4)$$

[13]Cohen (1962).

and for two-tailed tests:

$$n = \frac{(z_\beta - z_{\alpha/2})^2}{d^2}$$ (13.5)

To explore the use of these reduced formulas, consider the above example using a standardized effect size of 0.20 with $\alpha = .05$ and power $= .80$.

$$d = \frac{ES}{\sigma}$$

$$= \frac{20}{100}$$

$$= .20$$

For the one-tailed test, using formula 13.4, the required sample size for rejecting H_0: $\mu = 455$ is

$$n = \frac{(-.842 - 1.645)^2}{(.20)^2}$$

$$= \frac{(-2.487)^2}{(.20)^2}$$

$$= 154.63, \text{ or } 155$$

For the two-tailed test, using formula 13.5,

$$n = \frac{(-.842 - 1.96)^2}{(.20)^2}$$

$$= \frac{(-2.802)^2}{(.20)^2}$$

$$= 196.28, \text{ or } 197$$

As illustrated before, holding other factors constant, the sample size required for rejecting H_0 using a two-tailed test is greater than for a one-tailed test. Notice also that the sample size required to reject H_0 for $d = 0.20$ is less than that for $d = 0.10$. This illustrates the relationship between power and effect size for a fixed sample size, as well as the relationship between effect size and sample size when power is fixed; the smaller the effect size, the greater the required sample size.

To determine an appropriate sample size, the following factors must be considered:

1. The level of significance (σ).
2. The power of the test ($1 - \beta$).
3. The standardized effect size (d).
4. The directionality of the test (one- or two-tailed).

Standardized Effect Size: The Two-Sample Case

The process for determining the appropriate sample size for the two-sample case is the same as for the one-sample case; only the formulas are different. It is important to note, however, that the resulting sample size is the number of subjects that are to be assigned to *each* of the two groups under the assumption of equal sizes for the two groups. The formulas[14] for the two-sample case are as follows:

for one-tailed tests:

$$n = \frac{2\sigma^2(z_\beta - z_\alpha)^2}{(ES)^2}$$

or for a standardized effect size (d):

$$n = \frac{2(z_\beta - z_\alpha)^2}{d^2} \tag{13.6}$$

for two-tailed tests:

$$n = \frac{2\sigma^2(z_\beta - z_{\alpha/2})^2}{(ES)^2}$$

or for a standardized effect size (d):

$$n = \frac{2(z_\beta - z_{\alpha/2})^2}{d^2} \tag{13.7}$$

Consider again the example from Chapter 11 involving the mathematics achievement of two groups of third-grade students exposed to different teaching methods/environments. Recall that, in calculating the standard error of the difference ($s_{\overline{X}_1 - \overline{X}_2}$), the pooled estimate of the population variance (s^2) was computed and found to be equal to 128.07. Correspondingly, the estimated standard deviation is $\sqrt{128.07} = 11.32$. With this estimate, let us determine the sample size necessary for a 6-point difference in the mean mathematics achievement scores to be

---

[14]For the derivation of the formulas, see Guilford and Fruchter.

statistically significant. The four factors considered are:

1. *The level of significance.* We will use $\alpha = .05$ and a one-tailed test of H_0: $\mu_1 - \mu_2 = 0$. Thus, the critical value is -1.645.

2. *The power of the test.* Using $\alpha = .05$ and a 4:1 ratio of β to α, we establish power to be .80.

3. *The population error variance (σ^2).* We will use $s^2 = 128.07$ as the estimated population variance; the estimated standard deviation is $\sqrt{128.07} = 11.32$.

4. *The effect size.* Based on the previous assumption that a 6-point difference is substantial, we will use $ES = -6$ and $d = -6/11.32 = -0.530$.

Using formula 13.6 to find the necessary sample size (notice the signs of the values for z_α and z_β),

$$n = \frac{2[.842 - (-1.645)]^2}{(-.530)^2}$$

$$= \frac{2(2.487)^2}{(-.530)^2}$$

$$= 44.038, \text{ or } 45$$

Thus, 45 subjects are required for *each* group.

As might be expected, the process of determining the appropriate sample size for a two-tailed test is the same as for a one-tailed test; the only difference is that we use formula 13.7. For the above example, using $\alpha = .05$ and power $= .80$, the values for z_α and z_β in the first step are $+0.842$ and -1.96, respectively.

$$n = \frac{2[.842 - (-1.96)]^2}{(-.530)^2}$$

$$= \frac{2(2.802)^2}{(-.530)^2}$$

$$= 55.900, \text{ or } 56$$

Thus, 56 subjects are required for *each* group.

It is also possible to determine the sample size necessary to detect a standardized effect size (d) without using the estimated standard deviation. For this example, suppose we want to determine the sample size for detecting a standardized effect size of 0.50. Applying formulas 13.6 and 13.7 for this standardized effect size,[15] the required sample sizes for the one-tailed and two-tailed tests are

---

[15]Since the effect size is squared in the formula, the sign can be ignored.

as follows:

For the one-tailed test,

$$n = \frac{2[.842 - (-1.645)]^2}{(.50)^2}$$

$$= \frac{2(2.487)^2}{(.50)^2}$$

$$= 49.481, \text{ or } 50$$

Thus, 50 subjects are required for *each* group.

For the two-tailed test,

$$n = \frac{2[.842 - (-1.96)]^2}{(.50)^2}$$

$$= \frac{2(2.802)^2}{(.50)^2}$$

$$= 62.810, \text{ or } 63$$

Thus, 63 subjects are required for *each* group.

Tables for Sample Size Determination

As shown above, the process of determining the appropriate sample size can be very time consuming. To expedite the process, we have included tables in the Appendix that were developed using the methods outlined above, including the iterative process when small sample sizes are found in the first step. Table C.11 is for the one-sample case; Table C.13 is for the k-sample case where $k \geq 2$. In this chapter, we have been concerned only with determining the appropriate sample size for $k = 1$ and $k = 2$. In the next chapter, we will discuss methods for determining the sample size when the number of groups is greater than 2.

To use Table C.11 in determining the appropriate sample size for the one-sample case, values for the following must be specified:

1. The level of significance (α).
2. The power of the test ($1 - \beta$).
3. The directional nature of H_a (one-tailed or two-tailed test).
4. The standardized effect size ($d = \text{ES}/\sigma$).

For the one-sample case example, we used the following values and found the appropriate sample size to be 619:

1. $\alpha = .05$.
2. Power $= .80$.
3. One-tailed test.
4. $d = 0.10$.

Using Table C.11 for these values, we find the tabled value to be 619. Additional examples for using Table C.11 have been included in the exercises at the end of the chapter. *A note of caution:* there may be slight discrepancies between the computed values from using formulas 13.4 and 13.5 and the table values. The differences are a result only of rounding error.

For the two-sample case, we use Table C.13; however, the table contains values for two-tailed tests *only*. In order to use Table C.13, values for the following must be specified:

1. The level of significance (α).

2. The power of the test ($1 - \beta$).

3. The number of groups (k).

4. The standardized effect size ($d = \text{ES}/\sigma$).

First consider the example in which the standardized effect size was set at 0.50 with

1. $\alpha = .05$.

2. Power $= .80$.

3. $k = 2$.

From Table C.13, we find the required sample size to be 62; the calculated value using formula 13.7 is 62.810. The difference is due to rounding error.

Now use Table C.13 for the standardized effect size of 0.530, which corresponds to a 6-point difference between μ_1 and μ_2. (Notice that we have ignored the sign of the effect size, because the value is squared in the formulas.) Using formula 13.7 with this standardized effect size, the resulting sample size is 56 for *each* group. Notice that there is *not* a table value for this specific effect size; it is necessary to estimate the value of the appropriate sample size from the table. The nearest table values are for $d = 0.50$ and $d = 0.75$: 62 and 29, respectively. For $d = -0.530$ (in absolute value), our estimated sample size is approximately 56 — the value computed above. Thus, the sample size for *each* group needed to reject H_0 for this effect size (-0.530), given the a priori values for the level of significance (.05) and power (.80), is 56.

Effect Size: The More Important Question

Now that we have defined power and discussed the procedures for determining power and sample size, we can discuss what is considered to be the more important question: what is an appropriate effect size? In answering this question, we must reconsider Cohen's definition of effect size: the "degree to which a phenomenon exists." In addition to this definition, Cohen[16] goes on to define qualitatively "small," "medium," and "large" effect sizes in standardized units.

$$\text{Small} = .25\sigma$$

$$\text{Medium} = .50\sigma$$

$$\text{Large} = 1.00\sigma$$

Although these qualitative classifications are initially pleasing, Cohen acknowledges the ambiguity inherent in them:

> These qualitative adjectives . . . may not be reasonably descriptive in any specific area. Thus what a sociologist may consider a small effect may well be appraised as medium by a clinical psychologist.[17]

Other authors have also tried to classify effect sizes. For example, Feldt contrasts his classifications with those proposed by Cohen:

> For experiments involving two to four treatments $[2 \leq k \leq 4]$, Cohen's "large" effect corresponds to a range of means of $.80\sigma$ or more. A "medium" effect involves a range of $.50\sigma$ or more. A "small" effect represents a range of $.20\sigma$ or more. Clearly our "large" effect is even smaller than Cohen's "medium" effect.[18]

So now the behavioral science researcher, initially concerned with the question of an appropriate sample size, is faced with an even more perplexing question: that of an appropriate effect size. And, as Cook and Campbell point out,

> The problem of specifying magnitudes is . . . sometimes one of "consciousness," for the issue may simply not be considered in designing the research. Alternatively, it may be silently considered by some persons but never brought to the level of discussion for fear different parties to the research may disagree on the level of effect required to conclude a treatment has made a significant, practical difference.[19]

Therefore, we encourage the reader to recognize the complexity of the question of determining the sample size, its relationship to α and to the power of the test, and most important, its relationship to the effect size.

---

[16]Cohen (1965).

[17]Cohen (1965), p. 278.

[18]Feldt, p. 224.

[19]T. D. Cook and D. T. Campbell, *Quasi-experimentation: Design and Analysis Issues for Field Studies* (Chicago: Rand McNally, 1979), p. 46.

Answering the question regarding an appropriate effect size must precede answering the question regarding an appropriate sample size.

Summary

This chapter has involved an extensive discussion of the procedures for determining an appropriate sample size. Although the process might initially appear to be rather simple, it is actually very complex because many factors must be considered simultaneously. One of the major factors is the power of the statistical test, defined as the probability of rejecting the null hypothesis when it is false. Recall that the standard research convention (discussed in previous chapters) involves stating and testing the null hypothesis in hopes of rejecting it. Thus, the researcher seeks to use the most powerful test in order to reject the null hypothesis and provide a rationale for the tenability of the alternative hypothesis (H_a).

The relationship between the size of the sample (n) and the power of the statistical test $(1 - \beta)$ is that the larger the sample, the more powerful the test. Other factors affecting power are the level of significance and the directional nature of H_a. Using the sampling distributions of both H_0 and H_a, holding all other factors constant, increasing the level of significance increases the power of the test. Similarly, one-tailed tests are more powerful than two-tailed tests. Table 13.1 summarizes the computational formulas for determining sample size.

TABLE 13.1
Summary of Formulas for Determining Sample Size

		One-Sample Case	Two-Sample Case
Effect Size (ES)	One-Tailed Test of H_0	$n = \dfrac{\sigma^2(z_\beta - z_\alpha)^2}{(ES)^2}$	$n = \dfrac{2\sigma^2(z_\beta - z_\alpha)^2}{(ES)^2}$
	Two-Tailed Test of H_0	$n = \dfrac{\sigma^2(z_\beta - z_{\alpha/2})^2}{(ES)^2}$	$n = \dfrac{2\sigma^2(z_\beta - z_{\alpha/2})^2}{(ES)^2}$
Standardized Effect Size (d)	One-Tailed Test of H_0	$n = \dfrac{(z_\beta - z_\alpha)^2}{d^2}$	$n = \dfrac{2(z_\beta - z_\alpha)^2}{d^2}$
	Two-Tailed Test of H_0	$n = \dfrac{(z_\beta - z_{\alpha/2})^2}{d^2}$	$n = \dfrac{2(z_\beta - z_{\alpha/2})^2}{d^2}$

Understanding the concept of statistical power is a prerequisite to understanding the complex process of sample size determination. The process does *not* involve the actual calculation of the power of the test, but rather the determination of an acceptable power, a priori, along with the level of significance and the directional nature of H_a. However, the most important factor to be considered is the effect size (ES), or the degree to which the phenomenon exists. Thus, the sample size question is not simply — how large should the sample be? — but rather: how large should the sample be *in order to detect a specific effect size?* The specific effect size is predetermined by the researcher. The answer to this latter question requires giving consideration not only to the level of significance and the power of the test, but also to the effect size.

Exercises

1. The president of a major educational foundation is investigating the average salary of faculty members in major state universities across the United States. He decides to test the null hypothesis H_0: $\mu = 24,500$ and is concerned with making a Type II error only if the average salary is *less* than $600 *above* this hypothesized value. In this case, the effect size is set at 600; the alternative hypothesis is H_a: $\mu = 25,100$. Assume that the population standard deviation (σ) equals $4,800.

   a. For $\alpha = .05$ and $n = 144$, determine the power of *both* a one-tailed and a two-tailed test of the null hypothesis H_0: $\mu = 24,500$ against the alternative hypothesis H_a: $\mu = 25,100$.

   b. For the above example, consider only a one-tailed test. For $n = 144$, determine the power of the test of H_0: $\mu = 24,500$ against H_a: $\mu = 25,100$ for $\alpha = .01$ and $\alpha = .10$.

   c. For a one-tailed test and $\alpha = .05$, determine the power of the test of H_0: $\mu = 24,500$ against H_a: $\mu = 25,100$ for $n = 400$.

   d. Consider the effect size equal to 1,000; that is, testing H_0: $\mu = 24,500$ against H_a: $\mu = 25,500$. Find the power of a one-tailed test for $\alpha = .05$ and $n = 144$.

2. Suppose a researcher is testing the null hypothesis H_0: $\mu = \mu_0$, where μ_0 is some specified value. Determine the power of the test of this hypothesis (using $\alpha = .05$) for the following:

   a. Assume that the effect size equals 1 standard error above the value of μ_0; that is,

   H_0: $\mu = \mu_0$
   H_a: $\mu = \mu_0 + 1\sigma_{\overline{X}}$

   Notice that the standard error contains the value for n; the sample size is fixed.

   b. Similarly, find the power of the test when the effect size is 2 standard errors above the value of μ_0; that is,

   H_0: $\mu = \mu_0$
   H_a: $\mu = \mu_0 + 2\sigma_{\overline{X}}$

    c. Now determine the power of the test when the effect size is 3 standard errors above the value of μ_0; that is,

$$H_0: \mu = \mu_0$$
$$H_a: \mu = \mu_0 + 3\sigma_{\overline{X}}$$

*3. Researchers are interested in the perception of, and attitudes towards, physical fitness of adults over 40 years of age. Thay have an extensive inventory for measuring such perceptions and attitudes. They decide to test the null hypothesis $H_0: \mu = 100$ and are concerned about a Type II error if the average score is more than five points above the hypothesized value of 100. As such, the alternative hypothesis becomes $H_a: \mu = 105$; the standard deviation on the inventory is 24.

    a. Suppose the researchers select a random sample of size 128. Determine the power of a one-tailed test of the null hypothesis $H_0: \mu = 100$ against the alternative hypothesis $H_a: \mu = 105$ for $\alpha = .05$.

    b. Suppose the sample size is doubled to 256 and all other conditions remain the same. How will this affect the power of the test?

4. An industrial psychologist is studying the productivity of employees in small industries with end-of-the-year pay incentives and those in small industries without such incentives. For this study, the psychologist randomly selects 100 employees from both types of industries and determines their productivity using a standardized instrument. From the data, the pooled estimate of the population variance (s^2) is found to be 250. Therefore, the standard error of the difference is as follows:

$$s_{\overline{X}_1 - \overline{X}_2} = \sqrt{250\left(\frac{1}{100} + \frac{1}{100}\right)} \qquad \text{(see formula 11.1)}$$

$$= 2.236$$

    a. For $\alpha = .05$ and effect size equal to 5.0, determine the power for *both* a one-tailed and a two-tailed test of the null hypothesis. The null and alternative hypotheses are

$$H_0: \mu_1 - \mu_2 = 0$$
$$H_a: \mu_1 - \mu_2 = 5.0$$

    b. For effect size (ES) equal to 5.0 and a one-tailed test, determine the power for $\alpha = .01$ and $\alpha = .10$.

    c. Now for $n_1 = n_2 = 250$, determine the standard error of the difference and the power of a one-tailed test for $\alpha = .05$ and ES $= 5.0$.

    d. Now for the original study with $n_1 = n_2 = 100$, let the effect size (ES) equal 3.0. Find the power of a one-tailed test at $\alpha = .05$.

5. The chairman of the Federal Aviation Administration (FAA) has asked a training subcommittee to investigate the current regulations for issuing licenses to sailplane pilots. The null hypothesis to be tested is that the average number of hours of solo flights before awarding a license (regardless of the length of the flights) equals 40; that is, $H_0: \mu = 40$. Assume that the standard deviation in the population (σ) is 12.5. The subcommittee is concerned with making a Type II error only if the average number of flights is *more* than 2 flights *below* this hypothesized value. Thus, the effect size (ES) equals 2, and the null and alternative hypotheses are as follows:

$$H_0: \mu = 40$$
$$H_a: \mu = 38$$

a. For $\alpha = .05$ and $n = 140$, determine the power of *both* a one-tailed and a two-tailed test of the null hypothesis $H_0: \mu = 40$ against the alternative hypothesis $H_a: \mu = 38$.

b. For the above example, consider only a one-tailed test. For $n = 140$, determine the power of the test of $H_0: \mu = 40$ against $H_a: \mu = 38$ for $\alpha = .01$ and $\alpha = .10$.

c. For a one-tailed test and $\alpha = .05$, determine the power of the test of $H_0: \mu = 40$ against $H_a: \mu = 38$ for $n = 220$.

d. Consider the effect size (ES) equal to 1.5, testing $H_0: \mu = 40$ against $H_a: \mu = 38.5$. Find the power of a one-tailed test for $\alpha = .05$ and $n = 140$.

6. A family sociologist is interested in the difference in marital satisfaction between working and nonworking wives. Using a standardized marital satisfaction inventory with a standard deviation (σ) of 25, determine the power of the test of the null hypothesis $H_0: \mu_1 - \mu_2 = 0$ under the following conditions:

a. For $n_1 = n_2 = 75$, the standard error of the difference is

$$s_{\bar{X}_1 - \bar{X}_2} = \sqrt{(25)^2 \left(\frac{1}{75} + \frac{1}{75} \right)} \qquad \text{(see formula 11.1)}$$

$$= 4.08$$

For $\alpha = .05$, determine the power for *both* a one-tailed and a two-tailed test of the null hypothesis $H_0: \mu_1 - \mu_2 = 0$ against the alternative hypothesis $H_a: \mu_1 - \mu_2 = 7.5$ (the effect size equals 7.5).

b. For effect size equal to 7.5 and a one-tailed test, determine the power for $\alpha = .01$ and $\alpha = .10$.

c. Now let $n_1 = n_2 = 120$, determine the standard error of the difference, and determine the power of a one-tailed test for $\alpha = .05$ and $ES = 5.0$.

d. Now for the original study, in which $n_1 = n_2 = 75$, let the effect size equal 3.5. Find the power of a one-tailed test at $\alpha = .05$.

7. In Exercise 1, the president of the educational foundation was testing the null hypothesis $H_0: \mu = 24,500$. Using $\sigma = 4,800$, determine the appropriate sample size given the following conditions. Check your answers with Table C.10.

a. $\alpha = .05$, $\beta = .20$, $ES = 600$, one-tailed *and* two-tailed tests.

b. $\alpha = .05$, $\beta = .20$, $ES = 1,200$, one-tailed test only.

c. $\alpha = .01$, $\beta = .05$, $ES = 600$, one-tailed *and* two-tailed tests.

d. $\alpha = .01$, $\beta = .05$, $ES = 1,200$, one-tailed test only.

8. In Exercise 4, the industrial psychologist was testing the null hypothesis H_0: $\mu_1 - \mu_2 = 0$. Using $s^2 = 250$, determine the appropriate sample size given the following conditions. For the two-tailed tests, check your answers with Table C.11.

   a. $\alpha = .05$, $\beta = .20$, ES $= 12.5$, one-tailed *and* two-tailed tests.

   b. $\alpha = .05$, $\beta = .20$, ES $= 8.5$, two-tailed test only.

   c. $\alpha = .01$, $\beta = .05$, ES $= 12.5$, one-tailed *and* two-tailed tests.

   d. $\alpha = .01$, $\beta = .05$, ES $= 8.5$, two-tailed test only.

9. In Exercise 5, the chairman of the FAA was testing the null hypothesis H_0: $\mu = 40$. Determine the appropriate sample size given the following conditions. Notice that a value for the standardized effect size (d) is given instead of an actual effect size (ES). Check your answers with Table C.10.

   a. $\alpha = .05$, $\beta = .20$, $d = 0.20$, one-tailed *and* two-tailed tests.

   b. $\alpha = .05$, $\beta = .20$, $d = 0.50$, one-tailed test only.

   c. $\alpha = .01$, $\beta = .05$, $d = 0.20$, one-tailed *and* two-tailed tests.

   d. $\alpha = .01$, $\beta = .05$, $d = 0.50$, one-tailed test only.

10. In Exercise 6, the family sociologist was testing the null hypothesis H_0: $\mu_1 - \mu_2 = 0$. Determine the appropriate sample size given the following conditions. Notice that a value for the standardized effect size (d) is given instead of an actual effect size (ES). For the two-tailed tests, check your answers with Table C.11.

   a. $\alpha = .05$, $\beta = .20$, $d = 0.25$, one-tailed *and* two-tailed tests.

   b. $\alpha = .05$, $\beta = .20$, $d = 0.40$, two-tailed test only.

   c. $\alpha = .01$, $\beta = .05$, $d = 0.25$, one-tailed *and* two-tailed tests.

   d. $\alpha = .01$, $\beta = .05$, $d = 0.40$, two-tailed test only.

14

Hypothesis Testing, *K*-Sample Case: Analysis of Variance, One-Way Classification

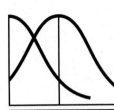

Key Concepts

Type I error rate
Levels of an independent variable
Total variation
Within-groups variation
Between-groups variation
Linear model
Sum of squares
Mean squares

Pooled estimate of the population variance
Expected mean squares
F ratio
F distributions
Homogeneity of variance
Repeated measures
Residual variation

So far, we have discussed statistical procedures that can be used to test hypotheses or to estimate specific parameters. These procedures have involved data for the one-sample case (Chapters 8 through 10) and the two-sample case (Chapters 11 and 12). In this chapter, we will present a procedure for testing the hypothesis that K population means are equal, where $K > 2$. This procedure is called analysis of variance (ANOVA).

For example, a psychologist is studying the effect of method of reinforcement on the number of trials needed to complete a learning task. There are five different methods of reinforcement, and 31 subjects are randomly assigned to the five

TABLE 14.1
Scores, Means, and Variances of the Comparison Groups and the Total Group

	Method					
	1	*2*	*3*	*4*	*5*	
	1	8	7	9	10	
	4	6	6	10	12	
	3	7	4	8	9	
	2	4	9	6	11	
	5	3	8	5	8	
	1	5	5			
	6		7			
			5			Total
n_k	7	6	8	5	5	$N = 31$
$\overline{X}_k$	3.14	5.50	6.38	7.60	10.00	$\overline{X} = 6.26$
s_k^2	3.81	3.50	2.84	4.30	2.50	$s^2 = 7.93$

methods.[1] After the experimental treatment, each subject completes the learning task; the number of trials needed to complete the task are found in Table 14.1. The means ($\overline{X}$) and variances (s^2) for the groups are also presented, along with the mean and variance for all the subjects combined. This latter mean will be called the *grand mean* in subsequent sections of this chapter.

To test whether the pairs of sample means differ by more than we would expect due to chance, we might initially conduct a series of *t* tests on the *K* sample means (see Chapter 11). For our example, this approach would require ten separate *t* tests in order to compare all possible pairs of means.[2] With various available computer software packages, these *t* tests could be computed easily and quickly. However, this approach has a major problem.

Problems with Multiple *t* Tests

Using multiple *t* tests for testing the hypothesis that *K* population means are equal is not appropriate. When more than one *t* test is run, each at a specified level of significance (such as $\alpha = .05$), the probability of making one or more Type I errors

---

[1] Note that, for this example, we have used unequal numbers of subjects in each of the five groups. In the exercises at the end of the chapter and in the next chapter, we will include examples with equal numbers of subjects in the groups.

[2] The number of *t* tests needed to compare all possible pairs of means would be $K(K - 1)/2$, where *K* = number of means.

in the series of t tests is greater than α. The **Type I error rate** is determined as follows:[3]

$$1 - (1 - \alpha)^c$$

where

$\alpha =$ level of significance for each separate t test

$c =$ number of independent t tests

Suppose the psychologist in our example conducts each of the ten independent t tests using $\alpha = .05$ for each test. The Type I error rate would be:

$$1 - (1 - .05)^{10} = .40$$

In other words, the probability of making at least one Type I error in the comparisons of the five group means is *not* 0.05, as one might expect. Instead, the probability of committing a Type I error is 0.40.

The problem with computing multiple independent t tests for comparing K sample means is that, as the number of t tests increases, the Type I error rate increases.

With this problem in mind, R. A. Fisher developed the procedure known as analysis of variance (ANOVA). This procedure allows us to test the hypothesis of equality of K population means while maintaining the Type I error rate at the pre-established α level for the entire set of comparisons.

The Variables in ANOVA

In any ANOVA, there are two kinds of variables: independent and dependent. In our example, the psychologist is interested in determining whether the five methods of reinforcement are equally effective. The variable that forms the groupings is called the *independent variable*. In this example, *method* is the independent variable, with five *levels* of that variable. In ANOVA, one-way classification, only one independent variable is considered, but there can be two, three, six, or (theoretically) any finite number of levels of the independent variable. However, the procedure is still only a one-way ANOVA, since there is only one independent variable.

[3] We will discuss the Type I error rate more thoroughly in Chapter 15. For a more extensive discussion see G. A. F. Seber, *Linear Regression Analysis* (New York: John Wiley & Sons, 1977), pp. 125–126.

One-way ANOVA involves the analysis of one independent variable with two or more levels.

From our earlier discussions of types of variables, the dependent variable was defined as the variable that is, or is presumed to be, the result of manipulating the independent variable. In ANOVA, hypotheses are formulated about the means of the groups on the dependent variable. In our example, the subjects' scores (the number of trials to complete the task) represent the dependent variable. The actual scores are found in Table 14.1, and they are symbolically depicted in Table 14.2. The symbol X_{11} refers to the score of the first subject in method 1, X_{21} to the score of the second subject in method 1, and so on for the n_k subjects in each of the five methods. Thus, the first subscript on the X indicates the subject number, and the second subscript indicates the method or group number.

Changes in the dependent variable in ANOVA are, or are presumed to be, the result of changes in the independent variable.

The use of two subscripts allows us to use summation notation (see Chapter 1) to indicate the addition of scores in more than one group. For example, the notation

$$\sum_{i=1}^{n_k} X_{i2}$$

TABLE 14.2
Data Layout of One-Way ANOVA for Five Method Categories

	Method				
	1	*2*	*3*	*4*	*5*
	X_{11}	X_{12}	X_{13}	X_{14}	X_{15}
	X_{21}	X_{22}	X_{23}	X_{24}	X_{25}
	X_{31}	X_{32}	X_{33}	X_{34}	X_{35}
	X_{41}	X_{42}	X_{43}	X_{44}	X_{45}
	X_{51}	X_{52}	X_{53}	X_{54}	X_{55}
	X_{61}	X_{62}	X_{63}		
	X_{71}		X_{73}		
			X_{83}		
n_k	7	6	8	5	5

indicates that we sum the n_k scores in method 2. The notation for summing *all* scores in all K groups is

$$\sum_{k=1}^{K} \sum_{i=1}^{n_k} X_{ik}$$

Concepts Underlying ANOVA

The null hypothesis tested in ANOVA is that the population means from which the K samples are selected are equal. Symbolically,

$$H_0: \mu_1 = \mu_2 = \cdots = \mu_K$$

where K is the number of levels of the independent variable. In our example, the null hypothesis is $H_0: \mu_1 = \mu_2 = \mu_3 = \mu_4 = \mu_5$. Therefore, the null hypothesis is that the population means for the five methods of reinforcement are equal. The alternative hypothesis (H_a) is that at least one population mean differs from the other population means.

$$H_a: \mu_i \neq \mu_k \text{ for some } i, k$$

The procedure for testing this null hypothesis against the alternative hypothesis is precisely what its name implies: analysis of the variance of the scores on the dependent variable. In this analysis, the total variation of the scores is partitioned into two components. The two component parts are (1) the variation of the scores within the K groups and (2) the variation between the group means and the mean of the total group (grand mean).

Intuitive Approach to Partitioning the Variation

The means and variances for the five groups in our example are given in Table 14.1, along with the mean and variance for the total group (disregarding group membership). The distributions of scores for these five groups are illustrated in Figure 14.1. As will be discussed, one of the assumptions underlying ANOVA is that the populations from which the samples are selected are normally distributed for the dependent variable. Thus, the distributions in Figure 14.1 are drawn as normal distributions centered around the group means.

Consider the variance for the total group (disregarding group membership). The computation involves summing the squared deviations from the grand mean for each subject and dividing by $n - 1$. From Table 14.1, we see that

$$\frac{SS}{n-1} = \frac{\Sigma(X - \overline{X})^2}{n-1} = \frac{237.94}{30} = 7.93$$

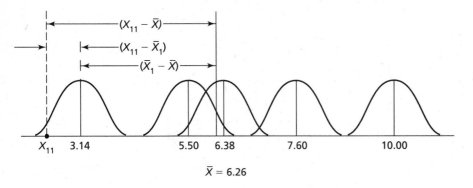

FIGURE 14.1
Partitioning the variation into between-groups and within-groups components

For subject 1 in group 1 (X_{11}), the deviation score ($X_{11} - \overline{X}$) is this subject's contribution to the total variation of the scores. Note in Figure 14.1 that this deviation score can be partitioned into two component parts. The first component is the deviation of X_{11} from the mean of group 1 ($X_{11} - \overline{X}_1$). The second component is the deviation of the group 1 mean from the grand mean ($\overline{X}_1 - \overline{X}$). The first component reflects the differences observed among subjects exposed to the same treatment. This variation among all subjects within a group is called the **within-groups variation**; we will denote this variation as s_W^2. It is assumed that within-groups variation of a similar magnitude exists in each of the five groups. This variation within any one group is a function of the specific subjects selected at random for the group or allocated at random to the group. Therefore, we can attribute variation within a group to random sampling fluctuation.

The second component has to do with the differences among group means. Even if there were absolutely no treatment effect, it would be unlikely that the sample means for the five groups would be identical. A more reasonable expectation is that the group means will differ, even without treatment, simply as a function of the individual differences of the subjects. For example, if we randomly reassign the subjects to different groups before treatment, we should expect somewhat different group means. Thus, we would expect the group means to vary somewhat due to the random selection (assignment) process in the formation of the groups.

If, in addition, different treatments that do have an effect on the dependent variable are applied to the different groups, we can expect even larger differences among the group means. Thus, the **between-groups variation**, denoted as s_B^2, reflects variation due to treatments plus variation attributable to the random process by which subjects are selected and assigned to groups.

In one-way ANOVA, the total variance can be partitioned into two sources: (1) variation of scores *within* groups and (2) variation *between* the group means and the grand mean. Both sources reflect variation due to random sampling. In addition, the between-groups variation reflects variation due to differential treatment effects.

Suppose that the null hypothesis is true; that is, $\mu_1 = \mu_2 = \mu_3 = \mu_4 = \mu_5$. In this case, the sample distributions might look like those in Figure 14.2A, even though the population distributions are identical. Recall that the within-groups variation (s_W^2) is that source of natural variation attributable to individual differences and random sampling fluctuation. Thus, we can consider this variance as an estimate of the population variance, which we will call *error variance*, or σ_e^2. It is error variance because it is not controlled.

Notice that, in Figure 14.2A, there are slight differences among the group means, even though the null hypothesis is true. Recall that the between-groups variation (s_B^2) contains variation due to differential treatment effects plus variation due to random sampling. Since we have assumed that the null hypothesis is true, there is no treatment effect, and s_B^2 is also an estimate of σ_e^2. Therefore, the ratio of these two estimates (s_B^2/s_W^2) is approximately 1.00. (Due to random sampling, we would not expect this ratio to be exactly equal to 1.00.)

Now suppose the null hypothesis is false, or $\mu_i \neq \mu_k$ for some i, k. This is illustrated graphically in Figure 14.2B. The between-groups variation would again be an estimate of σ_e^2 plus variation due to the differences among population means as a result of differential treatment effects. We denote this additional source of variation as σ_t^2. In this case, the ratio of s_B^2/s_W^2 is estimated to be:

$$\frac{\sigma_e^2 + \sigma_t^2}{\sigma_e^2}$$

As can be seen when the null hypothesis is not true, the ratio of s_B^2 to s_W^2 is expected to be greater than 1.00. Thus we can ask the question: how much can this ratio of variances depart from 1.00 before we can no longer attribute the between-groups variance to random sampling fluctuation?

The test statistic in ANOVA is the ratio of the between-groups variation to the within-groups variation. The specific procedures are similar to those for testing the hypothesis of no difference between the variances of two independent samples (see Chapter 10). In ANOVA, however, if the null hypothesis is true ($H_0: \mu_1 = \mu_2 = \mu_3 = \mu_4 = \mu_5$), this is equivalent to saying that the between-groups variation is only an estimate of error variance. In other words, if the five population means are equal, then the variance among sample means is merely error variance.

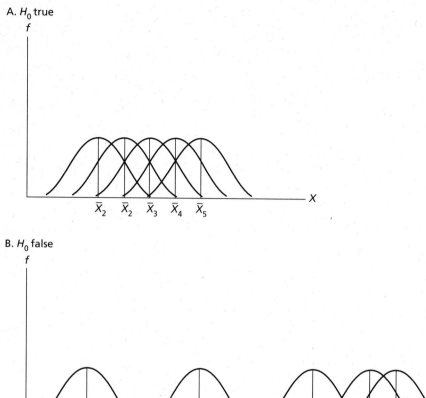

A. H_0 true

B. H_0 false

FIGURE 14.2
Illustration of sample data distributions when H_0 is true (A) and H_0 is false (B)

The null hypothesis in ANOVA is tested using the ratio of the two variance estimates—s_B^2, the between-groups variance, and s_W^2, the within-groups variance, or s_B^2/s_W^2.

The Linear Model

We have just considered one-way ANOVA in terms of partitioning the total variation of all sample scores into between-groups and within-groups variation. Another approach is to consider the composition of an individual score. Suppose we have

K groups or levels of the independent variable and consider the kth group. An individual score of a subject on the dependent variable can be expressed in terms of three additive components:

$$X_{ik} = \mu + \alpha_k + e_{ik} \tag{14.1}$$

where

$X_{ik} = i$th score in the kth group

$\mu =$ grand mean for the population

$\alpha_k = \mu_k - \mu =$ effect of belonging to group k

$e_{ik} =$ random error associated with this score

Notice that, in the above equation, the right side contains parameters μ and α_k plus e_{ik} — the random error. Thus, the **linear model** is an estimation of the components of one score in the population. In ANOVA, as in all inferential statistical procedures, parameters are unknown and are estimated from the sample statistics. The e_{ik}s (random errors) are assumed to be normally distributed, with a mean of zero for each of the groups. The variance of these errors across the groups is σ_e^2, the population or error variance.

The application of the estimated linear model to the X_{11} score in our example is illustrated in Figure 14.3. Notice the similarity between this figure and Figure 14.1. We are merely expressing the deviations as terms in the linear model. The deviation of the score from the group mean is the random error component for person 1 in group 1. The grand mean is the best estimate of the population mean, so it is designated as μ. The deviation of the group mean from the grand mean is our best estimate of the effect of being in group 1, so it is designated α_1.

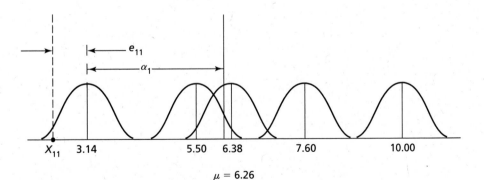

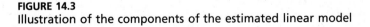

FIGURE 14.3
Illustration of the components of the estimated linear model

> The linear model in one-way ANOVA represents a single score in the population and contains three components: (1) μ, (2) α_k, and (3) e_{ik}.

Partitioning the Sum of Squares

Although the analysis of variance has been described as a process by which we partition the total variation, the procedure actually begins by partitioning the sum of the squared deviations around the grand mean, $\Sigma(X - \overline{X})^2$. Consider again the deviation score for subject 1 in group 1 and the partitioning of that deviation score into the two component parts:

$$(X_{11} - \overline{X}) = (X_{11} - \overline{X}_1) + (\overline{X}_1 - \overline{X})$$

To generalize the partitioning, we now want to consider the partitioning of the deviation score of the ith subject in the kth group, or:

$$(X_{ik} - \overline{X}) = (X_{ik} - \overline{X}_k) + (\overline{X}_k - \overline{X}) \tag{14.2}$$

In order to arrive at the total sum of squared deviations, we must square each deviation score and then sum the squared deviations across all subjects in all groups. Now, squaring both sides of equation 14.2,

$$(X_{ik} - \overline{X})^2 = [(X_{ik} - \overline{X}_k) + (\overline{X}_k - \overline{X})]^2$$

$$= (X_{ik} - \overline{X}_k)^2 + 2(X_{ik} - \overline{X}_k)(\overline{X}_k - \overline{X}) + (\overline{X}_k - \overline{X})^2$$

Summing these squared deviations across the n_k subjects in the kth group, the total sum of squares for this group is

$$\sum_{i=1}^{n_k}(X_{ik} - \overline{X})^2 = \sum_{i=1}^{n_k}(X_{ik} - \overline{X}_k)^2 + 2(\overline{X}_k - \overline{X})\sum_{i=1}^{n_k}(X_{ik} - \overline{X}_k) + \sum_{i=1}^{n_k}(\overline{X}_k - \overline{X})^2$$

Notice that, in the second term on the right side of the equation, 2 and $(\overline{X}_k - \overline{X})$ are both constants and thus appear in front of the summation sign. Since

$$\sum_{i=1}^{n_k}(X_{ik} - \overline{X}_k) = 0$$

for any of the groups (see Chapter 3, "Properties of the Mean"), the second term on the right side drops out of the equation. In addition, the third term on the right side of the equation $(\overline{X}_k - \overline{X})^2$ is a constant. Thus,

$$\sum_{i=1}^{n_k}(\overline{X}_k - \overline{X})^2 = n_k(\overline{X}_k - \overline{X})^2$$

The total sum of squares for the kth group is therefore

$$\sum_{i=1}^{n_k}(X_{ik} - \overline{X})^2 = \sum_{i=1}^{n_k}(X_{ik} - \overline{X}_k)^2 + n_k(\overline{X}_k - \overline{X})^2$$

Now, if we sum these sums of squares across the K groups, we have:

$$\sum_{k=1}^{K} \sum_{i=1}^{n_k} (X_{ik} - \overline{X})^2 = \sum_{k=1}^{K} \sum_{i=1}^{n_k} (X_{ik} - \overline{X}_k)^2 + \sum_{k=1}^{K} n_k (\overline{X}_k - \overline{X})^2 \qquad (14.3)$$

or

Total sum of squares = within sum of squares + between sum of squares

Equation 14.3 illustrates the partitioning of the total sum of squares (SS_T) into the within-groups sum of squares (SS_W) plus the between-groups sum of squares (SS_B). SS_W is the sum of the squared deviations of the original scores from the respective group means summed over the K groups. SS_B is the weighted sum of squared deviations of the group means from the grand mean, as well as the measure of the group effects. If there are group effects, the group means will differ substantially from the grand mean and SS_B will be substantially greater than zero.

In one-way ANOVA, the total sum of squares (SS_T) is partitioned into two components: (1) the sum of squares within groups (SS_W) and (2) the sum of squares between groups (SS_B).

Note that, since SS_B is based on group means that are sample means from a normal distribution, they reflect K means of a sampling distribution of means (see Chapters 7 and 8). Recall that this distribution of sample means is normally distributed, with the mean equal to μ (the mean of the population from which the samples were drawn). Hence, if we hypothesize the group means to be equal ($\mu_1 = \mu_2 = \cdots = \mu_K$), they must necessarily equal the population mean. Thus, under the null hypothesis, small observed differences between the sample means ($\overline{X}_k$) and the grand mean ($\overline{X}$) are attributable to sampling fluctuation. However, sizable differences between the group means and the grand mean, resulting in a substantial SS_B, are an indication that the null hypothesis may not be true.

Two Estimates of the Population Variance

The null hypothesis in ANOVA is

$$H_0: \mu_1 = \mu_2 = \cdots = \mu_K$$

In terms of the linear model, the null hypothesis could also be written

$$H_0: \alpha_1 = \alpha_2 = \cdots = \alpha_K = 0$$

That is, there is no group effect. The null hypothesis is not rejected when the difference between the sample means is attributable only to random sampling fluctuation. In ANOVA, the null hypothesis is tested using the ratio of two variance estimates

(s_B^2/s_W^2). So far, we do not have these variance estimates; we have only SS_B and SS_W. Recall from Chapter 3 that the sum of squared deviations is the numerator of the estimate of the population variance. Dividing this sum of squares by the degrees of freedom associated with this estimate $(n-1)$, we have the unbiased estimate of the population variance. In ANOVA, the between-groups and within-groups variance estimates are found by dividing SS_B and SS_W by the degrees of freedom associated with each of these estimates. The resulting between-groups and within-groups variance estimates are commonly called **mean squares** and are denoted MS_B and MS_W, respectively.

First, let us consider SS_B. It is calculated using the deviations of the group means from the grand mean, weighted for the number of observations in each group. Because there are K groups and the grand mean is calculated, there are $K-1$ degrees of freedom associated with the between-groups mean square (MS_B), or

$$MS_B = \frac{SS_B}{K-1} \tag{14.4}$$

But what does MS_B estimate? If H_0 is true, it estimates error variance. If H_0 is not true, it contains an additional component of variation due to the differences between the population means. To illustrate what MS_B estimates, consider again the sampling distribution of the mean. This distribution is normally distributed, with a mean equal to μ and standard deviation equal to $\sigma_{\overline{X}}$ (standard error of the mean):

$$\sigma_{\overline{X}} = \frac{\sigma_e}{\sqrt{n}}$$

Squaring both sides of the equation, we have:

$$\sigma_{\overline{X}}^2 = \frac{\sigma_e^2}{n}$$

where

$\sigma_e^2 = $ error variance

$n = $ sample size of each of the k samples (assuming equal sample sizes)

Algebraically manipulating this expression, we find:

$$\sigma_e^2 = n\sigma_{\overline{X}}^2$$

We use $s_{\overline{X}}^2$ to estimate $\sigma_{\overline{X}}^2$:

$$s_{\overline{X}}^2 = \frac{\Sigma(\overline{X}_k - \overline{X})^2}{K-1}$$

where

$\overline{X}_k = $ sample mean

$\overline{X} = $ grand mean

$K = $ number of groups

Therefore, under the null hypothesis, we can estimate σ_e^2 from the sample means and the grand mean:

$$MS_B = \frac{\Sigma n(\overline{X}_k - \overline{X})^2}{K - 1} = ns_{\overline{X}}^2 \xrightarrow{\text{estimates}} \sigma_e^2$$

Now consider SS_W. Recall that $s_1^2, s_2^2, \cdots, s_k^2$ are all estimates of σ_e^2. Hence, the **pooled estimate of the population variance** is

$$\frac{(n_1 - 1)s_1^2 + (n_2 - 1)s_2^2 + \ldots + (n_K - 1)s_K^2}{(n_1 - 1) + (n_2 - 1) + \cdots + (n_K - 1)}$$

This is the same as the sum of squared deviations within each group, pooled across groups and then divided by the sum of the degrees of freedom for each group. Therefore,

$$\frac{\Sigma(X_{i1} - \overline{X}_1)^2 + \Sigma(X_{i2} - \overline{X}_2)^2 + \cdots + \Sigma(\overline{X}_{iK} - \overline{X}_K)^2}{(n_1 - 1) + (n_2 - 1) + \cdots + (n_K - 1)} \xrightarrow{\text{estimates}} \sigma_e^2$$

Notice that the numerator of this expression is SS_W and the total degrees of freedom associated with this estimate are

$$(n_1 - 1) + (n_2 - 1) + \cdots + (n_K - 1) = N - K$$

where

$N = n_1 + n_2 + \cdots + n_K$ (the total number of observations)

$K =$ number of groups

Therefore,

$$MS_W = \frac{SS_W}{N - K} \xrightarrow{\text{estimates}} \sigma_e^2 \qquad (14.5)$$

A mean square (MS) is a variance estimate. The degrees of freedom associated with the two estimates of the population variance are $K-1$ and $N - K$ for MS_B and MS_W, respectively.

Testing the Null Hypothesis

At this point, let us briefly review the chain of reasoning for testing the null hypothesis in ANOVA. The data for ANOVA consist of scores for K (two or more) random samples from the K levels of the independent variable. In an experiment, the different experimental treatments are the levels of the independent variable.

After the experimental treatment, the samples exposed to the different experimental treatments reflect respective populations, and the sample means are thus estimates of the means of the respective populations.

The computed mean squares are actually statistics and have counterparts in the population, called **expected mean squares**. The expected mean squares [E(MS)] describe the variance components that are theoretically contained in the calculated mean squares. We have dealt intuitively with the concept of the components of the expected mean squares in the discussion of partitioning the total variation. We can now extend that discussion and determine the values of E(MS) using the concept of the linear model in ANOVA. The actual derivation of E(MS) involves the use of expected values, a mathematical concept beyond the scope of this book.[4]

Expected Mean Square Between: E(MS_B)

Even though the sum of α_k always equals zero [$\Sigma\alpha_k = \Sigma(\mu_k - \mu) = 0$], the sum of α_k^2 equals zero only when the null hypothesis is true ($\alpha_1 = \alpha_2 = \cdots = \alpha_K = 0$). When the null hypothesis is true, there is no group effect and MS_B is an estimate of σ_e^2. If there is at least one α_k not equal to zero, MS_B also contains a component of variance due to the sum of α_k^2, and the expected mean square between would be

$$E(MS_B) = \sigma_e^2 + \frac{n\Sigma\alpha_k^2}{K-1} \tag{14.6}$$

That is, $E(MS_B)$ equals the population error variance plus n times the variation due to treatment effects (where n is the sample size for each group).

Expected Mean Square Within: E(MS_W)

The expected value of the mean square within equals σ_e^2. The reasoning for this is quite straightforward. The within-groups variance for each group is a separate, independent estimate of the population variance. Pooling these separate variances across the K groups actually results in a more precise estimate of σ_e^2, since it is based on more observations. Hence,

$$E(MS_W) = \sigma_e^2 \tag{14.7}$$

---

[4]For the derivation of E(MS), see R. E. Kirk, *Experimental Design: Procedures for the Behavioral Sciences*, 2d ed. (Belmont, Calif.: Wadsworth, 1982), pp. 64–72.

The expected mean squares are parameters that are estimated by the observed mean squares. $E(MS_B)$ is the sum of two parameters: one estimates the variation among subjects in the population; the other estimates the variance due to treatment effects. $E(MS_W)$ estimates only the variation among subjects in the population.

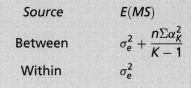

Source	$E(MS)$
Between	$\sigma_e^2 + \dfrac{n\Sigma\alpha_K^2}{K-1}$
Within	σ_e^2

The F Ratio

In Chapter 11, we indicated that the difference between two variances could be tested by forming a ratio of the variances and using the F distribution as the sampling distribution. We can now follow the same procedure, using the ratio of the two expected mean squares:

$$\frac{E(MS_B)}{E(MS_W)} = \frac{\sigma_e^2 + \dfrac{n\Sigma\alpha_k^2}{K-1}}{\sigma_e^2} \tag{14.8}$$

Therefore, the sample estimates of these values are

$$F = \frac{MS_B}{MS_W} \tag{14.9}$$

When the null hypothesis is true ($\alpha_1 = \alpha_2 = \cdots = \alpha_K = 0$), we would expect F to be $\sigma_e^2/\sigma_e^2 = 1$. Note that the observed mean squares merely estimate the parameters in $E(MS)$ and that these estimates may be larger or smaller than the corresponding parameters. Therefore, it is possible to have an observed F ratio less than 1.00, even though conceptually the ratio of $E(MS)$ cannot be less than 1.00.

Summary Table for ANOVA

The conventional way to report the results of ANOVA is in a standard tabular format. This format, often called the *summary table for ANOVA*, is illustrated in Table 14.3 for the general case. The variance estimates, or mean squares, are determined by dividing the sum of squares by the respective degrees of freedom. The F ratio is then determined by dividing the between-groups mean square (MS_B)

TABLE 14.3
Summary ANOVA: The General Case for One-Way Classification

Source of Variation	Sum of Squares	Degrees of Freedom	Variance Estimate (Mean Square)	F ratio
Between	$\sum_{K=1}^{K} n_K(\overline{X}_K - \overline{X})^2$	$K - 1$	$MS_B = \dfrac{SS_B}{K - 1}$	$\dfrac{MS_B}{MS_W}$
Within	$\sum_{k=1}^{K} \sum_{i=1}^{n_K} (X_{ik} - \overline{X}_K)^2$	$N - K$	$MS_W = \dfrac{SS_W}{N - K}$	
Total*	$\sum_{k=1}^{K} \sum_{i=1}^{n_K} (X_{ik} - \overline{X})^2$	$N - 1$		

*Note that the MS for the total is not computed.

by the within-groups mean square (MS_W):

$$F = \frac{MS_B}{MS_W}$$

Computational Formulas for Sums of Squares

In Chapter 3, we defined the sample variance of a set of scores as the sum of squared deviations divided by $n - 1$; then we proceeded to derive a computational formula. Analogously, the computational formulas for SS_B, SS_W, and SS_T can be derived. These formulas are presented below. The formulas are more convenient when the data are in raw score form. First of all, to simplify the formulas, it is convenient to denote the sum of all the scores in the kth group as T_k.

$$T_k = \sum_{i=1}^{n_k} X_{ik} \tag{14.10}$$

Second, denote the sum of all observations in all K groups as T.

$$T = \sum_{k=1}^{K} \sum_{i=1}^{n_k} X_{ik} \tag{14.11}$$

Using this notation, the computational formulas for SS_B, SS_W, and SS_T become for between groups:

$$SS_B = \sum_{k=1}^{K} n_k (\overline{X}_k - \overline{X})^2 \tag{14.12}$$

$$= \sum_{k=1}^{K} \frac{T_k^2}{n_k} - \frac{T^2}{N}$$

for within groups:

$$SS_W = \sum_{k=1}^{K} \sum_{i=1}^{n_k} (X_{ik} - \overline{X}_k)^2 \tag{14.13}$$

$$= \sum_{k=1}^{K} \sum_{i=1}^{n_k} X_{ik}^2 - \sum_{k=1}^{k} \frac{T_k^2}{n_k}$$

for the total:

$$SS_T = \sum_{k=1}^{K} \sum_{i=1}^{n_k} (X_{ik} - \overline{X})^2 \tag{14.14}$$

$$= \sum_{k=1}^{K} \sum_{i=1}^{n_k} X_{ik}^2 - \frac{T^2}{N}$$

Testing the Null Hypothesis in the Example

We are at last ready to use ANOVA to test the psychologist's hypothesis that the five methods of reinforcement are equally effective; that there are no differences in the average performance of subjects who experience the five methods. In preceding chapters, we included a series of steps that comprise a logical procedure for testing hypotheses; a similar series of steps can be used in ANOVA.

Step 1: State the Hypotheses

$$H_0: \mu_1 = \mu_2 = \mu_3 = \mu_4 = \mu_5$$
$$H_a: \mu_i \neq \mu_k \text{ for some } i, k$$

In this example, the psychologist hypothesizes that the mean performance on the learning task is the same for each reinforcement method. Suppose the psychologist wants to test this hypothesis at the .05 level of significance.

Step 2: Set the Criterion for Rejecting H_0

The test statistic for one-way ANOVA is the F ratio defined as MS_B/MS_W. The sampling distribution of the F ratio is the F distribution. There is a family of F distributions, each one a function of the degrees of freedom associated with the two variance estimates. In ANOVA, there are $K - 1$ degrees of freedom associated with MS_B and $N - K$ degrees of freedom associated with MS_W. Thus, the sampling distribution of the F ratio MS_B/MS_W is the F distribution with $K - 1$ and $N - K$

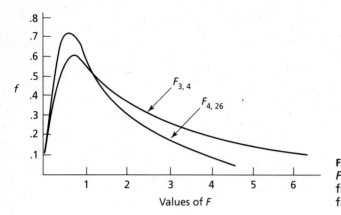

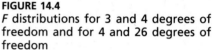

FIGURE 14.4
F distributions for 3 and 4 degrees of freedom and for 4 and 26 degrees of freedom

degrees of freedom, or $F_{(K-1,\ N-K)}$. Table C.5 contains critical values of F for a wide range of degrees of freedom. The degrees of freedom for MS_B are read across the column headings from left to right; the degrees of freedom for MS_W are read down the rows on the left side. The intersection of the row for $N - K$ degrees of freedom and the column for $K - 1$ degrees of freedom is the critical value of F.

The F distributions for 3 and 4 degrees of freedom and for 4 and 26 degrees of freedom are illustrated in Figure 14.4. In ANOVA, the alternative hypothesis does not specify the direction of the group differences; however, the region of rejection in the F distribution is only in the right-hand tail, because the observed F ratio is a ratio of two non-negative values, the mean squares. Large discrepancies among the sample means yield a large mean square among the groups, regardless of the direction of the differences. Because these discrepancies $(\overline{X}_k - \overline{X})$ are squared, direction is lost.

In our example, there are $K - 1 = 4$ degrees of freedom associated with MS_B and $N - K = 26$ degrees of freedom associated with MS_W. The critical value of F for 4 and 26 degrees of freedom, from Table C.5, for $\alpha = .05$ is 2.74; the critical value for $\alpha = .01$ is 4.14. Since the psychologist sets α at .05, the null hypothesis will be rejected if the calculated F exceeds 2.74.

Step 3: Compute the Test Statistic

The data are presented again in Table 14.4, along with the intermediate calculations that are used in the computational formulas for the sums of squares. The calculations of the sums of squares are provided in Table 14.5, and the summary table for ANOVA is shown in Table 14.6. The test statistic is F, the ratio of MS_B and MS_W. For this example,

$$F = \frac{37.63}{3.36} = 11.20$$

TABLE 14.4
Use of Computational Formulas in ANOVA, One-Way Classification

	1	2	3	4	5	
						Method
	1	8	7	9	10	
	4	6	6	10	12	
	3	7	4	8	9	
	2	4	9	6	11	
	5	3	8	5	8	
	1	5	5			
	6		7			
			5			
n_k	7	6	8	5	5	$N = 31$
T_k	22	33	51	38	50	$T = 194; \dfrac{T^2}{N} = \dfrac{(194)^2}{31} = 1{,}214.06$
$\bar{X}_k$	3.14	5.50	6.38	7.60	10.00	$\bar{X} = 6.26$
$\displaystyle\sum_{i=1}^{n_k} X_{ik}^2$	92	199	345	306	510	$\displaystyle\sum_{k=1}^{K}\sum_{i=1}^{n_k} X_{ik}^2 = 1{,}452$
$\dfrac{T_k^2}{n_k}$	69.14	181.50	325.13	288.80	500.00	$\displaystyle\sum_{k=1}^{K}\dfrac{T_k^2}{n_k} = 1{,}364.57$

TABLE 14.5
Calculations of Sums of Squares in ANOVA, One-Way Classification

$$SS_B = \frac{(22)^2}{7} + \frac{(33)^2}{6} + \frac{(51)^2}{8} + \frac{(38)^2}{5} + \frac{(50)^2}{5} - \frac{(194)^2}{31}$$

$$= (69.14 + 181.50 + 325.13 + 288.80 + 500.00) - 1{,}214.06$$

$$= 1{,}364.57 - 1{,}214.06 = \underline{150.51}$$

$$SS_W = 1{,}452.00 - 1{,}364.57 = \underline{87.43}$$

$$SS_T = 1{,}452.00 - 1{,}214.06 = \underline{237.94}$$

TABLE 14.6
Summary ANOVA for Example

Source	Summary ANOVA				
	SS	df	MS	F	F_{cv}
Between	150.51	4	37.63	11.20	2.74
Within	87.43	26	3.36		
Total	237.94	30			

The usual procedure is to report the entire summary table for ANOVA, including the sums of squares, degrees of freedom, and mean squares, as well as the *F* ratio.

Step 4: Interpret the Results

This step includes deciding whether to reject the null hypothesis and then providing some clarification about what that decision means in the context of the variables and the setting. The decision about the null hypothesis is made by comparing the test statistic, the computed *F* ratio, with the critical value from the table. If the computed *F* ratio exceeds the critical value, the hypothesis is rejected; if not, the hypothesis is not rejected. In this example, since the observed value of the test statistic, the *F* ratio, exceeds the critical value at the .05 level of significance, the null hypothesis is rejected. In terms of probability, the probability that the observed differences in sample means would have occurred by chance if the null hypothesis is true (the population means are equal) is less than .05. The conclusion is that not all of the population means are equal. Of course, at this point we do not know which means differ significantly. We will be able to determine this using techniques developed in Chapter 15.

Computer Example: The Survey of High School Students Data Set

For the survey of high school students data set ($n = 500$), consider the mathematics achievement (MATHACH) scores as the dependent variable and the recoded father's education (FAED) classification as the independent variable. The recoded classification was:

$\quad$ 1 = Less than HS
$\quad$ 2 = HS graduate
$\quad$ 3 = Some Vocational or College
$\quad$ 4 = College graduate or more.

The summary data for the four groups are as follows:

	n	$\overline{X}$	s
Group 1	132	10.8687	6.1866
Group 2	129	12.4419	6.4488
Group 3	105	13.2191	6.2831
Group 4	134	15.8309	6.4947

The ANOVA (ONEWAY) procedure from SPSS was used to generate the computer printout for this analysis (see Table 14.7).

Step 1: State the Hypotheses

For this example, the null hypothesis is that there is no difference in MATHACH scores between those students classified by the various levels of father's education

TABLE 14.7
SPSS Printout for One-Way ANOVA for the Survey of High School Students Data Set

Variable MATHACH $\quad$ math test score
By Variable FAED $\quad$ Father's education

Analysis of Variance

Source	D.F.	Sum of Squares	Mean Squares	F Ratio	F Prob.
Between Groups	3	1713.8734	571.2911	14.1307	.0000
Within Groups	496	20052.8090	40.4291		
Total	499	21766.6824			

(FAED). Symbolically,

$$H_0: \mu_1 = \mu_2 = \mu_3 = \mu_4$$

$$H_a: \mu_i \neq \mu_k \text{ for some } i, k$$

The level of significance set a priori for this analysis is $\alpha = .05$.

Step 2: Set the Criterion for Rejecting H_0

The test statistic for one-way ANOVA is the F ratio defined as MS_B/MS_W; the sampling distribution of the F ratio is the F distribution with $K - 1$ and $N - K$ degrees of freedom. In this example, there are $K - 1 = 4 - 1 = 3$ df associated with MS_B and $N - K = 500 - 4 = 496$ df associated with MS_W. Thus, the critical value for F with 3 and 496 degrees of freedom, from Table C.5 and for $\alpha = .05$, is $F_{cv} = 2.60$. The null hypothesis will be rejected if the calculated value of F exceeds the critical value.

Step 3: Compute the Test Statistic

The Summary ANOVA for this example is found in Table 14.7. The test statistic, F, was calculated as follows:

$$F = \frac{MS_B}{MS_W}$$

$$= \frac{571.2911}{40.4291}$$

$$= 14.1307$$

Step 4: Interpret the Results

For this example, the calculated value of F (14.1307) exceeds the critical value for F ($F_{cv} = 2.60$); therefore, the null hypothesis, $H_0: \mu_1 = \mu_2 = \mu_3 = \mu_4$, is rejected. The probability that the observed differences would have occurred by chance if this null hypothesis were true (i.e., the population means are equal) is less than .05 (the level of significance set a priori). Note in the printout, "F Prob" $= .0000$, which implies that the probability of such an occurrence is actually less than .0001. The conclusion is that not all the population means are equal. At this point, we do not know which means differ significantly. This determination will be made using the techniques developed in Chapter 15.

Assumptions Underlying ANOVA

We have been alluding to certain assumptions relevant to ANOVA. The importance of the assumptions being met lies in the appropriate use of the F distribution as the sampling distribution for testing the null hypothesis. The three primary assumptions are:

1. The observations are random and independent samples from the populations. The null hypothesis actually says that the samples come from populations that have the same mean. The samples must be random and independent if they are to be representative of the populations.

2. The distributions of the populations from which the samples are selected are normal. This assumption implies that the dependent variable is normally distributed (a theoretical requirement of the underlying distribution, the F distribution) in each of the populations.

3. The variances of the distributions in the populations are equal. We mentioned this assumption earlier; it is called the assumption of **homogeneity of variance**. This assumption, along with the normality assumption and the null hypothesis, provides that the distributions in the populations have the same shapes, means, and variances; that is, they are the same population.

How do we know whether or not these assumptions are met in a specific situation? First, random sampling and independence are technical concerns related to the way the samples are selected. The normality assumption can be checked with a "goodness-of-fit" test,[5] which tests whether the sample distributions are within random sampling fluctuation of normality. The homogeneity of variance assumption can also be tested using the sample data. In Chapter 11, a test for H_0: $\sigma_1^2 = \sigma_2^2$ was introduced. However, we would not want to compare all sample variances in combinations of two, for the same reason that we would not do multiple t tests of H_0: $\mu_1 = \mu_2 = \cdots = \mu_K$. One of the more commonly used tests is Bartlett's test for homogeneity of variance,[6] which tests the null hypothesis H_0: $\sigma_1^2 = \sigma_2^2 = \cdots = \sigma_K^2$. (The computation for this test is not presented here, but it should be noted that this test is very sensitive to non-normality.)

What happens if one or more of the assumptions underlying the ANOVA are not met? Generally, failure to meet these assumptions changes the Type I error rates. For example, instead of operating at the designated level of significance, the actual Type I error rate may be greater or less than α, depending on which assumptions are violated. An excellent treatment of this topic is "Consequences of Failure to Meet the Assumptions Underlying the Use of Analysis of Variance and

---

[5]This test uses the chi-square distribution discussed in Chapter 21.

[6]M. S. Bartlett, "Properties of Sufficiency and Statistical Tests," *Proceedings of the Royal Society of London* 160, Series A (1937): 268–282.

Covariance" by Glass *et al.*[7] This article reports the results of a number of studies in which the assumptions were systematically violated and the effect on the Type I error rate was observed. Briefly, some of the useful findings were as follows:

1. When the populations sampled are not normal, the effect on the Type I error rate is minimal.

2. If the population variances differ, there may be a serious problem when sample sizes are unequal. If the larger variance is associated with the larger sample, the *F* test will be too conservative. If the smaller variance is associated with the larger sample, the *F* test will be too liberal. (If the α level is .05, "conservative" means that the actual rate is less than .05.) If the sample sizes are equal, the effect of heterogeneity of variances on the Type I error is minimal.

In other words, the effects of violating the assumptions vary somewhat with the specific assumptions violated. If a statistical procedure is little affected by violating an assumption, the procedure is said to be robust with respect to that assumption.

> ANOVA is robust with respect to violations of the assumptions, except in the case of unequal variances with unequal sample sizes.

A Measure of Association (ω^2)

Rejecting the null hypothesis in ANOVA indicates that there are significant differences among the sample means, a greater difference than would be expected on the basis of chance. However, with large sample sizes, these statistically significant differences may have little practical significance. This is similar to our previous discussions of correlation coefficients in which a correlation of .15 might be significantly different from zero with a very large sample size yet indicate a weak relationship between the variables.

A measure of the strength of the association between the independent and dependent variables in ANOVA is ω^2, omega squared.[8] Omega squared indicates the proportion of the variance in the dependent variable that is accounted for by the levels of the independent variable. This is analogous to the coefficient of determination (r^2) that was developed in Chapter 5. The formula for omega

---

[7]G. V Glass et al., "Consequences of Failure to Meet the Assumptions Underlying the Use of Analysis of Variance and Covariance," *Review of Educational Research* 42, 3 (1972): 237–288.

[8]W. L. Hays, *Statistics* (New York: Holt, Rinehart and Winston, 1981), p. 382.

squared is:

$$\omega^2 = \frac{SS_B - (K-1)MS_W}{SS_T + MS_W} \tag{14.15}$$

Applying formula 14.15 to data in our example, we get

$$\omega^2 = \frac{150.51 - (5-1)(3.36)}{237.94 + 3.36} = .57$$

This means that the five methods of reinforcement, the independent variable in the ANOVA, account for about 57 percent of the variance in the number of trials necessary to complete a learning task, the dependent variable. The psychologist would interpret this as a strong association between reinforcement and learning rate.

But we should state some cautions about the use of ω^2. First, this statistic is most meaningful when the levels of the independent variable have been carefully chosen. Equal units along a quantitative dimension — for example, drug dosage (5 mg, 15 mg, and 25 mg) — give a cleaner measure of association than arbitrarily chosen values, such as 2 mg, 7 mg, and 23 mg. Groupings formed on the basis of nominal variables should be similarly reasonable. Second, the value of ω^2 is specific to the levels of the independent variable that are used in the ANOVA. A change in the levels, the comparison of different groups, could of course lead to a change in the calculated value of ω^2.

Omega squared can provide some useful information to help the researcher understand the relationship between the independent and dependent variables in the ANOVA. It can be particularly helpful in tempering the conclusions in situations in which there is considerable statistical power. A modest ω^2 may be found, even though the F ratio in the ANOVA is very large. In such a case, the researcher should caution others about overinterpreting the significance of results.[9]

When used appropriately, ω^2 can provide useful information about the association between the dependent variable and the levels of the independent variable.

The Relationship of ANOVA to the t Test

A researcher is interested in comparing the average number of telephone calls received each week by male college sophomores in dormitories and in fraternity

[9]G. V Glass and A. R. Hakstian, "Measures of Association in Comparative Experiments: Their Development and Interpretation," *American Educational Research Journal* 6 (1969): 403–414, and S. E. Maxwell, C. J. Camp, and R. D. Avery, "Measures of Strength of Association: A Comparative Examination," *Journal of Applied Psychology* 66 (1981): 525–534.

houses. The researcher randomly selects 10 sophomores from each group and determines the number of telephone calls received in the past week. It is assumed that the week in which the study is conducted is representative of the total school year. The living arrangement is the independent variable and has two levels (dormitory versus fraternity house). The dependent variable is number of telephone calls received. These data are found in Table 14.8.

In testing the null hypothesis H_0: $\mu_1 = \mu_2$ for independent samples, the researcher can use either the t test or ANOVA; the two procedures are equivalent for testing this hypothesis. The calculation of the t test (using formula 11.6) follows:

$$t = \frac{(3.50 - 6.30) - 0}{\sqrt{\dfrac{(10 - 1)(19.61) + (10 - 1)(24.01)}{10 + 10 - 2}\left(\dfrac{1}{10} + \dfrac{1}{10}\right)}}$$

$$= \frac{-2.80}{\sqrt{4.36}}$$

$$= -1.34$$

The critical values of t for 18 degrees of freedom, assuming a two-tailed test at $\alpha = .05$, are ± 2.101. Because our observed value of t does not exceed the critical value, the null hypothesis is not rejected. The researcher concludes that there is no

TABLE 14.8
Data Used to Illustrate the Relationship Between ANOVA and the *t* Test

	Dormitory Students	Fraternity Students	
	2	4	
	6	6	
	3	4	
	2	2	
	4	12	
	0	16	
	2	4	
	15	10	
	1	0	
	0	5	
			Totals
ΣX	35	63	98
ΣX^2	299	613	912
$\overline{X}$	3.5	6.3	
s^2	19.61	24.01	

TABLE 14.9
Summary ANOVA for Telephone Calls Example

		Summary ANOVA			
Source	SS	df	MS	F	F_{cv}
Between	39. 20	1	39. 20	1. 80	4. 41
Within	392. 60	18	21. 81		
Total	431. 80	19			

difference in the average number of telephone calls received by men in the two populations.

Now consider ANOVA for the same data. The sums of squares are computed as follows:

$$SS_B = \left[\frac{(35)^2}{10} + \frac{(63)^2}{10} \right] - \frac{(98)^2}{20} = 519.40 - 480.20 = 39.20$$

$$SS_W = 912 - \left[\frac{(35)^2}{10} + \frac{(63)^2}{10} \right] = 912 - 519.40- = 392.60$$

$$SS_T = 912 - (98)^2/20 = 912 - 480.20 = 431.80$$

The summary table is shown in Table 14.9. Since the observed value of the F ratio does not exceed the critical value of F for 1 and 18 degrees of freedom at the .05 level of significance, the null hypothesis would not be rejected. Note that, for the observed test statistic and the critical value of the test statistic, the following relationships hold:

$$t^2 = F \qquad (1.34^2 = 1.80)$$
$$t_{cv}^2 = F_{cv} \qquad (2.101^2 = 4.41)$$

For testing H_0: $\mu_1 = \mu_2$ versus H_a: $\mu_1 \neq \mu_2$, $F = t^2$. That is, ANOVA and the t test for independent samples give identical results.

Determining the Sample Size

In the last chapter, we discussed the procedures for determining sample size for the one-sample case and the two-sample case. We now expand that discussion to include the k-sample case. While the specific procedures for the k-sample case are analogous to those previously discussed, they are more complex and thus beyond the scope of this book. For specific details, see Hays (1981) or Kirk (1982). However, we have provided tables that can be used for the k-sample case in determining appropriate sample sizes.

To illustrate the use of these tables, recall from the last chapter that four factors must be considered in determining an appropriate sample size; they are:

1. The level of significance (α).

2. The power of the test ($1 - \beta$).

3. The population error variance (σ_e^2).

4. The effect size (ES).

Regarding the first two factors, we indicated in Chapter 13 that Type I errors are generally considered more serious than Type II errors, and we recommended using a 4:1 ratio of β to α. If, a priori, the researcher establishes α at .05, the corresponding β is .20, and thus the required power ($1 - \beta$) is .80. For the last two factors, the effect size is defined as the *minimum treatment effect*. It is the standardized difference ($d = \text{ES}/\sigma$) between the two most extreme groups. For example, consider an experiment in which a psychologist studies the effects of four different drug dosages on the performance of rats running a maze. Assuming that the researcher has done previous research in this area and is knowledgeable about the research literature, a standardized effect size (d) can be established. For this example, assume the researcher uses $d = 0.75$. By setting the standardized effect size at 0.75, the researcher implies that such a difference is needed in order for the rejection of the null hypothesis to be both statistically significant and practically important. The sample size question thus becomes: how large must the sample be in order to detect a difference between the two extreme groups of .75 standard deviations ($d = 0.75$) for $\alpha = .05$ and power $= .80$? Using Table C.12 for the following values:

$\alpha = .05$

power $= .80$

$d = 0.75$

$k = 4$

we find that the sample size required for *each* of the four groups is 40.

Determining the sample size for the *k*-sample case requires considering the level of significance (α), the power of the test ($1 - \beta$), and the effect size (ES).

ANOVA: Simple Repeated Measures

Repeated-measures designs involve measuring an individual two or more times on the dependent variable. For example, a researcher may test the same sample of individuals under different conditions or at different times. These people's scores comprise dependent samples. Learning experiments often involve measuring the same people's performance solving a problem under five different conditions. Such a situation is analogous to and an extension of the design in which the dependent t test was applied (see Chapter 11).

In repeated-measures analysis, scores for the same individual are dependent, whereas the scores for different individuals are independent. Accordingly, the partitioning of the variation in ANOVA needs to be adjusted so that appropriate F ratios can be computed. The total sum of squares (SS_T) is partitioned into three components: (1) the variation among individuals (SS_I); (2) the variation among test occasions (SS_O); and (3) the remaining variation, which is called the residual variation (SS_{Res}). The mean squares for these sources of variation are computed, as before, by dividing the sums of squares by their appropriate degrees of freedom. The mean square for the residual variation ($MS_{Res} = SS_{Res}/df_{Res}$) is used as the error term (the denominator of the F ratio) for testing the effect of test occasion, which is the effect of primary interest. It must be noted that there is no appropriate error term for testing the effect of differences among the individuals.

> In simple repeated-measures ANOVA, test occasion is the effect of primary interest. Additionally, there is *no* appropriate error term for testing the effect of differences among individuals.

The formulas for the sums of squares in repeated-measures ANOVA are as follows:

$$SS_T = \sum_{k=1}^{K} \sum_{i=1}^{n} X_{ik}^2 - \frac{T^2}{N} \tag{14.16}$$

where

$\displaystyle\sum_{k=1}^{K} \sum_{i=1}^{n} X_{ik}^2 =$ squared scores summed across all individuals and all test occasions

$K =$ number of test occasions

$n =$ number of individuals

$T =$ sum of all scores summed across all individuals and all test occasions

$N = nK =$ total number of scores

$$SS_I = \sum_{i=1}^{n} \left(\frac{T_i^2}{K} \right) - \frac{T^2}{N} \tag{14.17}$$

where

T_i = sum of observations for the *i*th individual

K = number of test occasions

$$SS_O = \sum_{k=1}^{K} \left(\frac{T_k^2}{n} \right) - \frac{T^2}{N} \tag{14.18}$$

where

T_k = sum of observations in the *k*th test occasion

n = number of individuals

$$SS_{Res} = SS_T - SS_I - SS_O \tag{14.19}$$

The degrees of freedom associated with each of these sums of squares are as follows:

$$df_T = N - 1$$
$$df_I = n - 1$$
$$df_O = K - 1$$
$$df_{Res} = (K - 1)(n - 1)$$

As indicated earlier, the effect of interest is the test occasion and is tested using the following F ratio:

$$F_O = \frac{MS_O}{MS_{Res}}$$

The summary ANOVA table for the general case of a simple repeated-measures ANOVA is found in Table 14.10.

TABLE 14.10
Summary ANOVA Table for a Repeated-Measures Analysis

Source	SS	df	MS	F
Individuals	SS_I	$n - 1$	$SS_I/(n - 1)$	
Occasions	SS_O	$K - 1$	$SS_O/(K - 1)$	MS_O/MS_{Res}
Residual	SS_{Res}	$(K - 1)(n - 1)$	$SS_{Res}/(K - 1)(n - 1)$	
Total	SS_T	$N - 1$		

In simple repeated-measures ANOVA, the total variation is partitioned into three components: (1) variation among individuals; (2) variation among test occasions; and (3) residual variation.

Suppose data are collected from a sample of ten individuals on three different test occasions, each involving a different treatment. Each individual is exposed to a different order of treatment. The dependent variable is performance on a learning task under the various treatments. (The data, with preliminary calculations, are found in Table 14.11.) This investigation is interested primarily in the possible differences among treatments. Now consider this example using the four steps of hypothesis testing.

Step 1: State the Hypotheses

The null hypothesis for this example would be that there is no difference among the treatments; that is,

$$H_0: \mu_1 = \mu_2 = \mu_3$$

where the μ_Ks represent the population means for the test occasions. The alternative hypothesis is

$$H_a: \mu_i \neq \mu_K \quad \text{for some } i, k$$

Step 2: Set the Criterion for Rejecting H_0

As discussed for the other ANOVAs, the test statistic in repeated-measures ANOVA is the F ratio—that is, the ratio of the mean square for the test occasion (MS_O) and the mean square residual (MS_{Res}). For this test, the underlying distribution is the F distribution with $K - 1 = 2$ degrees of freedom and $(n - 1)(K - 1) = (9)(2) = 18$ degrees of freedom. Using $\alpha = .05$, the critical value of F for this test statistic from Table C.5 in the Appendix is 3.55.

Step 3: Compute the Test Statistic

Using the data in Table 14.11, and formulas 14.16 to 14.19, the sums of squares are calculated as follows:

$$SS_T = 4,054 - \frac{(320)^2}{30}$$

$$= 640.67$$

TABLE 14.11
Scores for Ten Persons on Three Test Occasions

Person	Test 1	Test 2	Test 3	T_i	T_i^2/K
A	6	12	18	36	432.00
B	9	14	16	39	507.00
C	4	8	15	27	243.00
D	3	10	12	25	208.33
E	1	6	10	17	96.33
F	7	15	20	42	588.00
G	8	8	15	31	320.33
H	9	11	18	38	481.33
I	8	12	13	33	363.00
J	6	10	16	32	341.33
					$\sum\limits_{i=1}^{n} \dfrac{T_i^2}{K} = 3{,}580.66$
n	10	10	10	$N = 30$	
T_K	61	106	153	$T = 320$	
$\bar{X}_K$	6.1	10.6	15.3	$\dfrac{T^2}{N} = 3{,}413.33$	
$\sum\limits_{i=1}^{n} X_{ik}^2$	437	1,194	2,423	$\sum\limits_{k=1}^{K}\sum\limits_{i=1}^{n} X_{ik}^2 = 4{,}054$	
$\dfrac{T_K^2}{n}$	372.10	1,123.60	2,340.90	$\sum\limits_{k=1}^{K}\left(\dfrac{T_K^2}{n}\right) = 3{,}836.60$	

$$SS_I = \frac{(36)^2}{3} + \frac{(39)^2}{3} + \cdots + \frac{(32)^2}{3} - \frac{(320)^2}{30}$$

$$= 167.33$$

$$SS_O = \frac{(61)^2}{10} + \frac{(106)^2}{10} + \frac{(153)^2}{10} - \frac{(320)^2}{30}$$

$$= 423.27$$

$$SS_{Res} = 640.67 - 167.33 - 423.27$$

$$= 50.07$$

Using these sums of squares and the corresponding degrees of freedom, the summary ANOVA table is completed (see Table 14.12). Notice that the F ratio for test occasions equals 76.13.

Step 4: Interpret the Results

The calculated F ratio for test occasions (F_O) equals 76.13 and exceeds the critical value of F for 2 and 18 degrees of freedom; therefore, the null hypothesis is rejected. The conclusion is that the population means for the three test occasions are not equal. In terms of probability, the probability that the observed differences in the means of the test occasions would have occurred by chance if the null hypothesis were true (the population means are equal) is less than .05. Since the null hypothesis is rejected, it is necessary to conduct a post hoc multiple comparison analysis in order to determine which pairs or combinations of means differ (see Chapter 15).

Assumptions for Repeated-Measures ANOVA

There are several assumptions that underlie the use of simple repeated-measures ANOVA.

1. The sample was randomly selected from the population.

2. The dependent variable is normally distributed in the population.

TABLE 14.12
Summary Table for Repeated-Measures ANOVA

Source	SS	df	MS	F	F_{cv}
Individuals	167.33	9	18.59		
Occasions	423.27	2	211.64	76.13*	3.55
Residual	50.07	18	2.78		
Total	640.67	29			

*$p < .05$.

3. The population variances for the test occasions are equal.

4. The population correlation coefficients between pairs of test occasion scores are equal.

When the last two assumptions are not met, the Type I error rate can be seriously affected. However, we can make an appropriate correction by changing the degrees of freedom from $K - 1$ and $(n - 1)(K - 1)$ to 1 and $n - 1$, respectively. Such a change for the above example would result in the critical value of F being 5.12 (for df = 1, 9) rather than 3.55 (for df = 2, 18). This change provides a conservative test of the null hypothesis when assumptions 3 and 4 are not met.

Summary

The analysis of variance (ANOVA), one-way classification, is the method for testing the null hypothesis $H_0: \mu_1 = \mu_2 = \cdots = \mu_K$, that is, the equality of K population means. The t test for two independent samples is a special case of ANOVA for $K = 2$. Using one-way ANOVA, the equality of all population means can be tested simultaneously while maintaining the pre-established Type I error rate.

The test statistic in ANOVA (F) is the ratio of two variance estimates. The two variance estimates are called the mean square between groups (MS_B) and the mean square within groups (MS_W). MS_B is an estimate of the population or error variance plus the variation attributable to the treatment effects or differences between population means, while MS_W is an estimate of only the error variance.

The actual computational procedures for determining MS_B and MS_W involve computing the sum of squared deviations of the group means from the grand mean (sum of squares between groups: SS_B) and the sum of squared deviations of the individual scores from the group means (sum of squares within groups: SS_W). MS_B and MS_W are then determined by dividing SS_B and SS_W by the degrees of freedom associated with the two estimates: $K - 1$ and $N - K$, respectively. Thus, $MS_B = SS_B/(K - 1)$ and $MS_W = SS_W/(N - K)$. The actual computational procedures are summarized in Table 14.13, along with the expected mean squares E(MS).

The sampling distribution of the F ratio (MS_B/MS_W) is the F distribution with $K - 1$ and $N - K$ degrees of freedom. When this F ratio exceeds the critical value of F, the null hypothesis is rejected, and we conclude that the population means are not all equal.

Omega squared (ω^2) is a measure of association, which indicates the proportion of variation in the dependent variable that is accounted for by differences in the levels of the independent variable. We also discussed the process of determining sample size for the K-sample case. As for a two-sample case, the sample size for each of the K groups is dependent on α, power, and the effect size, all identified a priori by the researcher.

Repeated-measures ANOVA is similar in many respects to one-way ANOVA. The total sum of squares in repeated-measures ANOVA is partitioned differently

TABLE 14.13
Summary of Concepts and Procedures for ANOVA, One-Way Classification

Source	SS	Computational Formula	df	MS	E(MS)	F
Between	$\sum\limits_{k=1}^{K} n_k(\bar{X}_k - \bar{X})^2$	$\sum\limits_{k=1}^{K}\left(\dfrac{T_k^2}{n_k}\right) - \dfrac{T^2}{N}$	$K-1$	$\dfrac{SS_B}{K-1}$	$\sigma_e^2 + \dfrac{n\sum \alpha_k^2}{K-1}$	$\dfrac{MS_B}{MS_W}$
Within	$\sum\limits_{k=1}^{K}\sum\limits_{i=1}^{n_k}(X_{ik} - \bar{X}_k)^2$	$\sum\limits_{k=1}^{K}\sum\limits_{i=1}^{n_k}X_{ik}^2 - \sum\limits_{k=1}^{K}\left(\dfrac{T_k^2}{n_k}\right)$	$N-K$	$\dfrac{SS_W}{N-K}$	σ_e^2	
Total	$\sum\limits_{k=1}^{K}\sum\limits_{i=1}^{n_k}(X_{ik} - \bar{X})^2$	$\sum\limits_{k=1}^{K}\sum\limits_{i=1}^{n_k}X_{ik}^2 - \dfrac{T^2}{N}$	$N-1$			

TABLE 14.14
Summary of Computational Procedures for Repeated-Measures ANOVA

Source	SS	df	MS	F
Individuals	$\sum\limits_{i=1}^{n}\left(\dfrac{T_i^2}{K}\right) - \dfrac{T^2}{N}$	$(n-1)$	$SS_I/(n-1)$	
Occasions	$\sum\limits_{k=1}^{K}\left(\dfrac{T_k^2}{n}\right) - \dfrac{T^2}{N}$	$(K-1)$	$SS_O/(K-1)$	MS_O/MS_{Res}
Residual	$SS_T - SS_I - SS_O$	$(n-1)(K-1)$	$SS_{Res}/(n-1)(K-1)$	
Total	$\sum\limits_{k=1}^{K}\sum\limits_{i=1}^{n}X_{ik}^2 - \dfrac{T^2}{N}$	$N-1$		

in order to control for the fact that scores for each individual are not independent. SS_T (total sum of squares) is partitioned into SS_I (sum of squares for individuals), SS_O (sum of squares for test occasions), and SS_{Res} (residual sum of squares). The F ratio for testing the null hypothesis of no differences between test occasions is MS_O/MS_{Res}. The computational procedures for repeated-measures ANOVA are summarized in Table 14.14.

Exercises

1. A sociologist is interested in whether or not level of parental occupation affects anxiety scores among ninth-grade students. The data below are scores on a standardized anxiety test for students with parents in three different occupational levels. Test the appropriate hypothesis using $\alpha = .05$.

Occupational level

1	2	3
8	23	21
6	11	21
4	17	22
12	16	18
16	6	14
17	14	21
12	15	9
10	19	11
11	10	
13		

a. State the hypotheses.

b. Set the criterion for rejecting H_0.

c. Complete the ANOVA and compute the test statistic.

d. Compute ω^2.

e. Interpret the results.

*2. A psychology experiment is conducted using 65 young, adult males enrolled in an introductory psychology course. The independent variable is "solution strategy" for solving a concept attainment problem. The solution strategies, labeled below as A, B, C, D, and E, involve the extent to which the individual is taught to be conservative in determining the characterisitecs involved in the problem. The dependent variable is the time in minutes required to complete the problem. Thirteen males are randomly assigned to each strategy; each male solves the problem. The means for each strategy and a partially completed Summary ANOVA Table are given below. Complete the Summary ANOVA Table and answer the following questions for testing the appropriate hypothesis using $\alpha = .05$.

Solution strategy	A	B	C	D	E
Time to complete task	2.40	2.23	3.79	5.91	4.61

Summary ANOVA

Source	SS	df	MS	F	F_{cv}
Between	105.41				
Within					
Total	347.45				

a. State the hypotheses.

b. Set the criterion for rejecting H_0.

c. Complete the ANOVA and compute the test statistic.

d. Compute ω^2.

e. Interpret the results.

3. A cognitive psychologist designs a research study to investigate different problem-solving strategies. Subjects are randomly assigned to one of five different groups; each group is taught to use a different problem-solving strategy. After the training, the subjects are given a series of problems to solve using the strategies they have learned; the data below are times the subjects spent solving the problems. Test the appropriate hypothesis using $\alpha = .05$.

Problem-solving strategy

1	2	3	4	5
32	30	85	38	53
41	39	76	29	43
53	52	70	21	47
67	64	64		52
48	51			67
39	37			
44	44			

    a. State the hypotheses.

    b. Set the criterion for rejecting H_0.

    c. Complete the ANOVA and compute the test statistic.

    d. Compute ω^2.

    e. Interpret the results.

4. A marketing specialist is assessing the effectiveness of two types of formats for TV commercials. Twenty-four regular TV watchers are randomly selected and then assigned to watch one of the two formats of a commercial for an imported car. The following data are the scores on a measure for assessing the impact of commercials. Test the appropriate hypothesis using $\alpha = .05$.

Commercial format

A	B
21	38
33	32
40	56
37	58
22	45
25	46
39	57
28	49
34	43
25	41
27	55

    a. State the hypotheses.

    b. Set the criterion for rejecting H_0.

    c. Complete the ANOVA and compute the test statistic.

d. Compute ω^2.

e. Interpret the results.

5. For the data in Exercise 4, test the appropriate hypothesis using the independent *t* test (see Chapter 11) and $\alpha = .05$.

   a. State the hypotheses.

   b. Set the criterion for rejecting H_0.

   c. Compute the test statistic.

   d. Interpret the results.

   e. Show that $t^2 = F$ and $T_{cv}^2 = F_{cv}$

6. Suppose the cognitive psychologist in Exercise 3 replicates the study with eight subjects in each group. The following is a partially completed summary ANOVA table for this study. Complete the summary ANOVA table and answer the following questions for testing the appropriate hypothesis using $\alpha = .01$.

		Summary ANOVA			
Source	SS	df	MS	F	F_{cv}
Between	95.80				
Within			3.66		
Total					

   a. State the hypotheses.

   b. Set the criterion for rejecting H_0.

   c. Complete the ANOVA and compute the test statistic.

   d. Compute ω^2.

   e. Interpret the results.

7. Compare the formulas for the pooled estimate of the population variance from Chapter 11 (see formulas 11.3, 11.4, and 11.5) and the formula for the mean square within groups (MS_W). Discuss the similarity of the formulas.

8. A university biology professor has developed four methods for teaching the senior-level class in microbiology. At the start of the year, the professor randomly assigns students to the four groups. Attrition takes place during the year, so that some of the groups lose more members than others. The following are the final exam scores for each group. Test the appropriate hypothesis using $\alpha = .05$.

Method

1	2	3	4
61	72	82	93
65	78	58	84
76	67	64	80
69	91	76	75
78	80	70	
71	86		
	79		

 a. State the hypotheses.

 b. Set the criterion for rejecting H_0.

 c. Complete the ANOVA and compute the test statistic.

 d. Compute ω^2.

 e. Interpret the results.

9. For Exercise 8, suppose the biology professor computes all possible t tests rather than the ANOVA using $\alpha = .05$.

 a. How many t tests would have to be computed?

 b. Compute the Type I error rate.

10. The following data are for three groups of subjects who are exposed to three different methods of reinforcement. The dependent variable is the number of correctly identified nonsense syllables. Test the appropriate hypothesis using $\alpha = .05$.

Method

A	B	C
20	19	24
18	12	35
14	13	26
16	17	29
13	15	31
19	20	25
22	24	28

 a. State the hypotheses.

 b. Set the criterion for rejecting H_0.

 c. Complete the ANOVA and compute the test statistic.

 d. Compute ω^2.

 e. Interpret the results.

11. Suppose, for Exercise 10, only means and variances are available for the groups (see data below). Using the definitions of SS_B and SS_W, complete the summary ANOVA for the following data using $\alpha = .01$.

Method

	A	B	C
n_k	9	9	9
$\overline{X}_k$	22.4	29.8	37.5
s_k^2	10.8	11.4	10.6

 a. State the hypotheses.

 b. Set the criterion for rejecting H_0.

 c. Complete the ANOVA and compute the test statistic.

d. Compute ω^2.

e. Interpret the results.

12. Consider a single-factor design (one independent variable) with four levels of the factor. Use Table C.12 to determine the appropriate sample size, given the following conditions:

   a. $\alpha = .05$, $\beta = .20$, $d = 1.00$

   b. $\alpha = .05$, $\beta = .20$, $d = 0.85$

   c. $\alpha = .01$, $\beta = .05$, $d = 1.00$

   d. $\alpha = .01$, $\beta = .05$, $d = 0.85$

13. Consider a single-factor design (one independent variable) with three levels of the factor. Use Table C.12 to determine the appropriate sample size, given the following conditions:

   a. $\alpha = .05$, $\beta = .20$, $d = 0.75$

   b. $\alpha = .05$, $\beta = .20$, $d = 1.15$

   c. $\alpha = .01$, $\beta = .05$, $d = 0.75$

   d. $\alpha = .01$, $\beta = .05$, $d = 1.15$

14. A researcher is studying the performance of six subjects over five trials on a learning task. The data are given below. Complete the ANOVA using $\alpha = .01$.

   a. State the hypotheses.

   b. Set the criterion for rejecting H_0.

   c. Complete the ANOVA and compute the test statistic.

			Trial		
Subject	1	2	3	4	5
A	7	6	9	11	12
B	6	5	6	9	8
C	7	9	11	11	13
D	5	5	5	6	6
E	7	8	9	9	11
F	6	6	7	11	13

15. A school principal traces reading comprehension scores on a standardized test for a random sample of dyslexic students across three years. The data are given below. Complete the ANOVA using $\alpha = .05$.

Student	Third grade	Fourth grade	Fifth grade
A	2.8	3.2	4.5
B	2.6	4.0	5.1
C	3.1	4.3	5.0
D	3.8	4.9	5.7
E	2.5	3.1	4.4
F	2.4	3.1	3.9
G	3.2	3.8	4.3
H	3.0	3.6	4.4

a. State the hypotheses.

b. Set the criterion for rejecting H_0.

c. Complete the ANOVA and compute the test statistic.

d. Interpret the results.

16. In a pilot study of weight loss among obese middle-aged women, a research team of behavioral psychologists, physicians, and dietitians monitors the diet and weight loss of ten subjects. The following data are the subjects' weight losses across the four weeks of the pilot study. Complete the ANOVA using $\alpha = .05$.

Subject	Week 1	Week 2	Week 3	Week 4
1	5.0	3.5	3.5	2.5
2	4.5	3.0	1.0	1.5
3	3.0	3.0	2.0	2.0
4	5.5	5.0	4.5	2.0
5	7.0	5.0	2.0	2.0
6	2.0	3.0	3.0	1.0
7	4.5	4.5	2.5	1.0
8	5.0	4.5	2.5	1.0
9	5.5	3.5	2.5	2.0
10	6.0	6.0	5.0	3.0

a. State the hypotheses.

b. Set the criterion for rejecting H_0.

c. Complete the ANOVA and compute the test statistic.

d. Interpret the results.

15

Multiple-Comparison Procedures

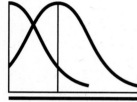

I n the previous chapter, we presented analysis of variance, with one-way classification, as the procedure for testing the null hypothesis that K population means are simultaneously equal. The null hypothesis is rejected when the observed F ratio exceeds the critical value; we can then conclude that there exists some difference among the means. Using ANOVA, this hypothesis is tested while maintaining the Type I error rate at the pre-established level. This is sometimes called the *omnibus test*. After the hypothesis is rejected, we are still faced with the problem of deciding which pairs or combinations of means are not equal. Thus, the omnibus test in ANOVA is often only the first test in analyzing a set of data. The procedures discussed in this chapter were developed to determine which means differ significantly after a significant F ratio has been found in the ANOVA. These procedures are called **post hoc multiple-comparison tests**. We will discuss additional procedures that are alternatives to ANOVA when we have specific hypotheses to be tested, rather than the null hypothesis in ANOVA. For example, we might want to determine whether the mean of three experimental groups differs from the mean of a single control group. We might want to determine the trend in the effects of different levels of the independent variable, or we might want to test a directional hypothesis. These procedures are called **planned**, or **a priori**, **comparisons**, and they are done in place of the omnibus F test of ANOVA.

Six multiple-comparison procedures are discussed in this chapter. Although many multiple-comparison procedures have been developed,[1] we think that these six procedures should suffice for most research situations. The six procedures to be discussed are illustrated in Figure 15.1, along with their appropriate contexts. The choice of procedure depends, first of all, on whether the hypothesis will require a specifically planned test or an ANOVA will be used, followed by post hoc multiple-comparison procedures, if the null hypothesis is rejected. Second, if we choose the post hoc procedures, the choice of procedure will depend on whether the group sizes for the levels of the independent variable are equal or unequal and on the relative statistical power of the procedure.

Figure 15.1 shows that either the *Tukey* or the *Newman-Keuls method* is used to identify which pairs of means differ following a significant F ratio in the ANOVA when the group sizes are equal. When the group sizes are unequal, the *Tukey/Kramer (TK) method* is used. The *Scheffé method* is used when combinations of means, rather than simply pairs of means, are contrasted. For situations in

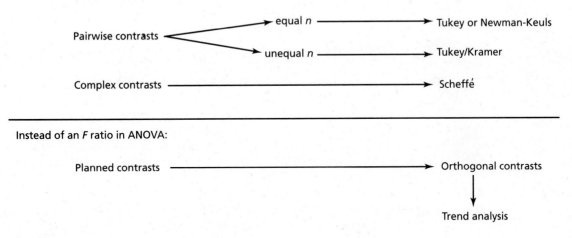

FIGURE 15.1
Multiple-comparison procedures presented in this chapter

---

[1] For a detailed discussion of the various multiple-comparison procedures, see G. R. Hancock and A. J. Klockars, "The Quest for α: Developments in Multiple Comparisons in Multiple Comparison Procedures in the Quarter Century Since Games (1971)," *Review of Educational Research* 66 (1996): 269–306; M. R. Stoline, "The Status of Multiple Comparisons: Simultaneous Estimation of All Pairwise Comparisons in One-Way ANOVA Designs," *American Statistician* 35 (1981):134–141; and R. E. Kirk, Chapter 3.

which hypotheses other than the general hypothesis of ANOVA are defined a priori, complex contrasts — *orthogonal contrasts* and *trend analysis* — can be used.

Post Hoc Multiple-Comparison Tests

When the null hypothesis in the ANOVA is rejected, we want to know whether the significant F ratio is due to differences between pairs of means or perhaps to some more complex combination of means. The post hoc tests to be discussed in this chapter were developed to maintain the a priori Type I error rate when computing a series of comparisons following the rejection of the null hypothesis in the ANOVA. Instead of the usual interpretation of the Type I error when testing a single hypothesis, these post hoc tests are usually described in terms of their *comparisonwise* error rate or *experimentwise* error rate.

> Post hoc multiple-comparison tests maintain the Type I error rate at α when a series of comparisons is made among sample means.

Type I Error Rates

The **comparisonwise error rate** is defined simply as α, or the level of significance, for each comparison. To control for this error rate requires only that each test be conducted at the a priori α level. For example, in the ANOVA using five groups discussed in Chapter 14, there are ten different pairs of means that could be compared: group 1 with group 2, group 1 with group 3, and so on. If we test each of these pairwise comparisons at $\alpha = .05$, the probability of a Type I error for *each* comparison is .05.

In contrast, the **experimentwise error rate** (α_E) is the probability of making a Type I error for the set of all possible comparisons. When all of the pairwise comparisons are independent,[2] the formula for defining α_E is

$$\alpha_E = 1 - (1 - \alpha)^c \tag{15.1}$$

where

$\alpha = $ comparisonwise error rate for each comparison

$c = $ number of comparisons

[2] See G. A. F. Seber, pp. 125–126.

Formula 15.1 is true only when the c comparisons are independent. In ANOVA, the c comparisons are not independent and α_E would be less than $1 - (1 - \alpha)^c$. For this situation, α_E can be approximated by

$$\alpha_E' = c(\alpha) \tag{15.2}$$

This approximation provides us with a simple procedure for determining the α necessary for each comparison in order to maintain α_E at the a priori level of significance across all comparisons. By dividing α_E by the number of comparisons (c), we find the comparisonwise error rate for testing each of the pairwise comparisons. In our example, if $\alpha_E = .05$, the comparisonwise error rate is $.05/10 = .005$. Note that, even though this procedure controls the experimentwise error rate, it is extremely conservative and may result in no significant comparisons even when the F ratio of the ANOVA is significant.

An experimentwise error rate is defined as the probability of making at least one Type I error for the set of all possible comparisons in an experiment. A comparisonwise error rate is the probability of making a Type I error for any of the comparisons.

A less conservative procedure is to use one of the post hoc multiple-comparison tests that maintain α_E at the a priori α level by using the **studentized range (Q) distributions**, rather than the t distributions, as the sampling distributions. The Tukey and the Newman-Keuls methods for equal group sizes and the Tukey/Kramer method for unequal group sizes all use the Q distributions as the sampling distributions. Whereas the Tukey and the Tukey/Kramer methods maintain α_E at the a priori α level, the Newman-Keuls has an α_E that is somewhat higher than the a priori α level; however, the α_E is substantially less than $c\alpha$.

The Q distributions were developed to determine the minimum difference between the largest and smallest means in a set of K sample means that is necessary to reject the hypothesis that the corresponding population means are equal. By dividing this difference by $\sqrt{MS_W/n}$, the Q distributions are analogous to the t distributions that we used earlier. In the development of the Q distributions, the number of groups varied but the number of observations in each of the K groups was the same. That is, the n in the expression is the sample size for each of the K groups and is the same for all groups. The 95th and 99th percentile points for the Q distributions are found in Table C.8; these are the critical values for the multiple-comparison tests at the .05 and .01 levels of significance, respectively. As might be expected, these critical values are dependent on the number of groups compared (K) and the degrees of freedom associated with the estimate of the population variance (MS_W).

> The Tukey, Tukey/Kramer, and Newman-Keuls methods use the Q distributions as the sampling distributions in order to control the experimentwise error rate.

The Tukey Method

The **Tukey method**, often called the *HSD (honestly significant difference) test*, is designed to make all pairwise comparisons while maintaining the experimentwise error rate (α_E) at the pre-established α level.[3] The null hypothesis tested for each pairwise comparison is

$$H_0: \mu_i = \mu_k \quad \text{for} \quad i \neq k$$

That is, each pair of population means is equal. The test statistic is Q, defined as follows:

$$Q = \frac{\overline{X}_i - \overline{X}_k}{\sqrt{MS_W/n}} \tag{15.3}$$

This test statistic is analogous to the t statistic found in Chapter 11. The only difference is in the use of the Q distributions as the sampling distributions.

Suppose a psychologist randomly selects 20 subjects from a population and then randomly assigns them to four different sets of motivating instructions (the levels of the independent variable). The dependent variable is the number of trials needed to successfully complete a complex motor task. The data are found in Table 15.1. Notice that the F ratio (8.87) exceeds the critical value of F at the .05 level; thus, we would reject the null hypothesis $(H_0: \mu_1 = \mu_2 = \mu_3 = \mu_4)$.

The calculation of the Q statistic for each of the possible pairwise comparisons is illustrated in Table 15.2. For computational convenience, the means of the groups are ranked from low to high.[4] (This step is *not* necessary for the Tukey method; however, it will be necessary for the Newman-Keuls method.) The second step involves finding the differences between the respective group means; that is, determining the numerator of the Q statistic. For example, $12.40 - 8.40 = 4.00$, $13.60 - 8.40 = 5.20$, and so on. The final step is to divide these differences by $\sqrt{MS_W/n}$. The critical value of the Q statistic is found by using Table C.8. In the table, r = the number of means being compared. For this example, $r = 4$. The *error df* are the degrees of freedom associated with MS_W; for this example,

---

[3] The Tukey method can also be used for complex comparisons when the group sizes are equal; however, we recommend the Scheffé method for complex comparisons and will discuss it later in the chapter.

[4] For all post hoc procedures, it is convenient to rank the means from low to high in computing the test statistic. Although such a strategy is not necessary for some of the procedures, it is consistent with the strategy used in SAS and SPSS.

TABLE 15.1
Data to illustrate the Tukey and Newman-Keuls Methods

	Group			
	1	*2*	*3*	*4*
	10	6	16	9
	16	12	18	12
	12	9	14	14
	12	8	20	11
	18	7	17	16
ΣX	68	42	85	62
$\overline{X}_k$	13.60	8.40	17.00	12.40

		Summary ANOVA			
Source	SS	df	MS	F	$F_{cv\,(.05)}$
Between	188.95	3	62.98	8.87	3.24
Within	113.60	16	7.10		
Total	302.55	19			

TABLE 15.2
Calculation of Q Using the Tukey Method

			$(\overline{X}_i - \overline{X}_k)$		
Group 2	8.40				
Group 4	12.40	4.00			
Group 1	13.60	5.20	1.20		
Group 3	17.00	8.60	4.60	3.40	
			Q		
Group 2					
Group 4		3.67			
Group 1		4.77*	1.10		
Group 3		7.89*	4.22*	3.12	

*$p < .05$ ($Q_{cv} = 4.05$ for df $= 16$).

$df_W = 16$. Thus, the critical values for Q for $r = 4$ and $df_W = 16$ are 4.05 and 5.19 for $\alpha = .05$ and $\alpha = .01$, respectively. Assuming $\alpha = .05$, the null hypotheses $H_0: \mu_1 = \mu_2$, $H_0: \mu_2 = \mu_3$, and $H_0: \mu_1 = \mu_3$ are rejected; the remaining null hypotheses are not rejected. The psychologist can then conclude that the population means for motivating instructions 1 and 3 are different from the population mean for motivating instruction 2 in terms of the number of trials necessary to complete the complex motor task. Also, the population means for 1 and 3 differ.

The Tukey method uses the Q distributions to maintain the Type I error rate at α for all possible pairwise comparisons.

The Newman-Keuls Method

The **Newman-Keuls method** is similar to the Tukey method except that it considers the differences between the ranked means. This method is based on a stairstep or layer approach, in which the critical values change depending on the range in the set of means being considered. When the observed sample means are ranked from low to high, the Newman-Keuls method requires smaller critical Q values for means that are adjacent and larger Q values for more widely separated means.

Consider again the data in Table 15.1; the calculation of the Q statistics for the Newman-Keuls method is presented in Table 15.3. For the Newman-Keuls method, it is necessary to rank the means from low to high and then compute the differences between the pairs of means. As with the Tukey method, the Q statistics are found by dividing these differences by $\sqrt{MS_W/n}$.

The critical values for Q are found in Table C.8. The *error df* is again $df_W = 16$, but the r is the number of steps between ordered means and will vary depending on which pair of means is being compared. Consider Table 15.3. When the highest mean ($\overline{X}_3 = 17.00$) is compared with the lowest mean ($\overline{X}_2 = 8.40$), $r = 4$ since

TABLE 15.3
The Q Statistics and Respective Critical Values Using the Newman-Keuls Method

Group 2	8.40			$(\overline{X}_i - \overline{X}_k)$
Group 4	12.40	4.00		
Group 1	13.60	5.20	1.20	
Group 3	17.00	8.60	4.60	3.40
Group 2				Q
Group 4		3.67*		
Group 1		4.77*	1.10	
Group 3		7.89*	4.22*	3.12*
		Critical Values of Q for $\alpha = .05$		
		3.00		
		3.65	3.00	
		4.05	3.65	3.00

*$p < .05$.

these are the extreme means of the four groups. When $\overline{X}_3$ is compared with $\overline{X}_4$ (the next lowest mean), $r = 3$. Finally, when $\overline{X}_3$ is compared with the adjacent mean $(\overline{X}_1)$, $r = 2$. Determining r for the remaining rows of Table 15.3 follows the same strategy. Using the various values for r, the critical values for the Q statistics for the Newman-Keuls method (assuming $\alpha = .05$) are found in Table 15.3.

The comparison of the observed values of the Q statistics with the critical values *must follow a rigid procedure*, as follows:

1. Go to the bottom row and test the difference between the largest and the smallest mean. The calculated Q (7.23) exceeds the critical value (4.05), so we continue across that row and compare the calculated Qs with the corresponding critical values of Q until either a nonsignificant Q is found or the row is completed.

2. Proceed up to the next row, comparing the next highest mean $(\overline{X}_1)$ with the smallest mean $(\overline{X}_2)$. Continue across the row and compare the calculated Qs with the corresponding critical values of Q until either a nonsignificant Q is found or the row is completed.

3. Proceed up to the next row and continue the process.

From the data in Table 15.3, we can conclude that the population mean of group 2 is different from the population means of the other three groups and that the population mean for group 3 is different from the population mean for group 4 and group 1. The remaining group means do not differ significantly, and the differences are attributable to sampling fluctuation.

> The Newman-Keuls method uses various Q distributions in the pairwise comparisons. The critical value for adjacent means is smaller than for more widely separated means.

Comparison of the Tukey and Newman-Keuls Methods

The Tukey and Newman-Keuls methods are both appropriate for pairwise comparisons following a significant F ratio in the ANOVA when the group sizes are equal. Sometimes, as in the above example, more statistically significant differences are found using the Newman-Keuls method than using the Tukey method. For this reason, the Newman-Keuls method is the more powerful test statistically. This advantage is offset by the fact that the experimentwise Type I error rate (α_E) for the Newman-Keuls is somewhat greater than the a priori α level set in the original ANOVA, but substantially less than if multiple t tests were conducted. For the Tukey procedure, α_E is maintained at the a priori α level.

When conducting post hoc tests with equal group sizes, both the Tukey and the Newman-Keuls methods are appropriate. The test statistic for both is Q, and the underlying distribution is the studentized range distribution for the appropriate degrees of freedom. Of the two methods, the Newman-Keuls is statistically more powerful. However, the Tukey method maintains the experimentwise Type I error rate at the a priori α level.

Post Hoc Tests for Unequal n: The Tukey/Kramer (TK) Method

As mentioned, the Tukey and Newman-Keuls methods are appropriate when the group sizes are equal. However, when the group sizes differ, a modification of the Tukey method is available; it is called the **Tukey/Kramer (TK) method**.[5] The method involves a modification in the denominator of the Q statistic. The formula for Q in the TK method is as follows:

$$Q = \frac{\overline{X}_i - \overline{X}_k}{\sqrt{MS_W \left(\dfrac{1/n_i + 1/n_k}{2} \right)}} \qquad (15.4)$$

Suppose a counselor receives clients from four different sources: the schools, the courts, parents, and "other persons." It is suspected that clients from the different sources have different levels of self-esteem. Random samples from each of four sources (populations) are selected and administered a measure of self-esteem during the initial counseling session. Thus, there are four levels of the independent variable, and the dependent variable is scores on the self-esteem inventory. An ANOVA is computed; the results are found in Table 15.4. Since the critical value of F for 3 and 29 degrees of freedom is 2.93 (see Table C.5), the computed F ratio is statistically significant, and the null hypothesis ($H_0: \mu_1 = \mu_2 = \mu_3 = \mu_4$) is rejected. However, at this point it is not known which means differ significantly.

The calculated values of the Q statistic for the TK method are found in Table 15.5. The critical value for the Qs with $r = 4$ and $df_W = 29$ is 3.86. These data lead to the following conclusions:

1. The population mean of the school-referred clients (S) is different from the population means of clients referred by parents (P) and others (O).

2. The population mean of the court-referred clients (C) is different from the population means of clients referred by parents (P) and others (O).

[5] See Stoline.

TABLE 15.4
Data for the Counselor Example

Group	Schools	Courts	Parents	Other
$\overline{X}_k$	21.86	23.00	17.29	16.80
n_k	7	9	7	10

		Summary ANOVA		

Source	SS	df	MS	F	$F_{cv(.05)}$
Between	255.45	3	85.15	10.47	2.93
Within	235.89	29	8.13		
Total	491.34	32			

TABLE 15.5
Calculation of Q Using the TK Procedure

$$Q_{s-c} = \frac{21.86 - 23.00}{\sqrt{8.13\left(\dfrac{1/7 + 1/9}{2}\right)}} = -1.12 \qquad Q_{c-p} = \frac{23.00 - 17.29}{\sqrt{8.13\left(\dfrac{1/9 + 1/7}{2}\right)}} = 5.62^*$$

$$Q_{s-p} = \frac{21.86 - 17.29}{\sqrt{8.13\left(\dfrac{1/7 + 1/7}{2}\right)}} = 4.24^* \qquad Q_{c-o} = \frac{23.00 - 16.80}{\sqrt{8.13\left(\dfrac{1/9 + 1/10}{2}\right)}} = 6.69^*$$

$$Q_{s-o} = \frac{21.86 - 16.80}{\sqrt{8.13\left(\dfrac{1/7 + 1/10}{2}\right)}} = 5.09^* \qquad Q_{p-o} = \frac{17.29 - 16.80}{\sqrt{8.13\left(\dfrac{1/7 + 1/10}{2}\right)}} = .49$$

$^*p < .05.$

Computer Exercise: The Survey of High School Seniors Data Set

In Chapter 14, we considered the differences in mathematics achievement scores (MATHACH) for students classified by father's education (FAED), i.e.,

1 = Less than HS

2 = HS graduate

3 = Some vocational or college

4 = College graduate or more

The Summary ANOVA table for this example is again repeated in Table 15.6. Recall that the null hypothesis, H_0: $\mu_1 = \mu_2 = \mu_3 = \mu_4$, was rejected since the

TABLE 15.6
SPSS Printout for One-Way ANOVA for the Survey of High School Seniors Data Set

```
           Variable  MATHACH     math test score
        By Variable  FAED        Father's education
                                  Analysis of Variance

                              Sum of        Mean        F        F
            Source       D.F.  Squares       Squares    Ratio    Prob.
Between Groups             3   1713.8734     571.2911   14.1307   .0000
Within Groups           496   20052.8090    40.4291
Total                   499   21766.6824
```

calculated value of F (14.1307) exceeded the critical value ($F_{cv} = 2.60$). We concluded that, in the population, the means for the four groups differed.

In order to determine which sample means differ significantly, we calculated both the Tukey HSD tests and the Newman-Keuls tests; these results are found in Table 15.7. Note that for both methods, the means for the groups are ordered from low to high. Note also that the actual Q statistics are not computed; only the * are used to denote which group means differ significantly.

For both methods, the null hypothesis tested for each pairwise comparison is

$$H_0: \mu_i = \mu_k \text{ for some } i, k$$

That is, each pair of population means is equal. The test statistics for both methods are defined as Q. For this example, we used the Tukey/Kramer (TK) formula for unequal n (see formula 15.4). Below we illustrate the calculation for the difference between Group 1 (Less than HS) and Group 2 (HS graduate).

$$Q = \frac{\overline{X}_i - \overline{X}_k}{\sqrt{MS_W \frac{1/n_i + 1/n_k}{2}}}$$

$$= \frac{12.4419 - 10.8687}{\sqrt{40.1307 \frac{1/129 + 1/132}{2}}}$$

$$= 2.8367$$

Note that the critical values for Q are given for both procedures; these values are interpolated from those given in Table C.8 in the Appendix.

For this example, note that the calculated value ($Q = 2.8367$) does *not* exceed the critical value for the Tukey method ($Q_{cv} = 3.65$) but it does for the Newman-Keuls method ($Q_{cv} = 2.81$). This again illustrates that the Newman-Keuls method is more powerful than the Tukey method. This additional power, however, is offset by the fact that the Type I error rate (α_E) is somewhat higher for the Newman-Keuls method than for the Tukey method.

TABLE 15.7
SPSS Printout for the Tukey and Newman-Keuls Post Hoc Tests

```
          Variable  MATHACH    math test score
      By Variable  FAED        Father's education


Multiple Range Tests: Tukey-HSD test with significance level .050
The difference between two means is significant if
    MEAN(J)- MEAN(I) >= 4.4961 * RANGE * SQRT(1/N(I)+1/N(J))
    with the following value(s) for RANGE: 3.65

    (*) Indicates the significant differences which are shown in the lower triangle
                            G  G  G  G
                            r  r  r  r
                            p  p  p  p

                            1  2  3  4
Mean      FEAD
10.8687   Grp 1
12.4419   Grp 2
13.2191   Grp 3                  *
15.8309   Grp 4               *  *  *



          Variable  MATHACH    math test score
      By Variable  FAED        Father's education


Multiple Range Tests: Student-Newman-Keuls test with significance level .050

The difference between two means is significant if
    MEAN(J)- MEAN(I) >= 4.4961 * RANGE * SQRT(1/N(I)+1/N(J))
    with the following value(s) for RANGE:

Step        2     3     4
RANGE    2.81  3.34  3.65


    (*) Indicates the significant differences which are shown in the lower triangle
                            G  G  G  G
                            r  r  r  r
                            p  p  p  p

                            1  2  3  4
Mean      FEAD
10.8687   Grp 1
12.4419   Grp 2               *
13.2191   Grp 3               *
15.8309   Grp 4               *  *  *
```

The data in Table 15.7 indicate that the results for the Tukey and the Newman-Keuls methods are the same with the exception of the difference between Group 1 and Group 2. These data lead to the following conclusions:

Group 4 (College graduate or more) has a significantly higher MATHACH mean than all other groups;

Group 3 (Some vocational or college) had a significantly higher mean than Group 1 (Less than HS); and

for the Newman-Keuls method only, Group 2 (HS graduate) had a significantly higher mean than Group 1 (Less than HS)

Post Hoc Test for Complex Comparisons: The Scheffé Method

Many times there are research settings in which a researcher is interested in testing hypotheses that are more complex than simple differences between pairs of means. One might, for example, want to know whether two experimental groups (considered together) differ from a control group. The **Scheffé method** is a versatile procedure that can be used to test these complex hypotheses.[6] With the Scheffé method, each hypothesis is stated in terms of a linear combination of coefficients and means; this combination is referred to as a **contrast**. In general, the form of the null hypothesis is as follows:

$$H_0: \sum_{k=1}^{K} C_k \mu_k = 0 \tag{15.5}$$

where

$$\sum_{k=1}^{K} C_k = 0$$

That is, we add (or subtract) the products of the means multiplied by these coefficients. The only restriction is that, for each hypothesis, the sum of the coefficients must equal zero. For any contrast, the coefficients (C_k) are nonzero only for the population means under consideration in the hypothesis. For any population mean not included in the hypothesis, $C_k = 0$.

Suppose the counselor in our earlier example is interested in determining whether clients referred by courts differ in self-esteem from clients referred by all other sources. This hypothesis can be written as

$$H_0: \mu_2 = \frac{\mu_1 + \mu_3 + \mu_4}{3}$$

This hypothesis can also be expressed in terms of a contrast.

$$H_0: (1/3)\mu_1 + (-1)\mu_2 + (1/3)\mu_3 + (1/3)\mu_4 = 0$$

[6]The Tukey method can also be used for complex comparisons when the group sizes are equal; however, we recommend the Scheffé method for complex comparisons.

Notice that the sum of the coefficients equals zero $(1/3 - 1 + 1/3 + 1/3 = 0)$. By multiplying each of the coefficients by 3, the fractional coefficients become integer values, and the null hypothesis becomes

$$H_0: (1)\mu_1 + (-3)\mu_2 + (1)\mu_3 + (1)\mu_4 = 0$$

While both sets of coefficients are appropriate, it is often easier to use the integer coefficients rather than the fractional ones.

When there are unequal group sizes, as in our example, it is necessary to adjust the coefficients.[7] For the example, C_1, C_3, and C_4 need to be changed. Consider the adjusted coefficient, C_1; it is computed by dividing the sample size for group 1 by the sum of the sample sizes for groups 1, 3, and 4. The adjusted coefficients for C_3 and C_4 are similarly computed. Since group 2 is being compared to the other three groups, C_2 is determined by dividing the sample size for group 2 by itself and multiplying by -1 in order to form the contrast. The adjusted coefficients for this example are computed as follows:

$$C_1 = \frac{7}{7+7+10} = .29$$

$$C_2 = (-1)\frac{9}{9} = -1$$

$$C_3 = \frac{7}{7+7+10} = .29$$

$$C_4 = \frac{10}{7+7+10} = .42$$

Notice that the values of these adjusted contrasts sum to zero.
The test statistic for the Scheffé method is:

$$F = \frac{\left(\Sigma C_k \bar{X}_k\right)^2}{(MS_W)\left[\Sigma(C_k^2/n_k)\right]} \tag{15.6}$$

The sampling distribution for this test statistic is the F distribution, but the critical value is *not* read directly from Table C.5. Instead, the critical value is determined by multiplying the critical value of F used in the ANOVA by the factor $K - 1$, where K is the number of groups. This multiplication results in an increase in the critical value and is the primary reason why the Scheffé method is a conservative method.

[7]For a more thorough discussion of adjusting the coefficients for unequal group sizes, see E. J. Pedhazur, *Multiple Regression in the Behavioral Sciences: Explanation and Prediction*, 2d ed. (New York: Holt, Rinehart and Winston, 1982), pp. 324–328.

TABLE 15.8
Computation of a Complex Contrast Using Means from Four Groups

Group	$\bar{X}_k$	n_k	Coefficients
Schools	21.86	7	.29
Courts	23.00	9	−1.00
Parents	17.29	7	.29
Other	16.80	10	.42

$$F = \frac{[.29(21.86) - 23.00 + .29(17.29) + .42(16.80)]^2}{8.13\left(\dfrac{.29^2}{7} + \dfrac{(-1)^2}{9} + \dfrac{.29^2}{7} + \dfrac{.42^2}{10}\right)}$$

$$F = \frac{21.07}{1.24} = 16.96^*$$

$^*p < .05.$

The calculation of the test statistic for this example is provided in Table 15.8. The critical value is determined as follows:

$$(K - 1)F_{K-1,N-K}$$

Using the .05 level of significance, the critical value is

$$(4 - 1)2.93 = 8.79$$

Since the observed value of the test statistic exceeds the critical value, the null hypothesis is rejected at the .05 level of significance. Thus, it would be concluded that the population of clients referred by the courts has a self-esteem mean that is different from the combined mean of the remaining three populations.

The Scheffé method is the most versatile and at the same time the most conservative post hoc multiple-comparison procedure; it is recommended for complex contrasts.

A Priori Planned Comparisons

In planning a research study, a researcher often determines a priori a specific set of hypotheses to test. The tests of these hypotheses are called *a priori*, or *planned*, *comparisons* and are performed instead of the overall test of the null hypothesis of

ANOVA. We will develop two of these procedures: planned orthogonal contrasts and trend analysis.

Planned Orthogonal Contrasts

In the Scheffé method, a contrast is defined as a linear combination of coefficients and means, such that the sum of the coefficients is zero. **Planned orthogonal contrasts** are a special set of complex contrasts, which are established before the data are collected. By definition, two contrasts are said to be orthogonal if knowledge of the outcome of one contrast in no way helps to predict the outcome of the second contrast. In other words, two contrasts are orthogonal if the respective tests are independent. Thus, both the comparisonwise error rate and the experimentwise error rate are equal to α.

> Planned orthogonal contrasts are established a priori by the researcher. The hypotheses reflected by these contrasts are independent, and thus the experimentwise error rate is maintained at α.

The criterion for determining whether two contrasts are orthogonal is that the sum of the cross-products of corresponding coefficients of the contrasts must equal zero:

$$\sum_{k=1}^{K} C_{1k} C_{2k} = 0$$

For example, in a four-group case, a researcher might be interested in testing the following hypotheses:

$$H_{0_1}: \frac{\mu_1 + \mu_2}{2} - \mu_3 = 0$$

$$H_{0_2}: \frac{\mu_1 + \mu_2}{2} - \frac{\mu_3 + \mu_4}{2} = 0$$

$$H_{0_3}: \mu_1 - \mu_2 = 0$$

The integer coefficients for these contrasts would be:

	μ_1	μ_2	μ_3	μ_4
H_{0_1}:	1	1	-2	0
H_{0_2}:	1	1	-1	-1
H_{0_3}:	1	-1	0	0

Consider the coefficients for H_{0_1} and H_{0_2}; the sum of the cross-products is $(1)(1) + (1)(1) + (-2)(-1) + (0)(-1) = 4$. Thus, these two contrasts are *not* orthogonal. Now consider the coefficients for H_{0_1} and H_{0_3}; the sum of the cross-products is $(1)(1) + (1)(-1) + (-2)(0) + (0)(0) = 0$. Finally, the cross-products of the coefficients for H_{0_2} and H_{0_3} are $(1)(1) + (1)(-1) + (-1)(0) + (-1)(0) = 0$. These latter two sets of contrasts are orthogonal.

There is a limitation on the number of orthogonal contrasts that can be generated for a set of data. If there are K levels of the independent variable, there are only $K - 1$ contrasts of means in any one set of contrasts. This corresponds to the $K - 1$ degrees of freedom associated with MS_B in the ANOVA. Of course, there can be several sets of $K - 1$ orthogonal contrasts. For our four-group design, one set would be

1	-1	0	0	$H_0: \mu_1 = \mu_2$
0	0	1	-1	$H_0: \mu_3 = \mu_4$
1	1	-1	-1	$H_0: \mu_1 + \mu_2 = \mu_3 + \mu_4$

If these contrasts reflect the three null hypotheses of interest to the researcher, the planned orthogonal contrasts will provide a powerful test of them.

> For a K-group design, there are only $K - 1$ contrasts that are orthogonal. Two contrasts are orthogonal if the sum of the cross-products of the corresponding coefficients equals zero.

The null hypotheses for orthogonal contrasts, as in the Scheffé method, are written as the linear combination of coefficients and means; the general form of the null hypothesis is

$$H_0: \sum_{k=1}^{K} C_k \mu_k = 0$$

The test statistic for orthogonal contrasts is again F and is computed using formula 15.6:

$$F = \frac{(\Sigma C_k \overline{X}_k)^2}{(MS_W)[\Sigma(C_k^2/n_k)]} \tag{15.6}$$

The critical value of F for each orthogonal contrast is read directly from Table C.5 for the appropriate degrees of freedom. The degrees of freedom are 1 and $N - K$ for each contrast. Note that this critical value is *not* inflated by the multiplication by $K - 1$.

> There are 1 and $N - K$ degrees of freedom associated with the test of each contrast. The critical value is read directly from Table C.5 and is not inflated by multiplying by $K - 1$ as in the Scheffé method.

An Example

An educational psychologist is investigating the effects of induced anxiety on scores in a card-sorting task. The researcher selects 60 subjects randomly from the population and assigns them randomly to four groups with 15 subjects in each group. The groups are then given different anxiety-producing treatments while performing the card-sorting task. The four different treatments are the levels of induced anxiety and are assumed to be equally spaced on a continuum from low to high. The data for this example are found in Table 15.9. The null hypotheses of interest to the researcher, each to be tested at $\alpha = .05$, are:

$$H_0: \mu_1 = \mu_4$$
$$H_0: \mu_2 = \mu_3$$
$$H_0: \mu_1 + \mu_4 = \mu_2 + \mu_3$$

TABLE 15.9
Calculation of the Test Statistic Using Planned Orthogonal Contrasts

	Group			
	1	*2*	*3*	*4*
$\overline{X}_k$	16.47	23.73	25.27	15.80
n_k	15	15	15	15
$C_1: \mu_1 = \mu_4$	1	0	0	−1
$C_2: \mu_2 = \mu_3$	0	1	−1	0
$C_3: \mu_1 + \mu_4 = \mu_2 + \mu_3$	1	−1	−1	1

Summary ANOVA

Source	SS	df	MS	F	F_{cv}
Between	1071.00	3	357.00	44.40	2.78
Within	450.00	56	8.04		
Total	1521.00	59			

$$F_1 = \frac{(16.47 - 15.80)^2}{8.04\left(\dfrac{1}{15} + \dfrac{1}{15}\right)} = .42$$

$$F_2 = \frac{(23.73 - 25.27)^2}{8.04\left(\dfrac{1}{15} + \dfrac{1}{15}\right)} = 2.21$$

$$F_3 = \frac{(16.47 - 23.73 - 25.27 + 15.80)^2}{8.04\left(\dfrac{1}{15} + \dfrac{1}{15} + \dfrac{1}{15} + \dfrac{1}{15}\right)} = 130.55^*$$

*$p < .05$.

Of course there has to be a strong theoretical rationale for these hypotheses to justify the use of planned comparisons.

The F statistics for these contrasts are in Table 15.9. The data indicate that the first two hypotheses are not rejected. For both hypotheses, the differences between the pairs of sample means are well within what would be expected from random sampling fluctuation. The third hypothesis is rejected at the .05 level because the observed value of F exceeds the critical value. The data indicate that the mean scores from the middle levels of anxiety-producing commands are significantly different from the mean scores of the more extreme levels.

Relationship of Orthogonal Contrasts and the Between-Groups Sum of Squares (SS_B)   As indicated, orthogonal contrasts involve independent tests of hypotheses. We have already noted that there are $K - 1$ possible orthogonal contrasts, reflecting the $K - 1$ degrees of freedom associated with MS_B in the ANOVA, and that each contrast is allocated 1 degree of freedom. Because each of these contrasts deals with independent information, the between-groups sum of squares (SS_B) is distributed across the $K - 1$ contrasts. In other words, the sum of SS for all contrasts equals SS_B. As an illustration, consider the definition of the sum of squares for each orthogonal contrast:

$$SS_{C_k} = \frac{(\Sigma C_k \overline{X}_k)^2}{\Sigma(C_k^2/n_k)} \tag{15.7}$$

Since each contrast has 1 degree of freedom, $SS_{C_k} = MS_{C_k}$, or the mean square equals the sum of squares. The test statistic for each contrast is defined by formula 15.6; this formula can also be expressed as

$$F_k = \frac{(\Sigma C_k \overline{X}_k)^2/\Sigma(C_k^2/n_k)}{MS_W} = \frac{MS_{C_k}}{MS_W} \tag{15.8}$$

Therefore,

$$MS_{C_k} = SS_{C_k} = (F_k)(MS_W)$$

Computing the sum of squares of each contrast,

$$SS_{C_1} = (F_1)(MS_W) = (.42)(8.04) = 3.38$$

$$SS_{C_2} = (F_2)(MS_W) = (2.21)(8.04) = 17.77$$

$$SS_{C_3} = (F_3)(MS_W) = (130.55)(8.04) = 1,049.62$$

Summing the sum of squares,

$$SS_B = (3.38) + (17.77) + (1,049.62)$$

$$= 1,070.77$$

The difference between this value and SS_B in the summary ANOVA table is due to rounding error.

The between-groups sum of squares (SS_B) equals the sum of the sum of squares for the orthogonal contrasts:

$$SS_B = \sum_{k=1}^{K-1} SS_{C_k}$$

One-tailed Tests Using Orthogonal Contrasts   One useful aspect of planned orthogonal contrasts is that we can do one-tailed tests by converting the F statistic to a t statistic. Recall from Chapter 14 that a one-way ANOVA for two groups is equivalent to a t test for independent samples. The relationship is that, for 1 degree of freedom in the numerator of F, $t = \sqrt{F}$. Because the test statistic for each orthogonal contrast is the F ratio with 1 and $N - K$ degrees of freedom, it is possible to convert the observed F ratio to a t value by taking the square root of F. Then, when a one-tailed t test is desired, the critical value for a one-tailed test can be used. For the same α level, a smaller critical value is required for statistical significance; thus, the statistical power of the orthogonal contrasts is enhanced.

Trend Analysis

In the previous examples of ANOVA, the levels of the independent variable were simply a nominal classification, such as the different sets of motivating instructions (Table 15.1) and the client/counselor categories (Table 15.4). Such an independent variable is called a *qualitative variable*. For such variables, data analysis is limited to testing the overall null hypothesis of ANOVA with subsequent post hoc comparisons, or planned contrasts. However, if the levels of the independent variable represent some logical order representing different amounts of a single common variable, the data analysis can be extended to examine the functional relationship between the levels of the independent variable and the group means on the dependent variable. For example, the levels of the independent variable could be (1) levels of sleep deprivation, (2) different dosages of a certain drug, or (3) lengths of practice time for a learning task. In these situations, the independent variable is called a *quantitative variable*.

Generally, in a research design with a qualitative variable, the researcher is interested principally in determining the differences among the K groups. However, in a research design with a quantitative variable, if the intervals between the levels of the independent variable can be specified, the researcher can investigate more penetrating questions regarding the trends in the data. For example, **trend analysis** allows the researcher to answer such questions as

1. Do the means of the treatment groups increase (decrease) in linear fashion with an increase in the level of the independent variable?

2. Is the trend linear or nonlinear?

3. If the trend is nonlinear, what degree equation (polynomial) is required to fit the data?

> When the independent variable in ANOVA is quantitative, the functional relationship between the levels of the independent variable and the dependent variable can be investigated through trend analysis.

Consider an example in which the levels of the independent variable are levels of sleep deprivation and the dependent variable is performance on a complex psychomotor task. The group means for the different levels of sleep deprivation are illustrated in Figure 15.2A. The data indicate a linear trend; higher levels of sleep deprivation are associated with lower performance on the task. This linear trend can be defined by a first-degree equation, $Y = bX + a$.

Another example of a quantitative variable involves lengths of practice time (independent variable) and resulting performance on a specified learning task (dependent variable). The group means on the learning task for various lengths of practice time are illustrated in Figure 15.2B. In this example, the trend is nonlinear. Performance on the learning task increases with length of practice until the final level, when performance decreases. The data might indicate that the final length of practice time is too long and that fatigue becomes a factor in performance at this final level. The nonlinear trend is quadratic and can be defined by a second-degree equation, $Y = b_1 X^2 + b_2 X + a$.

Cubic and quartic trends are illustrated in Figures 15.2C and 15.2D. These trends are defined by third- and fourth-degree equations, respectively. Because the trends can be represented by these equations, or polynomials, the functional relationship between a quantitative independent variable and the dependent variable can be investigated through trend analysis. In trend analysis, the researcher usually is not interested in determining the exact polynomial that describes the relationship. Rather, the purpose is to determine whether the trend departs significantly from the linear and, if so, whether the trend is quadratic, cubic, and so on. The procedures used for trend analysis in this chapter assume that the levels of the independent variable are equally spaced and that the group sizes are equal. These conditions

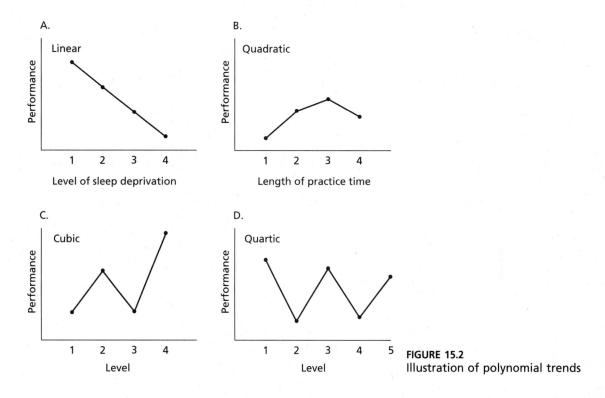

FIGURE 15.2
Illustration of polynomial trends

are not necessary for trend analysis, but the computation is greatly simplified when they are present.[8]

Orthogonal Polynomials: Contrasts Used in Trend Analysis

The procedures used in trend analysis are similar to those for planned orthogonal comparisons. Trend analysis is actually a special case of orthogonal contrasts, which involves the use of a set of specific coefficients called the *coefficients for orthogonal polynomials*. (The procedure for determining these coefficients is beyond the scope of this book.) As is the case for planned orthogonal contrasts, there are $K - 1$ orthogonal polynomial contrasts. For an independent variable with three levels, we can test for linear and quadratic trends. For an independent variable with four levels, we can test for linear, quadratic, and cubic trends. For an independent variable with five levels, we can test for linear, quadratic, cubic, and quartic trends. Even though there are $K - 1$ possible orthogonal polynomial contrasts, we would rarely test for a

---

[8]For a more thorough discussion of adjusting the coefficients for unequal spacing between the levels of the independent variable, see G. Keppel, *Design and Analysis: A Researcher's Handbook*, 2d ed. (Englewood Cliffs, N. J.: Prentice-Hall, 1982), pp. 629–634.

trend higher than the quartic regardless of the number of levels of the independent variable.

The coefficients for orthogonal polynomial contrasts are found in Table 15.10 and Table C.9; a more exhaustive list of coefficients is found in Kirk.[9] The test statistic for orthogonal polynomial contrasts is again the F statistic and is computed using formula 15.6:

$$F = \frac{(\sum C_k \overline{X}_k)^2}{(MS_W)[\sum(C_k^2/n_k)]}$$

The critical value of F is read directly from Table C.5 for 1 and $N - K$ degrees of freedom.

Usually, the tests for the respective trends are made sequentially; that is, we test for the linear trend, then for the quadratic trend, then for the cubic trend, and so on. To illustrate, consider again the example of the educational psychologist who is interested in the effect of anxiety on performance on a card-sorting task. The levels of induced anxiety (the independent variable) are assumed to be equally spaced on a continuum from low to high (an essential assumption for conducting trend analysis). The data and the computation of the test statistics are found in Table 15.11. Notice that $MS_W = 8.04$ (see Table 15.9) and that the critical value for each test for 1 and 56 degrees of freedom is 4.02.

The data indicate that the tests for the linear and cubic trends are not significant ($F = 0.02$ and $F = 2.61$, respectively). However, the test for a quadratic trend is significant ($F = 130.55$). Therefore, the educational psychologist can conclude that the trend in the data is quadratic and that the relationship between the independent

TABLE 15.10
Coefficients for Orthogonal Polynomials

$K = 3$					
Linear	−1	0	1		
Quadratic	1	−2	1		
$K = 4$					
Linear	−3	−1	1	3	
Quadratic	1	−1	−1	1	
Cubic	−1	3	−3	1	
$K = 5$					
Linear	−2	−1	0	1	2
Quadratic	2	−1	−2	−1	2
Cubic	−1	2	0	−2	1
Quartic	1	−4	6	−4	1

[9] See Kirk, p. 830.

TABLE 15.11
Calculation of Test Statistics for Trend Analysis
Using Orthogonal Polynomial Contrasts

	Group			
	1	2	3	4
$\overline{X}_k$	16.47	23.73	25.27	15.80
n_k	15	15	15	15
Linear	−3	−1	1	3
Quadratic	1	−1	−1	1
Cubic	−1	3	−3	1

$$F_{linear} = \frac{[(-3)(16.47) + (-1)(23.73) + (1)(25.27) + (3)(15.80)]^2}{8.04\left(\dfrac{9}{15} + \dfrac{1}{15} + \dfrac{1}{15} + \dfrac{9}{15}\right)} = .02$$

$$F_{quadratic} = \frac{[(1)(16.47) + (-1)(23.73) + (-1)(25.27) + (1)(15.80)]^2}{8.04\left(\dfrac{1}{15} + \dfrac{1}{15} + \dfrac{1}{15} + \dfrac{1}{15}\right)} = 130.55^*$$

$$F_{cubic} = \frac{[(-1)(16.47) + (3)(23.73) + (-3)(25.27) + (1)(15.80)]^2}{8.04\left(\dfrac{1}{15} + \dfrac{9}{15} + \dfrac{9}{15} + \dfrac{1}{15}\right)} = 2.61$$

*$p < .05$.

variable (induced anxiety) and the dependent variable (performance on the card-sorting task) can be described by a second-order polynomial. This trend is illustrated in a graph of the group means, in Figure 15.3.

We should note that more than one trend may be statistically significant. For example, in a four-group design, a significant cubic trend is sometimes accompanied by a linear trend. Such a situation is illustrated in Figure 15.4. The figure illustrates that there is a general linear increase over the levels of the independent variable. The slight decrease from level 2 to level 3 results in the cubic trend also being significant. Although linear and quadratic trends are investigated most often, there may be theoretically derived expectations of higher-order trends for some research studies. For these higher-order trends, it is important to inspect the graph of the means as well as to compute the test statistic.

In trend analysis with more than two levels of the independent variable, more than one contrast can be statistically significant.

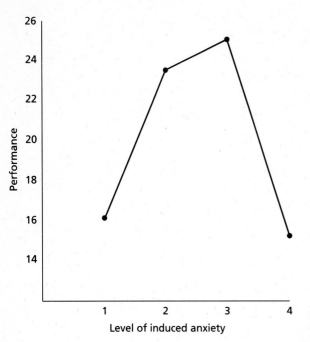

FIGURE 15.3
Quadratic polynomial trend for sample data

Summary

The multiple-comparison procedures discussed in this chapter are only a few of many procedures that have been developed; however, these methods should be sufficient for most research situations. The Tukey and Newman-Keuls methods are recommended for pairwise comparisons with equal group sizes, whereas the Tukey/Kramer (TK) is recommended for pairwise comparisons when group sizes are unequal. The Scheffé method is recommended for complex contrasts. When meaningful hypotheses can be determined before the analysis, planned contrasts including orthogonal contrasts and trend analysis can be used in place of the ANOVA. The formulas for the test statistics for each of these methods are summarized in Table 15.12.

The Tukey, Newman-Keuls, and Tukey/Kramer (TK) methods use the Q distributions as the sampling distributions, whereas the Scheffé method, orthogonal contrasts, and trend analysis use the F distributions. It must be emphasized that orthogonal contrasts and trend analysis are used only when the specific contrasts reflect hypotheses that are meaningful to a given research study. Some sets of orthogonal contrasts may not reflect meaningful hypotheses. On the other hand, some meaningful hypotheses may be reflected by contrasts that are not orthogonal.

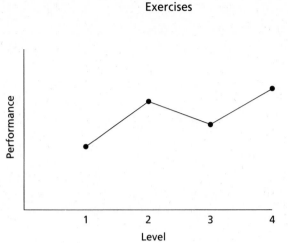

FIGURE 15.4
Example illustrating both a significant linear trend
and a cubic polynomial trend

Exercises

1. A company hires an industrial psychologist to determine the job satisfaction of salespeople who have been employed for varying lengths of time. Length of employment is classified as follows: (1) 1 to 2 years, (2) 3 to 5 years, (3) 5 to 10 years, and (4) more than 10 years. The following data are job satisfaction scores for the salespeople. Test the appropriate hypothesis using $\alpha = .05$.

Length of employment					*Length of employment*			
1	*2*	*3*	*4*		*1*	*2*	*3*	*4*
71	77	83	76		74	70	86	73
74	74	77	69		72	74	79	72
75	81	80	64		76	81	86	68
68	75	76	76		73	76	82	71

   a. State the hypotheses.

   b. Set the criterion for rejecting H_0.

   c. Complete the ANOVA and compute the test statistic.

   d. Compute ω^2.

   e. Use the Tukey method to detect the differences among the four groups.

   f. Compute the Newman-Keuls procedure. Compare the results with those for part e.

   g. Interpret the results of the ANOVA and the Tukey method.

*2. Consider the data in Exercise 2 in Chapter 14, in which the effects of various "solution strategies" upon solving a concept attainment problem were studied. Recall that 13 subjects were assigned to each of the five solution strategies; the dependent measure was the time in minutes to complete a concept attainment problem. The means for each strategy and the Summary ANOVA are given below.

Solution strategy	A	B	C	D	E
Time to complete task	2.40	2.23	3.79	5.91	4.61

TABLE 15.12
Summary Table of Multiple-Comparison Procedures

Procedure	Hypothesis Tested	To Be Used	Test Statistic	Underlying Distribution
Tukey	$H_0: \mu_i = \mu_k$	Pairwise contrasts with equal n_k	$Q = \dfrac{\bar{X}_i - \bar{X}_k}{\sqrt{MS_W/n}}$	Studentized range (Q) distributions (r ranges) with $N - K$ degrees of freedom
Newman-Keuls	$H_0: \mu_i = \mu_k$	Pairwise contrasts with equal n_k	$Q = \dfrac{\bar{X}_i - \bar{X}_k}{\sqrt{MS_W/n}}$	Studentized range (Q) distributions (varying ranges) with $N - K$ degrees of freedom
Tukey/Kramer	$H_0: \mu_i = \mu_k$	Pairwise contrasts with unequal n_k	$Q = \dfrac{\bar{X}_i - \bar{X}_k}{\sqrt{MS_W\left(\dfrac{1/n_i + 1/n_k}{2}\right)}}$	Studentized range (Q) distributions (r ranges) with $N - K$ degrees of freedom
Scheffé	$H_0: \sum c_k\mu_k = 0$	Pairwise contrasts with unequal n_{ki}; complex contrasts	$F = \dfrac{\left(\sum c_k\bar{X}_k\right)^2}{(MS_W)\sum\left(\dfrac{c_k^2}{n_k}\right)}$	$(K-1)$ (F distribution) with $K - 1, N - K$ degrees of freedom
Planned orthogonal contrasts	$H_0: \sum c_k\mu_k = 0$	Planned orthogonal contrasts	$F = \dfrac{\left(\sum c_k\bar{X}_k\right)^2}{(MS_W)\sum\left(\dfrac{c_k^2}{n_k}\right)}$	F distribution with $1, N - K$ degrees of freedom
Trend analysis	$H_0: \sum c_k\mu_k = 0$	Orthogonal polynomial contrasts	$F = \dfrac{\left(\sum c_k\bar{X}_k\right)^2}{(MS_W)\sum\left(\dfrac{c_k^2}{n_k}\right)}$	F distribution with $1, N - K$ degrees of freedom

Summary ANOVA

Source	SS	df	MS	F	F_{cv}
Between	105.41	4	26.35	6.54	2.53
Within	242.04	60	4.03		
Total	347.45	64			

    a. Use the Tukey method to detect the differences among the five groups.

    b. Interpret the results of the ANOVA and the Tukey method.

3. Consider the data in Exercise 10 in Chapter 14, in which the effect of three different methods of reinforcement on correct identification of nonsense syllables was studied. The means and standard deviations for the three groups and the summary ANOVA are given below.

Method

	A	B	C
n_k	7	7	7
$\overline{X}_k$	17.43	17.14	28.29
s_k	3.26	4.22	3.82

Summary ANOVA

Source	SS	df	MS	F	F_{cv}
Between	564.95	2	282.48	19.71	3.55
Within	258.00	18	14.33		
Total	822.95	20			

    a. Use the Tukey method to detect the differences among the three groups.

    b. Compute the Newman-Keuls procedure. Compare the results with those for part a.

    c. Interpret the results of the ANOVA and the Tukey method.

4. Consider the data in Exercise 1 in Chapter 14, in which a sociologist was investigating the effects of level of parental occupation on anxiety scores among ninth-grade students. The means and standard deviations for the three occupation levels and the summary ANOVA are given below.

Occupation level

	1	2	3
n_k	10	9	8
$\overline{X}_k$	10.90	14.56	17.13
s_k	4.09	5.08	5.11

Summary ANOVA					
Source	SS	df	MS	F	F_{cv}
Between	176.96	2	88.48	3.93	3.40
Within	540.00	24	22.50		
Total	716.96	26			

    a. Use the Tukey/Kramer method to detect the differences among the three groups.

    b. Interpret the results of the ANOVA and the Tukey/Kramer method.

5. Consider the data in Exercise 3 in Chapter 14, in which the cognitive psychologist was studying the effects of different problem-solving strategies. The means and standard deviations for the five groups and the summary ANOVA are given below.

Problem-solving strategy

	1	2	3	4	5
n_k	7	7	4	3	5
$\overline{X}_k$	46.29	45.29	73.75	29.33	52.40
s_k	11.31	11.34	8.96	8.50	9.10

Summary ANOVA					
Source	SS	df	MS	F	F_{cv}
Between	3810.99	4	952.75	8.87	2.84
Within	2255.47	21	107.40		
Total	6066.46	25			

    a. Use the Tukey/Kramer method to detect the differences among the five groups.

    b. Interpret the results of the ANOVA and the Tukey/Kramer method.

6. The following data were obtained from a university research study designed to measure the impact of employment level on job satisfaction. A composite index of job satisfaction was developed, taking into consideration such factors as reaction to employment supervision, adaptation to work-group environment, acceptance of personnel politics, and so on.

University Position

	Faculty	Admin.	Secretarial	Classified
n_k	21	21	21	21
$\overline{X}_k$	35.2	42.6	39.1	36.7

Summary ANOVA					
Source	*SS*	*df*	*MS*	*F*	*F_{cv}*
Between	517.5	3	172.500	29.417	2.74
Within	469.1	80	5.864		
Total	986.6	83			

Use the Scheffé method to test the following comparisons, using $\alpha = .05$.

   a. Faculty members versus secretarial staff.

   b. Classified staff versus faculty and secretarial staff.

   c. Administrators versus all other employees.

7. Use the properties of planned orthogonal contrasts with the data from Exercise 6 to test the following contrasts, using $\alpha = .05$. Illustrate how these contrasts are orthogonal.

   a. $H_0: \mu_1 = \mu_2$
      $H_a: \mu_1 \neq \mu_2$

   b. $H_0: \mu_3 = \mu_4$
      $H_a: \mu_3 \neq \mu_4$

   c. $H_0: \mu_1 + \mu_2 = \mu_3 + \mu_4$
      $H_a: \mu_1 + \mu_2 \neq \mu_3 + \mu_4$

8. For a four-group design like that in Exercise 6, what comparisonwise error rate is required to maintain the experimentwise error rate at 0.01 for all possible post hoc pairwise comparisons?

9. An exercise physiologist is studying the effects of drug dosage on delayed muscle soreness following a 30-minute standardized step-test exercise. The researcher assigns 48 subjects to four different groups. After the exercise, the subjects are given either a placebo or one of three levels of a drug. The following data are the mean levels of perceived pain for the four groups, measured six hours after the exercise. Use the coefficients provided in Table 15.10 to test for a linear, quadratic, and/or cubic trend. Use $\alpha = .01$. The within-groups mean square (MS_W) for these data is 14.36.

Drug Dosage				
	Placebo	*4 ml*	*8 ml*	*12 ml*
n_k	12	12	12	12
$\overline{X}_k$	82.5	63.4	56.6	57.2

10. Consider the data in Exercise 8 in Chapter 14, in which the biology professor was studying the effects of different methods of teaching microbiology. The means and standard deviations for the four groups and the summary ANOVA are given below.

	Teaching methods			
	1	*2*	*3*	*4*
n_k	6	7	5	4
$\overline{X}_k$	70.00	79.00	70.00	83.00
s_k	6.45	8.04	9.49	7.62

Summary ANOVA

Source	SS	df	MS	F	F_{cv}
Between	641.86	3	213.95	3.41	3.16
Within	1,130.00	18	62.78		
Total	1,771.86	21			

    a. Use the Tukey/Kramer method to detect the differences among the four groups.

    b. Interpret the results of the ANOVA and the Tukey/Kramer method.

11. For Exercise 1, in which the industrial psychologist is studying the effects of length of employment on job satisfaction, use the coefficients provided in Table 15.10 to test for a linear, a quadratic, and/or a cubic trend. Use $\alpha = .01$.

n_k	8	8	8	8
$\overline{X}_k$	72.88	76.00	81.13	71.13
s_k^2	6.41	13.71	14.41	16.70

16

Analysis of Variance, Two-Way Classification

Key Concepts

Factor
Factorial design
Interaction
Main effects
Multiple comparisons
Ordinal interaction
Disordinal interaction

Test of simple effects
Reduction in error variance
Fixed-effects model
Random-effects model
Mixed-effects model
Expected mean squares
Disproportionate cell frequencies

I n Chapter 14, we discussed analysis of variance (ANOVA), one-way classification, as the procedure for testing the null hypothesis that the population means for the K levels of a single independent variable are equal. In this chapter, we extend that discussion to include the procedure for testing null hypotheses when two independent variables are included simultaneously in the same design. We will limit the discussion in this chapter to analyses with two independent variables, called ANOVA, two-way classification, or simply two-way ANOVA.

The Two-Factor Design

Initially in this chapter, we will consider what is called a *two-factor, fixed-effects model*,[1] that is, a two-factor design in which the levels of *both* independent variables

[1] In a later section of this chapter, we will discuss random-effects and mixed-effects models.

TABLE 16.1
Two-Factor Design with Difficulty of Maze and Amount of Reward as the Independent Variables

		Levels of the Second Independent Variable (Reward)		
		b_1 (small)	*b_2 (medium)*	*b_3 (large)*
Levels of the	a_1 (simple)			
First Independent	a_2 (moderate)			
Variable (Maze)	a_3 (difficult)			

are of particular interest to the researcher. They are not randomly selected from a larger population of levels. We will also restrict the discussion to those two-factor designs in which there is an equal number of observations (*n*) in each cell of the design.[2]

For example, an experimental psychologist is interested in determining the time necessary for hungry rats to complete three mazes of different difficulty using three levels of reward. The three levels of maze difficulty are (1) simple, (2) moderate, and (3) difficult; the levels of reward are (1) small, (2) medium, and (3) large. This two-factor design is illustrated in Table 16.1. Notice that, with three levels for each independent variable, there are $3 \times 3 = 9$ possible treatment combinations; that is, simple maze/small reward, simple maze/medium reward, and so on. For this study, the psychologist can take 45 laboratory rats of the same age and breed and randomly assign 5 rats to each of the nine cells of the design.

The two-way ANOVA procedures for analyzing the data from this two-factor design are just the logical extension of the procedures for analyzing data from a one-factor design. The two-way ANOVA will be discussed as we discussed one-way ANOVA. We will first consider the partitioning of the total variation into the different component parts, then present the computational formulas, and finally discuss the linear model for two-way ANOVA.

In analysis of variance (ANOVA), two-way classification, two independent variables or factors are analyzed simultaneously in a single analysis.

[2]In a later section of this chapter, we will briefly introduce two-way ANOVA procedures when the numbers of observations in the cells are not equal.

Advantages of a Factorial Design

A researcher may decide to study the effect of K levels of a single independent variable (or factor) on a dependent variable. Such a study assumes that effects of other potentially important variables are the same over the K levels of this independent variable. An alternative is to investigate the effects of one independent variable on the dependent variable, in conjunction with one or more additional independent variables. This would be a **factorial design**.

There are several advantages to using a factorial design; one is efficiency. Consider the above example. With simultaneous analysis of the two independent variables, we are in essence carrying out two separate research studies concurrently. In addition to investigating how different levels of the two independent variables affect the dependent variable, we can test whether levels of one independent variable affect the dependent variable in the same way across the levels of the second independent variable. If the effect is not the same, we say there is an **interaction** between the two independent variables. Thus, with a two-factor design, we can study the effects of the individual independent variables, called the **main effects**, as well as the **interaction effect**.

Another advantage is control over a second variable by including it in the design as an independent variable. For example, an educational researcher might be interested in the mathematics achievement of students taught with three different teaching methods. If the researcher also has the students' IQ scores, a second independent variable can be included in the design. Suppose the researcher divides the IQ scores into four categories: (1) lower than 85, (2) 85 to 99, (3) 100 to 114, and (4) 115 or higher. This design, a 3×4 factorial, is illustrated in Table 16.2. Using this design, the researcher can now determine the effects of the various teaching methods, the effects of IQ, and the possible interaction between teaching method and IQ.

The third advantage of factorial designs was mentioned in connection with the other two advantages but deserves special attention. With factorial designs, it is possible to investigate the interaction of two or more independent variables. Since,

TABLE 16.2
Two-Factor Design with Teaching Method and IQ as Independent Variables

		Level of IQ			
		b_1 (<85)	b_2 (85–99)	b_3 (100–114)	b_4 (≥115)
Types of	a_1 (method 1)				
Teaching	a_2 (method 2)				
Method	a_3 (method 3)				

in the real world of research, the effect of a single independent variable is rarely unaffected by one or more other independent variables, the study of interaction among the independent variables may be the more important objective of an investigation. If only single-factor studies were conducted, the study of interaction among independent variables would be impossible.

The advantages of factorial designs are (1) efficiency, (2) control over additional variables, and (3) the study of the interaction among independent variables.

The Variables in Two-Way ANOVA

As indicated in Chapter 14, there are two kinds of variables in any ANOVA: independent and dependent. In two-way ANOVA, there are two independent variables and a single dependent variable. Changes in the dependent variable are, or are presumed to be, the result of changes in the independent variables. The subjects' scores on the dependent variable assigned to the various combinations of these levels are illustrated in the data layout of Table 16.3. The symbol X_{111} refers to the score on the dependent variable of the first subject assigned to the first level of the first independent variable and the first level of the second independent variable. Generally, the first independent variable is the variable in the rows, and the second variable is in the columns. Thus, X_{ijk} would be the score of the ith subject in the jth row and the kth column. X_{423} would be the score of the fourth subject in the second row and the third column. As previously mentioned, we will assume that there is the same number of observations (n) in each cell of the factorial; the total number of observations in the design is the sum of all the observations in all the cells, denoted N. Since there are JK cells, the total number of observations would be $N = nJK$.

We will use a new symbol called the *dot subscript* to denote the means of the rows and columns. For example, $\overline{X}_{1.}$ is the mean of all observations in the first row of the data matrix averaged across all columns; $\overline{X}_{.1}$ is the mean of all observations in the first column of the data matrix averaged across the rows. The mean of the observations in the first cell of the matrix corresponding to the first row and first column is denoted $\overline{X}_{11}$. In general terms, $\overline{X}_{jk}$ is the cell mean for the jth row and the kth column, $\overline{X}_{j.}$ is the row mean for the jth row, and $\overline{X}_{.k}$ is the column mean for the kth column.

TABLE 16.3
Data Layout for Two-Way ANOVA

		Levels of Second Independent Variable			
		b_1	b_2		b_k
	a_1	X_{111} X_{211} . . . X_{n11}	X_{112} X_{212} . . . X_{n12}	$\cdots$ $\cdots$ $\cdots$ $\cdots$ $\cdots$	X_{11k} X_{21k} . . . X_{n1k}	$\overline{X}_{1\cdot}$
	a_2	X_{121} X_{221} . . . X_{n21}	X_{122} X_{222} . . . X_{n22}	$\cdots$ $\cdots$ $\cdots$ $\cdots$	X_{12k} X_{22k} . . . X_{n2k}	$\overline{X}_{2\cdot}$
Levels of First Independent Variable		. . .	. . .	$\cdots$ $\cdots$ $\cdots$	. . .	
	a_j	X_{1j1} X_{2j1} . . . X_{nj1}	X_{1j2} X_{2j2} . . . X_{nj2}	$\cdots$ $\cdots$ $\cdots$ $\cdots$	X_{1jk} X_{2jk} . . . X_{njk}	$\overline{X}_{j\cdot}$
		$\overline{X}_{\cdot1}$	$\overline{X}_{\cdot2}$	$\cdots$	$\overline{X}_{\cdot k}$	$\overline{X}$

Partitioning the Variation in Two-Way ANOVA

In our discussion of one-way ANOVA, the partitioning of the total variation of scores on the dependent variable had two components: the variation within the groups (s_W^2) and the variation between groups (s_B^2). Under the null hypothesis, we indicated that both variances were estimates of the population variance (σ_e^2). Thus, the ratio s_B^2/s_W^2 is approximately equal to 1.00 if the null hypothesis is true. However, when the null hypothesis is false, s_B^2 is an estimate of σ_e^2 *plus* an estimate of the variance due to the differences among the population means. Thus, in testing the null hypothesis in ANOVA, we consider how much this ratio can differ from 1.00 before we can no longer attribute the between-groups variance to only random sampling fluctuation and conclude that this variance is due to random sampling fluctuation *plus* the differences among the population means.

Partitioning the total variation of scores on the dependent variable for two-way ANOVA is similar to that for one-way ANOVA. But rather than partitioning the variance into only two components, we partition the variation into four components. The first component is the within-cell variance (s_W^2), which corresponds to the within-groups variance in one-way ANOVA. This component is the source of natural variation attributed to differences due to random sampling fluctuation and is an estimate of the population variance (σ_e^2).

The other three components are analogous to the between-groups variance. They are (1) the variation among the J row means, (2) the variation among the K column means, and (3) the variation due to the interaction between the two independent variables. Consider, first of all, the variance among the J row means (s_J^2). The null hypothesis for the J row population means is

$$H_0: \mu_1. = \mu_2. = \cdots = \mu_J.$$

That is, there is no difference among the J row means. If the null hypothesis is true, the interpretation of s_J^2 is exactly the same as s_B^2 in one-way ANOVA; namely, that s_J^2 is an estimate of σ_e^2. Therefore, the ratio of the variance among the sample means for the J rows and the within-cell variance (s_J^2/s_W^2) is approximately 1.00. If, however, the null hypothesis for the row means is false, s_J^2 is an estimate of σ_e^2 *plus* an estimate of the variation in the J row means as a result of differential effects of the independent variable on the rows. Using the concept of expected mean squares (E[MS]) introduced in Chapter 14, E(MS) for the rows is

$$\sigma_e^2 + \frac{nK \sum \alpha_j^2}{J - 1}$$

Thus, when the ratio s_J^2/s_W^2 is greater than 1.00, we ask the question: how much can this ratio of variances depart from 1.00 before we no longer attribute the between-row variance to only random sampling fluctuation and conclude that the variance is due to sampling fluctuation *plus* the difference among the population means for the J rows?

The null hypothesis for columns is

$$H_0: \mu_{.1} = \mu_{.2} = \cdots = \mu_{.K}$$

That is, there is no difference among the K column population means. As is the case for the row means, the variance among the sample means for the K columns (s_K^2) is an estimate of random variance (σ_e^2), when H_0 is true, and we would expect the ratio s_K^2/s_W^2 to be approximately 1.00. However, when H_0 is false, s_K^2 is an estimate of σ_e^2 *plus* the variation due to the differences in the K column population means. The expected mean square for the columns is

$$\sigma_e^2 + \frac{nJ \sum \beta_k^2}{K - 1}$$

Thus, when H_0 is false, the ratio s_K^2/s_W^2 is greater than 1.00.

The null hypothesis for the interaction[3] is

$$H_0: \text{ all } (\mu_{jk} - \mu_{j.} - \mu_{.k} + \mu) = 0$$

or

$$H_0: \text{ all } \alpha\beta \text{ effects } = 0$$

That is, there is no difference in the JK cell means that cannot be explained by the differences among the row means, the column means, or both. For example, if we find a significant difference among adjacent row means, under the null hypothesis for interaction, we will find the same difference among all pairs of adjacent row means for each column. On the other hand, if the null hypothesis for interaction is false, we will find differences among the cell means that cannot be explained by differences among the J row means, the K column means, or both. (We will expand the discussion of interaction in a later section.) As is the case for both the between-row variance and the between-column variance, the variation among the sample means in the JK cells (s_{JK}^2) is an estimate of σ_e^2 when the null hypothesis is true. However, when the null hypothesis is false, s_{JK}^2 is an estimate of the error variance plus the variation due to the differences among cell means that cannot be attributed to either the between-row variance or the between-column variance. The expected mean square for the interaction is

$$\sigma_e^2 + \frac{n \sum (\alpha\beta)_{jk}^2}{(J-1)(K-1)}$$

Thus, if the null hypothesis is true, the ratio s_{JK}^2 / s_W^2 will be approximately 1.00; however, when the null hypothesis is false, the ratio will be greater than 1.00.

In two-way ANOVA, the total variation is partitioned into four components:

1. The within-cell variation.

2. The variation among the J row means.

3. The variation among the K column means.

4. The variation due to interaction.

---

[3]There are equivalent statements, in symbolic form, for the null hypothesis of the interaction; these equivalent statements are

$$H_0: \text{ all } (\mu_{jk} - \mu_{j.} - \mu_{.k} + \mu) = 0$$

$$H_0: \text{ all } \alpha\beta_{jk} = 0$$

$$H_0: \sum_j^J \sum_k^K (\mu_{jk} - \mu_{j.} - \mu_{.k} + \mu)^2 = 0$$

$$H_0: \sum_j^J \sum_k^K (\alpha\beta_{jk})^2 = 0$$

Partitioning the Sum of Squares

As is the case for one-way ANOVA, two-way ANOVA involves partitioning the total variation of the scores on the dependent variable. However, the process actually begins with partitioning the total sum of squared deviations around the grand mean, symbolized $\sum (X_{ijk} - \overline{X})^2$. Consider the deviation score for the ith subject in the jth row and the kth column and the partitioning of that score into two components:

$$(X_{ijk} - \overline{X}) = (X_{ijk} - \overline{X}_{jk}) + (\overline{X}_{jk} - \overline{X}) \tag{16.1}$$

The first term on the right-hand side of the equation reflects the within-cell component; the second term reflects the between-cell component. The between-cell component can be further partitioned into three components:

$$
\begin{aligned}
(\overline{X}_{jk} - \overline{X}) &= (\overline{X}_{j\cdot} - \overline{X}) + (\overline{X}_{\cdot k} - \overline{X}) \\
&\quad + \{(\overline{X}_{jk} - \overline{X}) - [(\overline{X}_{j\cdot} - \overline{X}) + (\overline{X}_{\cdot k} - \overline{X})]\} \\
&= (\overline{X}_{j\cdot} - \overline{X}) + (\overline{X}_{\cdot k} - \overline{X}) + (\overline{X}_{jk} - \overline{X}_{j\cdot} - \overline{X}_{\cdot k} + \overline{X})
\end{aligned}
$$

These three terms reflect (1) the between-rows component, (2) the between-columns component, and (3) the between-cells component after the first two components have been accounted for. Therefore, the partitioning of the deviation score would be

$$
\begin{aligned}
(X_{ijk} - \overline{X}) &= (X_{ijk} - \overline{X}_{jk}) + (\overline{X}_{j\cdot} - \overline{X}) + (\overline{X}_{\cdot k} - \overline{X}) \\
&\quad + (\overline{X}_{jk} - \overline{X}_{j\cdot} - \overline{X}_{\cdot k} + \overline{X})
\end{aligned}
\tag{16.2}
$$

Squaring both sides of the equation and summing over the n subjects in each cell (assuming equal numbers of subjects in the cells) across the J rows and the K columns, we have

$$
\begin{aligned}
\sum_{k=1}^{K} \sum_{j=1}^{J} \sum_{i=1}^{n} (X_{ijk} - \overline{X})^2 &= \sum_{k=1}^{K} \sum_{j=1}^{J} \sum_{i=1}^{n} [(X_{ijk} - \overline{X}_{jk}) + (\overline{X}_{j\cdot} - \overline{X}) \\
&\quad + (\overline{X}_{\cdot k} - \overline{X}) + (\overline{X}_{jk} - \overline{X}_{j\cdot} - \overline{X}_{\cdot k} + \overline{X})]^2
\end{aligned}
$$

When squaring and summing over all terms on the right-hand side of the equation, all but four terms drop out of the equation and we have

$$\sum_{k=1}^{K} \sum_{j=1}^{J} \sum_{i=1}^{n} (X_{ijk} - \overline{X})^2 \qquad \text{Total sum of squares: } (\text{SS}_\text{T})$$

$$= \sum_{k=1}^{K} \sum_{j=1}^{J} \sum_{i=1}^{n} (X_{ijk} - \overline{X}_{jk})^2 \qquad = \text{within-cell sum of squares: } (\text{SS}_\text{W})$$

$$+ nK \sum_{j=1}^{J} (\overline{X}_{j\cdot} - \overline{X})^2 \qquad + \text{row sum of squares: } (\text{SS}_J)$$

$$+ nJ \sum_{k=1}^{K} (\overline{X}_{\cdot k} - \overline{X})^2 \qquad\qquad + \text{column sum of squares: } (SS_K)$$

$$+ n \sum_{k=1}^{K} \sum_{j=1}^{J} (\overline{X}_{jk} - \overline{X}_{j\cdot} - \overline{X}_{\cdot k} + \overline{X})^2 \quad + \text{interaction sum of squares: } (SS_{JK})$$

$$(16.3)$$

Thus, the total sum of squares (SS_T) can be partitioned into four components: (1) the within-cell sum of squares (SS_W), (2) the row sum of squares (SS_J), (3) the column sum of squares (SS_K), and (4) the interaction sum of squares (SS_{JK}). That is,

$$SS_T = SS_W + SS_J + SS_K + SS_{JK}$$

Earlier, in Chapter 14, we stated that the process used to obtain the variance estimates, called *mean squares*, from the sum of squares was to divide the sum of squares by the degrees of freedom associated with each variance estimate. In two-way ANOVA, the degrees of freedom are as follows:

Rows	$(J - 1)$
Columns	$(K - 1)$
Interactions	$(J - 1)(K - 1)$
Within-cell	$JK(n - 1)$

Thus, the mean squares are

$$MS_J = SS_J/(J - 1)$$
$$MS_K = SS_K/(K - 1)$$
$$MS_{JK} = SS_{JK}/[(J - 1)(K - 1)] \qquad (16.4)$$
$$MS_W = SS_W/[JK(n - 1)]$$

The variance estimates in two-way ANOVA are called mean squares. They are calculated by dividing the sum of squares for the four components by the respective degrees of freedom.

Testing the Null Hypothesis

Three null hypotheses are tested in two-way ANOVA. The first two are concerned with the effects of the two independent variables considered individually:

$$H_0: \mu_1. = \mu_2. = \cdots = \mu_J.$$

$$H_0: \mu_{\cdot 1} = \mu_{\cdot 2} = \cdots = \mu_{\cdot K}$$

That is, the J row population means are simultaneously equal, as are the K column population means. The tests of these two null hypotheses are usually referred to as tests of the *main effects*. The alternative hypothesis, against which these null hypotheses are tested, is that at least one pair or combination of means differs. For both hypotheses, the test statistic is the F ratio:

$$Rows: F_J = MS_J/MS_W$$

$$Columns: F_K = MS_K/MS_W$$

The sampling distributions of these test statistics are F distributions with degrees of freedom equal to $J-1, JK(n-1)$ and $K-1, JK(n-1)$, respectively. If either F ratio exceeds the critical value of its F distribution found in Table C.5 in the Appendix, the corresponding null hypothesis is rejected.

The null hypothesis for the interaction is:

$$H_0: \text{ all } (\mu_{jk} - \mu_{j\cdot} - \mu_{\cdot k} + \mu) = 0$$

That is, there are no differences among the JK cell population means that cannot be explained by the row effect, the column effect, or both. The alternative hypothesis is that there are differences among the cell population means that cannot be attributable to the main effects. In other words, there is an interaction between the two independent variables. The test statistic for this hypothesis is also an F ratio:

$$Interaction: F_{JK} = MS_{JK}/MS_W$$

The sampling distribution of this F ratio is the F distribution with df $= (J-1)(K-1)$ and $JK(n-1)$.

The tabular format for reporting the results of a two-way ANOVA is similar to that for one-way ANOVA. This format is illustrated for the general case in Table 16.4. The variance estimates (mean squares) are determined by dividing the sum of squares by the respective degrees of freedom. The three F ratios are then determined by dividing MS_J, MS_K, and MS_{JK} by MS_W.

> The test statistic for the three null hypotheses of two-way ANOVA is the F ratio. The sampling distributions for these test statistics are F distributions with the appropriate degrees of freedom.

Computational Formulas for Sums of Squares

The formulas for the sums of squares in Table 16.4 are in the deviation form rather than the computational form. The computational formulas can be derived in the same way as in Chapter 14 for one-way ANOVA. First, consider the following

TABLE 16.4
Summary ANOVA: The General Case for Two-Way Classification

Source of Variation	Sum of Squares (SS)	Degrees of Freedom (df)	Variance Estimate or Mean Square (MS)
Rows	$nK\sum_{j=1}^{J}(\overline{X}_{j\cdot} - \overline{X})^2$	$J-1$	$\dfrac{SS_J}{J-1} = MS_J$
Columns	$nJ\sum_{k=1}^{K}(\overline{X}_{\cdot k} - \overline{X})^2$	$K-1$	$\dfrac{SS_K}{K-1} = MS_K$
Interaction	$n\sum_{k=1}^{K}\sum_{j=1}^{J}(\overline{X}_{jk} - \overline{X}_{j\cdot} - \overline{X}_{\cdot k} + \overline{X})^2$	$(J-1)(K-1)$	$\dfrac{SS_{JK}}{(J-1)(K-1)} = MS_{JK}$
Within Cell	$\sum_{k=1}^{K}\sum_{j=1}^{J}\sum_{i=1}^{n}(X_{ijk} - \overline{X}_{jk})^2$	$JK(n-1)$	$\dfrac{SS_W}{JK(n-1)} = MS_W$
Total	$\sum_{k=1}^{K}\sum_{j=1}^{J}\sum_{i=1}^{n}(X_{ijk} - \overline{X})^2$	$N-1$	

notation for the sum of the observations in each of the cells, denoted T_{jk}:

$$T_{jk} = \sum_{i=1}^{n} X_{ijk}$$

Second, we denote the sum of the observations in each of the J rows as $T_{j\cdot}$ and the sum of observations in each of the K columns as $T_{\cdot k}$.

$$T_{j\cdot} = \sum_{k=1}^{K}\sum_{i=1}^{n} X_{ijk}$$

$$T_{\cdot k} = \sum_{j=1}^{J}\sum_{i=1}^{n} X_{ijk}$$

Finally, the sum of all observations in all JK cells is denoted by:

$$T = \sum_{k=1}^{K}\sum_{j=1}^{J}\sum_{i=1}^{n} X_{ijk}$$

Using this notation, the computational formulas for the sum of squares are as follows:

Rows:

$$SS_J = nK\sum_{j=1}^{J}(\overline{X}_{j\cdot} - \overline{X})^2$$

$$= \frac{1}{nK}\sum_{j=1}^{J} T_{j\cdot}^2 - \frac{T^2}{N} \tag{16.5}$$

Columns:

$$SS_K = nJ \sum_{k=1}^{K} (\overline{X}_{\cdot k} - \overline{X})^2$$

$$= \frac{1}{nJ} \sum_{k=1}^{K} T_{\cdot k}^2 - \frac{T^2}{N} \tag{16.6}$$

Interaction:

$$SS_{JK} = n \sum_{k=1}^{K} \sum_{j=1}^{J} (\overline{X}_{jk} - \overline{X}_{j\cdot} - \overline{X}_{\cdot k} + \overline{X})^2$$

$$= \frac{1}{n} \sum_{k=1}^{K} \sum_{j=1}^{J} T_{jk}^2 - \frac{1}{nK} \sum_{j=1}^{J} T_{j\cdot}^2$$

$$- \frac{1}{nJ} \sum_{k=1}^{K} T_{\cdot k}^2 + \frac{T^2}{N}$$

$$= \frac{1}{n} \sum_{k=1}^{K} \sum_{j=1}^{J} T_{jk}^2 - SS_J - SS_K - \frac{T^2}{N} \tag{16.7}$$

Within cell:

$$SS_W = \sum_{k=1}^{K} \sum_{j=1}^{J} \sum_{i=1}^{n} (X_{ijk} - \overline{X}_{jk})^2$$

$$= \sum_{k=1}^{K} \sum_{j=1}^{J} \sum_{i=1}^{n} X_{ijk}^2 - \frac{1}{n} \sum_{k=1}^{K} \sum_{j=1}^{J} T_{jk}^2 \tag{16.8}$$

Total:

$$SS_T = \sum_{k=1}^{K} \sum_{j=1}^{J} \sum_{i=1}^{n} (X_{ijk} - \overline{X})^2$$

$$= \sum_{k=1}^{K} \sum_{j=1}^{J} \sum_{i=1}^{n} X_{ijk}^2 - \frac{T^2}{N} \tag{16.9}$$

Suppose a researcher is interested in the effects of the length of an exercise program on the flexibility of male and female subjects. The researcher randomly selects 24 male undergraduate students and randomly assigns them to one of three exercise programs. The training programs are one, two, and three weeks in length. The researcher also selects 24 female undergraduate students and assigns them to one of the same three exercise programs. At the end of the three programs, each subject's score on a flexibility measure is recorded. In this study, the two independent variables are gender and length of the exercise program. Gender of the subject is the variable arbitrarily placed in the rows; length of the program is placed in the columns. The dependent variable is the score on the flexibility measure. The data for this study are found in Table 16.5. At this point, consider the four steps for hypothesis testing that we have used throughout the book.

TABLE 16.5
Data for Two-Way ANOVA Example

		Length of Program					
		1 Week		2 Weeks		3 Weeks	
	Females	32	28	28	26	36	46
		27	23	31	33	47	39
		22	25	24	27	42	43
		19	21	25	25	35	40
Gender							
	Males	18	16	27	25	24	26
		22	19	31	32	27	30
		20	24	27	26	33	32
		25	31	25	24	25	29

Step 1: State the Hypotheses

For two-way ANOVA, we have three null hypotheses.

for rows:

> H_{0_1}: In the population, the mean for males equals the mean for females.
>
> H_{0_1}: $\mu_1. = \mu_2.$

for columns:

> H_{0_2}: In the population, the means for the three exercise programs are equal.
>
> H_{0_2}: $\mu._1 = \mu._2 = \mu._3$

for interaction:

> H_{0_3}: In the population, there is no interaction between gender of subject and length of exercise program.
>
> H_{0_3}: all $(\mu_{jk} - \mu_j. - \mu._k + \mu) = 0$

Step 2: Set the Criterion for Rejecting H_0

As discussed, the test statistic in two-way ANOVA is the F ratio, the ratio of mean squares for the main effects or the interaction to the mean square within cells. In our example, the F ratio for the row main effect is

$$F_J = \mathrm{MS}_J / \mathrm{MS}_W$$

The sampling distribution for this test statistic is the F distribution with 1 and 42 degrees of freedom. Using $\alpha = .05$, we see that the critical value for this test statistic, from Table C.5, is 4.07.

The test statistics for the column main effect and the interaction are as follows:

$$F_K = \mathrm{MS}_K / \mathrm{MS}_W$$

$$F_{JK} = \mathrm{MS}_{JK} / \mathrm{MS}_W$$

For both of these test statistics, the sampling distribution is the F distribution with 2 and 42 degrees of freedom. From Table C.5, we see that the critical value for these test statistics is 3.22.

Step 3: Compute the Test Statistic

A summary of the initial calculations is found in Table 16.6. Using these initial calculations with formulas 16.5 through 16.9, the sums of squares are computed as follows:

$$\mathrm{SS}_J = \frac{1}{nK} \sum_{j=1}^{J} T_{j\cdot}^2 - \frac{T^2}{N} = \frac{935,460}{(8)(3)} - \frac{(1,362)^2}{48}$$

$$= 38,977.50 - 38,646.75 = \underline{330.75}$$

$$\mathrm{SS}_K = \frac{1}{nJ} \sum_{k=1}^{K} T_{\cdot k}^2 - \frac{T^2}{N} = \frac{635,396}{(8)(2)} - \frac{(1,362)^2}{48}$$

$$= 39,712.25 - 38,646.75 = \underline{1,065.50}$$

$$\mathrm{SS}_{JK} = \frac{1}{n} \sum_{k=1}^{K} \sum_{j=1}^{J} T_{jk}^2 - \frac{1}{nK} \sum_{k=1}^{K} T_{j\cdot}^2 - \frac{1}{nJ} \sum_{k=1}^{K} T_{\cdot k}^2 + \frac{T^2}{N}$$

$$= \frac{323,144}{8} - \frac{935,460}{(8)(3)} - \frac{635,396}{(8)(2)} + \frac{(1,362)^2}{48}$$

$$= 40,393.00 - 38,977.50 - 39,712.25 + 38,646.75$$

$$= \underline{350.00}$$

$$\mathrm{SS}_W = \sum_{k=1}^{K} \sum_{j=1}^{J} \sum_{i=1}^{n} X_{ijk}^2 - \frac{1}{n} \sum_{k=1}^{K} \sum_{j=1}^{J} T_{jk}^2$$

$$= 41,014.00 - 40,393.00 = \underline{621.00}$$

$$\mathrm{SS}_T = \sum_{k=1}^{K} \sum_{j=1}^{J} \sum_{i=1}^{n} X_{ijk}^2 - \frac{T^2}{N}$$

$$= 41,014.00 - 38,646.75 = \underline{2,367.25}$$

TABLE 16.6
Summary of Calculations for Two-Way ANOVA Example

		Length of Program			
		1 Week	*2 Weeks*	*3 Weeks*	
Gender	Females	$T_{11} = 197.0$	$T_{12} = 219.0$	$T_{13} = 328.0$	$T_{1.} = 744.0$
		$\overline{X}_{11} = 24.63$	$\overline{X}_{12} = 27.38$	$\overline{X}_{13} = 41.00$	$\overline{X}_{1.} = 31.0$
		$\Sigma X_i^2 = 4,977.0$	$\Sigma X_i^2 = 6,065.0$	$\Sigma X_i^2 = 13,580.0$	$\Sigma X_i^2 = 24,622.0$
	Males	$T_{21} = 175.0$	$T_{22} = 217.0$	$T_{23} = 226.0$	$T_{2.} = 618.0$
		$\overline{X}_{21} = 21.88$	$\overline{X}_{22} = 27.13$	$\overline{X}_{23} = 28.25$	$\overline{X}_{2.} = 25.75$
		$\Sigma X_i^2 = 3,987.0$	$\Sigma X_i^2 = 5,945.0$	$\Sigma X_i^2 = 6,460.0$	$\Sigma X_i^2 = 16,392.0$
		$T_{.1} = 372.0$	$T_{.2} = 436.0$	$T_{.3} = 554.0$	$T = 1,362$
		$\overline{X}_{.1} = 23.25$	$\overline{X}_{.2} = 27.25$	$\overline{X}_{.3} = 34.63$	$\overline{X} = 28.38$
		$\Sigma X_i^2 = 8,964.0$	$\Sigma X_i^2 = 12,010.0$	$\Sigma X_i^2 = 20,040.0$	$\Sigma\Sigma\Sigma X_i^2 = 41,014.0$

$$\sum_{j=1}^{J} T_{j.}^2 = (744)^2 + (618)^2 = 935,460.0$$

$$\sum_{k=1}^{K} T_{.k}^2 = (372)^2 + (436)^2 + (554)^2 = 635,396.0$$

$$\sum_{k=1}^{K}\sum_{j=1}^{J} T_{jk}^2 = (197)^2 + (219)^2 + (328)^2 + (175)^2 + (217)^2 + (226)^2 = 323,144.0$$

$$T = \sum_{k=1}^{K}\sum_{j=1}^{J}\sum_{i=1}^{n} X_{ijk} = 1,362.0$$

$$\sum_{k=1}^{K}\sum_{j=1}^{J}\sum_{i=1}^{n} X_{ijk}^2 = 41,014.0$$

The summary ANOVA for this example is found in Table 16.7; notice that the F ratios are

for rows:
$$F_J = 330.75/14.79 = 22.36$$

for columns:
$$F_K = 532.75/14.79 = 36.02$$

for interaction:
$$F_{JK} = 175.00/14.79 = 11.83$$

TABLE 16.7
Summary Table for Two-Way ANOVA Example

Source	SS	df	MS	F	F_{cv}
Rows (gender)	330.75	1	330.75	22.36	4.07
Columns (length)	1,065.50	2	532.75	36.02	3.22
Interaction	350.00	2	175.00	11.83	3.22
Within cell	621.00	42	14.79		
Total	2,367.25	47			

Step 4: Interpret the Results

Since all three F ratios exceed their respective critical values, all three null hypotheses are rejected. For the row main effect, the conclusion is that, in the population, the mean for males (μ_1.) differs from the mean for females (μ_2.). For the column main effect, the conclusion is that, in the population, the means for the three exercise programs differ. The probability statement associated with both of these conclusions is

> The probability that the observed differences among the sample means would have occurred by chance, if in fact the null hypothesis were true, is less than .05.

In order to determine which pairs or combinations of means differ for the column main effect, multiple-comparison procedures, analogous to those discussed in Chapter 15, would be computed. These procedures will be illustrated later in this chapter.

For the interaction, the conclusion is that, in the population, there is an interaction between the length of exercise program and subject's gender. The probability statement associated with this conclusion is:

> The probability that the observed differences among the cell means (not attributable to the row or column main effects) would have occurred by chance, if in fact the null hypothesis were true, is less than .05.

There are also post hoc procedures that can be applied when a significant interaction is found; these procedures will be discussed later in the chapter.

The Meaning of Main Effects

In the exercise example, there are two independent variables: gender of subject and length of exercise program. The tests of the two hypotheses related to these independent variables have been referred to as the tests of the main effects. The data in Table 16.7 indicate that the tests of both main effects are statistically significant and thus that both null hypotheses are rejected. Notice that, in the interpretation of the row main effect, the row means are averaged across the K columns.

For the column main effect, the column means are averaged across the J rows. In the example, the significant row main effect indicates that the difference between the mean for the males, averaged across the three levels of exercise program ($\overline{X}_1. = 25.75$), and the mean for the females, averaged across the three levels ($\overline{X}_2. = 31.00$), is too great to attribute to sampling fluctuation if the null hypothesis is true. Therefore, the null hypothesis is rejected.

Similarly for the column main effect, the significant F ratio indicates that the differences among the means for the three levels of exercise program, averaged over both male and female subjects, are too great to attribute to sampling fluctuation if H_0 is true. Since there are three levels of this independent variable, a post hoc multiple-comparison procedure must be employed in order to determine which pairs or combinations of means differ.

> The tests of the hypotheses on the two independent variables are called tests of the main effects. The hypotheses refer to the J row population means averaged across the columns and the K column population means averaged across the rows.

Multiple-Comparison Procedures for the Main Effects

As is the case for one-way ANOVA, the tests of hypotheses in two-way ANOVA are only the first steps in the analysis of a data set. When a significant F ratio is found for either of the main effects, when there are more than two levels, and when the levels are fixed,[4] the researcher is still left with the problem of deciding what led to the rejection of the null hypothesis, or which pairs or combinations of means differ. In this section, we will apply multiple-comparison procedures to a two-factor design. Rather than deal with each of the previously discussed methods, we will illustrate only the use of the Tukey method. The other methods can be applied to two-factor designs by incorporating the same minor adjustments made here for the Tukey method.

In our example, we can readily assume that both independent variables, gender and length of exercise program, have fixed levels. As we have seen, the F ratio for gender is significant and, since there are only two levels, the researcher can conclude that female subjects combined over the three exercise programs differ in flexibility from male subjects. For the independent variable, length of exercise program, there is also a significant F ratio. However, since there are three levels,

---

[4]If a null hypothesis for a random effect is rejected, there is no need for a multiple-comparisons test, since the researcher is interested in generalizing to the population of levels — not to the specific levels selected for the study.

a post hoc test is necessary in order to determine which pair or combination of means differs. In applying the Tukey method to this example, we will test the null hypotheses relating to all possible pairwise comparisons of the column means (levels of exercise program) combined across the rows (male and female subjects). The three null hypotheses tested are

$$H_{0_1}: \mu_{\cdot 1} = \mu_{\cdot 2}$$

$$H_{0_2}: \mu_{\cdot 1} = \mu_{\cdot 3}$$

$$H_{0_3}: \mu_{\cdot 2} = \mu_{\cdot 3}$$

The test statistic for the Tukey method is Q, defined as follows:

$$Q = \frac{\overline{X}_i - \overline{X}_k}{\sqrt{MS_W/n'}} \tag{16.10}$$

where

$\overline{X}_i$ = sample mean for the ith group

$\overline{X}_k$ = sample mean for the kth group

MS_W = within-cell mean square from the summary ANOVA

n' = number of observations in each of the groups summed over the appropriate cells

The calculation of the Q statistic for each of the null hypotheses is found in Table 16.8. Notice that the differences between the respective sample means are determined and then divided by $\sqrt{MS_W/n'}$. In this example, the column means are based on $n' = nJ = (8)(2) = 16$ observations; thus, $\sqrt{MS_W/n'} = \sqrt{14.79/16} = 0.96$. The critical values for the Q statistics are found in Table C.8. For ranges = 3 and

TABLE 16.8
Data for Tukey Method for Two-Way ANOVA

Exercise Program	$\overline{X}_i$	$(\overline{X}_i - \overline{X}_k)$		Q	
One Week	23. 25				
Two Weeks	27. 25	4. 00		4. 17*	
Three Weeks	34. 63	11. 38	7. 38	11. 85**	7. 69**

$$Q = \frac{\overline{X}_i - \overline{X}_k}{\sqrt{MS_W/n'}}$$

*$p < .05$ $Q_{cv} = 3.44$ for df = 42.

**$p < .01$ $Q_{cv} = 4.37$ for df = 42.

$df_w = 42$, the critical values are 3.44 and 4.37 for $\alpha = .05$ and .01, respectively. Assuming $\alpha = .05$, all three null hypotheses are rejected. The researcher can conclude that subjects in the three-week program are more flexible than subjects in both one-week and two-week programs. Further, subjects in the two-week program are more flexible than those in the one-week program.

The application of the Tukey method in two-way ANOVA is the same as in one-way ANOVA. The test statistic is

$$Q = \frac{\overline{X}_i - \overline{X}_k}{\sqrt{MS_W/n'}}$$

where n' is the number of observations in each of the groups summed over the appropriate cells.

The Meaning of Interaction

In the introductory sections of this chapter, we indicated that an interaction between the independent variables is present in a two-factor design when the effect of the levels of the first independent variable is *not* the same across the levels of the second independent variable. One procedure for examining an interaction is to plot the cell means. The scale of the dependent variable is placed on the vertical (Y) axis; the levels of one of the independent variables are equally spaced on the horizontal (X) axis. Consider the data for our example; the plot of the cell means is found in Figure 16.1. Notice that we place the three levels of the exercise program on the X axis. The decision about which variable to place on the X axis is arbitrary. Regardless of which variable is placed on this axis, the interpretation of the interaction is the same.

If there is no interaction between the two independent variables, the lines connecting the cell means will be parallel or nearly parallel. Notice that, for these data, the lines are not parallel. For female subjects, the mean of the flexibility scores for subjects in the three-week program is much greater than the means for subjects in either the two-week or the one-week program. There is only a slight difference in the means for subjects in the two-week and one-week programs. For male subjects, the data indicate that there is little difference among the means for any of the groups, with the means for the two-week and three-week groups somewhat higher than the mean for the one-week group. Therefore, the data indicate

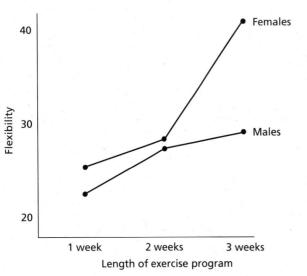

FIGURE 16.1
Plot of interaction for sample data

that the effect of the length of exercise program is not the same for male and female subjects; that is, there is an interaction between gender of subject and length of program.

General Characteristics of Interaction Plots

When a nonsignificant interaction is found in a two-way ANOVA, the lines connecting the cell means in the interaction plot will be nearly parallel (within sampling fluctuation). In our example, if the interaction had been nonsignificant, the plot of the interaction may have looked like the plot in Figure 16.2A. Because the lines connecting the cell means are parallel, the effect of the various exercise programs is shown to be the same for both male and female subjects. Now consider Figure 16.2B. For both groups, the lines are *not* straight across the three levels of exercise program, but the line segments connecting the cell means are parallel. Such a pattern also illustrates a nonsignificant interaction between the two independent variables.

The plot of a significant interaction can have many patterns. Consider the plots of the cell means in Figures 16.2C, and 16.2D. Since a significant interaction is assumed in these two plots, the lines connecting the cell means are not parallel. In Figure 16.2C, the cell means for females are always greater than for males across the three exercise programs, and the lines do *not* intersect. This pattern is called an **ordinal interaction**. Now consider Figure 16.2D. Notice that the lines intersect within the plot. This pattern is called a **disordinal interaction**.

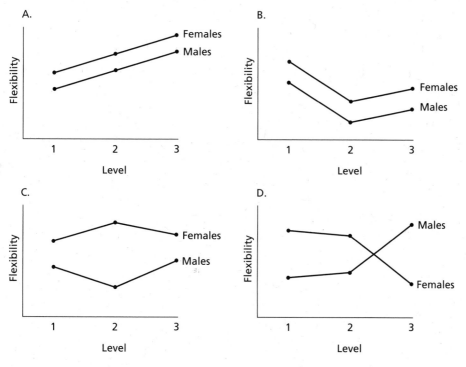

FIGURE 16.2
Plots of possible interactions

The significance of the interaction is determined in the ANOVA. A nonsignificant interaction is illustrated by nearly parallel lines (within sampling fluctuation) that connect the cell means. A significant interaction is ordinal when the lines do not intersect within the plot; an interaction is disordinal when they do intersect.

Tests of Simple Effects

When a significant F ratio is found in two-way ANOVA for either or both main effects, post hoc multiple-comparison tests are used to detect significant differences between pairs or combinations of row or column means. Similarly, when a significant F ratio is found for the interaction, a post hoc procedure, called the **test of simple effects**, is used in conjunction with the plotted cell means in interpreting the interaction. In our example, a significant interaction is found, and the conclusion is that the difference between the mean flexibility scores for males and females is *not* the same across the three exercise programs. The test of simple effects considers the differences among cell means within levels of the two independent variables. For example, using the test of simple effects, we can determine whether the means for the three exercise programs differ for the males and then we can determine whether the means differ for the females. Conceptually, we would be doing a one-way ANOVA for the males and a one-way ANOVA for the females. However, in computing the F ratio for each of these tests, we would use MS_W from the two-way ANOVA in the denominator.

In computing the sums of squares for the simple effects, we will first consider the cells in row 1 and then the cells in row 2; the computational formula for the simple effects for rows is

$$SS_k \text{ at row } j = \sum_{k=1}^{K} \frac{(T_{jk})^2}{n} - \frac{T_{j.}^2}{nK} \tag{16.11}$$

Using the data from Table 16.6,

$$
\begin{aligned}
SS_K \text{ at row 1} &= \frac{(197)^2}{8} + \frac{(219)^2}{8} + \frac{(328)^2}{8} - \frac{(744)^2}{24} \\
&= 24,294.25 - 23,064.00 \\
&= 1,230.25
\end{aligned}
$$

$$
\begin{aligned}
SS_K \text{ at row 2} &= \frac{(175)^2}{8} + \frac{(217)^2}{8} + \frac{(226)^2}{8} - \frac{(618)^2}{24} \\
&= 16,098.75 - 15,913.50 \\
&= 185.5
\end{aligned}
$$

Tests of simple effects can also be applied to the differences in the means of males and females at each level of exercise program. In computing the sums of squares for these simple effects, we will consider the cells in column 1, then the cells in column 2, and finally the cells in column 3. The formula for the simple effect for columns is:

$$SS_J \text{ at column } k = \sum_{j=1}^{J} \frac{(T_{jk})^2}{n} - \frac{T_{.k}^2}{nJ} \tag{16.12}$$

Using the data from Table 16.6,

$$
\begin{aligned}
SS_J \text{ at column 1} &= \frac{(197)^2}{8} + \frac{(175)^2}{8} - \frac{(372)^2}{16} \\
&= 8,679.25 - 8,649.00 \\
&= 30.25
\end{aligned}
$$

$$SS_J \text{ at column } 2 = \frac{(219)^2}{8} + \frac{(217)^2}{8} - \frac{(436)^2}{16}$$

$$= 11,881.25 - 11,881.00$$

$$= 0.25$$

$$SS_J \text{ at column } 3 = \frac{(328)^2}{8} + \frac{(226)^2}{8} - \frac{(554)^2}{16}$$

$$= 19,832.50 - 19,182.25$$

$$= 650.25$$

The expanded ANOVA table, which includes the sums of squares for the tests of simple effects, is in Table 16.9. Notice that, since we have three column means in each row, there are 2 degrees of freedom associated with the simple effects test in each row. Similarly, since there are two row means in each column, there is 1 degree of freedom associated with the simple effects test in each column. The F ratio for each simple effects test found in Table 16.9 is computed as follows:

$$F = \frac{MS_{\text{simple effect}}}{MS_W}$$

Notice that, in Table 16.9, there are three tests of simple main effects for the rows, one for each of the columns, and two tests of simple effects for the columns, one for each row. In determining the critical values of the F ratio for each of these tests of simple effects, the α level used in the initial tests of both row and column main effects is divided by the number of simple effects within the main effect. In our example, we used $\alpha = .05$ for the test of the row main effect (gender). Thus, the tests of the simple main effects for the rows at each column would use $\alpha = (.05)/3 = .017$. Similarly, the column main effect (program) was tested at

TABLE 16.9
Expanded Summary ANOVA Including Tests of Simple Effects

Source	SS	df	MS	F	F_{cv}
Rows (Gender)	330.75	1	330.75	22.36	4.07
Rows at column 1	30.25	1	30.25	2.05	6.73
Rows at column 2	0.25	1	0.25	0.02	6.73
Rows at column 3	650.25	1	650.25	43.97	6.73
Columns (Length)	1,065.50	2	532.75	36.02	3.22
Columns at row 1	1,230.25	2	615.13	41.59	4.43
Columns at row 2	185.25	2	92.63	6.26	4.43
Interaction	350.00	2	175.00	11.83	3.22
Within cell	621.00	42	14.79		
Total	2,367.25	47			

$\alpha = .05$; thus, the tests of the simple main effects for columns at each row would use $\alpha = (.05)/2 = .025$. By using this procedure, the overall α level is divided evenly across the respective simple effects.

The major problem with this strategy of dividing the α level for the main effects across the simple effects is in determining the critical value of the F ratio. Table C.5 does *not* contain the critical values of F for $\alpha = .017$ and $\alpha = .025$. To obtain these critical values, we must approximate them from the values for $\alpha = .05$ and $\alpha = .01$. For this example, F_{cv} for the row main effect (df = 1, 42) for $\alpha = .05$ was initially approximated to be 4.07. Similarly, F_{cv} (df = 1, 42) for $\alpha = .01$ is approximately 7.29. Therefore, F_{cv} for $\alpha = .017$ is between 4.07 and 7.29. To determine this value, first subtract .01 and .017 from .05 and then subtract 4.07 from 7.29.

$$
\begin{array}{rrr}
.05 & .050 & 7.29 \\
-.01 & -.017 & -4.07 \\
\hline
.04 & .033 & 3.22
\end{array}
$$

To determine F_{cv} for $\alpha = .017$, we must add $(.033/.04)(3.22)$ to 4.07.

$$F_{cv} = 4.07 + (.033/.04)(3.22)$$

$$= 4.07 + 2.66$$

$$= 6.73$$

F_{cv} for the simple effects for columns at each row are determined similarly. F_{cv} (df = 2, 42) for $\alpha = .05$ and $\alpha = .01$ equal 3.22 and 5.16, respectively. Therefore, F_{cv} for $\alpha = .025$ is between 3.22 and 5.16. Following the same procedure,

$$
\begin{array}{rrr}
.05 & .050 & 5.16 \\
-.01 & -.025 & -3.22 \\
\hline
.04 & .025 & 1.94
\end{array}
$$

Thus,

$$F_{cv} = 3.22 + (.025/.04)(1.94)$$

$$= 3.22 + 1.21$$

$$= 4.43$$

These two critical values are included in Table 16.9.

The results of the analysis of the simple effects for the rows at each column indicate that, for the three-week exercise program, there is a statistically significant difference between the mean flexibility scores of males ($\overline{X} = 28.25$) and of females ($\overline{X} = 41.00$). The results of the simple effects for the columns at each row indicate that, for *both* males and females, there are differences in the mean flexibility scores for each of the three exercise programs. To determine which means differ, the Tukey method, outlined in Chapter 15, would be applied to the cell means for both the males and females.

A Measure of Association (ω^2)

In Chapter 14, we introduced omega squared (ω^2) as a measure of association between the independent and dependent variables in ANOVA. As indicated in that discussion, the interpretation of ω^2 is analogous to the interpretation of r^2, the coefficient of determination; that is, ω^2 indicates the proportion of variance in the dependent variable that is accounted for by the levels of the independent variable. The use of ω^2 can be extended to two-way ANOVA; the formula is

$$\omega^2 = \frac{SS_{\text{effect}} - (df_{\text{effect}})(MS_W)}{SS_T + MS_W} \tag{16.13}$$

For the row main effect, the formula is

$$\omega^2 = \frac{SS_J - (J-1)(MS_W)}{SS_T + MS_W} \tag{16.14}$$

For the column main effect, the formula is

$$\omega^2 = \frac{SS_K - (K-1)(MS_W)}{SS_T + MS_W} \tag{16.15}$$

For the interaction, the formula is

$$\omega^2 = \frac{SS_{JK} - (J-1)(K-1)(MS_W)}{SS_T + MS_W} \tag{16.16}$$

Using the data from Table 16.7, ω^2 for the row and column main effects and the interaction are computed as follows:

for rows:

$$\omega^2 = \frac{330.75 - (1)(14.79)}{2,367.25 + 14.79} = .13$$

for columns:

$$\omega^2 = \frac{1,065.50 - (2)(14.79)}{2,367.25 + 14.79} = .44$$

for interaction:

$$\omega^2 = \frac{350.00 - (2)(14.79)}{2,367.25 + 14.79} = .13$$

These data indicate that approximately 13 percent of the variance of the scores on the flexibility measure can be attributed to the differences between males and females; 44 percent can be attributed to the differences in the length of the exercise programs; and 13 percent can be attributed to the interaction of the two independent variables.

Computer Exercise: Teaching Method and Reading Ability

Up to this point, the computer exercises included in the various chapters of the book have used the survey of high school seniors data set. However for this chapter, we cannot use this data set because we have restricted our discussion of two-way ANOVA to examples having equal cell size. So consider an example of a researcher who is interested in whether three methods of teaching are equally effective for students of differing reading levels. For the example, students from high, middle, and low reading ability levels are randomly assigned to the three different teaching strategies. The dependent variables are scores on a standardized test. The data for this example along with a summary of some of the initial calculations are found in Table 16.10; the ANOVA procedure from SPSS was used to generate the computer printout for this analysis (see Table 16.11).

Step 1: State the Hypotheses

For two-way ANOVA, there are three null hypotheses to be tested; one for the row effect (Reading level-ACH), one for the column effect (Teaching method-METHOD), and one for the interaction (ACH × METHOD). For the row effect

$$H_0: \mu_{1.} = \mu_{2.} = \mu_{3.}$$

$$H_a: \mu_{i.} \neq \mu_{j.} \text{ for some } i, j$$

For the column effect,

$$H_0: \mu_{.1} = \mu_{.2} = \mu_{.3}$$

$$H_a: \mu_{.i} \neq \mu_{.k} \text{ for some } i, k$$

For the interaction effect,

$$H_0: \text{ all } (\mu_{jk} - \mu_{j.} - \mu_{.k} + \mu) = 0$$

$$H_a: \text{ all } (\mu_{jk} - \mu_{j.} - \mu_{.k} + \mu) \neq 0$$

Step 2: Set the Criterion for Rejecting the H_0

The test statistic for all three hypotheses is the F ratio; the sampling distribution for these test statistics is the F distribution with appropriate degrees of freedom. The degrees of freedom associated with the main effects, the interaction, and the within cell mean squares are:

Reading Level (ACH)	$(J - 1) = (3 - 1) = 2$
Teaching Method (METHOD)	$(K - 1) = (3 - 1) = 2$
Interaction (ACH × METHOD)	$(J - 1)(K - 1) = (3 - 1)(3 - 1) = 4$
Within Cell	$JK(n - 1) = (3)(3)(4 - 1) = 27$

TABLE 16.10
Data Set and Initial Calculations for SPSS Example

	Teaching Method		
	1	2	3
High achievement	40	32	30
	36	29	35
	38	36	33
	36	35	38
Middle achievement	24	20	18
	21	29	26
	20	25	26
	20	30	25
Low achievement	22	20	18
	18	18	14
	16	19	19
	20	15	19

Methods

	1	2	3	
High	$T_{11} = 150$	$T_{12} = 132$	$T_{13} = 136$	$T_{1.} = 418$
	$\bar{X}_{11} = 37.5$	$\bar{X}_{12} = 33$	$\bar{X}_{13} = 34$	$\bar{X}_{1} = 34.83$
	$\sum X_i^2 = 5{,}636$	$\sum X_i^2 = 4{,}386$	$\sum X_i^2 = 4{,}658$	$\sum X_i^2 = 14{,}680$
Middle	$T_{21} = 85$	$T_{22} = 104$	$T_{23} = 95$	$T_{2.} = 284$
	$\bar{X}_{21} = 21.25$	$\bar{X}_{22} = 26$	$\bar{X}_{23} = 23.75$	$\bar{X}_{2.} = 23.67$
	$\sum X_i^2 = 1{,}817$	$\sum X_i^2 = 2{,}766$	$\sum X_i^2 = 2{,}301$	$\sum X_i^2 = 6{,}884$
Low	$T_{31} = 76$	$T_{32} = 72$	$T_{33} = 70$	$T_{3.} = 218$
	$\bar{X}_{31} = 19$	$\bar{X}_{32} = 18$	$\bar{X}_{33} = 17.5$	$\bar{X}_{3.} = 18.17$
	$\sum X_i^2 = 1{,}464$	$\sum X_i^2 = 1{,}310$	$\sum X_i^2 = 1{,}242$	$\sum X_i^2 = 4{,}016$
	$T_{.1} = 311$	$T_{.2} = 308$	$T_{.3} = 301$	$T = 920$
	$\bar{X}_{.1} = 25.92$	$\bar{X}_{.2} = 25.67$	$\bar{X}_{.3} = 25.08$	$\bar{X} = 25.56$
	$\sum X_i^2 = 8{,}917$	$\sum X_i^2 = 8{,}462$	$\sum X_i^2 = 8{,}201$	$\sum \sum \sum X_{ijk}^2 = 25{,}580$

With $\alpha = .05$, the critical values of the F ratios for the test of the main effects and the interaction are (see Table C.5 in the Appendix):

Reading level (ACH)	df $= 2, 27$	$F_{cv} = 3.35$
Teaching method (METHOD)	df $= 2, 27$	$F_{cv} = 3.35$
Interaction (ACH $\times$ METHOD)	df $= 4, 27$	$F_{cv} = 2.72$

TABLE 16.11
SPSS Printout for Two-Way ANOVA Computer Example

```
* * *   A N A L Y S I S   O F   V A R I A N C E   * * *
              SCORES
      by    ACH
            METHOD


            UNIQUE sums of squares
            All effects entered simultaneously
```

Source of Variation	Sum of Squares	DF	Mean Square	F	Sig of F
Main Effects	1735.278	4	433.819	48.103	.000
ACH	1730.889	2	865.444	95.963	.000
METHOD	4.389	2	2.194	.243	.786
2-Way Interactions	90.111	4	22.528	2.498	.066
ACH METHOD	90.111	4	22.528	2.498	.066
Explained	1825.389	8	228.174	25.301	.000
Residual	243.500	27	9.019		
Total	2068.889	35	59.111		

Step 3: Compute the Test Statistic

The F ratios for testing these three hypotheses are found in the computer printout in Table 16.11. Note the Summary ANOVA includes several additional lines that were not included in the Summary ANOVA tables presented previously, i.e., "Main Effects" and "Explained." These additional tests are used only occasionally in most behavioral science research. The test of the "Main Effects" involves simply summing the sum of square values for ACH and METHOD (1730.889 + 4.389 = 1735.278) and the associated degrees of freedom (2 + 2 = 4). The mean square is computed by dividing the summed value for the two main effects sum of squares by the summed value for the degrees of freedom (1735.278 ÷ 4 = 433.819). The F ratio for this line is determined by dividing this mean square by the MS_W (Residual), (433.819 ÷ 9.019 = 48.103).

Similarly, the "Explained" sum of squares is determined by summing the sum of squares values for ACH, METHOD, and ACH × METHOD (1730.889+4.389+ 90.111 = 1825.389); the associated degrees of freedom are also summed (2 + 2 + 4 = 8). The mean square for "Explained" (1825.389 ÷ 8 = 228.174) is divided by MS_W (Residual) to get the F ratio for this line (228.174 ÷ 9.019 = 25.301).

The specific F ratios of interest for this example are for ACH (row effect), METHOD (column effect), and ACH $\times$ METHOD (interaction effect).

$$F_J = 95.963$$
$$F_K = 0.243$$
$$F_{JK} = 2.498$$

Step 4: Interpret the Results

Note that only the F ratio for the row effect (ACH) ($F_J = 95.963$) exceeded the critical value ($F_{cv} = 3.35$). Thus, only the hypothesis for the row effect H_0: $\mu_{1.} = \mu_{2.} = \mu_{3.}$ is rejected. The associated probability statement is: the probability that the observed differences among the row means would have occurred by chance, if the null hypothesis is true, is less than .05. Note in the printout, "Sig of F" equals .000, which implies that the probability of such an occurrence is actually less than .001.

The conclusion is that not all the row means in the population are equal, i.e., the test scores for those students with differing reading levels were not the same. At this point, we do not know which groups differ significantly and we must conduct a post hoc multiple-comparison procedure (Tukey). The data for this procedure are found in Table 16.12. The test statistic for this post hoc procedure was defined by formula 16.10.

$$Q = \frac{\overline{X}_i - \overline{X}_k}{\sqrt{MS_W/n'}}$$

where $\sqrt{MS_W/n'} = \sqrt{9.019/12} = 0.867$.

As can be seen from the data in Table 16.12, the means for each group differ significantly from each other. We would conclude that the students with the "high" reading level had a significantly higher mean on the standardized test than did the group with the "middle" reading level, which in turn had a significantly higher mean than did the "low" group.

Since the F ratio for the column effect (METHOD) was not statistically significant, the conclusion is that the differences among the column means reflecting the

TABLE 16.12
Tukey Tests for Two-Way ANOVA Computer Example

Reading Level	$\overline{X}_i$	$(\overline{X}_i - \overline{X}_k)$				Q
Low	18.17					
Middle	23.67	5.50		6.34*		
High	34.28	16.66	11.16	19.22*	12.87*	

*$p < .05$ $Q_{CV} = 3.87$ for df $= 27$.

different teaching methods were not significantly different. Thus, there would be no need to conduct the post hoc procedure. Similarly, since the F ratio for the interaction effect (ACH × METHOD) was not statistically significant, the conclusion is that there was no difference in the effect of the various teaching methods across the levels of reading ability. Thus, there would be no need to plot the interaction means nor to conduct a test of simple effects.

Reduction in Error Variance in Two-Way ANOVA

As indicated earlier, one of the advantages of a factorial design is that we can identify and control an additional source of variation; that is, the variation associated with the second independent variable. This makes possible a **reduction in error variance** (MS_W). For example, suppose we analyze the data in Table 16.5 and consider only the effects of one of the independent variables: the length of the exercise program. Using the methods discussed in Chapter 14, we compute a one-way ANOVA for these data; the summary ANOVA is found in Table 16.13. Notice that, for the one-way ANOVA, $MS_B = 532.75$. This is the same value as for MS_K (column main effect) for the two-way ANOVA (see Table 16.7). However, notice that the MS_W is 28.93 for the one-way ANOVA and 14.79 for the two-way ANOVA. The consequence of the smaller MS_W for the two-way ANOVA is a smaller denominator in the F ratio and thus a larger F ratio. That is, with a larger F ratio, the rejection of the null hypothesis is more likely and the probability of making a Type II error is less. In this example, the F ratios for both the one-way and the two-way ANOVA are statistically significant, but the F ratio for the two-way ANOVA is nearly twice that for the one-way ANOVA. In other situations, it is possible for the effect of an independent variable to be insignificant in a one-way ANOVA but, through the control of the variation for the second independent variable, the effect could be significant in a two-way ANOVA.

The reduction of MS_W in two-way ANOVA through the control of a second independent variable is illustrated in Figure 16.3. The proportion of the variance due to differences among the lengths of the exercise programs is enclosed by the solid lines. The proportion of the variance due to gender and the interaction between length of program and gender are enclosed by broken lines. Recall that the total variance and the variance due to the length of exercise program were the same for

TABLE 16.13
Summary for One-Way ANOVA

Source	SS	df	MS	F	F_{CV}
Between (length of program)	1,065.50	2	532.75	18.42	3.21
Within	1,301.75	45	28.93		
Total	2,367.25	47			

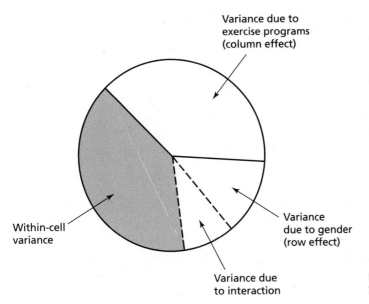

Variance due to
exercise programs
(column effect)

Within-cell
variance

Variance
due to gender
(row effect)

Variance due
to interaction

FIGURE 16.3
Schema of partitioning of variance

both the one-way and two-way ANOVAs, because the same scores were used in both. As can be readily seen, the addition of the second independent variable in this example reduces the error variance and thus enhances the statistical precision. Therefore, the researcher will be more likely to reject a false hypothesis.

If a second independent variable is used in the design of a research study, an additional source of variation is controlled and MS_W is reduced.

Assumptions in Two-Way ANOVA

In Chapter 14, we discussed the assumptions underlying the use of one-way ANOVA; the assumptions for two-way ANOVA are the same. The importance of the assumptions is in the use of the F distribution as the appropriate sampling distribution for testing the null hypothesis. The assumptions are:

1. The samples are independent, random samples from defined populations.

2. The scores on the dependent variable are normally distributed in the population.

3. The population variances in all cells of the factorial design are equal (homogeneity of variance).

The consequences of violating one or more of these assumptions were discussed in Chapter 14. As was the case for one-way ANOVA, two-way ANOVA is robust with respect to the violation of the assumptions, particularly when there are large *and* equal numbers of observations in each cell of the factorial.

The Linear Model for Two-Way ANOVA

As in one-way ANOVA, the observed score on the dependent variable for each subject in a two-way ANOVA can be expressed in terms of additive linear components. However, for two-way ANOVA, the linear model contains additional terms that represent the effects of the second independent variable and the interaction between the two independent variables. Thus, the score for a subject can be expressed as follows:

$$X_{ijk} = \mu + \alpha_j + \beta_k + (\alpha\beta)_{jk} + e_{ijk} \tag{16.17}$$

where

X_{ijk} = score of the ith subject in the jth row and the kth column

μ = grand mean of the population

α_j = $(\mu_j. - \mu)$, the effect of being in a particular level of the first independent variable

β_k = $(\mu._k - \mu)$, the effect of being in a particular level of the second independent variable

$(\alpha\beta)_{jk}$ = $\{(\mu_{jk} - \mu) - [(\mu_j. - \mu) + (\mu._k - \mu)]\} = (\mu_{jk} - \mu_j. - \mu._k + \mu)$, the interaction effect of being in a particular j and k combination after the main effects have been considered

e_{ijk} = random error associated with this score

The right side of equation 16.17 contains the parameters μ, α_j, β_k, and $(\alpha\beta)_{jk}$; thus, the linear model is a representation of the composition of a score in the population. As in one-way ANOVA, the errors, e_{ijk}, are assumed to be normally distributed, with a mean of zero and variance equal to σ_e^2, the population error variance.

The linear model for two-way ANOVA is an expression that represents an individual score as additive components that are associated with the main effects, the interaction effect, and random error.

Models for Two-Way ANOVA

In the exercise example, the independent variables (length of exercise program and gender) have fixed levels. That is, the levels of both independent variables were selected by the researcher because they were of particular interest. In other words, the levels did not constitute random samples from some larger population of levels. Such a factorial design is called a **fixed-effects model**, and generalizations can be made only to these levels.

Although most factorial designs used in behavioral science research are fixed-effects models, there are two other models. If the levels of an independent variable are randomly selected from a larger population of levels, that variable is called a *random effect*. A two-factor design in which both independent variables have random effects is called a **random-effects model**. With this model, results can be generalized to the population of levels from which the levels of the independent variables were randomly selected.

The remaining model for two-way ANOVA is the **mixed-effects model**, in which the levels of one of the independent variables are fixed and the levels of the second independent variable are random. In a mixed model, the results relative to the random effect can be generalized to the population of levels from which the levels were selected for the investigation; the results relative to the fixed effect can be generalized to the specific levels selected.

There are three models for a two-way ANOVA. In the *fixed-effects* model, the researcher selects specific levels of interest for both independent variables. In the *random-effects* model, the levels of both independent variables are randomly selected from populations of levels. In the *mixed-effects* model, one independent variable has fixed levels, and the second independent variable has random levels.

Expected Mean Squares for ANOVA Models

We indicated earlier that the three null hypotheses tested in two-way ANOVA are

for rows:

$$H_{0_1}: \mu_1. = \mu_2. = \mu_3. = \cdots = \mu_{J.}$$

for columns:

$$H_{0_2}: \mu_{\cdot 1} = \mu_{\cdot 2} = \mu_{\cdot 3} = \cdots = \mu_{\cdot K}$$

for interaction:

$$H_{0_3}: \text{ all } (\mu_{jk} - \mu_{j\cdot} - \mu_{\cdot k} + \mu) = 0$$

For testing these null hypotheses, we must consider the expected mean squares (E[MS]) in order to determine the denominator of the F ratio. E(MS) for row and column main effects and the interaction effects in fixed-effects, random-effects, and mixed-effects models are summarized in Table 16.14. The actual derivation of the expected mean squares for two-way ANOVA is beyond the scope of this book.

TABLE 16.14
Expected Mean Squares and Calculated F Ratios for Two-Way ANOVA

Source	Fixed-Effects Model (Rows Fixed, Columns Fixed)		Mixed-Effects Model (Rows Fixed, Columns Random)	
	E(MS)	Calculated F Ratio	E(MS)	Calculated F Ratio
Row	$\sigma_e^2 + \dfrac{nK\sum_{j=1}^{J}\alpha_{j.}^2}{J-1}$	$\dfrac{MS_J}{MS_W}$	$\sigma_e^2 + n\sigma_{jk}^2 + \dfrac{nK\sum_{j=1}^{J}\alpha_{j.}^2}{J-1}$	$\dfrac{MS_J}{MS_{JK}}$
Column	$\sigma_e^2 + \dfrac{nJ\sum_{k=1}^{K}\beta_{.k}^2}{K-1}$	$\dfrac{MS_K}{MS_W}$	$\sigma_e^2 + nJ\sigma_{.k}^2$	$\dfrac{MS_K}{MS_W}$
Interaction	$\sigma_e^2 + \dfrac{n\sum_{k=1}^{K}\sum_{j=1}^{J}(\alpha\beta)_{jk}^2}{(K-1)(J-1)}$	$\dfrac{MS_{JK}}{MS_W}$	$\sigma_e^2 + n\sigma_{jk}^2$	$\dfrac{MS_{JK}}{MS_W}$
Within	σ_e^2		σ_e^2	

Source	Random-Effects Model (Rows Random, Columns Random)	
	E(MS)	Calculated F Ratio
Row	$\sigma_e^2 + n\sigma_{jk}^2 + nK\sigma_{j.}^2$	$\dfrac{MS_J}{MS_{JK}}$
Column	$\sigma_e^2 + n\sigma_{jk}^2 + nJ\sigma_{.k}^2$	$\dfrac{MS_K}{MS_{JK}}$
Interaction	$\sigma_e^2 + n\sigma_{jk}^2$	$\dfrac{MS_{JK}}{MS_W}$
Within	σ_e^2	

In addition to E(MS) for the three models, Table 16.14 also contains the F ratios to be used in testing the null hypotheses. Consider the row effect in a fixed-effects model. The E(MS) in the denominator of the F ratio is σ_e^2; the E(MS) in the numerator of the F ratio contains σ_e^2 *plus* the component due to the differences between the J row means. The F ratios for the column effect and the interaction effect have comparable components in the respective numerators and denominators. Because these numerators contain only one additional component, the appropriate error term (the denominator of the F ratio) for testing the main effects and the interaction is MS_W.

Consider the random-effects model. Notice that, for both the row and the column effects, the E(MS) contain *two* components in addition to σ_e^2. One of these components is due to either the row or the column effect, and the other component is due to the interaction. The E(MS) for the interaction contains only one component in addition to σ_e^2. Thus, to test both the row and the column effects in the random-effects model, the mean square for interaction (MS_{JK}) is the appropriate error term. The appropriate error term for the interaction effect is MS_W.

Now consider the mixed-effects model with the row effect fixed and the column effect random. For the row effect, the E(MS) contains σ_e^2 plus two additional components of variance — one due to the interaction effect and one due to the row main effect. Thus, the appropriate error term for the row effect (the fixed effect in this mixed-effects model) is MS_{JK}. For both the column main effect and the interaction effect, the E(MS) contain only one source of variance in addition to σ_e^2. Thus, to test the column effect (the random effect in this mixed-effects model) and the interaction effect, the appropriate error term is MS_W.

As indicated above, the independent variables or factors in a factorial design generally have fixed levels. However, there are research studies that use factorial designs with one or more factors that have random levels. It is important to note that while the actual calculation of the sum of squares (SS), the degrees of freedom (df), and the mean squares (MS) remains the same regardless of whether the factors are fixed or random, the error terms (the denominators) used in the F ratios for testing the null hypotheses vary. The error terms used in fixed-effects, mixed-effects, and random-effects models are summarized below:

Fixed-Effects Model: The MS_W is the appropriate error term for testing both main effects and the interaction.

Mixed-Effects Model: The MS_{JK} is used to test the fixed main effect; the MS_W is used to test the random main effect and the interaction.

Random-Effects Model: The MS_{JK} is used to test both main effects; the MS_W is used to test the interaction.

Two-Way ANOVA with Disproportionate Cell Frequencies

Up to this point we have considered factorial designs that have had an equal number of observations in each cell of the design. However, unequal cell frequencies occur. The procedures discussed above for two-way ANOVA are appropriate only when the cell frequencies are either *equal* or *proportional*. By proportional we mean that the ratios of the cell sample sizes across one independent variable are equal to those of the cell sample sizes across the other independent variable. Consider the 2×2 factorial design below, in which n_{11} refers to the number of observations in the first cell of the data matrix.

$n_{11} = 20$	$n_{12} = 15$	$n_{1.} = 35$
$n_{21} = 40$	$n_{22} = 30$	$n_{2.} = 70$
$n_{.1} = 60$	$n_{.2} = 45$	

The ratio of the cell frequencies across the columns is $20/40 = 15/30 = 1/2$; the ratio across the rows is $20/15 = 40/30 = 4/3$. Thus, the cells are proportional, and the procedures discussed above for two-way ANOVA are appropriate.

Now consider a second example. For the 2×2 design below, the ratios of the cell frequencies across the columns are not equal ($30/10$ and $40/20$). Similarly, the ratios across the rows are not equal ($30/40$ and $10/20$). In this case, the cell frequencies are said to be disproportionate. If the previously discussed procedures for two-way ANOVA were applied, the results might be misleading in that greater emphasis would be given to cells with greater numbers of observations. Disproportionate cell frequencies also result in a correlation between the two independent variables, which interferes with the interpretation of their effect on the dependent variable.

$n_{11} = 30$	$n_{12} = 40$	$n_{1.} = 70$
$n_{21} = 10$	$n_{22} = 20$	$n_{2.} = 30$
$n_{.1} = 40$	$n_{.2} = 60$	

There are several procedures that can be applied when disproportionate cell frequencies occur, including regression analysis, hierarchical analysis, and unweighted means analysis; these procedures are beyond the scope of this book. The interested reader is referred to the Kirk reference in the Bibliography.

Determining the Sample Size

The process of determining sample size for each cell in a two-factor design is similar to that for a one-factor design with K levels. Consider the previous example

TABLE 16.15
Two-Factor Design with Teaching Method and IQ as Independent Variables

		Level of IQ			
		b_1 (<85)	b_2 (85–99)	b_3 (100–114)	b_4 (≥115)
Type of Teaching Method	a_1 (method 1)				
	a_2 (method 2)				
	a_3 (method 3)				

of the educational researcher who is interested in the mathematics achievement of students taught under three different teaching methods. In this example, the researcher also controlled for IQ levels; the design originally illustrated in Table 16.2 is repeated in Table 16.15. Assuming the researcher has done previous research in this area and knows the research literature well, a standardized effect size (d) can be established. Assume that the researcher uses $d = 0.75, \alpha = .05$, and power $= .80$; we can then use Appendix C.12 to determine the appropriate sample size. First, consider the independent variable, teaching method. With three levels and using $\alpha = .05$, power $= .80$, and $d = 0.75$, the sample size required for *each* level would be 35. Now consider the second independent variable, IQ, which has four levels. Using Appendix C.12, with $\alpha = .05$, power $= .80$, and $d = 0.75$, the sample size for *each* level of this variable is 40. For a two-factor design, it is necessary to satisfy the sample-size requirement for both independent variables, method and IQ. In order to meet the requirements for this 3×4 design, with $3 \times 4 = 12$ cells, the sample size for *each cell* would be 14. With 14 in each cell, the sample size for the three method levels would be $4 \times 14 = 56$, and for the four levels of IQ, it would be $3 \times 14 = 42$ (see Table 16.16). Recall from the last chapter that the power of the statistical test is directly related to the size of the sample; the larger the sample, the more powerful the test. Thus, with 42 in each level of IQ, the test of the hypothesis will be slightly more powerful than with just 40 in each level. Similarly, with 56 in each level of method, rather than 35, the test of the null hypothesis will be substantially more powerful.

For a two-factor design, the sample-size requirement for both independent variables must be satisfied. Such a strategy leads to slightly larger sample sizes for the levels of both variables and to more powerful statistical tests.

TABLE 16.16
Sample Size Requirements for the Two-Factor Design with Teaching Method and IQ as Independent Variables

		Level of IQ				
		b_1 (<85)	b_2 (85–99)	b_3 (100–114)	b_4 (≥115)	
Type of Teaching Method	a_1 (method 1)	14	14	14	14	56
	a_2 (method 2)	14	14	14	14	56
	a_3 (method 3)	14	14	14	14	56
		42	42	42	42	

Summary

In this chapter, we have discussed analysis of variance, two-way classification, as the procedure for testing hypotheses about means when two independent variables are considered simultaneously. Two-way ANOVA has several advantages over one-way ANOVA, namely: (1) *efficiency*, investigating the effects of two independent variables simultaneously; (2) *control*, taking into consideration the variation due to a second independent variable and thus enhancing the statistical precision; and (3) *interaction*, studying the interaction between the two independent variables.

The concept of partitioning the total variation and the computational procedures for two-way ANOVA are the logical extension of the procedures used for one-way ANOVA. The total variation is partitioned into four components: (1) variation due to the row main effect; (2) variation due to the column main effect; (3) variation due to the interaction effect after considering the main effects; and (4) error variance. The actual computational procedures are summarized in Table 16.17.

The null hypotheses tested in two-way ANOVA relate to the row effect, column effect, and interaction effect. The test statistic for all three hypotheses is the F ratio, which is the ratio of two mean squares (MS). While the MS are calculated in the ANOVA, the expected mean squares (E[MS]) determine which of the calculated mean squares are used in the numerator and denominator of the specific F ratio. The numerator contains the variation due to the effect being tested in the null hypothesis plus one or more additional sources of variation. The denominator contains only these latter sources of variation.

The sampling distribution of the F ratio is the F distribution for appropriate df values; two df values are required to identify the specific F distribution. As is the case for one-way ANOVA, when the observed F ratio exceeds the critical value of F, the respective null hypothesis is rejected. Following the rejection of the null hypothesis on either the row or column main effect, post hoc multiple-comparison procedures are applied. Similarly, when the null hypothesis for the interaction is rejected, post hoc test procedures, including plotting the cell means and tests of simple effects, would be applied.

TABLE 16.17
Summary of Computational Procedures for ANOVA, Two-Way Classification

	Row Effect	Column Effect	Interaction Effect	Within Cell	Total
Sum of Squares	$nK\sum_{j=1}^{J}(\bar{X}_{j\cdot} - \bar{X})^2$	$nJ\sum_{k=1}^{K}(\bar{X}_{\cdot k} - \bar{X})^2$	$n\sum_{k=1}^{K}\sum_{j=1}^{J}(\bar{X}_{jk} - \bar{X}_{j\cdot} - \bar{X}_{\cdot k} + \bar{X})^2$	$\sum_{k=1}^{K}\sum_{j=1}^{J}\sum_{i=1}^{n}(X_{ijk} - \bar{X}_{jk})^2$	$\sum_{k=1}^{K}\sum_{j=1}^{J}\sum_{i=1}^{n}(X_{ijk} - \bar{X})^2$
Computational Formula	$\frac{1}{nK}\sum_{j=1}^{J}T_{j\cdot}^2 - \frac{T^2}{N}$	$\frac{1}{nJ}\sum_{k=1}^{K}T_{\cdot k}^2 - \frac{T^2}{N}$	$\frac{1}{n}\sum_{k=1}^{K}\sum_{j=1}^{J}T_{jk}^2 - \frac{1}{nK}\sum_{j=1}^{J}T_{j\cdot}^2 - \frac{1}{nJ}\sum_{k=1}^{K}T_{\cdot k}^2 + \frac{T^2}{N}$	$\sum_{k=1}^{K}\sum_{j=1}^{J}\sum_{i=1}^{n}X_{ijk}^2 - \frac{1}{n}\sum_{k=1}^{K}\sum_{j=1}^{J}T_{jk}^2$	$\sum_{k=1}^{K}\sum_{j=1}^{J}\sum_{i=1}^{n}X_{ijk}^2 - \frac{T^2}{N}$
Degrees of Freedom	$J-1$	$K-1$	$(J-1)(K-1)$	$JK(n-1)$	$N-1$
Mean Square	$\frac{SS_J}{J-1}$	$\frac{SS_K}{K-1}$	$\frac{SS_{JK}}{(J-1)(K-1)}$	$\frac{SS_W}{JK(n-1)}$	
E(MS) Fixed-Effects Model	$\sigma_e^2 + \frac{nK\sum_{j=1}^{J}\alpha_{j\cdot}^2}{J-1}$	$\sigma_e^2 + \frac{nJ\sum_{k=1}^{K}\beta_{\cdot k}^2}{K-1}$	$\sigma_e^2 + \frac{n\sum_{k=1}^{K}\sum_{j=1}^{J}(\alpha\beta)_{jk}^2}{(J-1)(K-1)}$	σ_e^2	
E(MS) Mixed-Effects Model (Rows Fixed, Columns Random)	$\sigma_e^2 + n\sigma_{jk}^2 + \frac{nK\sum_{j=1}^{J}\alpha_{j\cdot}^2}{J-1}$	$\sigma_e^2 + nJ\sigma_k^2$	$\sigma_e^2 + n\sigma_{jk}^2$	σ_e^2	
E(MS) Random-Effects Model	$\sigma_e^2 + n\sigma_{jk}^2 + nK\sigma_j^2$	$\sigma_e^2 + n\sigma_{jk}^2 + nJ\sigma_k^2$	$\sigma_e^2 + n\sigma_{jk}^2$	σ_e^2	

The process for determining the sample size for a two-factor design was presented as an extension of the process for the one-factor case. However, using the strategy that satisfies the sample-size requirement for both factors results in sample sizes that are slightly larger for the levels of both. With larger sample sizes, the power of the statistical tests for these two factors is even greater.

Exercises

For each of the following exercises, complete the following:

1. State the hypotheses.

2. Set the criterion for rejecting each H_0.

3. Complete the ANOVA and compute the test statistic.

4. Compute ω^2 for the main effects and the interaction.

5. Plot the interaction (when significant).

6. Interpret the data.

1. A social psychologist is studying the effectiveness of three different reinforcement schedules on certain gender-role behaviors in boys and girls. The following data are the number of times that the desired behaviors were demonstrated (within a specified interval) following the treatment. Complete the ANOVA using $\alpha = .05$.

Reinforcement schedule

	1		*2*		*3*	
	32	30	34	40	40	35
Boys	36	38	36	31	42	38
	38	35	39	33	36	41
	25	29	40	38	43	36
Girls	27	26	29	34	41	39
	28	24	33	37	37	45

*2. An experiment is conducted with laboratory rats on the effects of different doses of a specific drug on physical performance. 48 rats of the same age, 24 female and 24 male, are randomly assigned to one of three levels of the drug, denoted D_1, D_2, and D_3, i.e., eight rats of each gender are randomly assigned to each drug dosage. After receiving the drug daily for one week, the rats are given a physical performance test; the scores on this test represent the dependent variable. The means for each cell of the design and a partially completed Summary ANOVA

Table are given below. Complete the Summary ANOVA Table and answer the above questions for testing the appropriate hypotheses using $\alpha = .05$.

	Drug Dosage			
	D_1	D_2	D_3	*Total*
Female	35.6	49.4	71.8	52.27
Male	55.2	92.2	110.0	85.8
Total	45.4	70.8	90.9	

Summary ANOVA					
Source	SS	df	MS	F	F_{cv}
Drug	14,832.5				
Gender	17,120.6				
Interaction					
Within	41,685.3				
Total	76,226.8				

3. The state superintendent of instruction asks the director of educational research to investigate differences in scores on a standardized teacher examination for senior education students majoring in the following subject areas:

   1. Vocational education
   2. Physical education
   3. English
   4. Mathematics

The following data are the examination scores for a random sample of 32 graduating seniors (16 males and 16 females). Complete the ANOVA using $\alpha = .05$.

	Major			
	Vocational education	*Physical education*	*English*	*Mathematics*
Men	32	34	34	32
	36	35	38	39
	32	36	40	38
	30	34	39	44
Women	28	35	36	35
	27	38	37	45
	30	33	40	48
	26	36	38	44

4. A family counselor is interested in the effects of three different counseling techniques on improving the self-esteem of clients from middle- and upper-class households. The data below are the mean self-esteem scores ($n = 8$ observations per cell) of clients in directive, combination, or nondirective therapy. Complete the ANOVA using $\alpha = .05$.

	Directive	Combination	Nondirective	Total
Upper	55.4	59.6	63.0	59.3
Middle	45.2	41.3	38.3	41.6
Total	50.3	50.3	50.7	

Summary ANOVA

Source	SS	df	MS	F	F_{cv}
Class			97.846		
Methods	8.47				
Interaction	412.84				
Within	923.80				
Total					

5. A behavioral psychologist is studying the effects of various levels of stress on performance of a complex physical task. The researcher also includes time of day (either 10:00 A.M. or 4:00 P.M.) in the design. The following data are the scores for the complex task. Complete the ANOVA using $\alpha = .05$.

	High stress	Low stress		High stress	Low stress
	15	16		22	20
	18	18		25	18
10 A.M.	14	12	4 P.M.	25	17
	12	15		19	20
	18	15		20	16

6. A researcher is interested in whether three methods of teaching are equally effective for students of different reading levels. Students from high, middle, and low reading levels are randomly assigned to the teaching methods. The following data are achievement scores. Complete the ANOVA using $\alpha = .05$.

	Teaching method					Teaching method		
	1	*2*	*3*			*1*	*2*	*3*
	40	32	30			22	20	18
High	36	29	35	*Low*		18	18	14
achievement	38	36	33	*achievement*		16	19	19
	36	35	38			20	15	19
	24	20	18					
Middle	21	29	26					
achievement	20	25	26					
	20	30	25					

7. In a large corporation, a survey of job satisfaction is given to separate random samples of labor and management representatives. These surveys are conducted at three different times during the year. Each time a new sample is selected. The following data are the job satisfaction scores for the samples at different times. Complete the ANOVA using $\alpha = .05$.

	Time					Time		
	1	*2*	*3*			*1*	*2*	*3*
	18	15	11			20	18	16
	18	17	15			20	17	16
Labor	15	18	14	*Management*		18	20	18
	17	14	15			15	15	12
	19	14	10			17	15	15

8. A political scientist who is interested in attitudes toward tax reform conducts a community survey of Republicans and Democrats of various ages. One analysis of particular interest is a comparison of young voters (ages 18 to 20) and older voters (ages 38 to 40). The following data are attitude scores for the survey respondents; the higher the score, the greater the concern for tax reform. Complete the ANOVA using $\alpha = .05$.

	Age group			
	18–20		*38–40*	
	4	4	10	10
	9	7	13	12
Democrats	7	5	9	11
	3	6	11	13
	5	2	15	8
	9	10	20	15
	9	8	22	19
Republicans	14	11	17	18
	12	10	16	20
	12	7	18	17

9. Consider a two-factor design with three levels for both factors. Use Table C.12 to determine the appropriate sample size for each cell, given the following conditions:

    a. $\alpha = .05, \beta = .20, d = 1.00$

    b. $\alpha = .05, \beta = .20, d = 0.60$

    c. $\alpha = .01, \beta = .05, d = 1.00$

    d. $\alpha = .01, \beta = .05, d = 0.60$

10. Consider a two-factor design with three levels for the first factor and four levels for the second factor. Use Table C.12 to determine the appropriate sample size for each cell, given the following conditions:

    a. $\alpha = .05, \beta = .20, d = 1.00$

    b. $\alpha = .05, \beta = .20, d = 0.65$

    c. $\alpha = .01, \beta = .05, d = 1.00$

    d. $\alpha = .01, \beta = .05, d = 0.65$

17

Linear Regression: Estimation and Hypothesis Testing

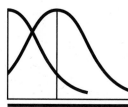

Key Concepts

Linear regression equation
Regression coefficient (*b*)
Regression constant (*a*)
Conditional distributions

Standard error of estimate
Homoscedasticity
Standard error of the regression coefficient

I n Chapter 7, we defined linear regression as the process of predicting or estimating scores on a *Y* variable based on knowledge of scores on an *X* variable;
in other words, the regression of *Y* on *X*. The term *linear* is included because
we assume that the relationship between *Y* and *X* variables can be represented in
a scatterplot with points tending to locate along a straight line. This straight line
is the linear regression line; it represents how, on the average, a change in the *X*
variable is associated with a change in the *Y* variable.

The discussion in Chapter 6 was limited to the descriptive aspects of linear regression: to the development and use of the linear regression equation in predicting
Y scores from *X* scores. We also introduced the concept of *conditional distributions* of *Y* scores. In this chapter, we will use these conditional distributions to
make probability statements about predicted *Y* scores.

Conditional Distributions

Consider the example from Chapter 6 in which we predicted creativity scores (Y) from logical reasoning scores (X). The data are repeated in Table 17.1. The regression equation for this example was

$$\hat{Y} = 0.65X + 5.28$$

Assume that Figure 17.1 represents the scatterplot of the creativity and logical reasoning scores with conditional distributions for $X_1 = 7$, $X_2 = 12$, and $X_3 = 17$. For the logical reasoning score $X_1 = 7$, we can find the corresponding predicted creativity score (Y_1) by using the previous regression equation.

TABLE 17.1
Data for Calculating the Regression Equation for Predicting Creativity Scores (Y) from Logical Reasoning Scores (X)

Student	X	Y	X²	Y²	XY
1	15	12	225	144	180
2	10	13	100	169	130
3	7	9	49	81	63
4	18	18	324	324	324
5	5	7	25	49	35
6	10	9	100	81	90
7	7	14	49	196	98
8	17	16	289	256	272
9	15	10	225	100	150
10	9	12	81	144	108
11	8	7	64	49	56
12	15	13	225	169	195
13	11	14	121	196	154
14	17	19	289	361	323
15	8	10	64	100	80
16	11	16	121	256	176
17	12	12	144	144	144
18	13	16	169	256	208
19	18	19	324	361	342
20	7	11	49	121	77
Σ	233	257	3037	3557	3205

$\overline{X} = 11.65$ $\overline{Y} = 12.85$
$SS_X = 322.55$ $SS_Y = 254.55$
$s_X = 4.12$ $s_Y = 3.66$
$r = .74$ $s_{XY} = 2.55$

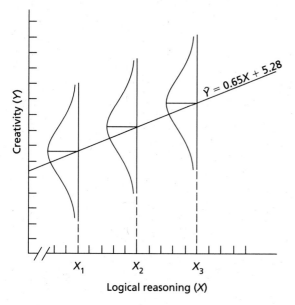

FIGURE 17.1
Conditional distributions of Y for X_1, X_2, and X_3

$$\hat{Y}_1 = 0.65(7) + 5.28$$

$$= 9.83$$

This predicted score is a point on the regression line. However, the predicted score is only an estimated creativity score ($\hat{Y}$) given a logical reasoning score of 7. There would actually be an entire distribution of creativity scores given the logical reasoning score of 7; in other words, the conditional distribution for $X_1 = 7$. The conditional distribution for $X_2 = 12$ and $X_3 = 17$ would be similarly defined, with the two predicted scores ($\hat{Y}_2$ and $\hat{Y}_3$) computed as follows:

$$\hat{Y}_2 = 0.65(12) + 5.28$$

$$= 13.08$$

$$\hat{Y}_3 = 0.65(17) + 5.28$$

$$= 16.33$$

As mentioned in Chapter 6, each of these distributions is assumed to be normal, with the mean of each equal to the predicted score ($\hat{Y}$) for the given X. The standard deviation for each of these distributions was defined as the *standard error of estimate*.

$$s_{Y \cdot X} = \sqrt{\frac{\sum e^2}{n - 2}}$$

(see formula 6.10)

or

$$s_{Y \cdot X} = s_Y \sqrt{1 - r^2} \sqrt{(n-1)/(n-2)} \qquad \text{(see formula 6.11)}$$

Using the data from Table 17.1 and formula 6.11, the standard error of estimate ($s_{Y \cdot X}$) for this example equals 2.55.

A second important assumption underlying linear regression is **homoscedasticity,** or the assumption that the standard deviations of conditional distributions are equal. Thus, with the assumptions of normality and homoscedasticity (as illustrated in Figure 17.1), we can make probability statements about predicted scores computed using the linear regression equation.

Predicted Scores and Conditional Distributions

In our example of predicting creativity scores from logical reasoning scores, the following questions might be asked:

If $X = 10$, what is the probability that Y will be greater than 14?

If $X = 13$, what is the probability that Y will be between 9 and 14?

The rationale for asking such questions must be couched in the realities of the situation and the decisions that need to be made. For example, the researcher might want to know the percentage of students with logical reasoning scores of 10 who will have creativity scores greater than 14. If this percentage is quite low, then perhaps these students should not be recommended for a special program for gifted children. To answer this question, we first need to compute the predicted creativity score for a logical reasoning score of 10. Using the above regression equation,

$$\hat{Y} = 0.65X + 5.28$$

$$= 0.65(10) + 5.28$$

$$= 11.78$$

Thus, the predicted creativity scores for students with logical reasoning scores of 10 are 11.78.

Now consider the conditional distribution of creativity scores for $X = 10$, illustrated in Figure 17.2. Recall that this distribution is assumed to be normal with the mean equal to the predicted score ($\hat{Y} = 11.78$), and under the assumption of homoscedasticity, the standard deviation of this conditional distribution equals 2.55. To find the percentage of students with logical reasoning scores of 10 who will have a creativity score greater than 14, we must determine the shaded area of the conditional distribution in Figure 17.2 by using the concepts of standard scores and the standard normal distribution. Recall the general formula for the standard score from Chapter 3.

$$z = \frac{X - \overline{X}}{s}$$

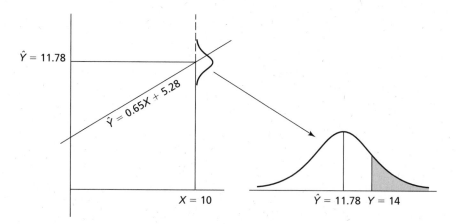

FIGURE 17.2
Conditional distribution for $X = 10$ with $\hat{Y} = 11.78$, illustrating the probability of a score ≥ 14

Applying this formula to the conditional distribution, the corresponding formula would be

$$z = \frac{Y - \hat{Y}}{s_{Y \cdot X}} \qquad (17.1)$$

For our example,

$$z = \frac{14 - 11.78}{2.55}$$

$$= 0.87$$

Thus, the percentage of Y scores that exceed 14 if $X = 10$ corresponds to the area under the standard normal curve beyond a z score of 0.87. Using Table C.1, we find that this area is 0.1922. If the criterion for admitting students to a special program for gifted students is a creativity score of 14 or greater, students with logical reasoning scores of 10 are *not* good candidates for the program.

The conditional distribution is used to determine the percentage of Y values associated with a specific value of X.

Confidence Intervals in Linear Regression

The concept of the conditional distribution can also be used to answer a more general question in our example:

If $X = 10$, what are likely values for Y?

The answer to this question involves using the concept of the confidence interval, introduced in Chapters 9 through 12. For this example, we want to determine a range of scores that has a .95 probability of including the creativity score of a student whose logical reasoning score is 10. This range of scores would be the 95 percent confidence interval (CI_{95}).

Consider the predicted creativity score ($\hat{Y}$), given a logical reasoning score of $X = 10$:

$$\hat{Y} = 0.65(10) + 5.28$$

$$= 11.78$$

The standard error of this predicted score is defined as follows:

$$s_{\hat{Y}} = s_{Y \cdot X} \sqrt{1 + \frac{1}{n} + \frac{(X - \overline{X})^2}{SS_X}} \qquad (17.2)$$

where

$$SS_X = (n - 1)s_X^2$$

Note that the standard error of a predicted score is a minimum when $X = \overline{X}$ and increases as X deviates from $\overline{X}$. This implies that the more the score on the predictor variable (X) deviates from the mean of the scores on the predictor variable ($\overline{X}$), the larger the standard error of the predicted score. For this example,

$$s_{\hat{Y}} = 2.55 \sqrt{1 + \frac{1}{20} + \frac{(10 - 11.65)^2}{322.55}}$$

$$= 2.55 \sqrt{1 + .05 + 0.0084}$$

$$= 2.62$$

The general formula for the confidence interval for predicted scores is given by

$$CI = \hat{Y} \pm (t_{cv})(s_{\hat{Y}}) \qquad (17.3)$$

where

$\hat{Y}$ = predicted score

t_{cv} = critical value of t for df = $n - 2$

$s_{\hat{Y}}$ = standard error of predicted scores

For this example, the 95 percent confidence interval (see Figure 17.3) would be computed as follows:

$$CI_{95} = 11.78 \pm (2.101)(2.62)$$

$$= 11.78 \pm 5.51$$

$$= (6.27, 17.29)$$

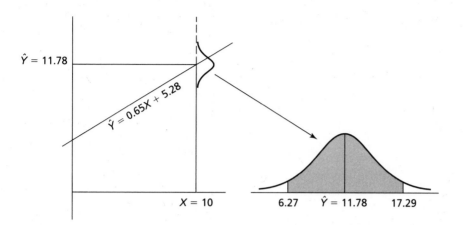

FIGURE 17.3
Conditional distribution for $X = 10$ with $\hat{Y} = 11.78$, illustrating the 95 percent confidence interval

Thus, for this example, we can say that we are 95 percent confident that, given a student's logical reasoning score (X) of 10, the interval from 6.27 to 17.29 will contain this student's creativity score (Y).

> CI_{95} is a range of scores for which the confidence is 95 percent that the interval will include the actual Y score.

Testing the Significance of the Regression Coefficient

Earlier we discussed the relationship between linear correlation and linear regression. In that discussion, we indicated that, if there were a perfect correlation between variables Y and X ($r = -1.00$ or $r = +1.00$), all points in the scatterplot would be on the linear regression line. In this case, we would be able to predict Y *perfectly* from X. However, when there is no correlation between Y and X ($r = 0$), it is impossible to fit a meaningful regression line to the points in the scatterplot. In fact, consider the formula for the regression coefficient (b) (see formula 17.4) and for the regression constant (a) (see formula 17.3) when $r = 0$.

$$b = r\frac{s_Y}{s_X}$$

$$= (0)\frac{s_Y}{s_X}$$

$$= 0$$

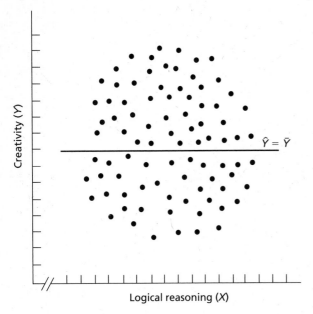

FIGURE 17.4
Regression line when $r = 0$; $\hat{Y} = \overline{Y}$

Thus, since $b = 0$,

$$a = \overline{Y} - b\overline{X}$$

$$= \overline{Y} - (0)\overline{X}$$

$$= \overline{Y}$$

That is, when the correlation coefficient (r) equals zero, the regression coefficient (b) equals zero and the regression constant (a) equals $\overline{Y}$.

Now consider the linear regression equation when there is no correlation between variables Y and X:

$$\hat{Y} = bX + a$$

Substituting the earlier values for b and a, we have

$$\hat{Y} = (0)X + \overline{Y}$$

$$= \overline{Y}$$

This regression equation is illustrated in Figure 17.4. Notice that, when $r = 0$, the regression equation is a horizontal line parallel to the X axis, which intercepts the Y axis at point $\overline{Y}$. That is, regardless of the value of X, the predicted score $(\hat{Y})$ will always be the mean of the Y scores $(\overline{Y})$.

Obviously, if there is no relationship between Y and X, the regression coefficient equals zero, and knowing an individual's X score will *not* help predict that individual's Y score. Thus, the issue becomes one of how different from zero the regression coefficient must be in order to statistically enhance the prediction or estimation of the Y scores. The answer to this question involves testing the regression coefficient for statistical significance. The process follows the logic of hypothesis testing developed in Chapters 7 through 12.

Step 1: State the Hypotheses

The null hypothesis to be tested is that the regression coefficient in the population (β) equals zero; the alternative hypothesis is that the population regression coefficient is *not* equal to zero, or

$$H_0: \beta = 0$$

$$H_a: \beta \neq 0$$

As before, the null hypothesis is established as the "straw man" to be rejected in order to show the tenability of the alternative. By rejecting the null hypothesis, the conclusion would be that the regression coefficient in the population is not equal to zero and thus that knowledge of scores on the X variable will enhance the prediction of scores on the Y variable. As we will see, rejecting $H_0: \beta = 0$ is the same as rejecting the null hypothesis $H_0: \rho = 0$ or the hypothesis of no relationship between X and Y in the population (see Chapter 10).

Step 2: Set the Criterion for Rejecting H_0

To illustrate the test of this hypothesis, consider our example of predicting creativity scores from logical reasoning scores. In this example, we assumed that we randomly selected 20 students from the population, gave each the creativity and logical reasoning tests, and computed the linear regression equation,

$$\hat{Y} = 0.65X + 5.28$$

For this sample, the regression coefficient (b) was equal to 0.65. Now consider the sampling distribution of the regression coefficient for all samples of size 20 selected from this population. (Recall the discussions of sampling distributions in Chapters 7 through 12.) It can be shown that the sampling distribution of the regression coefficient is the t distribution with $n - 2$ degrees of freedom.[4] The mean

---

[4] See E. J. Pedhazur, pp. 28–29.

of this distribution is established in the null hypothesis H_0: $\beta = 0$. The standard deviation of this distribution, called the **standard error of the regression coefficient** (s_b), is defined as follows:

$$s_b = \left(\frac{s_{Y \cdot X}}{\sqrt{SS_X}} \right) \tag{17.4}$$

where

$s_{Y \cdot X}$ = standard error of prediction or estimation

SS_X = sum of squares for the predictor variable (X), or $\Sigma(X - \overline{X})^2 = (n-1)s_X^2$

The general formula for the test statistic (see formula 10.1) will be used for testing this null hypothesis:

$$t = \frac{\text{statistic} - \text{parameter}}{\text{standard error of the statistic}}$$

For testing H_0: $\beta = 0$, the formula becomes

$$t = \frac{b - \beta}{s_b} \tag{17.5}$$

$$t = \frac{b - \beta}{\left(\dfrac{s_{Y \cdot X}}{\sqrt{SS_X}} \right)} \tag{17.6}$$

Assuming for this example that we will be testing H_0: $\beta = 0$ against H_a: $\beta \neq 0$ using a two-tailed test at $\alpha = .05$, the critical value of t (t_{cv}) for 18 degrees of freedom will be ± 2.101 (see Table C.3).

Step 3: Compute the Test Statistic

For the data in Table 17.1, the standard error of the regression coefficient (s_b) is computed as follows, using formula 17.4:

$$s_b = \left(\frac{s_{Y \cdot X}}{\sqrt{SS_X}} \right)$$

$$= \frac{2.55}{\sqrt{322.55}}$$

$$= \frac{2.55}{17.96}$$

$$= 0.14$$

Using formula 17.5, the test statistic is

$$t = \frac{b - \beta}{s_b}$$

$$= \frac{0.65 - 0}{0.14}$$

$$= 4.64$$

Step 4: Interpret the Results

Since the observed value of the test statistic ($t = 4.64$) exceeds the critical value ($t_{\text{cv}} = \pm 2.101$), the null hypothesis H_0: $\beta = 0$ would be rejected. The conclusion would be that the regression coefficient is different from zero, and that, therefore, knowledge of logical reasoning scores (X) will enhance the prediction of creativity scores (Y).

Computer Exercise: The Survey of High School Seniors Data Set

For the survey of high school seniors data set ($n = 500$), consider the mathematics achievement (MATHACH) scores as the criterion variable and the MOSAIC scores as the predictor variable. Now suppose we want to generate the regression equation for predicting MATHACH scores from MOSAIC scores and then to test the regression coefficient for statistical significance. The MULTIPLE RE-GRESSION procedure in SPSS was used to generate the computer printout for this analysis (see Table 17.2). Consider the computer printout in Table 17.2 and note the section denoted "Variables in the Equation." (We will discuss the upper section of the printout in Chapter 18.) The regression coefficient (b) is .207464 and the regression constant is 7.519532. Thus, we find the regression equation to be:

$$\hat{Y} = .207464X + 7.519532$$

Note that the standardized regression coefficient (Beta) is .291762. Thus, the regression equation in standard score form is:

$$z_{\hat{Y}} = .291762z_X$$

Step 1: State the Hypotheses

As in the previous example, the null hypothesis is that the regression coefficient in the population (β) equals zero. We will test this null hypothesis against the nondirectional alternative hypothesis and set the level of significance at $\alpha = .05$.

TABLE 17.2
SPSS Printout for Regression Exercise for the Survey of High School Seniors Data Set

```
          * * * *  M U L T I P L E    R E G R E S S I O N * * * *
Equation Number 1   Dependent Variable..    MATHACH


Variable(s) Entered on Step Number
    1..    MOSAIC


Multiple R            .29176
R Square              .08512
Adjusted R Square     .08329
Standard Error       6.32357


Analysis of Variance
                  DF     Sum of Squares    Mean Square
Regression         1         1852.88579     1852.88579
Residual         498        19913.79658       39.98754


 F =     46.33657    Signif F =   .0000


------------------ Variables in the Equation -------------


Variable              B        SE B       Beta      T     Sig T

MOSAIC             .207464    .030478    .291762   6.807   .0000
(Constant)        7.519532    .866934             8.674    .0000
```

Symbolically,

$$H_0: \beta = 0$$

$$H_a: \beta \neq 0$$

Step 2: Set the Criterion for Rejecting H_0

The general formula for the test statistic is used to test this null hypothesis.

$$t = \frac{\text{statistic} - \text{parameter}}{\text{standard error of the statistic}}$$

As was illustrated in the above example, the sampling distribution for this test statistic is the t distribution with $n - 2$ degrees of freedom. In this example, there are $n - 2 = 500 - 2 = 498$ degrees. Thus, the critical values from Table C.3 for $\alpha = .05$ are $t_{cv} = \pm 1.96$. If the calculated value for the test statistic exceeds the critical values, the null hypothesis would be rejected.

Step 3: Compute the Test Statistic

In the computer printout, there is a t-test for both the regression coefficient (b) and the regression constant (a). For most research studies in the behavioral sciences, we only concern ourselves with the test of the regression coefficient. The test statistic for this example is as follows:

$$t = \frac{b - \beta}{s_b}$$

Note from Table 17.2, the standard error of the regression coefficient $s_b = .030478$. Thus,

$$t = \frac{.207464 - 0}{.030478}$$

$$= 6.807$$

Step 4: Interpret the Results

Since the observed value of the test statistic ($t = 6.807$) exceeds the critical value ($t_{cv} = \pm 1.96$), the null hypothesis H_0: $\beta = 0$ would be rejected. The associated probability statement is that the probability that we would observe a regression coefficient of .207464 by chance if the null hypothesis is true is less than .05. In the printout, the value for "Sig T" is .0000, which implies that the probability of such occurrence is actually less than .0001. The conclusion would be that the variable MOSAIC is a statistically significant predictor of MATHACH scores, i.e., knowledge of MOSAIC scores enhances the prediction of MATHACH scores.

Testing H_0: $\beta = 0$ Versus Testing H_0: $\rho = 0$

In the above example of testing H_0: $\beta = 0$, by rejecting the null hypothesis, we are saying that knowledge of the X scores will enhance prediction of the Y scores. In essence, we are saying that there is a statistically significant relationship between the X and Y variables. That is, with a sample correlation of .74 between the creativity and logical reasoning scores, we would reject the null hypothesis H_0: $\rho = 0$ that the correlation in the population is zero.

In Chapter 10, we introduced the formula for testing the null hypothesis H_0: $\rho = 0$; this formula (10.7) was

$$t = r\sqrt{\frac{n - 2}{1 - r^2}}$$

Applying this formula to this example, the test statistic would be

$$t = .74\sqrt{\frac{18}{1 - (.74)^2}}$$

$$= .74\sqrt{39.79}$$

$$= 4.67$$

There are $n - 2$ degrees of freedom associated with this test statistic. For this example, with 18 degrees of freedom and $\alpha = .05$, the critical value of the test statistic (t_{cv}) would be ± 2.101. Thus, we would reject H_0: $\rho = 0$ and conclude that there is a statistically significant relationship between the creativity scores and the logical reasoning scores. Note that the difference between the two test statistics (those using formulas 17.5 and 10.7) is due only to rounding error and that the conclusions drawn from each are the same. That is, with a statistically significant relationship between the X and Y variables, knowledge of the X scores enhances prediction of the Y scores.

Summary

This chapter has extended the concept of linear regression initially discussed in Chapter 6 to include procedures for (1) making probability statements about predicted scores; (2) developing confidence intervals around predicted scores; and (3) testing hypotheses about the regression coefficients. For each of these procedures, we use the standard error of estimate ($s_{Y.X}$); this standard error is defined as the standard deviation of the errors of prediction. In the prediction of scores on a criterion variable (Y) from scores on a predictor variable (X), these errors of prediction are the differences between the observed scores and the predicted scores. These errors of prediction and consequently the standard error of estimate are related to the magnitude of the correlation between the Y and X variables. The larger the correlation coefficient, the better the prediction and the smaller the errors of prediction.

These errors of prediction can be illustrated using conditional distributions, which are defined as the distributions of scores on the Y variable for a given value of X. These conditional distributions are assumed to be normally distributed and homoscedastic; this latter assumption is that the standard deviations for all of these distributions are equal. With these two assumptions, it is possible to use the normal distribution to make probability statements and develop confidence intervals for predicted scores. The standard error used in confidence intervals for predicting an individual Y score is somewhat larger than the standard error for predicting a Y score for a group. The standard error for the former depends upon how far the X score for the individual is from the group mean on the X scores.

Testing the null hypothesis H_0: $\beta = 0$ was shown to be the same as testing the null hypothesis H_0: $\rho = 0$. If the latter is rejected, the conclusion is that the correlation in the population is not equal to zero. Analogously, if the former is rejected, the conclusion is that, in the population, the regression coefficient is not equal to zero. That is, with a statistically significant relationship between the Y and X variables, knowledge of X scores enhances the prediction of Y scores.

The concepts of linear regression with a single predictor variable can be extended to linear regression with more than one predictor variable. Although the logic underlying the development and use of the regression equation remains the same, the formulas for computing the regression coefficients and the tests of significance are more complicated. This process, called *multiple linear regression*, is discussed in the next chapter.

Exercises

1. In studying the relationship between verbal communication scores and marital satisfaction scores of women seeking marital counseling, a counseling psychologist collects the following data from 25 clients.

Client	Communication	Satisfaction	Client	Communication	Satisfaction
1	37	29	14	39	31
2	27	32	15	43	25
3	30	25	16	37	24
4	52	19	17	48	32
5	30	30	18	30	35
6	42	21	19	42	22
7	35	24	20	38	30
8	32	33	21	42	25
9	29	27	22	25	33
10	33	31	23	36	31
11	46	23	24	49	23
12	43	27	25	42	27
13	55	33			

$n = 25$ $\Sigma XY = 26,267$
$\Sigma X = 962$ $\Sigma Y = 692$
$\Sigma X^2 = 38,536$ $\Sigma Y^2 = 19,622$

   a. Plot the data in a scatterplot.

   b. Determine the regression equation for predicting marital satisfaction scores (Y) from communication scores (X).

   c. Draw the regression line on the scatterplot.

d. For a client with a communication score of 43, predict the marital satisfaction score.

e. Compute the standard error of estimate using formulas 6.11 and 6.12.

f. If clients have communication scores of 28, what percentage will have marital satisfaction scores greater than 25?

g. Develop CI_{95} for the predicted marital satisfaction score ($\hat{Y}$) given a communication score (X) of 33.

h. Compute the correlation between the communication scores and the marital satisfaction scores. For these data, test the null hypothesis $H_0: \rho = 0$ using formula 10.7.

i. Test the null hypothesis $H_0: \beta = 0$ using formula 17.6. Compare the results with part h of this exercise.

*2. Consider the data in Exercise 1 in Chapter 6, in which athletic trainers were using a short test of physical conditioning (X) to predict scores on a longer, more comprehensive test (Y). The regression equation was developed and equals:

$$Y = .635X + 1.222$$

The standard error of estimate, $s_{Y.X}$, was also computed and equals:

$$s_{Y.X} = .56$$

The summary data for these data were as follows:

$n = 24$	$\Sigma XY = 591.50$
$\Sigma X = 118$	$\Sigma Y = 104.3$
$X = 4.92$	$Y = 4.35$
$\Sigma X^2 = 704$	$\Sigma Y^2 = 510.01$
$SS_X = 123.83$	$SS_Y = 56.74$

a. If young adult males scored 5 on the short test, what is the probability that they would have a score of 5 or greater on the long test?

b. Develop the 95 percent confidence interval (CI_{95}) for the long test score for an individual has a score of 4 on the short test.

c. Test the null hypothesis that, in the population, the regression coefficient is zero, i.e., $H_0: \beta = 0$. Use $\alpha = .05$.

3. A sociologist is investigating the relationship between level of education (X) and income (Y). Data were collected from a randomly selected sample of 120 male shoppers at a local shopping mall. Each shopper was asked a series of questions, including number of years of formal education and unadjusted yearly income (recorded in thousands of dollars). Some initial data analyses were completed; these summary data were as follows:

$n = 120$	$r = .67$
$\Sigma X = 1,782$	$\Sigma Y = 1,854$
$s_X = 3.6$	$s_Y = 4.2$

a. Determine the regression equation for predicting income (Y) from level of education (X).

b. Using this regression equation, predict the income for a person with 13.5 years of formal education.

c. Compute the standard error of estimate using formulas 6.11 and 6.12.

d. For those with 15 years of formal education, what is the percentage that have an annual income greater than 18.5?

e. Develop CI_{95} for predicted annual income ($\hat{Y}$) given that the shopper indicates that he has had 16 years of formal education (X).

f. For the correlation between number of years of formal education and annual income, test the null hypothesis H_0: $\rho = 0$ using the formula 10.7.

g. Test the null hypothesis H_0: $\beta = 0$ using formula 17.6. Compare the results with part f of this exercise.

4. Consider the data in Exercise 4 of Chapter 6, in which an industrial psychologist is investigating the relationship between age (X) and efficiency rating (Y) for 20 employees of a small manufacturing company. The summary data for this example were:

$$n = 20 \qquad r = .64$$
$$\Sigma X = 796 \qquad \Sigma Y = 1{,}266$$
$$s_X = 9 \qquad s_Y = 12$$

a. If employees are 45 years old, what percentage have an efficiency rating between 58 and 65?

b. Develop CI_{95} for predicted efficiency rating ($\hat{Y}$) for an employee 35 years old (X).

c. For the correlation between age and efficiency rating, test the null hypothesis H_0: $\rho = 0$ using formula 12.7.

d. Test the null hypothesis H_0: $\beta = 0$ using formula 17.6. Compare the results with part c of this exercise.

5. Consider the data in Exercise 5 of Chapter 6, in which a researcher is investigating the relationship between the comprehensiveness (X) of a college or university and the scholarly productivity of its psychology department (Y). The summary data for this example were:

$$n = 18 \qquad \Sigma XY = 254{,}402$$
$$\Sigma X = 3{,}202 \qquad \Sigma Y = 1{,}406$$
$$\Sigma X^2 = 635{,}616 \qquad \Sigma Y^2 = 111{,}370$$

a. If institutions have comprehensiveness ratings of 170, what percentage have scholarly productivity ratings between 80 and 85?

b. Develop CI_{95} for the predicted scholarly productivity rating ($\hat{Y}$) given a comprehensiveness rating of 195 (X).

    c. Compute the correlation between comprehensiveness and scholarly productivity. For these data, test the null hypothesis H_0: $\rho = 0$ using formula 10.7.

    d. Test the null hypothesis H_0: $\beta = 0$ using formula 17.6. Compare the results with part c of this exercise.

6. Consider the data in Exercise 7 of Chapter 6, in which the dean is investigating the relationship between number of hours worked per week (X) and semester GPA (Y) for 25 students. The summary data for this example were:

$n = 25$ $\Sigma XY = 1,290.7$

$\Sigma X = 482$ $\Sigma Y = 69.2$

$\Sigma X^2 = 10,276$ $\Sigma Y^2 = 196.22$

    a. For students who worked 20 hours per week, what percentage have a GPA between 2.5 and 2.8?

    b. Develop CI_{95} for a predicted GPA ($\hat{Y}$) given that the student worked 23 hours per week.

    c. Compute the correlation between hours worked per week (X) and GPA (Y). For these data, test the null hypothesis H_0: $\rho = 0$ using formula 10.7.

    d. Test the null hypothesis H_0: $\beta = 0$ using formula 17.6. Compare the results with part c of this exercise.

18

Multiple Linear Regression

I n Chapter 5, we introduced the concept of the relationship between two variables, which is called *bivariate correlation*. In Chapters 6 and 17, we discussed bivariate linear regression using a predictor variable (X) and a criterion variable (Y). These concepts of bivariate correlation and linear regression can be extended to multiple correlation and multiple linear regression.[1] In **multiple correlation**, the relationship between the criterion variable (Y) and multiple predictor $(k \geq 2)$ variables $(X_1, X_2, \ldots, X_k)$ is determined, whereas in **multiple linear regression**, scores on the criterion variable (Y) are predicted using multiple predictor variables $(X_1, X_2, \ldots, X_k)$.

Concepts of Multiple Regression

In bivariate linear regression, a straight line was fitted to the scatterplot of points; the equation of this line, called the regression equation, has the form $\hat{Y} = bX + a$,

---

[1] The purpose of this chapter is to introduce the basic concepts of multiple regression analysis. For more extensive coverage of this topic, see E. J. Pedhazur.

where $\hat{Y}$ is the predicted value of the criterion variable for a given X value on the predictor variable. The regression coefficient (b) is the slope of the line, and the regression constant (a) is the Y intercept. In multiple linear regression, we again have a single criterion variable (Y), but we have k predictor variables ($k \geq 2$). These predictor variables are combined into an equation, called the *multiple regression equation*, which can be used to predict scores on the criterion variable ($\hat{Y}$) from scores on the predictor variables (X_i's). The general form of this equation is

$$\hat{Y} = b_1 X_1 + b_2 X_2 + \cdots + b_k X_k + a \qquad (18.1)$$

where the b's are the regression coefficients for the respective predictor variables and a is the regression constant.

> In multiple linear regression, we have a single criterion variable (Y) and multiple predictor variables (X_i). The multiple regression equation contains a regression coefficient (b_i) for each predictor variable and the regression constant (a).

The Geometry of Multiple Linear Regression

When developing the bivariate regression equation, we fitted the line ($\hat{Y} = bX + a$) to the scatterplot of points. The regression equation was determined using the **least-squares criterion**, which requires that $\Sigma(Y - \hat{Y})^2$ be minimized—that is, that the sum of the squared differences between the actual Y scores and the predicted Y scores be a minimum. It is relatively easy to visualize this regression equation fitted to a scatterplot in two dimensions. However, the geometry of multiple regression is more difficult to visualize. Consider the situation in which we have one criterion variable (Y) and two predictor variables (X_1 and X_2). This regression equation would have the form:

$$\hat{Y} = b_1 X_1 + b_2 X_2 + a \qquad (18.2)$$

In this situation, since there are three variables, each individual would have three scores, and these three scores could be represented by a point in three dimensions (see Figure 18.1).

The next step in developing the geometry of multiple linear regression would be to fit a plane to this mass of points in the three-dimensional space. This process is similar to fitting the regression line to the scatterplot in two-dimensional space using the least-squares criterion, that is, where $\Sigma(Y - \hat{Y})^2$ is a minimum. For the two predictor variable case, the predicted value ($\hat{Y}$) is computed using the regression equation in formula 18.2. The regression coefficient b_1 is the slope of this plane as it intercepts the Y, X_1 plane; the regression coefficient b_2 is the slope

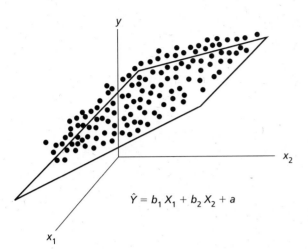

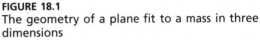

FIGURE 18.1
The geometry of a plane fit to a mass in three dimensions

of this plane as it intercepts the Y, X_2 plane. The regression constant (a) is the point where the regression plane intercepts the Y axis. The geometry of the two predictor variable case is illustrated in Figure 18.1.

Since we live in a three-dimensional world, we can extend the reasoning, but we cannot visualize the geometry, when there are more than two predictor variables. If we have three predictor variables (X_1, X_2, and X_3), we would fit a three-dimensional solid, called a **hyperplane**, to the mass of points in the four-dimensional space. Thus, for the case of k predictor variables, we would fit a k-dimensional hyperplane to the mass of points in $k + 1$-dimensional space. Regardless of the number of predictor variables, the least-squares criterion is used in the process of fitting the hyperplane to the mass of points.

In multiple regression with k predictor variables, a k-dimensional hyperplane is fitted to a mass of points in a $k + 1$-dimensional space using the least-squares criterion.

Multiple Regression in Standard Score Form

When using multiple regression in applied situations, the commonly used form of the regression equation is that in formula 18.1. This form is called the *raw score form*, because the actual scores of the predictor variables are used. However, in order to better conceptualize multiple regression, it might be helpful to consider the equation in *standard score form*. Suppose the scores on all the variables, the k predictors and the criterion, are transformed to standard scores (z scores) using the methods described in Chapter 3; that is, each variable is transformed into a standardized variable with mean equal to 0 and standard deviation (and variance) equal to 1. Geometrically, this changes the measurement scale of the axes to a standard scale and shifts the hyperplane so that it will go through the origin of the system (the point with coordinates $[0, 0, \ldots, 0]$). Using conventional notation, the regression equation becomes

$$z_{\hat{Y}} = \beta_1 z_1 + \beta_2 z_2 + \cdots + \beta_k z_k \tag{18.3}$$

The regression coefficients are now symbolized by the Greek letter beta (β) and are called **beta coefficients**. The regression constant has disappeared, because the intercept is zero.

Whether we use the regression equation in raw score or standard score form, the regression coefficients (b or β) and the regression constant (a, when using raw scores) are uniquely determined. The hyperplane is fitted by generating and solving a set of simultaneous equations in such a way that the least-squares criterion is met. The computational procedures for generating and solving these simultaneous equations are complex and beyond the scope of this book.[2] For most applied situations, we use the computer to solve these equations and determine the coefficients.

For the two-predictor variable case, the beta (β) coefficients can be found using the following formulas:

$$\beta_1 = \frac{r_{Y1} - (r_{Y2})(r_{12})}{1 - r_{12}^2}$$

$$\tag{18.4}$$

$$\beta_2 = \frac{r_{Y2} - (r_{Y1})(r_{12})}{1 - r_{12}^2}$$

where

r_{Y1} = correlation between the criterion variable (Y) and the first predictor variable (X_1)

r_{Y2} = correlation between the criterion variable (Y) and the second predictor variable (X_2)

r_{12} = the correlation between the two predictor variables (X_1 and X_2)

Multiple Regression in Raw Score Form

After solving for the beta coefficients, we can compute the b coefficients using the following:

$$b_1 = \beta_1 \left(\frac{s_Y}{s_1}\right)$$

$$\tag{18.5}$$

$$b_2 = \beta_2 \left(\frac{s_Y}{s_2}\right)$$

[2] See Pedhazur, pp. 51–55.

That is, the b coefficient for a specific predictor variable equals the β coefficient for that variable times the ratio of the standard deviation of the criterion variable (Y) to the standard deviation of that variable. The regression constant (a) is given by

$$a = \overline{Y} - \sum_{i=1}^{k} b_i \overline{X}_i \tag{18.6}$$

The raw score regression coefficients (*b*) can be determined from the standard score regression coefficients (β_1) using the following formula:

$$b_i = \beta_i \left(\frac{s_Y}{s_i} \right)$$

Example

To illustrate the strategy for conducting a multiple regression analysis, we propose the following four-step process.

Step 1. Determine the regression model. This step involves determining the regression coefficients and the regression constant.

Step 2. Determine the multiple correlation coefficient (R) and the proportion of shared variance (R^2). This step involves computing R and R^2, the *coefficient of determination.*

Step 3. Determine whether the multiple R is statistically significant. This step involves testing the null hypothesis H_0: $R_{\text{pop}} = 0$.

Step 4. Determine the significance of the predictor variables. This step involves testing the individual regression coefficients for statistical significance—in other words, testing the null hypothesis that the regression coefficients in the population equal zero (H_0: $\beta_i = 0$).

Suppose a researcher is trying to predict a posttest score (Y) from a pretest score (X_1) and the average number of correct responses on six training quizzes (X_2); the data for 15 individuals are given in Table 18.1. (We shall carry computations in this chapter four figures to the right of the decimal point because rounding errors can markedly affect the results in multiple regression.)

TABLE 18.1
Data for Two Predictor Example

	X_1	X_2	Y
	15	7.70	36
	22	8.20	39
	16	7.80	35
	19	9.30	43
	22	8.20	40
	20	8.80	42
	28	12.10	49
	14	8.00	38
	18	8.10	36
	21	11.20	44
	26	9.40	35
	14	10.30	43
	19	8.50	37
	22	7.60	41
	20	8.40	40
		$r_{Y1} = .3919$	
		$r_{Y2} = .7695$	
		$r_{12} = .4536$	
$\overline{X}$	19.7333	8.9067	39.8667
SS	230.9296	25.0896	215.7337
s	4.0614	1.3387	3.9255

Step 1: Determine the Regression Model Using formula 18.4, the β coefficients are

$$\beta_1 = \frac{.3919 - (.7695)(.4536)}{1 - .4536^2}$$

$$= 0.0540$$

$$\beta_2 = \frac{.7695 - (.3919)(.4536)}{1 - .4536^2}$$

$$= 0.7450$$

Thus, the regression equation in standard score form is

$$z_{\hat{Y}} = 0.0540z_1 + 0.7540z_2$$

Now, using formulas 18.5 and 18.6, the raw score regression coefficients and the regression constant are

$$b_1 = (0.0540)\frac{3.9255}{4.0614}$$

$$= 0.0521$$

$$b_2 = (0.7450)\frac{3.9255}{1.3387}$$

$$= 2.1846$$

$$a = 39.8667 - (0.0521)(19.7333) - (2.1846)(8.9067)$$

$$= 19.3810$$

Thus, the regression equation in raw score form is

$$\hat{Y} = 0.0521X_1 + 2.1846X_2 + 19.3810$$

Step 2: Determine R and R^2   The multiple correlation coefficient (R) is a Pearson product-moment correlation coefficient between the criterion variable (Y) and the predicted score on the criterion variable $(\hat{Y})$, which is a **linear combination** of the predictor variables (that is, $\hat{Y} = b_1X_1 + b_2X_2 + \cdots + b_kX_k + a$). For this example, the predicted scores for the 15 individuals have been computed using the above multiple regression equation and are found in Table 18.2. Using the raw score formula for the correlation coefficient (formula 5.3), we find the correlation between Y and $\hat{Y}$ to be .7710.

The multiple correlation coefficient (R) generally assumes the characteristics of a Pearson product-moment correlation coefficient, except that it takes on only positive values, from 0.0 to 1.0, inclusive. It can be expressed as the average cross-product of the two variables in standardized form divided by the product of their

TABLE 18.2

Observed (Y) and Predicted ($\hat{Y}$) Scores on the Criterion Variable for the Data of Table 18.1

Y	$\hat{Y}$	$Y - \hat{Y}$	$(Y - \hat{Y})^2$
36	36.9839	−.9839	.9681
39	38.4409	.5591	.3126
35	37.2545	−2.2545	5.0828
43	40.6877	2.3123	5.3467
40	38.4409	1.5591	2.4308
42	39.6475	2.3525	5.5343
49	47.2735	1.7265	2.9808
38	37.5872	.4128	.1704
36	38.0141	−2.0141	4.0566
44	44.9426	−.9426	.8885
35	41.2708	−6.2708	39.3229
43	42.6118	.3882	.1507
37	38.9400	−1.9400	3.7636
41	37.1302	3.8698	14.9754
40	38.7736	1.2264	1.5041

standard deviations; symbolically,

$$R_{Y \cdot 12 \cdots k} = \frac{\Sigma z_Y z_{\hat{y}}}{n s_{z_Y} \cdot s_{z_{\hat{y}}}} \tag{18.7}$$

By substituting into this expression for z_Y and $z_{\hat{y}}$ and simplifying[3] we get

$$R_{Y \cdot 12 \cdots k} = \sqrt{\beta_1 r_{Y1} + \beta_2 r_{Y2} + \cdots + \beta_k r_{Yk}} \tag{18.8}$$

That is, the multiple R equals the square root of the sum of the products of the beta coefficients multiplied by the correlations between the criterion variable and the respective predictor variable. For the data in Table 18.1,

$$R_{Y \cdot 12} = \sqrt{(0.0540)(0.3919) + (0.7450)(0.7695)}$$

$$= .7710$$

Thus, the correlation (R) between the criterion variable (Y) and the linear combination of the predictor variables (X_i) is .7710.

The square of the multiple correlation coefficient (R^2) is interpreted in the same way as the square of the bivariate correlation coefficient (r^2). That is, R^2 is the proportion of the variation in the criterion variable that can be attributed to the variation of the combined predictor variables. In this case, $.7710^2 = .5944$. Therefore, approximately 59 percent of the variation in the posttest scores can be attributed to the variation in the combination of the pretest scores and the scores on the training quizzes.

> Multiple R is the correlation coefficient between the scores on the criterion variable (Y) and the predicted scores for the criterion variable ($\hat{Y}$) using the linear combination of the predictor variables.

Step 3: Determine Whether the Multiple R is Statistically Significant   In Chapter 10, we tested the null hypothesis H_0: $\rho = 0$ for the bivariate case; that is, we tested that the correlation in the population was equal to zero. For multiple correlation, we can test the null hypothesis that the multiple correlation in the population equals zero.

$$H_0: R_{\text{pop}} = 0$$

The test statistic for this hypothesis is as follows:

$$F = \frac{R^2 / k}{(1 - R^2)/(n - k - 1)} \tag{18.9}$$

[3] $s_{z_{\hat{y}}}$ is the standard deviation of the distribution of standardized predicted scores. Although this distribution has a mean of zero, the standard deviation is not 1.0 unless $R = 0$.

where

k = number of predictor variables

The underlying distribution of this test statistic is the F distribution with k and $n - k - 1$ degrees of freedom. If the computed value of this test statistic exceeds the critical value of F for a given level of significance, the null hypothesis is rejected, and we can conclude that there is a nonzero relationship in the population between the criterion variable and the linear combination of the predictor variables.

For our example, the test statistic would be

$$F = \frac{(.7710)^2/2}{[1 - (.7710)^2]/(15 - 2 - 1)} = 8.7929$$

The critical value of F for 2 and 12 degrees of freedom at the .05 level of significance is 3.89. Thus, since the observed value of F exceeds the critical value, the null hypothesis is rejected. The associated probability statement would be: the probability that $R = .7710$ would have occurred by chance, if the null hypothesis were true, is less than .05. We would conclude that there is a nonzero relationship in the population between the criterion variable and the linear combination of the predictor variables.

Step 4: Determine the Significance of the Predictor Variables In Chapter 17, the regression coefficient (b) was tested for statistical significance (that is, testing the null hypothesis H_0: $\beta = 0$) using the following formula (see formula 17.5):

$$t = \frac{b - \beta}{s_b}$$

The same procedure can be applied to the regression coefficients in multiple regression; however, the procedure is more complicated. For testing the null hypothesis H_0: $\beta_i = 0$ for each of the regression coefficients,[4] the formula for the test statistic is

$$t = \frac{b_i}{s_{b_i}} \tag{18.10}$$

where

b_i = regression coefficient

s_{b_i} = standard error of the respective coefficient

The sampling distribution of this test statistic is the t distribution with $n - k - 1$ degrees of freedom.

In order to test the null hypotheses for the individual regression coefficients (b_i), we need to define and compute the standard error of estimate for multiple

---

[4]In multiple regression analysis, the beta coefficients (β_i) are the regression coefficients in standard score form. We also use the symbol β_i in the statement of the null hypothesis regarding the population regression coefficients. It is important to differentiate between the two uses of this symbol in multiple regression analysis.

regression. In Chapter 17, we defined the standard error of estimate as the standard deviation of the errors of prediction.

$$s_{Y \cdot X} = \sqrt{\frac{\Sigma(Y - \hat{Y})^2}{n - 2}} = \sqrt{\frac{SS_Y(1 - r^2)}{n - 2}}$$

where SS_Y = sum of squares for the criterion variable, or

$$\Sigma(Y_i - \overline{Y})^2 = (n - 1)s_Y^2$$

For multiple regression, the definition of the standard error of estimate is the logical extension of these formulas.

$$s_{Y \cdot 12 \cdots k} = \sqrt{\frac{\Sigma(Y - \hat{Y})^2}{n - k - 1}} = \sqrt{\frac{SS_Y(1 - R^2)}{n - k - 1}} \qquad (18.11)$$

For our example,

$$s_{Y \cdot 12} = \sqrt{(215.7337)[1 - (.7710)^2]/(15 - 2 - 1)}$$
$$= 2.7001$$

Now consider the formula for the standard error of the first regression coefficient (b_1)

$$s_{b_1} = \sqrt{\frac{s_{Y \cdot 12 \cdots k}^2}{SS_{X_1}(1 - R_{1 \cdot 23 \cdots k}^2)}} \qquad (18.12)$$

where

$s_{Y \cdot 12 \cdots k}^2$ = square of the standard error of estimate

SS_{X_1} = sum of squares for the first predictor variable, or $\Sigma(X_{i1} - \overline{X}_1)^2 = (n - 1)s_{X_1}^2$

$R_{1 \cdot 23 \cdots k}^2$ = square of the multiple R when X_1 is the criterion variable and X_2 through X_k are the predictor variables

Note that, for the standard error of each of the regression coefficients (b_i), we would use the appropriate SS_{X_i} in formula 18.12. Also, the multiple R would vary for the different standard errors. For example, in the formula for the standard error for b_2, we would use SS_{X_2} and $R_{2 \cdot 13 \cdots k}^2$. In general, the predictor variable associated with b_i serves as the criterion variable, with the remaining $k-1$ predictor variables serving as the predictor variables for this multiple R. The original criterion variable *does not* enter into the computation of the R^2 found in the denominator of formula 18.12.

For the two predictor variable case, the formulas for the standard errors of the two regression coefficients are:

$$s_{b_1} = \sqrt{\frac{s_{Y \cdot 12}^2}{SS_{X_1}(1 - r_{12}^2)}}$$

$$s_{b_2} = \sqrt{\frac{s_{Y \cdot 12}^2}{SS_{X_2}(1 - r_{12}^2)}}$$

(18.13)

For our example,

$$s_{b_1} = \sqrt{\frac{2.7001^2}{(230.9296)(1 - .4536^2)}}$$

$$= 0.1994$$

$$s_{b_2} = \sqrt{\frac{2.7001^2}{(25.0896)(1 - .4536^2)}}$$

$$= 0.6049$$

Using formula 18.10 to test the null hypothesis H_0: $\beta_i = 0$,

$$t = \frac{b_i}{s_{b_i}}$$

For the first predictor variable,

$$t = \frac{0.0521}{0.1994}$$

$$= 0.2613$$

For the second predictor variable,

$$t = \frac{2.1846}{0.6049}$$

$$= 3.6115$$

The critical value of t for 12 degrees of freedom, using $\alpha = .05$, is 2.179. Thus, the null hypothesis for the first predictor is not rejected; the null hypothesis for the second predictor is rejected. The conclusion is that the first predictor variable (X_1) does not contribute to the regression when used in combination with the second predictor variable (X_2). However, the second predictor variable (X_2) does contribute to the regression when used in combination with the first predictor variable (X_1). Therefore, we could use the simple bivariate regression equation with only the second predictor variable. It should be noted that if this were done, the new regression equation would need to be computed, since the regression coefficient (b) and the regression constant (a) would change.

Inspecting the data in Table 18.1 gives some insight into why X_1 does not contribute significantly to the regression when used in combination with X_2. Notice that X_2 has a high correlation (.7695) with the criterion variable (Y), whereas X_1 has only a modest correlation (.3919) with Y and a correlation of .4536 with X_2. Thus, most of the variance that X_1 shares with Y is already being accounted for by X_2 due to its high correlation with Y and modest correlation with X_1.

Caution should be exercised when interpreting regression weights. In this example, we used more than one predictor variable because we wanted to determine how well the *set* of predictor variables accounted for the variance in the criterion variable, not how well each predictor variable fared separately in the process. Also, we tested whether the regression coefficients differed from zero, not whether they differed from each other.

Selecting Predictor Variables

The practical problem in multiple linear regression is one of selecting an effective set of predictor variables, which will maximize the multiple R. We have defined R^2 as the proportion of variance in the criterion variable that can be attributed to the variance of the combined predictor variables.

Therefore, we want to identify predictor variables that are highly correlated with the criterion variable. But, since the variance in the criterion can be accounted for only once, we want predictor variables that account for different proportions of the variance in the criterion variable. Thus, the predictor variables should have low correlations among themselves. For example, if we have two predictor variables that are highly correlated with the criterion variable and are also highly correlated with each other, they explain the same variation in the criterion variable, and only one will contribute significantly to the regression when they are used in combination.

Therefore, the goal is to select predictor variables that are highly correlated with the criterion variable but have low correlation among themselves. (Sometimes a variable will increase R^2 even though it has a low correlation with the criterion and a high correlation with the other predictors; such a variable is called a *suppressor* variable. The important goal, though, is to maximize R^2.)

In multiple regression, predictor variables should be highly correlated with the criterion variable and have low correlations among themselves.

Number of Predictor Variables

When conducting research in the behavioral sciences using multiple regression, there may be a tendency to follow the "more-the-merrier" syndrome—that is, the tendency to continue including predictor variables as long as there seems to be an increase in the multiple R. In practice, however, it is simply not feasible to obtain measures on a large number of predictor variables that meet the goal of being highly correlated with the criterion variable and uncorrelated among themselves. Thus, including more than five or six predictor variables rarely produces a substantial increase in the multiple R.

One other point on the number of predictor variables is worth noting. In bivariate correlation with only two observations, the correlation will be either $+1.0$ or -1.0, since a straight line is determined by two points. Correspondingly, in multiple correlation, as the number of predictors approaches the number of observations $(k \rightarrow n)$, the multiple R increases. Thus, for studies with a small n and numerous predictors, the multiple R may be spuriously large simply as a result of the underlying geometry. In fact, if $k + 1 = n$, the multiple R equals 1.0. However, in studies in which the number of observations is large $(n > 100)$ and there are only five or six predictor variables, the researcher need not be concerned with this characteristic.

An adjustment in R^2 can be made when there are numerous predictor variables and a small number of observations.[5] The formula for this adjustment is

$$\text{Adjusted } R^2 = 1 - (1 - R^2)\left(\frac{n-1}{n-k-1}\right) \tag{18.14}$$

where

$R^2 = $ unadjusted R^2

$k = $ number of predictor variables

$n = $ number of observations

This formula provides a more conservative estimate of the percentage of variance in the criterion variable that can be attributed to the combined predictor variables.

The multiple correlation coefficient can be spuriously large if there are a large number of predictor variables and a small number of observations.

---

[5] See Pedhazur, p. 148.

Example: Four Predictor Variables

The mathematics for developing the multiple regression equation with more than two predictor variables requires matrix algebra and is beyond the scope of this book. We have included an example with four predictor variables but will discuss the solution using a printout generated from SPSS. In this example, the researcher has data from 200 individuals on a criterion variable (Y) and four predictor variables (X_1, X_2, X_3, and X_4). Table 18.3 contains the means and standard deviations for the variables, as well as the correlations between them. Table 18.4 contains the pertinent parts of the SPSS printout.[6] Now consider this example in the context of the four steps for multiple regression analysis.

Step 1: Determine the Regression Model

From Table 18.4, we find the raw score regression coefficients (b_i), the regression constant (a), and the standardized regression coefficients (β_i). These data indicate that the raw score regression equation is

$$\hat{Y} = 0.1762X_1 + 0.2320X_2 + 0.8101X_3 + 0.0488X_4 + 4.6418$$

The regression equation in standard score form is

$$z_{\hat{y}} = 0.1270z_1 + 0.1295z_2 + 0.5087z_3 + 0.0363z_4$$

TABLE 18.3
Data for Four Predictor Variable Example ($n = 200$)

	Y	X_1	X_2	X_3	X_4
Means	30	20	22	18	44
Standard deviations	8.6	6.2	4.8	5.4	6.4
Y	1.00	0.42	0.37	0.65	0.45
X_1	0.42	1.00	0.20	0.50	0.35
X_2	0.37	0.20	1.00	0.41	0.18
X_3	0.65	0.50	0.41	1.00	0.68
X_4	0.45	0.35	0.18	0.68	1.00

[6]The SPSS program for this example is found on page 494. Note that this example was generated using the means and standard deviations for each variable and the matrix of correlations among them. Note also that in the printout for this example, the computations have been carried out to more than four decimal places.

TABLE 18.4
Computer Solution for the Four Predictor Variable Example Using All Four Variables ($n = 200$)

```
Multiple R            .66952
R Square              .44826
Adjusted R Square     .43694
Standard Error       6.45318
```

Analysis of Variance

	DF	Sum of Squares	Mean Square
Regression	4	6597.54013	1649.38503
Residual	195	8120.49987	41.64359

```
F =      39.60718      Signif F =  .0000
```

---------------- Variables in the Equation --------------------

Variable	B	SE B	Beta	T	Sig T
X4	.048799	.098578	.036315	.495	.6211
X2	.232004	.105649	.129491	2.196	.0293
X1	.176225	.085208	.127046	2.068	.0399
X3	.810138	.134780	.508691	6.011	.0000
(Constant)	4.641777	3.957945		1.173	.2423

Step 2: Determine R and R^2

For this example, the multiple correlation coefficient (R) and the coefficient of determination (R^2) are

$$R = .66952$$

$$R^2 = .44826$$

This R^2 indicates that approximately 45 percent of the variance in the criterion variable is attributable to the variance of the combined predictor variables. Notice that the adjusted R^2 (.43694) is only slightly different from the unadjusted R^2. This is due to the relatively large number of observations and the small number of predictor variables.

Step 3: Determine Whether the Multiple R is Statistically Significant

For testing the null hypothesis H_0: $R_{pop} = 0$, we use the test statistic defined in formula 18.9.

$$F = \frac{R^2/k}{(1 - R^2)/(n - k - 1)}$$

SPSS Listing for Four Predictor Variable Example

TITLE CHAPTER 18—FOUR PREDICTOR VARIABLES
INPUT PROGRAM
NUMERIC Y X1 X2 X3 X4
. INPUT MATRIX FREE
END INPUT PROGRAM
REGRESSION READ CORR N MEANS STDDEV/
VARIABLES = Y X1 X2 X3 X4/
DEP=Y/ENTER X1 X2 X3 X4/
DEP=Y/ENTER X1 X2 X3
BEGIN DATA
30 20 22 18 44
8.6 6.2 4.8 5.4 6.4
1.00 0.42 0.37 0.65 0.45
0.42 1.00 0.20 0.50 0.35
0.37 0.20 1.00 0.41 0.18
0.65 0.50 0.41 1.00 0.68
0.45 0.35 0.18 0.68 1.00
200
END DATA
FINISH
//

Using the data from Table 18.4,

$$F = \frac{(.44826)/4}{(1 - .44826)/(200 - 4 - 1)}$$

$$= 39.60718$$

The underlying distribution of this test statistic is the F distribution with 4 and 195 degrees of freedom. Assuming $\alpha = .05$, we find the critical value to be 2.42 (see Table C.5 in the Appendix). Since the computed value of F exceeds the critical value, the null hypothesis H_0: $R_{pop} = 0$ would be rejected. The associated probability statement would be that the probability that $R = .66952$ would have occurred by chance, if the null hypothesis were true, is less than .05. Notice on the printout that "Signif F = .0000" indicates that the probability is actually less than .0001. Thus, we conclude that, in the population, the correlation between the criterion variable (Y) and the combined predictor variables (X_1, X_2, X_3, and X_4) is different from zero.

Step 4: Determine the Significance of the Predictor Variables

The procedure for testing H_0: $\beta_i = 0$ for each regression coefficient was discussed in the two predictor variable examples. The test statistic is

$$t = \frac{b_i}{s_{b_i}}$$

Consider the test of significance for the first predictor variable (X_1). The standard error for b_1 would be

$$s_{b_1} = \sqrt{\frac{s_{\hat{Y} \cdot 1234}^2}{SS_{X_1}(1 - R_{1 \cdot 234}^2)}}$$

where

$s_{\hat{Y} \cdot 1234}^2$ = square of the standard error of estimate

$\quad\quad = \dfrac{SS_Y(1 - R^2)}{n - k - 1}$ (see formula 18.11)

SS_{X_1} = sum of squares for the first predictor variable, or $\Sigma(X_{i1} - \overline{X}_1)^2$
$\quad\quad = (n - 1)s_{X_1}^2$

$R_{1 \cdot 234}^2$ = square of the multiple R when X_1 is the criterion variable and X_2 through X_4 are the predictor variables

For the data in Table 18.4,

$$t = \frac{0.176225}{0.085208} = 2.068$$

The critical value of this test statistic is found in Table C.3. Using $\alpha = .05$, the critical value of t for 195 degrees of freedom is 1.96. Since the computed value $(t = 2.068)$ exceeds this critical value, the null hypothesis would be rejected. The associated probability statement would be that the probability that $b_1 = 0.176225$ would have occurred by chance, if the null hypothesis were true, is less than .05. Notice on the printout that the "Sig T" of 0.0399 indicates that the probability is actually .0399. Thus, we conclude that, in the population, the first predictor variable is a significant contributor to the regression when used in combination with the other three predictor variables. Similarly, the t values for the other three predictor variables indicate that variables X_2 and X_3 are significant contributors to the regression, but that X_4 is not. Thus, X_4 can be dropped from the equation.

If X_4 is dropped from the equation, the entire solution must be redone using only the three remaining predictor variables; this solution is found in Table 18.5. For the remaining three predictor variables, the regression equation is

$$\hat{Y} = 0.1769X_1 + 0.2243X_2 + 0.8519X_3 + 6.1941$$

Testing the Difference Between Two Multiple R's

An alternative procedure for deciding on the number of predictor variables to retain in the regression equation is to test the difference between the multiple R with k_1 predictors and the multiple R with k_2 predictors where the k_2 predictors are a subset

TABLE 18.5
Computer Solution for the Four Predictor Variable Example Using Only X_1, X_2, and X_3 ($n = 200$)

```
Multiple R           .66901
R Square             .44757
Adjusted R Square    .43911
Standard Error      6.44074
```

Analysis of Variance

	DF	Sum of Squares	Mean Square
Regression	3	6587.33516	2195.77839
Residual	196	8130.70484	41.48319

F =   52.93177   Signif F =   .0000

---------------- Variables in the Equation --------------------

Variable	B	SE B	Beta	T	Sig T
X3	.851918	.104880	.534925	8.123	.0000
X2	.224281	.104290	.125180	2.151	.0327
X1	.176857	.085035	.127501	2.080	.0388
(Constant)	6.194149	2.410169		2.570	.0109

of the k_1 predictors.[7] The test statistic for this procedure is

$$F = \frac{(R_1^2 - R_2^2)/(k_1 - k_2)}{(1 - R_1^2)/(n - k_1 - 1)} \tag{18.15}$$

where

R_1^2 = multiple R for k_1 predictor variables

R_2^2 = multiple R for k_2 predictor variables

k_1 = number of predictor variables in R_1^2

k_2 = number of predictor variables in R_2^2

The underlying distribution of this test statistic is the F distribution with $k_1 - k_2$ and $n - k_1 - 1$ degrees of freedom.

In the preceding example, we could have tested the difference between the multiple R for the regression model with all four predictor variables and the multiple R for the regression model with the fourth predictor variable excluded. Using

---

[7]For a more general use of this F test, see Pedhazur, pp. 62–63.

formula 18.15 and the data in Tables 18.4 and 18.5

$$F = \frac{(.4483 - .4476)/(4 - 3)}{(1 - .4483)/(200 - 4 - 1)}$$

$$= \frac{0.0007/1}{0.5517/195}$$

$$= 0.2476$$

The critical value of this test statistic is found in Table C.5. Assuming that $\alpha = .05$, the critical value of F for 1 and 195 degrees of freedom is 3.84. Since the computed value ($F = 0.2476$) does not exceed this critical value, the null hypothesis is not rejected. Thus, we can conclude that the three predictor variables (X_1, X_2, and X_3) are as effective as all four predictor variables and that it is unnecessary to include X_4 as a predictor variable.

Computer Exercise: The Survey of High School Seniors Data Set

For the Survey of High School Seniors ($n = 500$), we will complete a two-step multiple regression analysis. In the first step, the mathematics achievement (MATH-ACH) is the criterion variable and the MOSAIC and VISUAL scores are the multiple predictor variables. In the second step, the variables father's education (FAED) and mother's education (MAED) (as originally coded in the date set — see Appendix B) are included as predictor variables. The means and standard deviation for all five variables as well as the correlation matrix are found in Table 18.6. The MULTIPLE REGRESSION procedure in SPSS was used to generate the computer printouts for the two steps; they are found in Table 18.7.

Step 1: Determine the Regression Model

In the first step of the proposed analysis, only the variables, MOSAIC and VISUAL, are entered into the regression mode. From Table 18.7, we find the regression coefficients (b_i), the regression constant (a), and the standardized regression coefficients (β) in the section "Variables in the Equation." For the first step, the raw score regression equation is:

$$\hat{Y} = .129528X_1 + .657378X_2 + 5.870055$$

The standard score regression equation is:

$$z_{\hat{Y}} = .182159z_1 + .386841z_2$$

Now consider the regression equations when all four predictor variables are entered into the regression model; the equations are

$$\hat{Y} = .115740X_1 + .620031X_2 + .481283X_3 + .084642X_4 + 5.870055$$

$$z_{\hat{Y}} = .162768z_1 + .364863z_2 + .186708z_3 + .029358z_4$$

TABLE 18.6
Summary Data for Multiple Regression Exercise for the Survey of High School Seniors Data Set

```
* * * *   M U L T I P L E   R E G R E S S I O N   * * * *


Listwise Deletion of Missing Data

          Mean  Std Dev  Label

MATHACH  13.098   6.605
MOSAIC   26.889   9.288
VISUAL    5.697   3.887
FAED      4.786   2.699  Father's education
MAED      4.202   2.291  Mother's education


N of Cases =   500


Correlation, 1-tailed Sig:

              MATHACH     MOSAIC     VISUAL       FAED       MAED

MATHACH       1.000        .292       .438       .275       .208
                .          .000       .000       .000       .000


MOSAIC         .292       1.000       .283       .111       .131
               .000         .         .000       .007       .002


VISUAL         .438        .283      1.000       .119       .135
               .000        .000         .        .004       .001


FAED           .275        .111       .119      1.000       .551
               .000        .007       .004         .        .000


MAED           .208        .131       .135       .551      1.000
               .000        .002       .001       .000         .
```

Step 2: Determine R and R²

When the variables, MOSAIC and VISUAL, are entered into the regression model, the multiple correlation coefficient (R) and the coefficient of determination (R^2) are:

$$R = .47197$$
$$R^2 = .22276$$

The R^2 indicates that over 22 percent of the variance in the MATHACH scores can be attributed to the variance of the combined predictor variables, MOSAIC and

TABLE 18.7
SPSS Printout for Multiple Regression Exercise for the Survey of High School Seniors Data Set

```
* * * *   M U L T I P L E   R E G R E S S I O N   * * * *

Equation Number 1    Dependent Variable..    MATHACH


Variable(s) Entered on Step Number
    1..     VISUAL
    2..     MOSAIC


Multiple R             .47197
R Square               .22276
Adjusted R Square      .21963
Standard Error        5.83440

Analysis of Variance
                     DF      Sum of Squares       Mean Square
Regression            2          4848.69680       2424.34840
Residual            497         16917.98557         34.04021

F =      71.22013       Signif F =   .0000


----------------- Variables in the Equation -------------------

Variable           B        SE B       Beta         T    Sig T
MOSAIC        .129528     .029321     .182159     4.418    .0000
VISUAL        .657378     .070074     .386841     9.381    .0000
(Constant)   5.870055     .818968                 7.168    .0000
```

VISUAL. Note that the adjusted R^2 is only slightly different from the unadjusted R^2 due to the larger number of observations and the small number of predictor variables.

For the second step in the analysis, with the variables FAED and MAED added to the regression model, the R and R^2 are:

$$R = .51728$$
$$R^2 = .26758$$

Note that with including the two additional predictor variables, the R^2 went from .22276 to .26758. This R^2 increase ($.265789 - .22276 = .04482$) indicates that the inclusion of both FAED and MAED accounts for an additional 4.5 percent of the variance in the MATHACH scores.

TABLE 18.7
Continued

```
          * * * *   M U L T I P L E   R E G R E S S I O N   * * * *

Equation Number 1    Dependent Variable..    MATHACH


Variable(s) Entered on Step Number
    1..    MAED      Mother's education
    2..    MOSAIC
    3..    VISUAL
    4..    FAED      Father's education


Multiple R          .51728
R Square            .26758
Adjusted R Square   .26166
Standard Error      5.67509

Analysis of Variance
                   DF       Sum of Squares       Mean Square
Regression          4          5824.38305        1456.09576
Residual          495         15942.29933          32.20667

F =      45.21101      Signif F =   .0000


------------------ Variables in the Equation --------------------

Variable            B         SE B       Beta        T   Sig T

MOSAIC         .115740     .028673    .162768     4.036  .0001
VISUAL         .620031     .068590    .364863     9.040  .0000
FAED           .481243     .113046    .196708     4.257  .0000
MAED           .084642     .133686    .029358      .633  .5269
(Constant)    3.794493     .894480               4.242  .0000
```

Step 3: Determine Whether the Multiple R is Statistically Significant

In the first step, we will test the multiple R for statistical significance when the variables, MOSAIC and VISUAL, are included in the model. In the second step, we will test the difference between the multiple R for the two variable model (MOSAIC and VISUAL) and the multiple R for the four variable model (MOSAIC, VISUAL, FAED, MAED). For the first test, the null hypothesis is H_0: $R_{pop} = 0$;

the test statistic was defined by formula 18.9. Using data in Table 18.7, the F value found in the "Analysis of Variance" section of the computer printout is computed as follows:

$$F = \frac{R^2/k}{(1 - R^2)/(N - k - 1)}$$

$$= \frac{.222762/2}{(1 - .22276)/(500 - 2 - 1)}$$

$$= 71.22013$$

The sampling distribution for the test statistic is the F distribution with k and $N - k - 1$ degrees of freedom. For this first step, $k = 2$ and $N - k - 1 = 500 - 2 - 1 = 497$. Thus the appropriate F distribution would be for 2 and 497 degrees of freedom. With $\alpha = .05$, the critical value of F from Table C.5 in the Appendix is $F_{cv} = 3.00$. Since the computed value (71.22013) exceeds the critical value (3.00), the null hypothesis, H_0: $R_{pop} = 0$, is rejected. The associated probability statement is: the probability that the observed $R = .47197$ would have occurred by chance, if in fact the null hypothesis is true, is less than .05. Note the "Signif of F" value in the printout is .0000. As before, this implies that the actual probability is less than .0001.

In the second step of the analysis, the question being asked is whether the addition of both FAED and MAED results in a statistically significant increase in the multiple R. The test statistic for testing the difference between two multiple Rs was defined by formula 18.15.

$$F = \frac{(R_1^2 - R_2^2)/(k_1 - k_2)}{(1 - R_1^2)/(N - k_1 - 1)}$$

where R_1^2 is the square of the multiple R for the model with the larger number of predictor variables (k_1) and R_2^2 is the square of the multiple R with the smaller number of predictor variables (k_2). For the data in Table 18.7,

$$F = \frac{(.26758 - .22272)/(4 - 2)}{(1 - .26758)/(500 - 4 - 1)}$$

$$= \frac{(.04482)/(2)}{(.73242)/(495)}$$

$$= \frac{.02241}{.00148}$$

$$= 15.1419$$

The sampling distribution for the test statistic is the F distribution with $k_1 - k_2$ and $N - k_1 - 1$ degrees of freedom. For this first step, $k_1 - k_2 = 4 - 2 = 2$ and $N - k_1 - 1 = 500 - 4 - 1 = 495$. Thus the appropriate F distribution would be for 2 and 495 degrees of freedom. With $\alpha = .05$, the critical value of F from Table C.5 in the Appendix is $F_{cv} = 3.00$. Since the computed value (15.1419) exceeds the critical value (3.00), the conclusion would be that the addition of FAED and MAED resulted in a statistically significant increase in the multiple R.

Step 4: Determine the Significance of the Predictor Variables

For this step, we can assume that either all four predictor variables were entered at the same time or that the first two variables (MOSAIC and VISUAL) were entered and then the next two variables (FAED and MAED) were entered. The question now becomes whether or not all four predictor variables contribute to the regression model. This question involves testing the null hypothesis H_0: $\beta_i = 0$ for each of the predictor variables. The test statistic for this hypothesis is:

$$t = \frac{b_i}{s_{b_i}}$$

For testing the null hypothesis for each predictor variable using this test statistic, the sampling distribution is the t distribution for $N - k_1 - 1$ degrees of freedom. Assuming $\alpha = .05$ and for df $= 500 - 4 - 1 = 495$, the critical value $t_{cv} = \pm 1.96$.

The regression coefficients, respective standard errors, and "Sig T" values are found in the printout.

$t_1 = 4.036$

$t_2 = 9.040$

$t_3 = 4.257$

$t_4 = 0.633$

Note that the t values for MOSAIC, VISUAL, and FAED exceeded the critical values; however, the t value for MAED did not. Thus, the mother's education level was not a significant contributor to the regression mode. Thus, we can conclude that the three variable (MOSAIC, VISUAL, and FAED) regression model is as effective as the four variable model, which included MAED.

Partial Correlation and Part Correlation

Partial correlation and part correlation coefficients are both Pearson correlation coefficients; both are related to multiple regression analysis.[8] A **partial correlation** is the correlation between two variables with the influence of a third variable removed from both. For example, a researcher could investigate the relationship between height and running speed with the influence of age removed. A **part correlation** is the correlation between two variables with the influence of a third variable

[8]For a more extensive discussion of partial and part correlations, see Pedhazur, pp. 97–133.

removed from only one of the variables. The same researcher could investigate the relationship between height and running speed with the influence of age removed from only the height variable.

In Table 18.8, we have data for five subjects on the three variables (X, Y, and Z). Assume that we want to determine the relationship between Y and Z with the effects of variable X partialed out. The process would involve determining the regression equations for predicting Y from X *and* for predicting Z from X. These two regression equations are included in Table 18.8. Next, the differences between the predicted Y scores and the actual Y scores ($e_Y = Y - \hat{Y}$) are determined, as well as the differences between the predicted Z scores and the actual Z scores ($e_Z = Z - \hat{Z}$). The partial correlation between Y and Z with the influence of X removed ($r_{YZ \cdot X}$) is actually the correlation between the e_Y scores and the e_Z scores. Notice that this partial correlation equals $-.0981$.

The computational formula for the partial correlation coefficient is

$$r_{YZ \cdot X} = \frac{r_{YZ} - r_{XY} r_{XZ}}{\sqrt{(1 - r_{XY}^2)(1 - r_{XZ}^2)}} \qquad (18.16)$$

For these data,

$$r_{YZ \cdot X} = \frac{.5055 - (.7838)(.7004)}{\sqrt{(1 - .7838^2)(1 - .7004^2)}}$$

$$= -.0981$$

Thus, removing the effects of X from both Y and Z reduces the correlation between Y and Z from .5055 to near zero.

Now consider the correlation between Y and Z, removing the effects of X from *only* Y; this is the part correlation $r_{Z(Y \cdot X)}$. It is the correlation between Z and e_Y. For these data, the part correlation is $-.0700$. The actual computational

TABLE 18.8
Data for Partial and Part Correlations

X	Y	Z	$\hat{Y}_X$	$e_y = Y - \hat{Y}_X$	$\hat{Z}_X$	$e_z = Z - \hat{Z}_X$
2	12	4	12.2982	−0.2982	5.4385	−1.4385
7	14	10	15.8947	−1.8947	8.8155	1.1845
8	18	8	16.6140	1.3860	9.4909	−1.4909
4	15	9	13.7368	1.2632	6.7893	2.2107
5	14	7	14.4561	−0.4561	7.4647	−0.4647

$r_{XY} = 0.7838$ $r_{XZ} = 0.7004$ $r_{YZ} = 0.5055$
$\hat{Y}_X = 0.7193X + 10.8596$
$\hat{Z}_X = 0.6754X + 4.0877$

formula for the part correlation is

$$r_{Z(Y \cdot X)} = \frac{r_{YZ} - r_{XY}r_{XZ}}{\sqrt{1 - r_{XY}^2}} \tag{18.17}$$

For these data,

$$r_{Z(Y \cdot X)} = \frac{.5055 - (.7838)(.7004)}{\sqrt{(1 - .7838^2)}}$$

$$= -.0700$$

Thus, removing the effects of X from only Y changes the correlation between Y and Z from .5055 to $-.0700$.

Partial and part correlation analyses are techniques by which two variables are correlated, with the influence of a third variable partialed out of both or only one of the variables.

Alternative Procedures for Multiple Regression

The multiple regression analysis we have just illustrated is called the **backward solution**. All predictor variables are initially entered into the regression model (called the *full model*), and then individual predictor variables are deleted if they do not make a significant contribution to the regression. There are two other approaches to selecting predictor variables.[9] The first of these is the **forward solution**, a procedure that involves entering the variables one at a time. The first variable selected for inclusion into the regression model is the predictor variable that has the highest correlation with the criterion variable. The next predictor variable selected is the one with the highest partial correlation with the criterion variable, with the effects of the first variable partialed out. This variable will result in the greatest increase in R^2—that is, the predictor variable that accounts for the greatest amount of the remaining variance in the criterion variable after the effect of the first predictor variable has been removed. The next predictor variable is similarly selected. The forward solution is terminated when the increase in R^2 is no longer statistically significant or all the predictor variables are included, whichever occurs first.

[9]For a more extensive discussion of backward, forward, and stepwise solutions, see Pedhazur, pp. 158–164.

In the backward solution, all predictor variables are initially included in the regression model, and individual predictor variables are deleted if they do not contribute to the regression. In the forward solution, the predictor variables are entered one at a time until the increase in R^2 is no longer statistically significant *or* until all predictor variables have been included in the regression model.

The other approach to multiple regression analysis is the **stepwise solution**, a variation of the forward solution. In the forward solution, when a predictor variable enters the regression model, it remains in the model regardless of whether it continues to contribute to the regression as other predictor variables are entered. In the stepwise solution, predictor variables are selected in a similar way; however, at each step after a new predictor variable is added to the model, a second significance test is conducted to determine the contribution of each of the previously selected predictor variables, as if it were the last variable entered. Therefore, it is possible for a predictor variable to be deleted if it loses its effectiveness as a predictor when considered in combination with newly entered predictors. As with the forward solution, the stepwise solution is terminated when all the predictor variables are entered *or* when the remaining predictor variables do not make a statistically significant contribution to the regression.

The stepwise solution is a variation of the forward solution. Predictor variables are entered one at a time but can be deleted if they do not contribute significantly to the regression when considered in combination with newly entered predictors.

Multiple Regression and Analysis of Variance

In the preceding examples of multiple regression analysis, we have considered only continuous criterion and predictor variables. In ANOVA, the independent variable is categorical with k levels, each representing group membership, with only the dependent (or criterion) variable continuous. It is possible, however, to create predictor variables that will represent the group membership in ANOVA and to use these predictor variables in multiple regression analysis.

In our discussion of multiple regression, we interpreted R^2 (the coefficient of determination) as the proportion of the variation of scores on the criterion variable that can be attributed to the variation of the scores on the linear combination of the predictor variables—in other words, the shared variance between the criterion variable and the combined predictor variables. We also tested the multiple R for statistical significance; that is, we tested H_0: $R_{pop} = 0$. In our discussion of ANOVA, we illustrated how the variation of scores on the criterion variable is partitioned into several sources, between-groups variation and within-groups variation. When we create predictor variables that represent the group membership in ANOVA and use these variables in multiple regression, the resulting R^2 is interpreted in the same way. It is the proportion of the variation of the scores on the criterion variable that can be attributed to the score on the linear combination of the predictor variables. But, in this case, the predictor variables represent the group membership in ANOVA. Testing the resulting multiple R for statistical significance is thus testing the null hypothesis of no differences in the population between the means of the groups.

To show the relationship between multiple regression and ANOVA, consider the data in Table 18.9; the summary ANOVA was generated using the methods outlined in Chapter 14 for testing the null hypothesis H_0: $\mu_1 = \mu_2 = \mu_3$. Since the computed value of F (46.875) exceeds the critical value (3.89, assuming that $\alpha = .05$), the null hypothesis is rejected.

This same result can be obtained with multiple regression analysis, using the predictor variables that were created to represent the group membership in ANOVA. In order to conduct the multiple regression analysis, however, the data must be rearranged and these predictor variables created; these data are found in Table 18.10. Notice that the scores on the criterion variable (Y) are listed in a single column and are not partitioned into groups. The first five scores are those for subjects in group I, the next five are those for subjects in group II, and the last five are those

TABLE 18.9
Data for Showing the Relationship Between ANOVA and Multiple Regression Analysis

I	II	III
18	14	7
19	15	9
20	15	10
21	15	11
22	16	13

Summary ANOVA

Source	SS	df	MS	F	F_{cv}
Between	250	2	125	46.875	3.89
Within	32	12	2.67		
Total	282	14			

for subjects in group III. The columns X_1 and X_2 are the predictor variables[10] indicating group membership. Observations for subjects in group I have a value of 1 on X_1. All other observations on X_1 are 0. That is, a subject in group I is given 1 on X_1. Similarly, observations for subjects in group II have a value of 1 on X_2. All other observations on X_2 are 0. That is, a subject in group II is given 1 on X_2. There is no need for a third predictor variable X_3 because a subject with 0 on both X_1 and X_2 must be in group III. (This is related to the concept of degrees of freedom in ANOVA; that is, there are as many predictor variables created as there are degrees of freedom between groups [$K - 1$].)

Using the data in Table 18.10, the multiple R can be computed and tested for statistical significance. Using formula 18.4, the β_i are

$$\beta_1 = \frac{.8154 - (.0000)(-.5000)}{1 - (-.50000)^2}$$

$$= 1.0872$$

$$\beta_2 = \frac{.0000 - (.8154)(-.5000)}{1 - (-.50000)^2}$$

$$= 0.5436$$

TABLE 18.10
ANOVA Data for Using Multiple Regression Analysis

X_1	X_2	Y
1	0	18
1	0	19
1	0	20
1	0	21
1	0	22
0	1	14
0	1	15
0	1	15
0	1	15
0	1	16
0	0	7
0	0	9
0	0	10
0	0	11
0	0	13

$r_{Y1} = .8154$
$r_{Y2} = .0000$
$r_{12} = -.5000$

[10]In multiple regression analysis, these variables are called *dummy variables*. See Pedhazur, pp. 274–289.

Using formula 18.8, the multiple R and R^2 are

$$R = \sqrt{(1.0872)(0.8154) + (0.5436)(0.0000)}$$

$$= .9415$$

$$R^2 = .8865$$

Testing this multiple R for statistical significance using formula 18.9,

$$F = \frac{.8865/2}{(1 - .8865)/12}$$

$$= 46.8636$$

Notice that this value is the same (within rounding error) as the F found in the ANOVA. So what is the relationship between multiple regression analysis and ANOVA? In multiple regression, R^2 is interpreted as the proportion of variation in the criterion variable that can be explained by the linear combination of the predictor variables. In ANOVA, the sum of squares between groups is the variation attributable to the differences between the groups. Thus,

$$R^2 \times SS_T = \text{explained sum of squares}$$

$$(.8865)(282) = 250$$

$$= SS_B$$

Correspondingly, in multiple regression, $1 - R^2$ is the proportion of unexplained or error variation, whereas SS_W is the random variation within the groups not accounted for by the differences between the groups. Thus,

$$(1 - R^2) \times SS_T = \text{unexplained (error) sum of squares}$$

$$(0.1135) \times (282) = 32$$

$$= SS_W$$

The relationship between multiple regression analysis and ANOVA is

$$R^2 \times SS_T = SS_B$$

$$(1 - R^2) \times SS_T = SS_W$$

Summary

In this chapter, we have discussed the concepts of multiple correlation and multiple regression. The multiple correlation coefficient, R, is the index of the relationship between the criterion variable (Y) and a linear combination of predictor variables $(X_1, X_2, \ldots, X_k)$. We also tested the multiple R for statistical significance, that is, tested the null hypothesis H_0: $R_{pop} = 0$. If this null hypothesis is rejected, we concluded that, in the population, the multiple R is different from zero. In other words, there is a statistically significant relationship between the criterion variable and the linear combination of predictor variables. Subsequently, we used the square of the multiple R (R^2, the coefficient of determination) to indicate the proportion of the variance in the criterion variable that could be attributed to the variance in the combined predictor variables.

The multiple regression equation is used in predicting scores on the criterion variable from scores on the predictor variables. This equation is developed by fitting a k-dimensional hyperplane to a mass of points in $k + 1$-dimensional space using the least-squares criterion. In this way, the variance of the distribution of errors in prediction is minimized.

In general, we want to select predictor variables that correlate highly with the criterion variable but have low correlations among themselves. With this restriction, including more than five or six predictor variables rarely produces a substantial increase in the multiple R. For those predictor variables included in the multiple regression equation, we determined the significance of each by testing the null hypothesis H_0: $\beta_i = 0$; that is, the individual regression coefficients in the population equal zero. When one or more of these hypotheses are rejected, we conclude that the corresponding predictor variables are significant contributors to the regression when used in combination with the other predictor variables.

In the last section, we illustrated the relationship between ANOVA and multiple regression analysis and showed that $R^2 \times SS_T = SS_B$ and that $(1 - R^2) \times SS_T = SS_W$. In the next chapter, we will use ANOVA together with multiple regression analysis in analysis of covariance (ANCOVA).

Exercises

*1. A social pyschologist at Children's Protective Services (CPS) works with youth 18 years of age and older who previously have been in custody at CPS, but now have been released. She decides to conduct a study on the extent to which these youths can live independently, measured by the *Independent Living Scale* (ILS), upon their release. The specific purpose of the study is to determine the relationship between this criterion variable (Y) and three predictor variables:

X_1 = Self-confidence Scale (SCS)
X_2 = Social Skills Inventory (SSI)
X_3 = Academic Aptitude Scale (AAS)

Data are available on the four scales for 50 youth who had previously been in custody. The data set is found in Appendix B; the summary data and the regression analysis are found below.

a. Determine the regression model (raw score and standard score form).

b. Determine R and R^2.

c. Determine whether the multiple R is statistically significant. (Test H_0: $R_{pop} = 0$. Use $\alpha = .05$.)

d. Determine the significance of the predictor variables. (Test H_0: $\beta_i = 0$. Use $\alpha = .05$.)

e. Interpret the results in terms of the effectiveness of using these three predictor variables in the regression equation.

Variable	Mean	Std Dev	Range	Minimum	Maximum	Valid N
X1	48.90	8.66	38.00	28.00	66.00	50
X3	49.52	7.61	36.00	35.00	71.00	50
X2	50.06	10.11	44.00	28.00	72.00	50
Y	51.60	10.08	45.00	27.00	72.00	50

	X1	X2	X3	Y
X1	1.0000 (50) P= .	.6318 (50) P= .000	.5620 (50) P= .000	.4858 (50) P= .000
X2	.6318 (50) P= .000	1.0000 (50) P= .	.5237 (50) P= .000	.6689 (50) P= .000
X3	.5620 (50) P= .000	.5237 (50) P= .000	1.0000 (50) P= .	.4768 (50) P= .000
Y	.4858 (50) P= .000	.6689 (50) P= .000	.4768 (50) P= .000	1.0000 (50) P= .000

(Coefficient / (Cases) / 2-tailed Significance)

```
  * * * *   M U L T I P L E   R E G R E S S I O N   * * * *
```

Listwise Deletion of Missing Data

Equation Number 1   Dependent Variable..   Y

Block Number 1. Method: Enter   X1   X2   X3

Variable(s) Entered on Step Number
     1..   X3
     2..   X2
     3..   X1

Multiple R	.68588
R Square	.47043
Adjusted R Square	.43589
Standard Error	7.57180

Analysis of Variance

	DF	Sum of Squares	Mean Square
Regression	3	2342.72282	790.90761
Residual	46	2637.27718	57.33211

F =   13.62077   Signif F = .0000

------------------ Variables in the Equation ------------------

Variable	B	SE B	Beta	T	Sig T
X1	.050583	.172023	.043450	.294	.7700
X2	.555965	.143116	.557349	3.885	.0003
X3	.212757	.178213	.160511	1.194	.2387
(Constant)	10.759157	7.626844		1.411	.1651

2. A business researcher is investigating the relationship between certain personality characteristics of small business owners and the businesses' growth potential over the next five years. The researcher randomly selects 15 small business owners, administers the "Speed and Impatience" (X_1) and "Job Involvement" (X_2) subscales of the Jenkins Activity Survey, and then determines the

growth potential (Y) for the businesses. The resulting data are as follows:

Company	Speed and impatience (X_1)	Job involvement (X_2)	Growth potential (Y)
1	4	3	5
2	2	5	2
3	5	7	7
4	4	5	6
5	9	10	8
6	9	8	10
7	7	5	8
8	3	4	7
9	4	5	3
10	2	3	3
11	5	7	7
12	4	7	4
13	1	4	2
14	5	8	6
15	5	5	5

$\Sigma X_1 = 69.0$	$\Sigma X_2 = 86.0$	$\Sigma Y = 83.0$
$\overline{X}_1 = 4.600$	$\overline{X}_2 = 5.733$	$\overline{Y} = 5.533$
$\Sigma X_1^2 = 393.0$	$\Sigma X_2^2 = 550.0$	$\Sigma Y^2 = 539.0$
$\Sigma X_1 X_2 = 444.0$	$\Sigma X_1 Y = 448.0$	$\Sigma X_2 Y = 514.0$
$SS_1 = 75.614$	$SS_2 = 56.956$	$SS_Y = 79.702$
$s_1 = 2.324$	$s_2 = 2.017$	$s_Y = 2.386$
$r_{12} = .738$	$r_{Y1} = .853$	$r_{Y2} = .566$

a. Determine the regression model (raw score and standard score form).

b. Determine R and R^2.

c. Determine whether the multiple R is statistically significant. (Test H_0: $R_{pop} = 0$. Use $\alpha = .05$.)

d. Determine the significance of the predictor variables. (Test H_0: $\beta_i = 0$. Use $\alpha = .05$.)

e. Determine partial and part correlation coefficients:

$r_{Y1 \cdot 2}$ $r_{Y2 \cdot 1}$

$r_{Y(1 \cdot 2)}$ $r_{Y(2 \cdot 1)}$

3. In a study of the relationship between measures of aggression (Y) and anger (X_1) and patience (X_2), a psychologist collects data from 100 subjects and computes the following summary statistics:

	Aggression (Y)	Anger (X_1)	Patience (X_2)	
$\overline{X}$	15.6	8.2	5.7	$r_{Y1} = .630$
s	3.4	2.4	1.6	$r_{Y2} = -.580$
SS	1,144.44	570.24	253.44	$r_{12} = -.180$

a. Determine the regression model (raw score and standard score form).

b. Determine R and R^2.

c. Determine whether the multiple R is statistically significant. (Test H_0: $R_{pop} = 0$. Use $\alpha = .05$.)

d. Determine the significance of the predictor variables. (Test H_0: $\beta_i = 0$. Use $\alpha = .05$.)

e. Determine partial and part correlation coefficients:

$r_{Y1 \cdot 2}$ $r_{Y2 \cdot 1}$

$r_{Y(1 \cdot 2)}$ $r_{Y(2 \cdot 1)}$

4. An educational researcher is studying the relationship between reading speed (X_1), aptitude (X_2), and achievement (Y). The following summary statistics are computed from data on 100 middle-school students. (Notice that the means and standard deviations are *not* given.)

$r_{12} = .450$ $r_{Y1} = .320$ $r_{Y2} = .610$

a. Determine the regression model (standard score form only).

b. Determine R and R^2.

c. Determine whether the multiple R is statistically significant. (Test H_0: $R_{pop} = 0$. Use $\alpha = .05$.)

d. Determine partial and part correlation coefficients:

$r_{Y1 \cdot 2}$ $r_{Y2 \cdot 1}$

$r_{Y(1 \cdot 2)}$ $r_{Y(2 \cdot 1)}$

5. For the following data, complete an ANOVA. (Use $\alpha = .05$.) Then, using the coding methods discussed in this chapter, complete a multiple regression analysis. Show that the F test in the ANOVA is the same as the F test in the multiple regression analysis.

I	II	III		I	II	III
8	10	9	ΣX	50	74	65
10	10	8	$\overline{X}$	8.333	12.333	10.833
10	12	14	ΣX^2	430	950	729
7	14	12	s	1.633	2.733	2.229
6	17	10				
9	11	12				

6. At a large state university, a new faculty rating system is initiated. The director of institutional research at the university is asked to investigate the faculty's reaction to this rating system and to determine the relationship between their reaction and their age and perception of their work environment. The following data are collected from 15 faculty members.

Faculty member	Reaction to rating (Y)	Age (X_1)	Perception of environment (X_2)
1	20	42	8
2	18	31	9
3	11	60	3
4	17	48	6
5	14	52	4
6	18	29	7
7	16	36	5
8	20	38	9
9	15	44	5
10	12	50	3
11	16	33	6
12	19	31	7
13	14	47	5
14	16	41	5
15	13	56	4

$\Sigma Y = 239.0$ $\Sigma X_1 = 638.0$ $\Sigma X_2 = 86.0$

$\overline{Y} = 15.933$ $\overline{X}_1 = 42.533$ $\overline{X}_2 = 5.733$

$\Sigma Y^2 = 3,917.0$ $\Sigma X_1^2 = 28,426.0$ $\Sigma X_2^2 = 546.0$

$\Sigma X_1 Y = 9,879.0$ $\Sigma X_2 Y = 1,441.0$ $\Sigma X_1 X_2 = 3,465.0$

$SS_Y = 108.899$ $SS_1 = 1,289.702$ $SS_2 = 52.908$

$s_Y = 2.789$ $s_1 = 9.598$ $s_2 = 1.944$

$r_{Y1} = -.764$ $r_{Y2} = .931$ $r_{12} = -.738$

a. Determine the regression model (raw score and standard score form).

b. Determine R and R^2.

c. Determine whether the multiple R is statistically significant. (Test H_0: $R_{pop} = 0$. Use $\alpha = .05$.)

d. Determine the significance of the predictor variables. (Test H_0: $\beta_i = 0$. Use $\alpha = .05$.)

e. Determine partial and part correlation coefficients:

$r_{Y1\cdot2}$ $r_{Y2\cdot1}$

$r_{Y(1\cdot2)}$ $r_{Y(2\cdot1)}$

7. An industrial psychologist employed at a large engineering company is asked to investigate the relationship between job satisfaction (Y) and productivity (X_1) and perseverance (X_2). Data are collected from 100 employees and summarized as follows:

	Satisfaction (Y)	Productivity (X_1)	Perseverance (X_2)	
$\overline{X}$	20.5	8.6	7.3	$r_{12} = .230$
s	5.3	1.4	1.7	$r_{Y1} = .770$
SS	2,780.91	194.04	286.11	$r_{Y2} = .670$

    a. Determine the regression model (raw score and standard score form).

    b. Determine R and R^2.

    c. Determine whether the multiple R is statistically significant. (Test H_0: $R_{pop} = 0$. Use $\alpha = .05$.)

    d. Determine the significance of the predictor variables. (Test H_0: $\beta_i = 0$. Use $\alpha = .05$.)

    e. Determine partial and part correlation coefficients:

$$r_{Y1 \cdot 2} \qquad r_{Y2 \cdot 1}$$

$$r_{Y(1 \cdot 2)} \qquad r_{Y(2 \cdot 1)}$$

8. A researcher used six predictor variables in an effort to identify persons who were at risk for coronary artery disease. He achieved an R^2 of .72, but was concerned because the sample size was only 25.

    a. What would be the adjusted value of R^2?

    b. Interpret R^2 and the adjusted R^2 in the context of this research.

    c. Is the unadjusted R^2 great enough to allow us to conclude that the population value of R does not equal zero?

    d. Does your answer to part c change your answer to part b? Explain.

9. A school psychologist tried to relate achievement scores with attendance data. She found $R = .55$ when predicting achievement from days absent and late. The R increased to .60 when the number of suspensions was added to the regression equation. Her sample size was 84.

    a. Was the increase in R significant at the .05 level?

    b. Would you recommend that the school psychologist include the number of suspensions in her regression equation? Explain.

10. The owners of a weight-loss clinic gathered data from 100 clients in order to predict the number of visits to the clinic in one-year's time. When they used age, gender, education, and initial weight as predictors, they found that $R = .46$. Later, they added income as a predictor, and R increased to .58.

    a. Was the increase in R significant at the .05 level?

    b. Was the correlation between income and number of visits positive or negative?

11. Suppose that a researcher found $R = .71$ when he used ten predictors and a sample size of five.

    a. Find the adjusted value of R^2.

    b. Interpret the results.

19

Analysis of Covariance

I n general, research is conducted for the purpose of explaining the effects of the independent variable on the dependent variable, and the purpose of research design is to provide a structure for the research. In the **research design**, the researcher identifies and controls independent variables that can help to explain the observed variation in the dependent variable. Since the research design is structured before the research begins, this method of control is called **experimental control**.

In addition to controlling and explaining variation through research design, it is also possible to use **statistical control** to explain variation in the dependent variable. Statistical control, used when experimental control is difficult, if not impossible, can be achieved by measuring one or more variables in addition to the independent variables of primary interest and by controlling the variation attributed to these variables through statistical analysis rather than through research design. The analytic procedure employed in this statistical control is **analysis of covariance (ANCOVA)**.[1]

---

[1] The purpose of this chapter is to introduce the basic concepts underlying analysis of covariance. For more in-depth coverage of ANCOVA, see J. D. Elashoff, "Analysis of Covariance: A Delicate Instrument," *American Educational Research Journal* 6 (1969): 383–402, and B. E. Huitema, *The Analysis of Covariance* (New York: Wiley, 1980).

Controlling and explaining variation in the dependent variable can be accomplished with either experimental control, using research design, or statistical control, using analysis of covariance.

Statistical Control Using ANCOVA

Analysis of covariance is used primarily as a procedure for the statistical control of an extraneous variable. ANCOVA, which combines regression analysis and analysis of variance (ANOVA), controls for the effects of this extraneous variable, called a **covariate**, by partitioning out the variation attributed to this additional variable. In this way, the researcher is better able to investigate the effects of the primary independent variables.

The researcher must carefully select the covariate. In order for ANCOVA to be effective, the covariate must be linearly related to the dependent variable. In addition, the covariate must be unaffected by other independent variables. For example, in an experiment, it must be unaffected by the manipulation of the experimental variable.

Consider an example in which a researcher is interested in the effects of three different teaching strategies on student achievement. In addition to randomly assigning students to groups that are exposed to the three different teaching strategies, suppose the researcher also has access to a recent IQ score for each student. In this example, teaching strategy would be the primary independent variable, the achievement measure would be the dependent variable, and IQ would be the covariate.

By statistically controlling for the variation attributed to the covariate, the researcher increases the precision of the research by reducing the error variance. This is illustrated in Figure 19.1. In this figure, the area inside the circle represents the total variation of the scores on the dependent variable. The proportion of the variation attributed to the treatment effects (different teaching strategies) is shown, along with the variation attributed to the covariate (IQ). Note that, if the effects of the covariate were not considered, the amount of error variance would be considerably larger. However, with ANCOVA, this variation can be controlled statistically and partitioned out of the error variance.

Using ANCOVA, the researcher can increase the precision of the research by partitioning out the variation attributed to the covariate, which results in a smaller error variance.

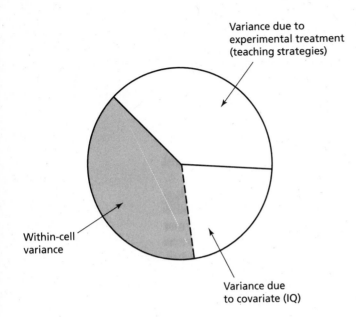

Variance due to
experimental treatment
(teaching strategies)

Within-cell
variance

Variance due
to covariate (IQ)

FIGURE 19.1
Schema of partitioning variation in
ANCOVA

Many authors[2] cite a second application of ANCOVA: when using intact groups of subjects. Such an application is used when treatments *can* be randomly assigned to groups but subjects *cannot* be randomly assigned to treatment groups. For such an experimental arrangement, there are two potential benefits of using ANCOVA. The first is an adjustment for preexisting differences that may exist among the intact groups prior to the research; the second is, as stated before, the increase in the precision of the research from reducing the error variance. Although ANCOVA can be useful in adjusting for preexisting differences among intact groups, it *must* be used with caution. In fact, Lord[3] concluded that, when subjects are not randomly assigned to treatment groups, there is no logical or statistical procedure that can properly allow for uncontrolled, preexisting differences among groups. Given this caution, the researcher must initially consider the differences among the groups on the covariate. If the differences are quite small, ANCOVA serves primarily to reduce the error variance, as discussed above. However, when the differences are large, and/or there is interaction between the covariate and the treatment, the use of ANCOVA is inappropriate and should *not* be used. We will discuss this fact later in the chapter, in the section on the assumptions underlying ANCOVA.

[2] T. D. Cook and D. T. Campbell, *Quasi-Experimentation: Design Analysis Issues for Field Settings* (Chicago: Rand-McNally, 1979), pp. 147–205.

[3] F. M. Lord, "Statistical Adjustment When Comparing Pre-existing Groups," *Psychological Bulletin* 72 (1969): 336–337.

When intact groups are used, ANCOVA can be used to partially adjust for preexisting differences among the groups. However, ANCOVA must be used with caution in such situations.

Example

The general notation of the research design for one-way ANCOVA with a single covariate is shown in Table 19.1. Notice that each subject in each treatment group has a score on the dependent variable (Y) of student achievement and the covariate (X) of intellectual ability. The means for each group on the dependent variable are denoted $\overline{Y}_1, \overline{Y}_2, \ldots, \overline{Y}_k$; the means for each group on the covariate are denoted $\overline{X}_1, \overline{X}_2, \ldots, \overline{X}_k$. To illustrate this notation for ANCOVA, consider the previous example of the effects of three different teaching strategies on student achievement while controlling for IQ. Suppose the researcher assigns 36 subjects randomly to the three teaching strategies, 12 to each group, and that IQ data are available for all subjects. Following the use of the three teaching strategies, a standardized achievement test is given to each subject; the resulting data are found in Table 19.2. Also included in this table are summary data that will be used in subsequent ANCOVA calculations. Because of the complexity of the calculations, we will initially discuss the results of the ANCOVA for this example using printouts from SAS and SPSS. The formulas used in ANCOVA and the actual calculations will be shown later in the chapter.

Preliminary Analyses In order to illustrate how ANCOVA increases the precision of the research by reducing error variance, we have computed an analysis of variance (ANOVA) on the dependent variable (Y = achievement scores) *without*

TABLE 19.1
General Data Layout for One-Way Analysis of Covariance

Group 1		Group 2		Group k	
Y_{11}	X_{11}	Y_{12}	X_{12}	Y_{1k}	X_{1k}
Y_{21}	X_{21}	Y_{22}	X_{22}	Y_{2k}	X_{2k}
Y_{31}	X_{31}	Y_{32}	X_{32}	Y_{3k}	X_{3k}
...	...	...	...	...	...
$Y_{n_1 1}$	$X_{n_1 1}$	$Y_{n_2 2}$	$X_{n_2 2}$	$Y_{n_k k}$	$X_{n_k k}$
$\overline{Y}_1$	$\overline{X}_1$	$\overline{Y}_2$	$\overline{X}_2$	$\overline{Y}_k$	$\overline{X}_k$

TABLE 19.2
Data for Analysis of Covariance Example

	Group 1		Group 2		Group 3		Total	
	Y	X	Y	X	Y	X		
	60	98	62	104	65	102		
	63	102	63	109	68	117		
	66	104	67	104	72	108		
	69	103	71	117	76	117		
	72	112	77	120	78	105		
	75	113	79	113	80	116		
	78	118	82	117	82	111		
	80	120	84	126	84	120		
	67	115	64	113	75	107		
	70	106	68	109	77	104		
	74	112	72	125	79	128		
	76	122	77	118	81	119		
n	12		12		12		36	
$\Sigma Y, \Sigma X$	850	1,325	866	1,375	917	1,354	2,633	4,054
$\overline{Y}, \overline{X}$	70.83	110.42	72.17	114.58	76.42	112.83	73.14	112.61
$\Sigma Y^2, \Sigma X^2$	60,620	146,959	63,126	158,135	70,429	153,458	194,175	458,552
ΣXY	94,305		99,665		103,690		297,660	
SS_Y, SS_X	411.67	656.92	629.67	582.92	354.92	681.67	1,600.31	2,026.56
r	.867		.719		.451		.641	

considering the covariate. The results of this analysis, using the procedures outlined in Chapter 14, are summarized in Table 19.3. Notice that the observed F ratio
($F = 2.41$) does not exceed the critical value of F ($F_{cv} = 3.29$) at the .05 level
of significance. Thus, the null hypothesis of no differences among the population
means is not rejected.

   Also for illustration, an ANOVA was computed for the IQ scores, again using
the procedures outlined in Chapter 14; the results are summarized in Table 19.4.
Notice again that the observed F ratio ($F = 0.90$) does not exceed the critical
value ($F_{cv} = 3.29$) at the .05 level of significance. Again, the null hypothesis of
no difference among the three population means is not rejected. Thus, it can be
concluded at this point of the analysis that there is no difference among the three
treatment groups on both the dependent variable and the covariate.

Step 1: State the Hypotheses   The null and alternative hypotheses of ANCOVA are
similar to those for ANOVA. Conceptually, however, these population means have
been adjusted for the covariate. Thus, in reality, the null hypothesis of ANCOVA
is of no difference among the adjusted population means.

TABLE 19.3
Summary ANOVA for Dependent Variable (Y = Achievement Scores)

Source	SS	df	MS	F	F_{cv}
Between	204. 056	2	102. 028	2. 411	3. 29
Within	1,396. 250	33	42. 311		
Total	1,600. 306	35			

	$\overline{X}$			s	
Group 1	70. 833			6. 118	
Group 2	72. 167			7. 566	
Group 3	76. 417			5. 680	
Grand mean	73. 14				

TABLE 19.4
Summary ANOVA for Covariate (X = Intellectual Ability)

Source	SS	df	MS	F	F_{cv}
Between	105. 056	2	52. 528	0. 902	3. 29
Within	1,921. 500	33	58. 227		
Total	2,026. 556	35			

	$\overline{X}$			s	
Group 1	110. 417			7. 728	
Group 2	114. 583			7. 279	
Group 3	112. 833			7. 872	
Grand mean	112. 61				

$$H_0: \mu_i' = \mu_2' = \mu_3'$$
$$H_a: \mu_i' \neq \mu_k' \quad \text{for some } i, k$$

Step 2: Set the Criterion for Rejecting H_0 The format of the summary table for ANCOVA is similar to that for ANOVA (see Table 19.5); the difference is that the values for the sums of squares and degrees of freedom have been adjusted for the effects of the covariate. The between-groups degrees of freedom are still $K - 1$, but the within-groups degrees of freedom and the total degrees of freedom are

522

19 ▪ Analysis of Covariance

TABLE 19.5
General Layout for Summary ANCOVA

Source	SS	df	MS	F	F_{cv}
Covariate	SS_{cov}	1			
Between	SS'_B	$K-1$			
Within	SS'_W	$N-K-1$			
Total	SS'_T	$N-1$			

$N - K - 1$ and $N - 1$, respectively. This reflects the loss of a degree of freedom when controlling for the covariate; this control places an additional restriction on the data. As we will see in the next step, the test statistic for ANCOVA (F) is the ratio of the adjusted between-groups mean square (MS'_B) to the adjusted within-groups mean square (MS'_W):

$$F = \frac{MS'_B}{MS'_W} \tag{19.1}$$

The underlying distribution of this test statistic is the F distribution with $K - 1$ and $N - K - 1$ degrees of freedom.

For this example, suppose the null hypothesis is being tested against the alternative hypothesis, using $\alpha = .05$. Thus, for $K - 1 = 3 - 1 = 2$ and $N - K - 1 = 36 - 3 - 1 = 32$ degrees of freedom, the critical value of the test statistic (F_{cv}) is 3.30 (see Table C.5). In order to reject the null hypothesis, the observed F ratio must exceed this critical value.

Step 3: Compute the Test Statistic The SAS and SPSS printouts of the results of the ANCOVA for this example are found in Table 19.6; these summary data are also shown in Table 19.7, using the format illustrated in Table 19.5. Even though we stated only one null hypothesis in Step 1, two test statistics (F) are found in Table 19.7. The first of these is for the covariate (IQ). Recall that, in selecting a covariate for the ANCOVA, it should be linearly related to the dependent variable. Thus, the null hypothesis tested with the F ratio for the covariate is that there is no relationship between the covariate (IQ) and the dependent variable (achievement):

$$H_0: \rho = 0$$
$$H_a: \rho \neq 0$$

The test statistic (F) for this hypothesis is the ratio of the mean square for the

TABLE 19.6
Summary ANCOVA for Example, SAS and SPSS Printouts

SAS

ANCOVA
General Linear Models Procedure

Dependent Variable: Score

Source	SS	df	MS	F	p > F
Model	843. 54150403	3	281. 18050134	11. 89	0. 0001
Error	756. 76405152	32	23. 64887661		
Corrected Total	1,600. 30555556	35			

Source	Type I SS	df	F	p > F	df
IQ	658. 20878490	1	27. 83	0. 0001	1
Group	185. 33271913	2	3. 92	0. 0300	2

SPSS

ANCOVA
Analysis of Variance

Score by Group with IQ

Source of Variation	SS	df	MS	F	Signif of F
Covariates	658. 209	1	658. 209	27. 833	0. 000
IQ	658. 209	1	658. 209	27. 833	0. 000
Main Effects	185. 333	2	92. 666	3. 918	0. 030
Group	185. 333	2	92. 666	3. 918	0. 030
Explained	843. 542	3	281. 181	11. 890	0. 000
Residual	756. 764	32	23. 649		
Total	1,600. 306	35	45. 723		

covariate (MS_{cov}) to the adjusted within-groups mean square (MS'_W):

$$F = \frac{MS_{cov}}{MS'_W}$$

$$= \frac{658.209}{23.649} \tag{19.2}$$

$$= 27.833$$

The critical value of this test statistic (F_{cv}) is found in Table C.5 for 1 and 32 degrees of freedom. Assuming the researcher has set the level of significance (α) at .05, F_{cv} would be approximately 4.15. Thus, since the observed value of

TABLE 19.7
Summary ANCOVA for Example

Source	SS	df	MS	F	F_{cv}
Covariate	658. 209	1	658. 209	27. 833	4. 15
Between	185. 333	2	92. 667	3. 918	3. 30
Within	756. 764	32	23. 649		
Total	1,600. 306	35			

the test statistic ($F = 27.833$) exceeds the critical value ($F_{cv} = 4.15$), the null hypothesis of no relationship between the covariate (IQ) and the dependent variable (achievement) is rejected.

The effect of this relationship between the covariate and the dependent variable in ANCOVA can be seen by comparing the within sum of squares in Table 19.3 for ANOVA, when the covariate is not considered in the analysis ($SS_W = 1,396.250$), with the adjusted within-groups sum of squares in Table 19.7 for ANCOVA ($SS'_W = 756.764$). This decrease reflects the partitioning of the effect of the covariate out of the within-cell variation, as illustrated in Figure 19.1.

Now consider the test of the null hypothesis that was stated in Step 1: $H_0: \mu'_1 = \mu'_2 = \mu'_3$. The test statistic for this hypothesis (F) is the ratio of the adjusted between-groups mean square (MS'_B) to the adjusted within-groups mean square (MS'_W):

$$F = \frac{MS'_B}{MS'_W}$$
$$= \frac{92.666}{23.649}$$
$$= 3.918$$

Step 4: Interpret the Results   Recall from the ANOVA on the achievement scores *without* adjusting for the covariate (see Table 19.3) that the observed F (2.411) did *not* exceed the critical value (F_{cv}) and that the null hypothesis of no difference among the "unadjusted" means ($H_0: \mu_1 = \mu_2 = \mu_3$) was *not* rejected. However, when the effects of the covariate were partitioned out in the ANCOVA, both the between-groups sum of squares and the within-groups sum of squares were adjusted; the resulting test statistic ($F = 3.918$) exceeded the critical value ($F_{cv} = 3.30$). Thus, the null hypothesis of no differences among the "adjusted" means ($H_0: \mu'_1 = \mu'_2 = \mu'_3$) is rejected. The conclusion at this point is that at least one pair or combination of adjusted means differs. In order to determine which pair or combination of means differs, a post hoc multiple comparison test must be conducted.

Post Hoc Multiple Comparison Test   The post hoc multiple comparison tests that can be used following an ANOVA were introduced in Chapter 15. While several different procedures were considered in that chapter, we will discuss only the Tukey procedure in this chapter. However, it is possible to use the other procedures, depending on the nature of the investigation as well as the nature of the data collected in the ANCOVA. For all these procedures, though, it is necessary to consider two adjustments that must be made before conducting post hoc tests following an ANCOVA.

The first adjustment is to the sample means; the formula for the adjusted means is as follows:

$$\overline{Y}'_k = \overline{Y}_k - b_W(\overline{X}_k - \overline{X})$$ (19.3)

where

$\overline{Y}'_k$ = adjusted group mean on the dependent variable

$\overline{Y}_k$ = unadjusted group mean on the dependent variable

b_W = "pooled" within-groups regression coefficient reflecting the correlation between the dependent variable and the covariate

$\overline{X}_k$ = group mean on the covariate

$\overline{X}$ = grand mean on the covariate

Again, because of the complexity of the calculation of b_W, we will discuss the results of the post hoc tests using the adjusted means available from the SAS and SPSS printouts (the actual calculation of b_W and the adjusted means for this example will be shown later in the chapter). However, consider the nature of the adjustment to the sample means, as reflected in formula 19.3. Assume, first of all, that there is no difference among the group means on the covariate; each of the group means equals the grand mean. In this case, the value $\overline{X}_k - \overline{X}$ is equal to zero for each group, and there will be no adjustment to the group means on the dependent variable. However, even with modest differences among the group means on the covariate, there will be an adjustment. For those groups with means less than the grand mean on the covariate, $\overline{X}_k - \overline{X}$ will be less than zero, and the adjusted mean ($\overline{Y}'_k$) will thus be *greater* than the unadjusted mean ($\overline{Y}_k$). In our example, the grand mean on the covariate (IQ) equals 112.61 (see Table 19.4). Since the mean on the covariate (IQ) for group 1 is less than the grand mean, the adjusted mean on the dependent variable (achievement) will be *greater* than the unadjusted mean for group 1. Conversely, for those groups with means greater than the grand mean on the covariate, $\overline{X}_k - \overline{X}$ will be positive, and the adjusted mean on the dependent variable will, thus, be *less* than the unadjusted mean. For this example, the adjusted means for both group 2 and group 3 will be less than the unadjusted means.

The adjusted means for this example are found in Table 19.8 along with the partial printouts from SAS and SPSS. To obtain the adjusted means using SAS, the LSMEANS option must be used.

To obtain the adjusted means using SPSS, the STATISTICS 1 option must be specified. Consider the partial SPSS printout. The "unadjusted" means (the

TABLE 19.8
Adjusted Means for SAS and SPSS Printouts

SAS	ANCOVA General Linear Models Procedure				

Least Squares Means

	Score	Prob> |T|		H_0: LSMEAN(I) = LSMEAN(J)	
Group	LSMEAN	I/J	1	2	3
1	72.0992931	1		0.6031	0.0445
2	71.0289054	2	0.6031		0.0128
3	76.2884682	3	0.0445	0.0128	

SPSS	ANCOVA Multiple Classification Analysis

Score by Group with IQ
Grand mean = 73. 14

Variable + Category	N	Unadjusted DEV'N ETA	Adjusted for Independents DEV'N BETA	Adjusted for Independents + Covariates DEV'N BETA
Group				
1	12	−2. 31		−1. 04
2	12	−0. 97		−2. 11
3	12	3. 28		3. 15
		0.36		0.34
Multiple R Squared	0.527			
Multiple R	0.726			

means before adjusting for the covariate) are found by adding to the grand mean ($\overline{Y} = 73.14$; see Table 19.3) the entries in the column labeled "Unadjusted DEV'N ETA."

$$\text{Group 1} \quad \overline{Y}_1 = 73.14 - 2.31 = 70.83$$

$$\text{Group 2} \quad \overline{Y}_2 = 73.14 - 0.97 = 72.17$$

$$\text{Group 3} \quad \overline{Y}_3 = 73.14 + 3.28 = 76.42$$

Similarly, the "adjusted" means are found by adding to the grand mean the entries in the column labled "Adjusted for Independents + Covariates DEV'N BETA."

$$\text{Group 1} \quad \overline{Y}'_1 = 73.14 - 1.04 = 72.10$$

$$\text{Group 2} \quad \overline{Y}'_2 = 73.14 - 2.11 = 71.03$$

$$\text{Group 3} \quad \overline{Y}'_3 = 73.14 + 3.15 = 76.29$$

Notice that $\overline{Y}'_1$ is greater than $\overline{Y}_1$, but both $\overline{Y}'_2$ and $\overline{Y}'_3$ are less than $\overline{Y}_2$ and $\overline{Y}_3$, respectively.

A second adjustment must be made before completing the Tukey procedure. As discussed in Chapter 15, the test statistic (Q) was defined as follows:

$$Q = \frac{\overline{X}_i - \overline{X}_k}{\sqrt{MS_W/n}}$$

where

$\overline{X}_i, \overline{X}_k$ = group means

MS_W = within-groups mean square from the ANOVA

n = number of observations in each group (assuming equal group sizes)

In ANCOVA, the MS'_W must be further adjusted to take into consideration the differences among the group means on the covariate; this adjustment is as follows:

$$MS''_W = MS'_W \left[1 + \frac{SS_{B(X)}}{(k-1)SS_{W(X)}} \right] \tag{19.4}$$

where

$SS_{B(X)}$ = between-groups sum of squares from ANOVA on the covariate

$SS_{W(X)}$ = within-groups sum of squares from ANOVA on the covariate

Using the summary data from Tables 19.4 and 19.7.

$$MS''_W = 23.649 \left[1 + \frac{105.056}{(3-1)1,921.500} \right]$$

$$= 23.649(1 + .027)$$

$$= 24.295$$

The Tukey procedure is used to make all pairwise comparisons of means while maintaining the experimentwise error rate (α_E) at the preestablished α level. For each of these pairwise comparisons, the null hypothesis is H_0: $\mu'_i = \mu'_k$, for $i \neq k$. In ANCOVA, these hypotheses are of no difference among the adjusted population means. The test statistic is again Q, defined as follows:

$$Q = \frac{\overline{Y}'_i - \overline{Y}'_k}{\sqrt{MS''_W/n}} \tag{19.5}$$

where

$\overline{Y}'_i, \overline{Y}'_k$ = pairs of adjusted sample means

MS''_W = adjusted MS'_W; see formula 19.4

The calculation of the Q statistic for each of the possible pairwise comparisons is found in Table 19.9. Again, for convenience, we have ranked the adjusted sample means from low to high, found the difference between these means $(\overline{Y}'_i - \overline{Y}'_k)$, and then found Q by dividing these differences by $\sqrt{MS''_W/n}$. The critical value for

TABLE 19.9
Tukey Post Hoc Tests for Example

	$\overline{Y}'_i$	$(\overline{Y}'_i - \overline{Y}'_j)$		Q	
$\overline{Y}'_2$	71.03	—	—	—	
$\overline{Y}'_1$	72.10	1.07	—	0.752	
$\overline{Y}'_3$	76.29	5.26	4.19	3.697*	2.944
$\sqrt{MS''_W/n} = \sqrt{24.295/12} = 1.423$					

*$Y < .05\,Q_{cv} \neq 3.48$

Q is found in Table C.8. For ranges equal to 3 and degrees of freedom for MS'_W equal to 32, we find Q_{cv} to be 3.48 for the .05 level of significance. As noted in Table 19.9, only the difference between the adjusted means for groups 2 and 3 ($Q = 3.697$) is statistically significant. The differences between the adjusted means for groups 1 and 2 and for groups 1 and 3 are not significant.

Computer Exercise: The Survey of High School Seniors Data Set

For the survey of high school seniors data set ($n = 500$), consider again the mathematics achievement (MATHACH) scores as the dependent variable and the recoded father's education classification (FAED) as the independent variable. Recall the recoded classification for FAED was:

1 = Less than HS

2 = HS graduate

3 = Some Vocational or College

4 = College graduate or more

The ANOVA for this example was completed in Chapters 14 and 15.

Now consider an extended analysis in which the variable, MOSAIC, is a covariate. The summary data for the four groups on both variables, MATHACH and MOSAIC, are as follows:

		MATHACH		MOSAIC	
	n	$\overline{Y}$	s	$\overline{X}$	s
Group 1	132	10.8687	6.1866	25.2955	9.9470
Group 2	129	12.4419	6.4488	27.1860	9.0683
Group 3	105	13.2191	6.2831	27.0095	10.4230
Group 4	134	15.8309	6.4947	28.0784	7.6214

The ANOVA procedure in SPSS was used to generate the printouts for the analyses to be considered. The one-way ANOVAs for both MATHACH and MOSAIC scores are found in Table 19.10; the ANCOVA for MATHACH scores with MOSAIC scores as the covariate is found in Table 19.11.

Based upon the data in the Summary ANOVA for the MATHACH scores (see Table 19.10), the null hypothesis, H_0: $\mu_1 = \mu_2 = \mu_3 = \mu_4$, was rejected; the conclusion was that significant differences exist in the means of the four groups on the MATHACH scores. In Chapter 15, we used the Tukey method to determine

TABLE 19.10
SPSS Printout for One-Way ANOVA for MATHACH and MOSAIC

```
- - - - - O N E W A Y - - - - -


    Variable   MATHACH     math test score
  By Variable  FAED        Father's education


                           Analysis of Variance


                       Sum of        Mean          F        F
       Source      D.F.   Squares     Squares     Ratio    Prob.

Between Groups      3    1713.8734    571.2911    14.1307   .0000
Within Groups      496   20052.8090    40.4291
Total              499   21766.6824
```

```
- - - - - O N E W A Y - - - - -


    Variable   MOSAIC      pattern test score
  By Variable  FAED        Father's education


                           Analysis of Variance


                       Sum of        Mean          F        F
       Source      D.F.   Squares     Squares     Ratio    Prob.

Between Groups      3     537.6596    179.2199    2.0910    .1005
Within Groups      496   42511.4299    85.7085
Total              499   43049.0895
```

TABLE 19.11
SPSS Printout for One-Way ANCOVA for MATHACH with MOSAIC as a Covariate

```
* * *   A N A L Y S I S   O F   V A R I A N C E   * * *

          MATHACH   math test score
   by     FAED      Father's education
   with   MOSAIC    pattern test score

          UNIQUE sums of squares
          All effects entered simultaneously
```

Source of Variation	Sum of Squares	DF	Mean Square	F	Sig of F
Covariates	1524.455	1	1524.455	40.727	.000
MOSAIC	1524.455	1	1524.455	40.727	.000
Main Effects	1385.442	3	461.814	12.338	.000
FAED	1385.442	3	461.814	12.338	.000
Explained	3238.328	4	809.582	21.629	.00
Residual	18528.354	495	37.431		
Total	21766.682	499	43.621		

that Group 4 differed from the other three groups and that Group 3 differed from Group 1. Note also in Table 19.10 that the null hypothesis, H_0: $\mu_1 = \mu_2 = \mu_3 = \mu_4$, was not rejected for the MOSAIC scores; the conclusion was that the differences between the group means on the MOSAIC scores were not statistically significant. Now consider the ANCOVA for the four groups with the MATHACH scores as the dependent variable and the MOSAIC scores as the covariate.

Step 1: State the Hypotheses

The null hypothesis for ANCOVA is that there are no differences in the population means on the MATHACH scores after they have been adjusted for the MOSAIC scores. The null and alternative hypotheses are:

$$H_0: \mu'_1 = \mu'_2 = \mu'_3 = \mu'_4$$

$$H_a: \mu'_i \neq \mu'_k \text{ for some } i, k$$

Step 2: Set the Criterion for Rejecting H_0

The test statistic for this one-way ANCOVA is the F ratio defined as the ratio of MS_B and MS_W, both of which have been adjusted for the covariate, i.e.,

$$F = \frac{MS'_B}{MS'_W}$$

The sampling distribution for this F ratio is the F distribution for $K-1$ and $N-K-1$ degrees of freedom. For this example, there are $K-1 = 4-1 = 3$ degrees of freedom associated with the MS'_B and $N-K-1 = 500-4-1 = 495$ degrees of freedom associated with the MS'_W. Thus, assuming $\alpha = .05$, the critical value of F from Table C.5, for 3 and 495 degrees of freedom, is $F_{cv} = 2.60$.

Step 3: Compute the Test Statistic

In ANCOVA, two test statistics are computed: one for testing the significance of the relationship between the covariate (MOSAIC) and the dependent variable (MATHACH); and one for testing the null hypothesis identified in Step 1, H_0: $\mu'_1 = \mu'_2 = \mu'_3 = \mu'_4$. For the first test statistic, the null hypothesis was that there is no relationship between the covariate and the dependent variable, i.e., H_0: $\rho = 0$. The test statistic is defined by formula 19.2, i.e.,

$$F = \frac{MS'_{cov}}{MS'_W}$$

For the data in Table 19.11, we find

$$F = \frac{1524.455}{37.431}$$

$$= 40.727$$

The critical value of F for this test statistic is identified in the F distribution for 1 and $N-K-1 = 500-4-1 = 495$ degrees of freedom. With $\alpha = .05$ and using Table C.5, $F_{cv} = 3.84$. Since the calculated F value ($F = 40.727$) exceeds the critical value ($F_{cv} = 3.84$), the null hypothesis of no relationship between the MOSAIC scores and the MATHACH scores is rejected (see Table 19.11). The conclusion is that since the four groups did not differ on the MOSAIC scores and since there is a statistically significant relationship between the MOSAIC scores and the MATHACH scores, the variable MOSAIC is an appropriate covariate.

The second test statistic was defined in Step 2,

$$F = \frac{MS'_B}{MS'_W}$$

and is used to test the null hypothesis defined in Step 1, $H_0: \mu'_1 = \mu'_2 = \mu'_3 = \mu'_4$. Note the data in the Summary ANCOVA,

$$F = \frac{461.814}{37.431}$$

$$= 12.338$$

Step 4: Interpret the Results

Since the calculated F value ($F = 12.338$) exceeds the critical value ($F_{cv} = 2.60$), the null hypothesis, $H_0: \mu'_1 = \mu'_2 = \mu'_3 = \mu'_4$, is rejected. The conclusion is that the adjusted means in the population are not equal. The associated probability statement would be: the probability that the observed differences among the adjusted means would have occurred by chance, if the null hypothesis is true, is less than .05. Note in the printout, "Sig of F" equals .000, which implies that the probability of such an occurrence is actually less than .001.

In order to identify which adjusted means differ, it is necessary to conduct the Tukey procedure for adjusted means. The adjusted means for the groups are computed using formula 19.3;

$$\overline{Y}'_k = \overline{Y}_k - b_W(\overline{X}_k - \overline{X})$$

The calculation of the adjusted means, using this formula, is illustrated in Table 19.12.

TABLE 19.12
SPSS Printout for Adjusted Means from the One-Way ANCOVA for MATHACH with MOSAIC as a Covariate

ANCOVA Multiple Classification Analysis

MATHACH by FAED with MOSAIC
Grand Mean = 13.10

Variable + Category	N	Unadjusted Dev'n	Eta	Adjusted for Independents + Covariates Dev'n	Beta
FAED					
1	132	−2.23		−1.93	
2	129	−.66		−.71	
3	105	.12		.10	
4	134	2.73		2.51	
			.28		.25
Multiple R Squared		.149			
Multiple R		.386			

TABLE 19.13
Tukey Tests for One-Way ANCOVA Computer Example

Group	Y_i	$(Y_i - Y_k)$			Q		
Group 1	11.17						
Group 2	12.39	1.22			2.27		
Group 3	13.20	2.03	0.81		3.58	1.42	
Group 4	15.61	4.44	3.22	2.41	8.35*	6.02*	4.27*

$* \, p < .05 \quad Q_{CV} = 3.68 \text{ for df} = 120$

$$
\begin{array}{ll}
\text{Group 1} & \overline{Y}'_1 = 13.10 - 1.93 = 11.17 \\
\text{Group 2} & \overline{Y}'_2 = 13.10 - 0.71 = 12.39 \\
\text{Group 3} & \overline{Y}'_3 = 13.10 + 0.10 = 13.20 \\
\text{Group 4} & \overline{Y}'_4 = 13.10 + 2.51 = 15.61
\end{array}
$$

The Tukey procedure is conducted using the Tukey/Kramer (T/K) formula for calculating the Q statistic.

$$
Q = \frac{\overline{Y}'_i - \overline{Y}'_k}{\sqrt{MS''_W \left(\dfrac{1/n_i + 1/n_k}{2} \right)}}
$$

Note that it is necessary to use the MS''_W that has been adjusted the second time (see formula 19.4). Using data from Tables 19.10 and 19.11, we compute MS''_W to be

$$
MS''_W = MS'_W \left(1 + \frac{SS_{B(X)}}{(K-1)SS_{W(X)}} \right)
$$

$$
= 37.431 \left(1 + \frac{537.6596}{(3)42,511.4299} \right)
$$

$$
= 37.589
$$

The data for the Tukey procedure are found in Table 19.13. As can be seen, the data indicate that the adjusted mean for Group 4 differs from the adjusted means for the other three groups.

The Linear Model

In Chapter 14, we presented the **linear model** for one-way ANOVA as

$$
Y_{ik} = \mu + \alpha_k + e_{ik} \tag{19.6}
$$

where

$Y_{ik} = i$th score in the kth group

$\mu = $ grand mean of the population

$\alpha_k = \mu_k - \mu = $ effect of belonging to group k

$e_{ik} = $ random error associated with this score

This model indicates that each subject's score (Y_{ik}) is a function of the grand mean (μ), the mean of the particular group to which the subject belongs ($\alpha_k = \mu_k - \mu$), and random error (e_{ik}). As indicated in Chapter 14, the random errors are assumed to be normally distributed, with a mean of zero for each group. The variance of these errors across the groups is defined as σ_e^2, the population error variance.

Now consider the linear model for one-way ANCOVA. As mentioned above, ANCOVA serves to control for additional variation by partitioning out variation from the error variance. Thus, by including a covariate in the linear model, the error would be partioned into two components; one would be the variation attributed to the covariate and the other would be the remaining error, or $e_{ik} = \text{cov} + e'_{ik}$. Thus, the linear model for ANCOVA, one-way classification, would be

$$Y_{ik} = \mu + \alpha_k + \text{cov} + e'_{ik} \qquad (19.7)$$

The term cov in the equation is actually

$$\text{cov} = \beta_W(X_{ik} - \mu_X) \qquad (19.8)$$

where

$\beta_W = $ regression coefficient representing the relationship between the dependent variable (Y) and the covariate (X)

$X_{ik} = $ subject's score on the covariate

$\mu_X = $ grand mean of the scores on the covariate

Combining formulas 19.7 and 19.8, we have

$$Y_{ik} = \mu + \alpha_k + \beta_W(X_{ik} - \mu_X) + e'_{ik} \qquad (19.9)$$

which can be expressed as follows:

$$Y_{ik} - \beta_W(X_{ik} - \mu_X) = \mu + \alpha_k + e'_{ik} \qquad (19.10)$$

Formula 19.10 illustrates the adjustment of the scores on the dependent variable when incorporating the covariate in the linear model. It also illustrates how regression analysis is incorporated in ANCOVA by adjusting the scores on the dependent variable; the nature of this adjustment will be discussed in a later section. Most importantly, formula 19.10 illustrates that ANCOVA can be thought of as an ANOVA using adjusted scores, in which the adjusted score is defined as $Y_{ik} - \beta_W(X_{ik} - \mu_X)$.

The linear model for one-way ANCOVA illustrates that this procedure is actually an ANOVA on adjusted scores.

Computing the Adjusted Sum of Squares for ANCOVA

In the chapter on one-way ANOVA, we discussed the partitioning of the *total* variation of scores on the dependent variable into two components, *between* and *within*, and provided the computational formulas for the total sum of squares (SS_T), the between-groups sum of squares (SS_B), and the within-groups sum of squares (SS_W). Analogously, in Chapter 17, which dealt with bivariate regression, we discussed the partitioning of the total variation into the variation of the predicted scores and the variation of the errors of estimation. In ANCOVA, we combine both of these concepts when adjusting SS_T, SS_B, and SS_W for the variance attributed to the covariate.[4]

First consider the adjusted total sum of squares (SS_T'). Recall from the chapters on correlation (Chapter 5) and regression (Chapters 6 and 17) that the coefficient of determination (r^2) was defined as the proportion of the total variation of Y that can be attributed to the variation in X. With this definition, SS_T' is defined as the total sum of squares after removing the variation attributed to the covariate:

$$SS_T' = SS_T(1 - r_T^2) \tag{19.11}$$

where

SS_T = total sum of squares from the ANOVA

r_T = correlation between all scores on the dependent variable and the covariate

From Tables 19.2 and 19.3,

$$SS_T' = 1,600.306(1 - .641^2)$$

$$= 1,600.306(.589)$$

$$= 942.580$$

The adjusted within-groups sum of squares (SS_W') is similarly defined:

$$SS_W' = SS_W(1 - r_W^2) \tag{19.12}$$

---

[4]The slight differences between the adjusted sum of squares computed above and those found in Tables 19.6 and 19.7 are due to rounding error.

where

> SS_W = within-groups sum of squares from the ANOVA
>
> r_W = pooled correlation coefficient between the scores on the dependent variable and the covariate

The calculation of the **pooled correlation coefficient**[5] is as follows:

$$r_W = \frac{\Sigma(n\Sigma XY - \Sigma X \Sigma Y)}{\sqrt{\{\Sigma\left[n\Sigma X^2 - (\Sigma X)^2\right]\}\{\Sigma\left[n\Sigma Y^2 - (\Sigma Y)^2\right]\}}} \qquad (19.13)$$

Using the data from Table 19.2, the numerator for formula 19.13 is computed as follows:

$$
\begin{aligned}
\Sigma(n\Sigma XY - \Sigma X \Sigma Y) &= [12(94,305) - (1,325)(850)] \\
&\quad + [12(99,665) - (1,375)(866)] \\
&\quad + [12(103,690) - (1,354)(917)] \\
&= 5,410 + 5,230 + 2,662 \\
&= 13,302
\end{aligned}
$$

Now consider the left-hand side of the denominator under the square-root sign.

$$
\begin{aligned}
\Sigma\left[n\Sigma X^2 - (\Sigma X)^2\right] &= [12(146,959) - (1,325)^2] \\
&\quad + [12(158,135) - (1,375)^2] \\
&\quad + [12(153,458) - (1,354)^2] \\
&= 7,883 + 6,995 + 8,180 \\
&= 23,058
\end{aligned}
$$

For the right-hand side of the denominator under the square-root sign,

$$
\begin{aligned}
\Sigma\left[n\Sigma Y^2 - (\Sigma Y)^2\right] &= [12(60,620) - (850)^2] \\
&\quad + [12(63,126) - (866)^2] \\
&\quad + [12(70,429) - (917)^2] \\
&= 4,940 + 7,556 + 4,259 \\
&= 16,755
\end{aligned}
$$

---

[5] Notice that formula 19.13 incorporates the general formula for the correlation coefficient but involves pooling values across the groups for both the numerator and the denominator. When there are unequal group sizes, the computational formulas must be adjusted. These adjustments are discussed in Huitema, pp. 26–30.

```
UNIVERSITY OF SOUTHERN MISSISSIPPI
           TEXTBOOK CENTER
           HATTIESBURG, MS


356768529  STINSON MARK RICHARD
-------------------------------------------
0860043U  GRADUATE TEXT          53.50
0000001   POLICY LETTER            .00
-------------------------------------------
          TOTAL PAID--->         53.50

EADLINE FOR RETURNING RENTAL BOOKS
S SAT. MAY 13, 2000 12 NOON

29  011400   14:26   HTY032  VISA/MC

SA/MC: 5291071488225333   EXP: 0700
```

```
      UNIVERSITY OF SOUTHERN MISSSISSIPPI
             TEXTBOOK CENTER
             HATTIESBURG, MS


0356768529  STINSON MARK RICHARD
-----------------------------------------
0860043U  GRADUATE TEXT          53.50
0000001   POLICY LETTER            .00
-----------------------------------------
          TOTAL PAID--->          53.50

DEADLINE FOR RETURNING RENTAL BOOKS
3 SAT. MAY 13, 2000 12 NOON
-----------------------------------------
29  011400   14:26   HTY032  VISA/MC

BA/MC: 5291071488225333   EXP: 0700
```

Thus, for this example,

$$r_{\text{W}} = \frac{13,302}{\sqrt{(23,058)(16,755)}}$$

$$= .677$$

Using the data from Table 19.3,

$$SS'_{\text{W}} = 1,396.25(1 - .677^2)$$

$$= 1,396.25(.542)$$

$$= 756.768$$

There is a corresponding formula for the adjusted between-groups sum of squares (SS'_{B}); however, the calculation is somewhat cumbersome. Therefore, we will compute SS'_{B} by subtracting SS'_{W} from SS'_{T}, as follows:

$$SS'_{\text{B}} = SS'_{\text{T}} - SS'_{\text{W}} \tag{19.14}$$

For these data,

$$SS'_{\text{B}} = 942.580 - 756.768$$

$$= 185.812$$

Finally, the sum of squares for the covariate is computed by subtracting the adjusted total sum of squares (SS'_{T}) from the total sum of squares (SS_{T}). For these data,

$$SS_{\text{cov}} = SS_{\text{T}} - SS'_{\text{T}}$$

$$= 1,600.306 - 942.580$$

$$= 657.726$$

Computing the Adjusted Means

The adjusted means in the ANCOVA for this example were presented and discussed earlier in the chapter; the formula for the adjusted means (see formula 19.3) is as follows:

$$Y'_k = \overline{Y}_k - b_{\text{W}}(\overline{X}_k - \overline{X})$$

where

$\overline{Y}'_k$ = adjusted group mean on the dependent variable

$\overline{Y}_k$ = unadjusted group mean on the dependent variable

b_{W} = pooled within-groups regression coefficient reflecting the correlation between the dependent variable and the covariate

$\overline{X}_k$ = group mean on the covariate

$\overline{X}$ = grand mean on the covariate

In order to use this formula, we must compute the pooled within-groups **regression coefficient** (b_W). First, consider the relationship between the correlation coefficient (r) and the regression coefficient (b) as discussed in Chapter 17 (see formula 17.4).

$$b = r\frac{s_Y}{s_X} = r\sqrt{\frac{s_Y^2}{s_X^2}}$$

Now recall that, in ANOVA, MS_W is the pooled estimate of the variance (s^2). Thus, using these two concepts, the formula for b_W is as follows:

$$b_W = r_W\sqrt{\frac{MS_{W(Y)}}{MS_{W(X)}}} \tag{19.15}$$

where

r_W = pooled correlation coefficient (see formula 19.13)

$MS_{W(Y)}$ = MS_W for the dependent variable in ANOVA (see Table 19.3)

$MS_{W(X)}$ = MS_W for the covariate in ANOVA (see Table 19.4)

For this example,

$$b_W = .677\sqrt{\frac{42.311}{58.227}}$$

$$= 0.577$$

Therefore, the adjusted means for the three groups would be

$$\overline{Y}'_k = Y_k - (0.577)(\overline{X}_k - 112.61)$$

$$\overline{Y}'_1 = 70.83 - (0.577)(110.42 - 112.61)$$

$$= 70.83 + 1.26$$

$$= 72.09$$

$$\overline{Y}'_2 = 72.17 - (0.577)(114.58 - 112.61)$$

$$= 72.17 - 1.14$$

$$= 71.03$$

$$\overline{Y}'_3 = 76.42 - (0.577)(112.83 - 112.61)$$

$$= 76.42 - 0.13$$

$$= 76.29$$

Assumptions for ANCOVA

In addition to the assumptions underlying ANOVA (see Chapter 14), there are two major assumptions that underlie the use of ANCOVA[6]; both concern the nature of the relationship between the dependent variable and the covariate. The first is that the relationship is linear. If the relationship is nonlinear, the adjustments made in the ANCOVA will be biased; the magnitude of this bias depends on the degree of departure from linearity, especially when there are substantial differences between the groups on the covariate. Thus, it is important for the researcher, in preliminary analyses, to investigate the nature of the relationship between the dependent variable and the covariate (by looking at a scatterplot of the data points), in addition to conducting an ANOVA on the covariate.

The second assumption has to do with the regression lines within each of the groups. We assume the relationship to be linear. Additionally, however, the regression lines for these individual groups are assumed to be parallel; in other words, they have the same slope. This assumption is often called **homogeneity of regression** or *parallelism* and is necessary in order to use the pooled within-groups regression coefficient (b_W) for adjusting the sample means. Failure to meet this assumption implies that there is an interaction between the covariate and the treatment.

A test of the homogeneity-of-regression assumption is a prerequisite to conducting ANCOVA; this test is outlined below. The procedure involves adjusting the sum of squares within each group, using the correlation coefficient between the dependent variable and the covariate for the respective groups (r_i), rather than the pooled correlation coefficient (r_W). These adjustments are made to the groups individually, and then the adjusted sums of squares are summed across the groups:

$$\text{SS}_{\text{hreg}} = \Sigma \text{SS}_i (1 - r_i^2) \tag{19.16}$$

where

$\text{SS}_i = \Sigma Y_i^2 - (\Sigma Y_i)^2 / n_i$ for each group

$r_i = $ correlation coefficient for each group

The null hypothesis for this assumption is $H_0: \beta_1 = \beta_2 = \cdots = \beta_k$. The β_i's are the population regression coefficients for each group. The test statistic is defined as follows:

$$F = \frac{(\text{SS}_W' - \text{SS}_{\text{hreg}})/(K - 1)}{\text{SS}_{\text{hreg}}/K(n - 2)} \tag{19.17}$$

---

[6]For a more thorough discussion of the assumptions underlying ANCOVA, see Huitema, and Cook and Campbell.

where

SS'_W = adjusted within-groups sum of squares using r_W

SS_{hreg} = adjusted within-groups sum of squares using the individual r_i and summing across the groups

The critical value for this test statistic is found in Table C.5 for $K - 1$ and $K(n - 2)$ degreees of freedom.

The data for testing the homogeneity-of-regression assumption for the example are found in Tables 19.2 and 19.7.

$$SS'_W = 756.77$$

$$SS_{hreg} = 411.67(1 - .867^2)$$

$$+ 629.67(1 - .719^2)$$

$$+ 354.92(1 - .451^2)$$

$$= 102.22 + 304.16 + 282.73$$

$$= 689.11$$

Thus,

$$F = \frac{(756.77 - 689.11)/(3 - 1)}{689.11/3(12 - 2)}$$

$$= 33.83/22.97$$

$$= 1.47$$

With a .05 level of significance, the critical value of the F ratio for this test of the homogeneity-of-regression assumption with 2 and 30 degrees of freedom is 3.32. Because the observed value of the test statistic does *not* exceed the critical value, the null hypothesis is not rejected. By not rejecting the null hypothesis, the homogeneity-of-regression assumption is met, and the researcher can proceed with the ANCOVA.

If the null hypothesis is rejected, the regression lines for the individual groups are not parallel, implying that there is a covariate-treatment interaction. In this case, the use of the pooled regression coefficient (b_W) will result in biased adjusted means. When this situation arises, alternative procedures are recommended.[7] Although most of these alternatives are beyond the scope of this book, one alternative to be considered would be for the researcher to categorize subjects into levels of the covariate and to use the covariate as a second factor in a factorial design. Such a strategy can lead to unequal cell sizes in the factorial design and thus can create other analytical problems. However, when the homogeneity-of-regression assumption is not met in an ANCOVA, such a strategy is considered a viable alternative.

[7] See the discussion of the Johnson-Neyman procedure in Huitema, pp. 270–297.

Summary

The purpose of this chapter was to introduce the basic concepts underlying one-way analysis of covariance. ANCOVA incorporates regression analysis and analysis of variance in statistically controlling for variation attributable to a covariate. In this statistical control, the total, between-groups, and within-groups sums of squares are reduced as a function of the relationship between the covariate and the dependent variable. In this way, ANCOVA serves to reduce the error variance and thus to increase the precision of testing the null hypothesis.

Although many authors cite a second function of ANCOVA—that of adjusting for preexisting differences among treatment groups—such statements need to be made and interpreted cautiously. Indeed, the group means can be adjusted for the covariate, but the use of these adjusted means is valid only when the assumptions of linearity and homogeneity of regression have been met. If these assumptions are not met, alternative procedures are available; most are beyond the scope of this chapter. When these assumptions are met, however, ANCOVA is a very useful statistical technique for controlling variation.

The procedures described in this chapter are limited to one-way ANCOVA. ANCOVA can also be incorporated into more complex designs, like those discussed in Chapter 16. In these more complex designs, it is even more important to test the assumptions that underlie ANCOVA as well as to interpret the results with caution.

Exercises

1. A behavioral psychologist is interested in whether goal-setting activities can be applied successfully in public school classrooms. The psychologist randomly selects 30 middle-school students and randomly assigns them to three different groups that are assigned different levels of goal difficulty. The dependent variable is performance on a verbal-fluency task. Before the study, each student is given a general-ability test, whose scores are used as the covariate. For the data below, use $\alpha = .05$ in the data analyses.

Group	Verbal fluency (Y)	General ability (X)
1	42	16
1	40	16
1	48	19
1	55	24
1	45	20
1	51	22
1	47	20
1	38	23
1	46	20
1	45	18

$n = 10$

$\Sigma Y = 457 \qquad \Sigma X = 198$

$\overline{Y} = 45.7 \qquad \overline{X} = 19.8$

$\Sigma Y^2 = 21,113 \qquad \Sigma X^2 = 3,986$

$\Sigma XY = 9,110$

$SS_Y = 228.10 \qquad SS_X = 65.60$

$r = .50$

(continued)

Group	Verbal fluency (Y)	General ability (X)
2	51	25
2	50	18
2	45	17
2	52	19
2	47	17
2	50	20
2	62	23
2	60	24
2	48	21
2	47	20
3	46	20
3	47	21
3	43	19
3	54	26
3	49	21
3	44	25
3	53	24
3	52	23
3	42	18
3	49	22

$n = 10$

$\Sigma Y = 512 \qquad \Sigma X = 204$
$\overline{Y} = 51.2 \qquad \overline{X} = 20.4$
$\Sigma Y^2 = 26,496 \qquad \Sigma X^2 = 4,234$
$\Sigma XY = 10,541$
$SS_Y = 281.60 \qquad SS_X = 72.40$
$r = .67$

$n = 10$

$\Sigma Y = 479 \qquad \Sigma X = 219$
$\overline{Y} = 47.9 \qquad \overline{X} = 21.9$
$\Sigma Y^2 = 23,105 \qquad \Sigma X^2 = 4,857$
$\Sigma XY = 10,559$
$SS_Y = 160.90 \qquad SS_X = 60.90$
$r = .70$

a. Complete an ANOVA on both the covariate (general ability) and the dependent variable (verbal fluency).

b. Test the assumption of homogeneity of regression.

c. Complete the ANCOVA.

d. Compute the adjusted means.

e. Conduct the Tukey post hoc test on these adjusted means.

*2. Suppose the social psychologist in Exercise 1 in Chapter 18 was interested in determining if age has an effect on the critierion variable, *Independent Living Scale* (ILS), after the youths have been released from Children's Protective Services (CPS). Of the 50 youth participating in the study, 10 are 20 years old, 20 are 19 years old, and 20 are 18 years old. The psychologist wants to determine the effect of age after controlling for scores on the Social Adjustment Scale (SAS). The data set for this exercise is found in Appendix B; the summary data and the ANCOVA analysis are found below. Set $\alpha = .05$.

a. Interpret the results of the ANOVA on the covariate, SAS.

b. Interpret the ANCOVA using ILS as the criterion variable and SAS as the covariate.

c. Conduct the Tukey post hoc test on the adjusted means.

Number of valid observations (listwise) =   50.00

Variable	Mean	Std Dev	Minimum	Maximum	Valid N
Y	51.60	10.08	27.00	72.00	50
COV	72.04	5.92	60.00	80.00	50

- - Correlation Coefficients - -

	COV	Y
COV	1.0000	.4087
	(50)	(50)
	P= .	P= .003
Y	.4087	1.0000
	(50)	(50)
	P= .003	P= .

(Coeffiecent / (Cases) / 2-tailed Significance)

* * *   C E L L   M E A N S * * *

                    Y
               by GROUP

Total Population

       51.60
    (50)

GROUP

	1	2	3
	63.00	53.30	44.20
	(10)	(20)	(20)

```
* * *  A N A L Y S I S   O F   V A R I A N C E  * * *

          COV
     by   GROUP

          UNIQUE sums of squares
          All effects entered simultaneously
```

Source of Variation	Sum of Squares	DF	Mean Square	F	Sig of F
Main Effects	148.120	2	74.060	2.220	.120
GROUP	148.120	2	74.060	2.220	.120
Explained	148.120	2	74.060	2.220	.120
Residual	1567.800	47	33.357		
Total	1715.920	49	35.019		

```
* * *  A N A L Y S I S   O F   V A R I A N C E  * * *

            Y
       by   GROUP
       with COV

          HIERARCHICAL sums of squares
          Covariates entered FIRST
```

Source of Variation	Sum of Squares	DF	Mean Square	F	Sig of F
Covariates	831.943	1	831.943	16.612	.000
COV	831.943	1	831.943	16.612	.000
Main Effects	1844.347	2	922.174	18.414	.000
GROUP	1844.347	2	922.174	18.414	.000
Explained	2676.290	3	892.097	17.813	.000
Residual	2303.710	46	50.081		
Total	4980.000	49	101.633		

```
Covariate    Raw Regression Coeffiecient
COV               .696
```

```
* * *  M U L T I P L E    C L A S S I F I C A T I O N    A N A L Y S I S  * * *
```

              Y
       by   GROUP
       with COV

Grand Mean =   51.60   Adjusted for
                                                   Independents
                                     Unadjusted   + Covariates

Variable + Category   N   Dev'n Eta   Dev'n   Beta

GROUP
    1   10   11.40   10.36
    2   20   1.70   1.53
    3   20   -7.40   -6.70
                                            .70   .64

Multiple R Squared   .537
Multiple R   .733

3. An exercise physiologist is interested in comparing the rate of weight gain under three different diet regimens: (a) macrobiotic, (b) low fiber, and (c) high fiber. The researcher selects 30 freshman women from a physical-fitness class and randomly assigns them to the three treatment groups. The measures of rate of weight gain are available, because these same subjects participated in a study in which they were weighed daily. The rate of weight gain was computed by taking a composite of each subject's weight change during a three-month period. The subject's initial weight was used as a covariate. For the following data, use $\alpha = .05$ in the data analyses.

Group	Rate of gain (Y)	Initial weight (X)
1	15	105
1	17	110
1	19	109
1	16	107
1	15	108
1	21	110
1	18	108
1	19	109

$n = 8$

$\Sigma Y = 140 \qquad \Sigma X = 866$

$\overline{Y} = 17.50 \qquad \overline{X} = 108.25$

$\Sigma Y^2 = 2,482 \qquad \Sigma X^2 = 93,764$

$\Sigma XY = 15,173$

$SS_Y = 32.00 \qquad SS_X = 19.50$

$r = .72$

(continued)

Group	Rate of gain (Y)	Initial weight (X)
2	21	109
2	22	112
2	18	107
2	21	109
2	17	105
2	21	109
2	20	110
2	21	110
3	23	111
3	24	110
3	26	112
3	20	108
3	24	111
3	21	109
3	24	111
3	23	111

$n = 8$

$\Sigma Y = 161$ $\Sigma X = 871$

$\overline{Y} = 20.13$ $\overline{X} = 108.88$

$\Sigma Y^2 = 3{,}261$ $\Sigma X^2 = 94{,}861$

$\Sigma XY = 17{,}552$

$SS_Y = 20.88$ $SS_X = 30.88$

$r = .91$

$n = 8$

$\Sigma Y = 185$ $\Sigma X = 883$

$\overline{Y} = 23.13$ $\overline{X} = 110.38$

$\Sigma Y^2 = 4{,}303$ $\Sigma X^2 = 97{,}473$

$\Sigma XY = 20{,}435$

$SS_Y = 24.88$ $SS_X = 11.88$

$r = .91$

a. Complete an ANOVA on both the covariate (initial weight) and the dependent variable (rate of gain).

b. Test the assumption of homogeneity of regression.

c. Complete the ANCOVA.

d. Compute the adjusted means.

e. Conduct the Tukey post hoc test on these adjusted means.

4. An educational psychologist is interested in comparing problem-solving skills of three different groups of adolescents: (a) those who have not reached Piaget's formal operations stage; (b) those who have just reached the formal operations stage; and (c) those who have been in the formal operations stage for more than six months. The dependent variable is the number of problems solved correctly in six trials of a problem-solving task. The covariate is a standardized measure of creativity. For the following data, use $\alpha = .05$ in the data analyses.

Group	Problem solving (Y)	Creativity (X)
1	143	205
1	159	219
1	153	213
1	155	223
1	145	210
1	156	220
1	145	219

$n = 7$

$\Sigma Y = 1{,}056$ $\Sigma X = 1{,}509$

$\overline{Y} = 150.86$ $\overline{X} = 215.57$

$\Sigma Y^2 = 159{,}550$ $\Sigma X^2 = 325{,}545$

$\Sigma XY = 227{,}815$

$SS_Y = 244.86$ $SS_X = 247.71$

$r = .70$

(continued)

Group	Problem solving (Y)	Creativity (X)		
2	150	219	$n = 7$	
2	151	218	$\Sigma Y = 1,068$	$\Sigma X = 1,543$
2	146	216	$\overline{Y} = 152.57$	$\overline{X} = 220.43$
2	149	221	$\Sigma Y^2 = 163,168$	$\Sigma X^2 = 340,187$
2	163	226		$\Sigma XY = 235,528$
2	159	223	$SS_Y = 221.71$	$SS_X = 65.71$
2	150	220		$r = .91$
3	151	229	$n = 7$	
3	142	218	$\Sigma Y = 1,021$	$\Sigma X = 1,533$
3	148	219	$\overline{Y} = 145.86$	$\overline{X} = 219.00$
3	141	218	$\Sigma Y^2 = 149,163$	$\Sigma X^2 = 335,939$
3	136	209		$\Sigma XY = 223,779$
3	149	218	$SS_Y = 242.86$	$SS_X = 212.00$
3	154	222		$r = .79$

a. Complete an ANOVA on both the covariate (creativity) and the dependent variable (problem solving).

b. Test the assumption of homogeneity of regression.

c. Complete the ANCOVA.

d. Compute the adjusted means.

e. Conduct the Tukey post hoc test on these adjusted means.

20

Other Correlation Coefficients

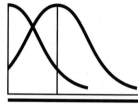

T he discussion of correlation in Chapter 5 considered the Pearson product-moment correlation coefficient for situations in which both variables are measured on quantitative (interval or ratio-level) scales. We introduced the Spearman rho coefficient as a special case of the Pearson for times when both variables are sets of rankings. In this chapter we will discuss other coefficients that can be computed when one or both of the variables are qualitative, or measured on ordinal or nominal scales. In fact, coefficients have been developed for use with virtually every combination of quantitative and qualitative variables.[1]

Scales of Measurement

One important consideration in determining the most appropriate correlation coefficient for assessing the relationship between two variables is the scale of measurement of the two variables. In Chapter 1, we defined quantitative and qualitative

[1] See A. M. Liebetrau, *Measures of Association* (Beverly Hills, Calif.: Sage Publications, 1983).

548

variables and the hierarchy of measurement scales. We will continue to use these classifications in this chapter.

Nominal Scale

In this chapter, we must distinguish between nominal variables with two levels of classification and those with more than two levels. Nominal variables with two classification levels are called **dichotomous variables**; the classification indicates the presence or absence of a particular characteristic. Presence of the characteristic is generally denoted by 1, and absence of the characteristic is denoted by 0. Typical examples of dichotomous variables are marital status (1 = married; 0 = unmarried) and smoking (1 = smoker; 0 = nonsmoker). For other variables, such as gender or political party affiliation, the assignment of 1 and 0 is arbitrary (1 = female and 0 = male; 1 = Republican and 0 = Democrat).

There are other nominal variables that have more than two classification levels, such as geographic region and religious preference. For these two variables, we could use the following classification scheme:

Geographic region	*Religious preference*
1 = Northeast	1 = Protestant
2 = Southeast	2 = Catholic
3 = Northwest	3 = Jewish
4 = Southwest	4 = No preference

For both of these variables, the assignment of a numeric value to the classification level is arbitrary, implying no order of importance.

Ordinal Scale

As we saw in Chapter 1, the ordinal scale is a scale of ranked values. The ranks are assigned based on the amount of the measured characteristic that is possessed by each individual in the group. Examples of the use of the ordinal scale include the ranking of the ten finalists in the Miss America contest and a grading system that incorporates A, B, C, D, and F grades.

A special case of the ordinal scale to be considered in this chapter is that in which the available data for a given variable classify individuals into two categories (a dichotomy), but in which the variable itself is assumed to have underlying continuity and to be normally distributed. For example, we might have data only on whether a child's IQ is greater than or equal to 100 (assigned the value 1) or less than 100 (assigned the value 0). Another example might be a psychological scale that classifies subjects as normal (1) or abnormal (0). For both of these examples, the actual measurement of the characteristic is assumed to have underlying continuity and to be normally distributed, but the available data are dichotomous. For these dichotomies, there is an implied order underlying the presence or absence of the characteristic.

Interval or Ratio Scale

In this chapter, we combine the interval and ratio scales in one category. Both of these scales exhibit equal precision in measurement, but the ratio scale has a known zero point. We will assume in this chapter that quantitative variables measured on interval or ratio scales are normally distributed, although in some situations, a distribution may only approximate the normal.

To illustrate the relationships among the interval or ratio scale, the ordinal scale, and the dichotomy with underlying continuity, consider the following distribution of IQ scores for ten students:

Student	IQ	Ranked IQ	Dichotomy
1	103	5	1
2	94	7	0
3	117	1	1
4	112	2	1
5	89	9	0
6	93	8	0
7	99	6	0
8	107	4	1
9	87	10	0
10	110	3	1

The actual IQ scores are given in the second column, and the rankings of these scores are listed in the third column. The fourth column contains the scores obtained when IQ scores greater than 100 are assigned the value 1 and those less than 100 are assigned the value 0. Recall that we are assuming IQ to be a quantitative variable measured on an interval scale and normally distributed. Thus, the fourth column illustrates these scores on a quantitative variable converted to a dichotomous variable with underlying continuity. Note that researchers do not indiscriminately convert scores on a quantitative variable to ranks or dichotomies. By doing so, they would be discarding or ignoring available data, and the resulting correlation coefficient might give a distorted picture of the relationship between the variables. However, there are situations in which available data may exist only in dichotomous form.

Correlation Coefficients for Quantitative and Qualitative Variables

Now that we have reconsidered the measurement scale hierarchy, we will define and discuss the correlation coefficients for the various combinations of quantitative and qualitative variables; these coefficients[2] are classified in Table 20.1. Notice that, by combining the interval and ratio scales, there are only three levels of measurement for each variable (X and Y) depicted in the table and only $3 \times 3 = 9$ different possible combinations of the variables. For some of these combinations, we will consider more than one coefficient. Also, because no distinction is made between correlating Y with X and correlating X with Y, three cells are denoted with numbers in parentheses. The coefficients for these cells would be the same as for those cells with numbers not in parentheses.

> The appropriate correlation coefficient depends on the scales of measurement of the two variables being correlated.

TABLE 20.1
Matrix Showing Correlation Coefficients Appropriate for Scales of Measurement for Variable *X* and Variable *Y*

		Variable X		
		Nominal	*Ordinal*	*Interval/Ratio*

Variable Y	*Nominal*	1 a. Phi (ϕ)
		   b. C coefficient
		   c. Cramer's *V*
		   d. λ and λ_Y
			(4)	(6)

	Ordinal	4 Rank-biserial	2 a. Tetrachoric b. Spearman ρ	(5)
	Interval/Ratio	6 Point-biserial	5 Biserial	3 Pearson *r*

[2]We have included coefficients that are commonly used for research in the behavioral sciences.

Pearson Product-Moment Correlation Coefficients

The correlation coefficient for cell 3 of Table 20.1 is the Pearson product-moment coefficient (r) that we discussed in Chapter 5. It is used when (1) both variables are measured on either an interval or a ratio scale and (2) the underlying distributions of both variables are normal. The raw score formula for the Pearson r was presented in Chapter 5; it is

$$r = \frac{n\Sigma XY - (\Sigma X)(\Sigma Y)}{\sqrt{\left[n\Sigma X^2 - (\Sigma X)^2\right]\left[n\Sigma Y^2 - (\Sigma Y)^2\right]}} \qquad (20.1)$$

Several special cases for this formula have been developed, producing correlation coefficients that can be used for certain combinations of measurement scales. These coefficients, as shown in Table 20.1, are the point-biserial (r_{Pb}), phi (ϕ), and the Spearman rho (ρ). In the following discussion, we present an example of r_{Pb} and ϕ and then calculate the correlation between the two variables using formula 20.1. Subsequently, we give the specific formulas for these coefficients, each one derived from formula 20.1. The correlation coefficient for each example is then recalculated using the specific formula.

Point-Biserial   The **point-biserial correlation coefficient (r_{pb})** is the special case of the Pearson r in which one variable is measured on an interval or ratio scale and the other variable is a nominal variable with two classification levels (a dichotomous variable). For example, suppose we want to correlate success on one item of a test (the dichotomy—either right or wrong) with the total score on the test. The resulting correlation coefficient is the index of the relationship between performance on one test item and performance on the test as a whole.

Consider scores for item 1 of the test (X) and the total test scores (Y) for ten students on a 20-item test; these data are found in Table 20.2, along with the preliminary calculations for using formula 20.1. For this example, we have assigned the value 1 to a correct response to item 1 and the value 0 to an incorrect response. Using formula 20.1 for calculating the correlation between the item scores and total test scores, we find

$$r = \frac{n\Sigma XY - (\Sigma X)(\Sigma Y)}{\sqrt{[n\Sigma X^2 - (\Sigma X)^2][n\Sigma Y^2 - (\Sigma Y)^2]}}$$

$$= \frac{10(59) - (5)(96)}{\sqrt{[10(5) - (5)^2][10(1016) - (96)^2]}}$$

$$= .716$$

Since X is a dichotomous variable with 0 and 1 as the only possible values, formula 20.1 can be simplified into the formula for the point-biserial correlation.

TABLE 20.2
Data for Calculating the Point-Biserial Correlation Coefficient

Subject	Item Score (X)	Test Score (Y)	X^2	Y^2	XY
A	1	10	1	100	10
B	1	12	1	144	12
C	1	16	1	256	16
D	1	10	1	100	10
E	1	11	1	121	11
F	0	7	0	49	0
G	0	6	0	36	0
H	0	11	0	121	0
I	0	8	0	64	0
J	0	5	0	25	0
Σ	5	96	5	1,016	59

The coefficient is denoted r_{pb}, and the formula is as follows:

$$r_{pb} = \frac{\overline{Y}_1 - \overline{Y}_0}{\sigma_Y}\sqrt{pq} \qquad (20.2)$$

where

$\overline{Y}_1 =$ mean of the Y scores for those individuals with X scores equal to 1

$\overline{Y}_0 =$ mean of the Y scores for those individuals with X scores equal to 0

$\sigma_Y =$ standard deviation of all Y scores; that is, $\sqrt{[\Sigma Y^2 - (\Sigma Y)^2/n]/n}$

$p =$ proportion of individuals with an X score of 1

$q =$ proportion of individuals with an X score of 0

For the data in Table 20.2,

$$\overline{Y}_1 = (10 + 12 + 16 + 10 + 11)/5 = 11.80$$

$$\overline{Y}_0 = (7 + 6 + 11 + 8 + 5)/5 = 7.40$$

$$\sigma_Y = \sqrt{[1016 - (96)^2/10]/10} = 3.07$$

$$p = q = 0.50$$

Using these data in formula 20.2, the point-biserial correlation coefficient is found to be

$$r_{pb} = \frac{11.80 - 7.40}{3.07}\sqrt{(0.50)(0.50)}$$

$$= .716$$

This is the same value that we found when we used formula 20.1 for the Pearson r.

For formula 20.2, consider the situation in which $\overline{Y}_1 = \overline{Y}_0$. This means that those subjects who had item 1 incorrect had the same mean score as those subjects who had item 1 correct. Thus, the numerator of formula 20.2 is zero, and $r_{pb} = 0.00$. If $\overline{Y}_1 - \overline{Y}_0$ had been negative, then r_{pb} would have been negative. In this example, this would have meant that subjects scoring high on the total test tended to answer item 1 incorrectly and that those with lower scores tended to answer the item correctly. For some dichotomous variables — gender, for example — the assignment of 1 and 0 is strictly arbitrary, and the sign of the r_{pb} depends on the meaning of that assignment. For example, suppose we want to determine the relationship between the student's gender and score on the same 20-item test, and we code females 1 and males 0. If the r_{pb} is positive, then the interpretation is that females tend to score higher than males on the test.

The point-biserial correlation coefficient is a special case of the Pearson r, in which one variable is a quantitative variable measured on an interval or ratio scale and the other is a nominal (dichotomous) variable. The formula is

$$r_{pb} = \frac{\overline{Y}_1 - \overline{Y}_0}{\sigma_Y} \sqrt{pq}$$

Phi(ϕ)   The special case of the Pearson r in which both variables are nominal dichotomous variables is the **phi (ϕ) coefficient**. Suppose we want to determine the relationship between gender and political party affiliation (Republican or Democrat). The data for this example are found in Table 20.3. For these data, we have arbitrarily assigned the values 1 and 0 as shown in the table. Applying formula 20.1 to these data,

$$r = \frac{(10)(4) - (5)(6)}{\sqrt{[(10)(5) - (5)^2][(10)(6) - (6)^2]}}$$

$$= .408$$

This coefficient indicates that there is a low positive relationship between gender and party affiliation. These results indicate that females tend to be affiliated with the Republican party, and males tend to be affiliated with the Democratic party. This direction is evidenced by the positive correlation, which indicates that scores of 1 tend to be associated with scores of 1 (1 = female, Republican) and zeros with zeros (0 = male, Democrat).

We can also arrange these data into a 2 × 2 contingency table (see Table 20.4). The numbers in each cell of the contingency table are frequencies; in this case, there are two Republican males, four Republican females, and so on. Table 20.5 shows the general form of the 2 × 2 contingency table. The letters *A, B, C,* and *D* represent the

TABLE 20.3
Data for Calculating the Phi (ϕ) Coefficient

Person	Gender (X)	Party (Y)	X^2	Y^2	XY
A	1	1	1	1	1
B	1	1	1	1	1
C	1	0	1	0	0
D	1	1	1	1	1
E	1	1	1	1	1
F	0	0	0	0	0
G	0	1	0	1	0
H	0	1	0	1	0
I	0	0	0	0	0
J	0	0	0	0	0
Σ	5	6	5	6	4

1 = female   1 = Republican
0 = male   0 = Democrat

TABLE 20.4
2 × 2 Contingency Table for Computing the Phi (ϕ) Coefficient

		Gender		
		Male (0)	Female (1)	Totals
Political Party	Republican (1)	2	4	6
	Democrat (0)	3	1	4
	Totals	5	5	10

TABLE 20.5
General Form of the 2 × 2 Contingency Table for Computing the Phi (ϕ) Coefficient

		Variable X		
		0	1	Totals
Variable Y	1	A	B	A + B
	0	C	D	C + D
	Totals	A + C	B + D	N

frequencies in each of the four cells. In the above example, the frequency in cell *A* reflects the number of males affiliated with the Republican party. Note that the order in which the frequencies are placed in the table is important in the interpretation of the phi coefficient; that is, the placement of the (0, 1) combinations in the table should be consistent with the initial assignments of 0 and 1 in Table 20.3.

Again, the formula for the Pearson r can be simplified for the case of the phi coefficient:

$$\phi = \frac{BC - AD}{\sqrt{(A + B)(C + D)(A + C)(B + D)}} \qquad (20.3)$$

Using the data from Table 20.4 in formula 20.3, the phi coefficient is calculated as follows:

$$\phi = \frac{(4)(3) - (2)(1)}{\sqrt{(2 + 4)(3 + 1)(2 + 3)(4 + 1)}}$$

$$= .408$$

This value for ϕ using formula 20.3 is the same as the correlation computed using the general formula for the Pearson r. It is important to note, however, that the maximum values of $+1.0$ and -1.0 are possible only when $(A + B) = (C + D)$ and $(A + C) = (B + D)$. Whenever these two pairs of sums are not equal, the maximum value[3] of ϕ will be less than $+1.0$ and greater than -1.0.

The phi (ϕ) coefficient is the special case of the Pearson r in which both variables are nominal dichotomies. The formula is

$$\phi = \frac{BC - AD}{\sqrt{(A + B)(C + D)(A + C)(B + D)}}$$

Nonproduct Moment Coefficients

The point-biserial, phi (ϕ), and Spearman rho (ρ) are all special cases of the Pearson product-moment correlation coefficient. We now turn our attention to measures of relationship other than product-moment coefficients.

Measures of Association There are several coefficients listed in cell 1 of Table 20.1. As we have seen, the phi (ϕ) coefficient is an appropriate coefficient to use when both variables being correlated are nominal variables with only two classification levels (dichotomous variables). However, ϕ is restricted to 2×2 contingency tables. But there are many research studies in which the variables under investigation are nominal variables with more than two classification levels. The resulting contingency table is therefore larger than 2×2; it may be 3×5, 4×3, and so on. For example, suppose a sociologist is investigating the relationship between level of education (X) and occupational choice (Y); the 4×5 contingency table for these data is found in Table 20.6.

---

[3]The procedure for determining the maximum value of ϕ for a given data set is found in J. B. Carroll, "The Nature of Data, or How to Choose a Correlation Coefficient," *Psychometrika* 26 (1961): 347–372.

TABLE 20.6
Data for Determining the Relationship Between Level of Education and Occupational Choice

	Level of Education					
	Less than HS	HS Graduate	Some College	College Graduate	Graduate Degree	Totals
Laborer/Farmer	347	128	84	37	5	601
Skilled Crafts	164	277	103	43	36	623
Sales/Clerical	30	77	217	147	80	551
Professional/Managerial	2	34	82	198	267	583
Total	543	516	486	425	388	2,358

For contingency tables larger than 2×2 (i levels of the variable placed on the rows and j levels of the variable placed on the columns), a variety of coefficients have been developed. These coefficients are called **measures of association**. Two of these are **Pearson's contingency coefficient (C)** and **Cramer's V coefficient**. The formula for the C coefficient is

$$C = \sqrt{\frac{\chi^2}{n + \chi^2}}$$

(20.4)

The formula for the V coefficient is

$$V = \sqrt{\frac{\chi^2}{n(q - 1)}}$$

(20.5)

where

q = the smaller of i and j

Both of these coefficients require knowledge of the χ^2 (chi-square) statistic, which we will discuss in Chapter 21. Thus, we will postpone further discussion of these coefficients until then.

Another commonly used measure of association is the Goodman and Kruskal **lambda (λ) coefficient**.[4] The formula is

$$\lambda = \frac{\sum_{j=1}^{J} n_{mj} + \sum_{i=1}^{I} n_{im} - n_{m+} - n_{+m}}{2n - n_{m+} - n_{+m}}$$

(20.6)

where

n_{mj} = largest frequency in the jth column

n_{im} = largest frequency in the ith row

---

[4] For a thorough discussion of λ as a measure of proportional reduction in error, see H. T. Reynolds, *Analysis of Nominal Data*, 2d ed. (Beverly Hills, Calif.: Sage Publications, 1984), pp. 49–57.

$$n_{m+} = \text{largest marginal row total}$$
$$n_{+m} = \text{largest marginal column total}$$
$$n = \text{number of observations}$$

For the data in Table 20.6,

$$\sum_{j=1}^{J} n_{mj} = (347 + 277 + 217 + 198 + 267) = 1,306$$

$$\sum_{i=1}^{I} n_{im} = (347 + 277 + 217 + 267) = 1,108$$

$$n_{m+} = 623$$
$$n_{+m} = 543$$
$$n = 2,358$$

Applying formula 20.6 to these data,

$$\lambda = \frac{1,306 + 1,108 - 623 - 543}{2(2,358) - 623 - 543}$$

$$= .352$$

This coefficient is called the *symmetric* version of λ; it is used when the researcher is unwilling to specify which variable (X or Y) can be considered the dependent variable. However, when the researcher can legitimately identify which variable can be considered the dependent variable, the asymmetric version of λ should be used. For the above example, the researcher might consider occupational choice the dependent variable. The *asymmetric* λ is denoted λ_Y and is defined as

$$\lambda_Y = \frac{\sum_{j=1}^{J} n_{mj} - n_{m+}}{n - n_{m+}} \tag{20.7}$$

where

$$n_{mj} = \text{largest frequency in the } j\text{th column}$$
$$n_{m+} = \text{largest marginal row total}$$
$$n = \text{number of observations}$$

For the data from Table 20.6,

$$\lambda = \frac{1,306 - 623}{2,358 - 623}$$

$$= 0.394$$

For most practical situations, the interpretation of the λ_Y and the λ coefficients is similar to interpreting any correlation coefficient.[5] For this example, we would conclude that there is a moderate relationship between level of education and occupational choice. The data in Table 20.6 indicate that those individuals with more education tend to have sales/clerical or professional/managerial positions, whereas those with less education tend to have laborer/farmer or skilled-crafts positions.

Biserial Correlation Coefficient   The **biserial correlation coefficient** (r_b) is similar to the point-biserial; its use is appropriate when one variable is quantitative and measured on the interval or ratio scale and the other variable is dichotomous but with underlying continuity. Suppose we want to determine the relationship between anxiety level and performance on a standardized achievement test. Achievement is measured on an interval scale, but the data on the anxiety measure are recorded only in terms of high and low anxiety. The data for this example are found in Table 20.7. Since the case can be made that anxiety, as a psychological construct, has underlying continuity and is normally distributed, the biserial is the appropriate coefficient to use. The formula for the biserial correlation coefficient is

$$r_b = \left(\frac{\overline{Y}_1 - \overline{Y}_0}{\sigma_Y} \right) \left(\frac{pq}{Y} \right) \tag{20.8}$$

where

$\overline{Y}_1 =$ mean of the Y scores for those individuals with X scores equal to 1

$\overline{Y}_0 =$ mean of the Y scores for those individuals with X scores equal to 0

$\sigma_Y =$ standard deviation of all Y scores, that is, $\sqrt{[\Sigma Y^2 - (\Sigma Y)^2/n]/n}$

$p =$ proportion of individuals with an X score of 1

$q =$ proportion of individuals with an X score of 0

$Y =$ ordinate (height) of the unit normal curve at the point of the division between the p and q proportions under the curve

Note the similarity between this formula and the formula for the point-biserial correlation. The major difference is the inclusion of Y, the ordinate of the unit normal curve. To illustrate the procedure for determining Y, consider the data in Table 20.7. Notice that, for the dichotomy representing anxiety level (which is assumed to be normally distributed), $p = 0.40$ and $q = 0.60$ (see Figure 20.1). First, we determine the z score for which 40 percent of the area under the unit normal curve is above this score and 60 percent of the area is below this score. From Table C.2, we find this z score to be 0.25.[6] Second, we must estimate the ordinate at this z score using values from Table C.1. For $z = 0.25$, the ordinate (Y) equals 0.3867.

---

[5]For a thorough discussion of the interpretation of λ, see Reynolds.

[6]For computational convenience, we have rounded the z score to two decimal places.

TABLE 20.7
Data for Calculating the Biserial
Correlation Coefficient

Anxiety Level (X) High = 1 Low = 0	Achievement Score (Y)
1	435
1	380
1	365
1	410
1	420
1	430
0	510
0	485
0	390
0	560
0	330
0	475
0	400
0	495
0	530

$$\Sigma Y = 6{,}615$$

$$\Sigma Y^2 = 2{,}977{,}725$$

$$\bar{Y}_1 = \Sigma Y_1/n_1 = 406.67$$

$$\bar{Y}_0 = \Sigma Y_0/n_0 = 463.89$$

$$\sigma_Y = \sqrt{\Sigma Y^2 - (\Sigma Y)^2/n} = 63.51$$

$$p = 6/15 = .40$$

$$q = 9/15 = .60$$

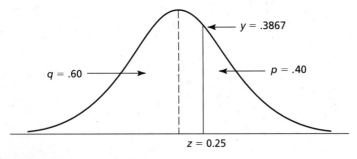

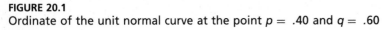

FIGURE 20.1
Ordinate of the unit normal curve at the point $p = .40$ and $q = .60$

For the data in Table 20.7,

$$\overline{Y}_1 = (435 + 380 + 365 + 410 + 420 + 430)/6 = 406.67$$
$$\overline{Y}_0 = (510 + 485 + 390 + 560 + 330 + 475 + 400 + 495 + 530)/9 = 463.89$$
$$\sigma_Y = \sqrt{[2,977,725 - (6,615)^2/15]/15} = 63.51$$
$$p = 6/15 = 0.40$$
$$q = 9/15 = 0.60$$
$$Y = 0.3867$$

Using these data in formula 20.8, the biserial correlation coefficient is found to be

$$r_b = \left(\frac{\overline{Y}_1 - \overline{Y}_0}{\sigma_Y}\right)\left(\frac{pq}{Y}\right)$$
$$= \left(\frac{406.67 - 463.89}{63.51}\right)\left[\frac{(0.40)(0.60)}{0.3867}\right]$$
$$= -.56$$

This coefficient indicates that there is a negative relationship between level of anxiety and performance on the standardized achievement test. More specifically, the data from these 15 individuals indicate that those with a high level of anxiety tended to have lower achievement scores and that those with lower anxiety levels tended to have higher achievement score.

Theoretically, the maximum values of the biserial correlation are $+1.0$ and -1.0. However, if the distribution of the continuous variable underlying the dichotomy departs appreciably from normal, or if the dividing point of the dichotomy departs appreciably from $p = q = 0.50$, the values of the biserial correlation can exceed ± 1. This is a rare occurrence, but if either of these conditions is present, a more appropriate coefficient should be used.

There is a similarity between the biserial and the point-biserial correlation coefficients, represented by the following expression:

$$r_b = r_{pb}\frac{\sqrt{pq}}{Y}$$

The factor $\sqrt{pq}/Y$ is always greater than 1; as a result the biserial is always greater than the point-biserial. The difference increases with the extremeness of the dichotomy. For example, when $p = q = 0.50$, $\sqrt{pq}/Y = 1.253$; but when either p or $q = 0.99$, $\sqrt{pq}/Y = 3.741$.

The biserial correlation coefficient is appropriate when one variable is quantitative and measured on the interval or ratio scale and the other variable is dichotomous but with underlying continuity. The formula is

$$r_b = \left(\frac{\overline{Y}_1 - \overline{Y}_0}{\sigma_Y} \right) \left(\frac{pq}{Y} \right)$$

Tetrachoric Correlation Coefficient   The tetrachoric correlation coefficient is similar to the phi (ϕ) coefficient because its use is appropriate when both variables are dichotomies. But, unlike the phi coefficient, the **tetrachoric correlation coefficient** (r_{tet}) is used when we can assume that both variables have underlying continuity and are normally distributed. For example, suppose we want to determine the relationship between two items on a symbolic reasoning test. These two items are assumed to measure constructs that are normally distributed in the population. However, the data are restricted to a correct response (1) and an incorrect response (0) to the items; hence, the data are in the form of dichotomies. The data in Table 20.8 represent the responses to the two items for 120 introductory psychology students.

The formula for the tetrachoric correlation coefficient is

$$r_{tet} = \cos \frac{180°}{1 + \sqrt{BC/AD}} \tag{20.9}$$

This formula, which is an approximation of a more complex one developed by Karl Pearson in 1900, requires some understanding of trigonometric functions. However, we have developed a table (Table C.13) to simplify the task of computing r_{tet}. This table is based on the BC/AD ratio, where A, B, C, and D are the observed cell frequencies of the 2×2 contingency table. If BC/AD is greater than 1, r_{tet} is read directly from Table C.13. For example, if $BC/AD = 2.99$, then $r_{tet} = +.41$. If, however, BC/AD is less than 1, then the AD/BC ratio is computed, and the value found in Table C.13 is assigned a negative value. For example, if $BC/AD = 0.137$, then $AD/BC = 7.322$ and $r_{tet} = -.66$.

For the data in Table 20.8, $BC/AD = 0.432$. Since this value is less than 1, we compute AD/BC; we find this ratio to be 2.316. Referring to Table C.13, we find that $r_{tet} = -.32$. This coefficient indicates that there is a low negative relationship between the responses on the two items. Exploring the data in Table 20.8 provides additional insight into the nature of this low negative relationship. As can be seen, students who had a correct response to the first item also tended to have a correct response to the second item. But notice that there was a larger percentage of students who had an incorrect response to the first item and a correct response to the second item, whereas very few students had both items incorrect.

TABLE 20.8
Data for Calculating the Tetrachoric Correlation Coefficient

		Item No. 1			
		Incorrect (0)	Correct (1)		
Item No. 2	Correct (1)	33	57	90	
	Incorrect (0)	6	24	30	
		39	81	120	

The limits of the values for the tetrachoric coefficient are $+1.0$ and -1.0, even when $(A+B) \neq (C+D)$ or $(A+C) \neq (B+D)$. Recall that this was not the case for the phi (ϕ) coefficient. This property, by itself, makes r_{tet} superior to ϕ as the index of relationship between two variables, if it is reasonable to assume that both dichotomous variables have underlying continuity and are normally distributed. Note also that the tetrachoric correlation coefficient, like ϕ, is restricted to 2×2 contingency tables. The logical extension of r_{tet} to larger contingency tables has been developed, but the computational procedures are beyond the scope of this book.[7]

> The tetrachoric correlation coefficient is appropriate when both dichotomous variables are assumed to have underlying continuity and to be normally distributed. The formula is
>
> $$r_{tet} = \cos \frac{180°}{1 + \sqrt{BC/AD}}$$

Rank-Biserial Correlation Coefficient   The **rank-biserial correlation coefficient** (r_{rb}) is similar to the point-biserial. For both coefficients, one variable is a nominal dichotomy, but for the rank-biserial, the other variable is a ranked variable. For example, suppose a researcher is interested in the relationship between the fact that an individual is at least a second-generation American (X) and socioeconomic status (Y); the data are found in Table 20.9. In this example, the X variable (immigration status) is considered a nominal dichotomy ($0 =$ less than second generation; $1 =$ second generation or greater). The data for the Y variable (socioeconomic status) are ranked with $1 =$ highest status; $2 =$ next highest status; and so on.

[7]H. O. Lancaster and M. A. Hamden, "Estimates of the Correlation Coefficient in Contingency Tables with Possibly Nonmetric Characters," *Psychometrika* 29 (1964): 383–391.

TABLE 20.9
Data for Calculating the Rank-Biserial
Correlation Coefficient

Person	Immigrating Generation (X)	Rank of Socio-economic Status (Y)
A	1	1
B	1	2
C	1	3
D	0	4
E	0	5
F	1	6
G	1	7
H	0	8
I	1	9
J	0	10
K	0	11
L	0	12

$n = 12$

$\overline{Y}_1 = 28/6 = 4.667$

$\overline{Y}_0 = 50/6 = 8.333$

The formula for the rank-biserial was originally developed by E. E. Cureton.[8] Glass refined the formula for use when there are no tied ranks.[9] This formula is

$$r_{rb} = \frac{2}{n}\left(\overline{Y}_1 - \overline{Y}_0\right) \tag{20.10}$$

where

$n =$ number of observations

$\overline{Y}_1 =$ mean rank for individuals with X scores equal to 1

$\overline{Y}_0 =$ mean rank for individuals with X scores equal to 0

Applying this formula to the data in Table 20.9, we find

$$r_{rb} = \frac{2}{12}\left(\frac{28}{6} - \frac{50}{6}\right)$$

$$= \frac{1}{6}(4.667 - 8.333)$$

$$= -.611$$

[8] E. E. Cureton, "Rank-Biserial Correlation," *Psychometrika* 21 (1956): 287–290, and "Rank-Biserial Correlation—When Ties Are Present," *Educational and Psychological Measurement* 28 (1968): 77–79.

[9] G. V Glass, "Note on the Rank-Biserial Correlation," *Educational and Psychological Measurement* 26 (1966): 623–631.

This negative coefficient indicates that those who are at least second-generation Americans tend to have higher socioeconomic status than those whose families immigrated more recently.

The maximum values for the rank-biserial correlation coefficient are $+1$ and -1. Thus, r_{rb} can be interpreted in the same way as any of the Pearson product-moment coefficients. Note, however, that formula 20.10 is appropriate for r_{rb} only when there are no tied ranks. When tied ranks do occur, the original formula developed by Cureton must be used.

> The rank-biserial correlation coefficient is appropriate when one variable is a nominal dichotomy and the other is a ranked variable. The formula is
>
> $$r_{rb} = \frac{2}{n}(\overline{Y}_1 - \overline{Y}_0)$$

Coefficient of Nonlinear Relationship: Eta (η)

The correlation coefficients we have discussed are measures of the linear relationship between two variables. These coefficients are appropriate when we can assume that the relationship in the population is linear, or the points in a scatterplot locate along a straight line. But, as noted in Chapter 5, not all variables are linearly related; some have a curvilinear relationship. This means that a straight line does not fit the points in the scatterplot, but that some type of curved line does. Some variables have a linear relationship over short intervals on the scale of measurement but do not maintain a linear relationship over long intervals. For example, the relationship between performance on a complex physical task and age is not linear. The very young do not perform well, but in the teens and twenties, performance increases as age increases. At a certain point — say, late twenties or early thirties — performance begins to decline as age increases. Thus, over a long time span, the relationship between the variables is not linear.

An index that does not assume a linear relationship between two variables is the **eta (η) coefficient**; it is sometimes called the *correlation ratio*. Eta should be used whenever a scatterplot suggests that there is a nonlinear relationship between two variables. For example, recall that, when we discussed the biserial correlation coefficient, we described the relationship between anxiety level and performance on a standardized achievement test. In that example, anxiety was measured over a limited span, and it had a linear relationship with the achievement variable. However, if anxiety is measured over a wide span, from very low to very high, its relationship to many performance and achievement variables may not be linear. Instead, people with either low or high anxiety levels might not perform well, whereas those with a moderate amount of anxiety might be the better performers. In other cases, a modulating, hills-and-valleys scatterplot fits the data for the relationship between anxiety and another variable. For example, consider the relationship between anxiety and scores on a complex concept-attainment task. The scatterplot for this example is found in Figure 20.2. As can be seen, subjects with low anxiety scores tend to have poor performance scores, and subjects with high anxiety scores tend to have good performance scores. However, between these extremes, the fit of the curve to the points tends to reverse itself.

Computation of the eta coefficient is more complex than that for any of the previously discussed correlation coefficients; the procedures are outlined in Table 20.10. The first step is to categorize one of the variables into intervals of equal width. It is somewhat arbitrary about which of the variables to categorize and the number of categories to use. We do not want an excessive number of categories, yet we do want to bring out the general shape of the scatterplot. A rule of thumb that works well for most variables is to have no fewer than 6 categories and no more than 12. For this example, we will categorize the test anxiety scores (X) into the following intervals: 1–3, 4–6, 7–9, 10–12, 13–15, and 16–18.

The next step is to calculate the mean performance score for the total group (grand mean = $\overline{Y}_t$ = 9.44) as well as for each of the categories of test anxiety (group mean = $\overline{Y}_k$; see Table 20.10). Then, for each individual, (1) determine the

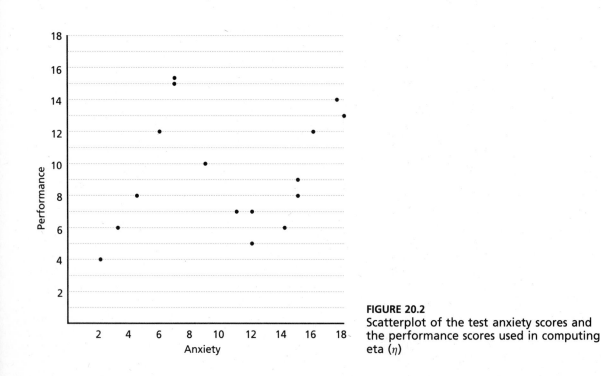

FIGURE 20.2
Scatterplot of the test anxiety scores and the performance scores used in computing eta (η)

TABLE 20.10
Data for Calculating the Eta (η) Coefficient

Subject	Anxiety Score (X)	Category of X	Performance Score (Y)	$\bar{Y}_k$	$(Y - \bar{Y}_k)$	$(Y - \bar{Y}_k)^2$	$(Y - \bar{Y}_t)$	$(Y - \bar{Y}_t)^2$
						Computations for Eta (η)		
1	2	(1–3)	4	5.00	−1.00	1.00	−5.44	29.59
2	3	(1–3)	6	5.00	1.00	1.00	−3.44	11.83
3	5	(4–6)	8	10.00	−2.00	4.00	−1.44	2.07
4	6	(4–6)	12	10.00	2.00	4.00	2.56	6.55
5	7	(7–9)	15	13.33	1.67	2.78	5.56	30.91
6	7	(7–9)	15	13.33	1.67	2.78	5.56	30.91
7	9	(7–9)	10	13.33	−3.33	11.09	0.56	0.31
8	11	(10–12)	7	6.33	0.67	0.45	−2.44	5.95
9	12	(10–12)	7	6.33	0.67	0.45	−2.44	5.95
10	12	(10–12)	5	6.33	−1.33	1.77	−4.44	19.71
11	14	(13–15)	6	7.67	−1.67	2.78	−3.44	11.83
12	15	(13–15)	8	7.67	0.33	0.11	−1.44	2.07
13	15	(13–15)	9	7.67	1.33	1.77	−0.44	0.19
14	16	(16–18)	12	13.00	−1.00	1.00	2.56	6.55
15	17	(16–18)	14	13.00	1.00	1.00	4.56	20.79
16	18	(16–18)	13	13.00	0.00	0.00	3.56	12.67
Σ	169		151		0.00	35.98	0.00	197.88

deviation score from the group mean $(\overline{Y}_k)$ and from the grand mean $(\overline{Y}_t)$; (2) square the deviation scores $[(Y - \overline{Y}_k)^2$ and $(Y - \overline{Y}_t)^2]$; and (3) sum the squared deviations. The formula for the eta coefficient is as follows:

$$\eta = \sqrt{1 - \frac{\Sigma(Y - \overline{Y}_k)^2}{\Sigma(Y - \overline{Y}_t)^2}} \qquad (20.11)$$

Using the data from Table 20.10,

$$\eta = \sqrt{1 - \frac{35.98}{197.88}}$$

$$= .91$$

The eta coefficient can also be computed using the procedures outlined in Chapter 14 for ANOVA. By categorizing the scores on one of the variables (X, the test anxiety scores), the categories can serve as the levels of this variable. In our example, there would be six levels. The other variable (Y, the scores on the concept-attainment task) can serve as the dependent or criterion variable. Using ANOVA procedures, the eta coefficient is defined as follows:

$$\eta = \sqrt{1 - \frac{SS_W}{SS_T}} \qquad (20.12)$$

$$= \sqrt{\frac{SS_B}{SS_T}}$$

The interpretation of the eta (η) coefficient is the same as for the Pearson r. That is, the square of the coefficient (η^2) is the proportion of the variance in the Y variable that can be attributed to the variance in the X variable. For this example, the eta (η) coefficient indicates a strong relationship between anxiety scores and performance on the concept-attainment task. Specifically, there is a strong nonlinear relationship, as illustrated in Figure 20.2. Had the Pearson r formula been applied to these data, we would have found $r = .24$. Thus, the use of the Pearson product-moment correlation coefficient for nonlinear data underestimates the relationship between the two variables and gives a distorted picture of the relationship.

There are other important differences between r and η. The range of values for η is from 0 to +1; the coefficient cannot be negative. It is also important to note that η^2 will never be smaller than r^2; that is, $0 \le r^2 \le \eta^2 \le 1$. Finally, if we correlate two variables using the Pearson r, there is only one r value. But, when two variables are correlated using the eta coefficient, there are two possible values of η, depending on which variable is categorized. If, in our example, we categorized the scores on the performance variable rather than the scores on the anxiety variable, then the value of η would have been different.

The eta (η) coefficient is the index of nonlinear relationship between two variables. The formula is

$$\eta = \sqrt{1 - \frac{\Sigma(Y - \overline{Y}_k)^2}{\Sigma(Y - \overline{Y}_t)^2}}$$

The use of the Pearson r when there is a nonlinear trend in the data will underestimate the relationship between the two variables.

Summary

The Pearson product-moment correlation coefficient (r) is by far the most frequently used correlation coefficient in behavioral science research. In this chapter, though, we have extended the discussion of correlation beyond the Pearson r. The various coefficients discussed require certain assumptions about the distributions of the variables being correlated. Within these assumptions is the nature of the variables; that is, quantitative variables (interval or ratio-scale measurement) versus qualitative variables (nominal or ordinal-scale measurement). Table 20.11 contains the formulas for the various coefficients discussed.

Exercises

1. A research psychologist is interested in knowing the relationship between scores on a finger-dexterity test and gender. The following data are from five males and five females who participated in the investigation. Compute the point-biserial correlation coefficient.

Male	Female
11	10
13	7
12	9
15	6
10	7

2. For the following data, compute the phi (ϕ) coefficient to determine the relationship between gender of respondent and whether the respondent is for or against an issue.

	Male	Female
For	20	30
Against	10	40

TABLE 20.11
Formulas for Correlation Coefficients

Pearson r	$r = \dfrac{n\Sigma XY - (\Sigma X)(\Sigma Y)}{\sqrt{[n\Sigma X^2 - (\Sigma X)^2][n\Sigma Y^2 - (\Sigma Y)^2]}}$	(20.1)
Point-biserial	$r_{pb} = \dfrac{\bar{Y}_1 - \bar{Y}_0}{\sigma_Y}\sqrt{pq}$	(20.2)
Phi (ϕ)	$\phi = \dfrac{BC - AD}{\sqrt{(A+B)(C+D)(A+C)(B+D)}}$	(20.3)
Spearman rho (ρ)	$\rho = 1 - \dfrac{6\Sigma d^2}{n(n^2 - 1)}$	(5.8)
Contingency (C)	$C = \sqrt{\dfrac{\chi^2}{n + \chi^2}}$	(20.4)
Cramer's V	$V = \sqrt{\dfrac{\chi^2}{n(q-1)}}$	(20.5)
Symmetric λ	$\lambda = \dfrac{\sum\limits_{j=1}^{J} n_{mj} + \sum\limits_{i=1}^{I} n_{im} - n_{im+} - n_{+m}}{2n - n_{m+} - n_{+m}}$	(20.6)
Asymmetric λ	$\lambda_Y = \dfrac{\sum\limits_{j=1}^{J} n_{mj} - n_{m+}}{n - n_{m+}}$	(20.7)
Biserial	$r_b = \left(\dfrac{\bar{Y}_1 - \bar{Y}_0}{\sigma_Y}\right)\left(\dfrac{pq}{Y}\right)$	(20.8)
Tetrachoric	$r_{tet} = \cos\dfrac{180°}{1 + \sqrt{BC/AD}}$	(20.9)
Rank-biserial	$r_{rb} = \dfrac{2}{n}(\bar{Y}_1 - \bar{Y}_0)$	(20.10)
Eta (η)	$\eta = \sqrt{1 - \dfrac{\Sigma(Y - \bar{Y}_k)^2}{\Sigma(Y - \bar{Y}_t)^2}}$	(20.11)

3. A psychologist administers the Rathus Assertiveness Scale to 200 students. He also administers a test measuring the degree of masculinity (high or low) to the same 200 students. For the following data, compute the tetrachoric correlation coefficient.

	High masc.	*Low masc.*
High assertiveness	82	40
Low assertiveness	23	55

4. A family therapist is investigating the relationship between the marital status of graduate students and their perceived level of life satisfaction (high or low). For the following data, compute the phi (ϕ) coefficient.

	High	Low
Single	21	34
Married	44	16

5. An educational psychologist thinks that left-handed people have poorer handwriting than those who are right-handed. An independent judge, without knowledge of the handedness of ten students, ranks their writing on a scale of 1 (poor) to 10 (good). Using the following data, compute the rank-biserial correlation coefficient.

Student	Left or right	Rank	Student	Left or right	Rank
1	L	1	6	R	6
2	R	2	7	L	7
3	L	3	8	R	8
4	L	4	9	L	9
5	R	5	10	R	10

6. The instructor of an introductory psychology class is interested in conducting an item analysis of the final examination. Compute the point-biserial correlation coefficient between the scores on the final examination and whether the student got item 1 correct. The data are as follows:

Student	Item 1 (1 = correct) (0 = incorrect)	Score
1	1	70
2	1	84
3	1	75
4	1	80
5	0	94
6	0	87
7	0	90
8	0	85
9	0	92
10	0	90

*7. A group of physical therapy students is administered an academic aptitude test upon entrance to the program. Near the end of the program, the students complete an internship. Although there are numerous measures of their performance taken during their internship, they are given a pass or fail grade upon completion. 100 students complete the internship; 62 pass and 38 fail. For those who passed the internship, the mean score on the academic aptitude test was 121.8; for those who failed, the mean score was 105.6. The standard deviation for the total group was 18.2. Compute the appropriate correlation coefficient in determining the relationship between academic aptitude and success in the internship.

8. For the following data, determine the biserial correlation coefficient between parents' income (over $15,000 = 1, under $15,000 = 0) and student's expected yearly income within 10 years.

Parents' income	Student's expected income	Parents' income	Student's expected income
1	$20,000	0	$18,000
1	15,000	0	20,000
1	18,000	0	12,000
1	17,500	0	10,000
1	22,000	0	8,000
0	12,000	0	14,000
0	15,000	0	15,000
0	10,000		

9. The governor of a large midwestern state is interested in the relationship between a person's geographic location and attitude toward marijuana use. For the following data, compute λ. Also, assuming that attitude toward marijuana use is the dependent variable, compute λ_Y.

		Geographic location				
Attitude	Rural	Small suburban	Large suburban	Small urban	Large urban	Total
Should be legalized	10	68	173	98	202	551
Should be misdemeanor	34	79	48	105	104	370
Should be felony	157	105	3	86	57	408
Total	201	252	224	289	363	1,329

10. The number of points scored per game by a certain high school basketball player appeared to be related to his perception of how important each game was to his team's success. The data below represent his rating of each game's importance and the number of points that he scored each game. The ratings were done on a six-point scale, with 1 meaning not important and 6 meaning very important.

Game	Perceived importance	Points	Game	Perceived importance	Points
1	2	5	9	6	5
2	1	4	10	2	5
3	3	8	11	3	7
4	3	10	12	4	18
5	5	12	13	4	15
6	1	3	14	6	6
7	2	8	15	5	10
8	5	12			

    a. Plot the data.

    b. Compute eta.

    c. Interpret the results in practical terms for this basketball player.

11. A nutrition consultant to a local public school was interested in finding whether students' choices of breakfast cereals correlated with their grade levels. She asked a random sample of 500 students about cereal preferences. Using data below, compute λ.

		Cereal		
Grade	Cheerios	Wheaties	Frosted Flakes	Trix
3	27	20	32	61
6	51	47	49	42
9	42	58	34	37

12. A counselor explored the relationship between family structure and children's self-confidence by testing 200 randomly chosen middle-school students on the XYZ Self-Confidence Inventory. He then computed summary statistics for (a) all 200 students, (b) the 80 students from single-parent households, and (c) the 120 students from two-parent households. The statistics are presented below. Compute the biserial correlation coefficient between the number of parents in the households and self-confidence.

$\overline{X}$ for all students = 61.02

s for all students = 8.15

$\overline{X}$ for students in single-parent households = 55.80

$\overline{X}$ for students in two-parent households = 64.50

13. A psychologist investigated the relationship between handedness and reaction time. Reaction time (RT) was defined as the time required to press a button when a bulb was lit. The data for a random sample of ten subjects are listed below. Calculate the rank-biserial correlation coefficient and interpret it in words.

RT	Rank	Handedness
.13	1	R
.15	2	R
.16	3	L
.18	4	R
.19	5	L
.26	6	R
.29	7	R
.30	8	R
.33	9	L
.47	10	R

21

Chi-Square (χ^2) Tests for Frequencies

Key Concepts

Nonparametric tests
χ^2 distribution
Observed frequency
Theoretical (expected) frequency
Goodness-of-fit test

Standardized residual
Test of homogeneity
Yates' correction for continuity
McNemar test
Stuart-Maxwell test

I n previous chapters, we discussed the use of the normal distribution, the Student's t distribution, and the F distribution to test various hypotheses concerning population parameters. In each test, certain assumptions were made about the parameters of the populations from which the samples were drawn; namely, assumptions of normality and homogeneity of variance for the respective populations. These tests are generally called *parametric tests*. But what do we do when the data for a research project do not meet parametric assumptions? This, of course, is not unusual in behavioral science research. **Nonparametric tests** have been developed for such data; these tests have less restrictive assumptions.[1]

> Nonparametric tests can be used when the parametric assumptions of normality and homogeneity of variance are not met.

[1] J. Gaito, "Measurement Scales and Statistics: Resurgence of an Old Misconception," *Psychological Bulletin* 87 (1980): 564–567.

The χ^2 Distribution

Before we begin the discussion of the various nonparametric tests,[2] we must define the χ^2 (chi-square) distribution. The most frequent use of the χ^2 **distribution** is in the analysis of nominal data.[3] In such analyses, we compare **observed frequencies** of occurrence with **theoretical** or **expected frequencies**. Observed frequencies are those that the researcher obtains empirically through direct observation; theoretical or expected frequencies are developed on the basis of some hypothesis. For example, in 200 flips of a coin, we would expect 100 heads and 100 tails. But what if we observed 92 heads and 108 tails? Would we reject the hypothesis that the coin is fair? Or would we attribute the difference between observed and expected frequencies to random fluctuation?

Consider another example. Suppose we hypothesize that we have an unbiased die. To test this hypothesis, we roll the die 300 times and observe the frequency of occurrence of each of the faces. Since we hypothesized that the die is unbiased, we expect that each face will occur 50 times. However, suppose we observe frequencies of occurrence as follows:

Face value	Occurrence
1	42
2	55
3	38
4	57
5	64
6	44

Again, what would we conclude? Is the die biased, or do we attribute the difference to random fluctuation?

Consider a third example. The president of a major university hypothesizes that at least 90 percent of the teaching and research faculty will favor a new university policy on consulting with private and public agencies within the state. Thus, for a random sample of 200 faculty members, the president would *expect* $0.90 \times 200 = 180$ to favor the new policy and $0.10 \times 200 = 20$ to oppose it. Suppose, however, for this sample, 168 faculty favor the new policy and 32 oppose it. Is the difference between observed and expected frequencies sufficient to reject the president's hypothesis that 90 percent would favor the policy? Or would the differences be attributed to chance fluctuation?

---

[2] For additional nonparametric tests, see M. Hollander and D. A. Wolfe, *Nonparametric Statistical Methods* (New York: Wiley, 1973), and L. A. Marascuilo and M. McSweeney, *Nonparametric and Distribution-free Methods for the Social Sciences* (Monterey, Calif.: Brooks/Cole, 1977).

[3] H. T. Reynolds, *Analysis of Nominal Data*, 2d ed. (Beverly Hills, Calif.: Sage Publications, 1984).

In each of these examples, the test statistic for comparing observed and expected frequencies is χ^2, defined as follows:

$$\chi^2 = \sum_{i=1}^{k} \frac{(O - E)^2}{E} \tag{21.1}$$

where

O = observed frequency

E = expected frequency

k = number of categories, groupings, or possible outcomes

The calculations of χ^2 for each of the three examples, using formula 21.1, are found in Tables 21.1, 21.2, and 21.3.

TABLE 21.1
Calculation of χ^2 for the Coin-Toss Example

	O	E	$O - E$	$(O - E)^2$	$(O - E)^2/E$
Heads	92	100	−8	64	.64
Tails	108	100	+8	64	.64
Total	200	200	0	—	$1.28 = \chi^2$

TABLE 21.2
Calculation of χ^2 for the Die Example

Face Value	O	E	$O - E$	$(O - E)^2$	$(O - E)^2/E$
1	42	50	−8	64	1.28
2	55	50	5	25	.50
3	38	50	−12	144	2.88
4	57	50	7	49	.98
5	64	50	14	196	3.92
6	44	50	−6	36	.72
Total	300	300	0	—	$10.28 = \chi^2$

TABLE 21.3
Calculation of χ^2 for the Consulting-Policy Example

	O	E	$O - E$	$(O - E)^2$	$(O - E)^2/E$
Favor	168	180	−12	144	.80
Oppose	32	20	+12	144	7.20
Total	200	200	0	—	$8.00 = \chi^2$

The theoretical sampling distribution of χ^2 can be generated.[4] As for the t distribution, there is a family of χ^2 distributions, each a function of the degrees of freedom associated with the number of categories in the sample data. As in the case of the t distributions, only a single df value is required to identify the specific χ^2 distribution. Unlike the t distributions, which are symmetrical, the theoretical sampling distributions of χ^2 are *positively skewed.* However, as the number of degrees of freedom associated with χ^2 increases, the respective sampling distribution approaches symmetry. The χ^2 distributions for 1, 3, 5, and 10 degrees of freedom are illustrated in Figure 21.1. Notice that all values of χ^2 are positive, ranging from zero to infinity.

Consider the degrees of freedom for each of the above examples. In the coin example, note that the *expected* frequencies in each of the two categories (heads or tails) are *not* independent. To obtain the expected frequency of tails (100), we need only to subtract the expected frequency of heads (100) from the total frequency (200), or $200 - 100 = 100$. Similarly, for the example of the new consulting policy, the expected number of faculty members who oppose it (20) can be found by subtracting the expected number who support it (180) from the total number in the sample (200), or $200 - 180 = 20$. Thus, given the expected frequency in one of the categories, the expected frequency in the other is readily determined. In other words, only the expected frequency in one of the two categories is free to vary; that is, there is only 1 degree of freedom associated with these examples.

For the die example, there are six possible categories of outcomes: the occurrence of the six faces. Under the assumption that the die is fair, we would *expect* that the frequency of occurrence of each of the six faces of the die would be 50. Note again that the expected frequencies in each of these categories are *not* independent. Once the expected frequency for five of the categories is known, the expected frequency of the sixth category is uniquely determined, since the total frequency

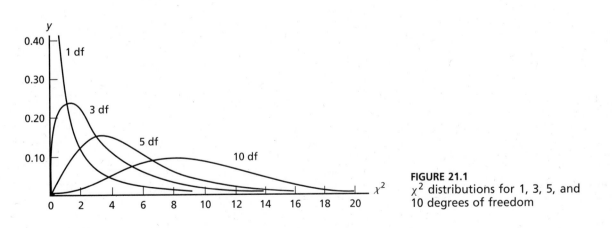

FIGURE 21.1
χ^2 distributions for 1, 3, 5, and 10 degrees of freedom

[4]For a more exhaustive discussion of the development of the χ^2 distribution, see L. Horowitz, *Elements of Statistics for Psychology and Education* (New York: McGraw-Hill, 1974), pp. 371–380.

equals 300. Thus, only the expected frequencies in five of the six categories are free to vary; there are only 5 degrees of freedom associated with this example.

> The categories are based on nominal-, ordinal-, interval-, and ratio-level data. The values in the categories are frequencies.

The Critical Values for the χ^2 Distribution

The use of the χ^2 distribution in hypothesis testing is analogous to the use of the t and F distributions. A null hypothesis is stated, a test statistic is computed, the observed value of the test statistic is compared to the critical value, and a decision is made whether or not to reject the null hypothesis. For the coin example, the null hypothesis is that the frequency of heads is equal to the frequency of tails. For the die example, the null hypothesis is that the frequency of occurrence of each of the six faces is the same. In general, it is not a requirement for the categories to have equal expected frequencies. For instance, in the example of the new consulting policy, the null hypothesis is that 90 percent of the faculty will support the new policy and 10 percent will not.

The critical values of χ^2 for 1 through 30 degrees of freedom are found in Table C.4. Fourteen different percentile points in each distribution are given, although in this chapter we are primarily interested in the more commonly used α levels, such as .05 and .01. For the coin and consulting-policy examples, the critical values of χ^2 for 1 degree of freedom, with $\alpha = .05$ and $\alpha = .01$, are 3.841 and 6.635, respectively. For the die example, the corresponding critical values of χ^2 for 5 degrees of freedom are 11.070 and 15.086. Although Table C.4 is sufficient for most research settings in the behavioral sciences, there are some situations in which the degrees of freedom associated with a χ^2 test are greater than 30. For these situations, the following expression has a sampling distribution that is approximately normal:

$$z = \sqrt{2\chi^2} - \sqrt{2df - 1} \qquad (21.2)$$

Since the sampling distribution of this formula is approximately normal, the critical values for the .05 and .01 levels of significance are 1.96 and 2.58, respectively.

> The χ^2 distributions comprise a family of distributions, each determined by a single degree-of-freedom value.

Now that we have seen the table of critical values for the χ^2 distribution, we can complete the examples. For the coin example, the null hypothesis is that the frequency of heads equals the frequency of tails. As we mentioned, since there are only two categories, once the expected value of the first category is determined, the second is uniquely determined. Thus, there is only 1 degree of freedom associated with this example. Assuming that the .05 level of significance is used in testing this null hypothesis, the critical value of χ^2 (χ^2_{cv}) is 3.841 (see Table C.4). Notice that, in Table 21.1, the calculated value of χ^2 is 1.28. Since this value does *not* exceed the critical value, the null hypothesis (the coin is fair) is *not* rejected; the differences between observed and expected frequencies are attributable to chance fluctuation.

For the example of the new consulting policy, the null hypothesis is that 90 percent of the faculty would support it and 10 percent would not. Again, since there are only two categories, there is 1 degree of freedom associated with the test of this hypothesis. Thus, assuming $\alpha = .05$, the χ^2_{cv} is 3.841. From Table 21.3, we see that the calculated value of χ^2 is 8.00; therefore, the null hypothesis is rejected. The conclusion is that the percentage of faculty supporting the new consulting policy is not .90.

For the die example, the null hypothesis is that the frequency of occurrence of each of the six faces is the same. With six categories, there are 5 degrees of freedom associated with the test of this hypothesis; the χ^2_{cv} for $\alpha = .05$ is 11.070. Using the data from Table 21.2, $\chi^2 = 10.28$. Since this calculated value is less than the critical value, the null hypothesis is *not* rejected. The conclusion is that the differences between the observed and expected frequencies in each of the six categories are attributable to chance fluctuation.

The null hypothesis is stated in words, not symbols.

χ^2 *Goodness-of-Fit Test*

The one-sample case for nominal data was illustrated in the above discussion of the χ^2 distributions. The one-sample case is also called the **goodness-of-fit test**. This terminology comes from the idea that the test indicates whether or not the observed frequencies are a good fit to the expected frequencies. *The fit is good when the observed frequencies are within random fluctuation of the expected frequencies and the computed χ^2 value is relatively small,* or less than the critical value of χ^2 for the appropriate degrees of freedom.

Suppose a sociologist believes that the distribution of people in various occupations in a specific geographic region is as follows:

20% agricultural workers
30% laborers
30% local, state, or federal government employees
15% professional and self-employed business people
 5% industrial managers

Step 1: State the hypotheses. In this study, the sociologist is testing the null hypothesis relative to the distribution of people in various occupations in the region; this hypothesized distribution is given above. The alternative hypothesis is that the distribution of people in various occupations differs from the hypothesized one. The test statistic for testing this null hypothesis is χ^2, defined by formula 21.1.

Step 2: Set the criterion for rejecting H_0. In this example, there are five categories of occupations. Thus, there are 4 degrees of freedom associated with the test of this null hypothesis. That is, once the expected frequencies in any four of the categories are determined, the expected frequency for the fifth category is uniquely determined. Assuming that the sociologist sets α at .05, the critical value of χ^2 (χ^2_{cv}) for 4 degrees of freedom is 9.488.

Step 3: Compute the test statistic. To test the null hypothesis for this example, suppose the sociologist randomly selects a sample of 864 workers in the region and categorizes them into their respective occupations; the observed distribution of persons in the various occupations is found in Table 21.4. The expected frequencies are found by multiplying the total number in the sample (864) by the respective hypothesized percentages. For agricultural workers, the expected frequency is $0.20 \times 864 = 172.80$. The expected frequencies for the other categories are computed similarly; they are found in Table 21.4, along with the calculation of the χ^2 value (49.91).

Step 4: Decide about H_0 and interpret the results. Since the calculated value ($\chi^2 = 49.91$) exceeds the critical value ($\chi^2_{cv} = 9.488$), the null hypothesis is rejected. The sociologist would conclude that the differences between the observed and expected frequencies in the five categories are too great to be attributed to sampling fluctuation. In other words, the sample

TABLE 21.4
Calculation of χ^2 for the (One-Sample Case) Occupation Data

Occupation	O	E	$O - E$	$(O - E)^2$	$(O - E)^2/E$	R
Agricultural workers	145	172.80	−27.80	772.84	4.47	−2.12
Industrial laborers	310	259.20	50.80	2,580.64	9.96	3.16
Government employees	305	259.20	45.80	2,097.64	8.09	2.84
Professionals/self-employed business people	78	129.60	−51.60	2,662.56	20.54	−4.53
Industrial managers	26	43.20	−17.20	295.84	6.85	−2.62
Total	864	864.00	0	—	$49.91 = \chi^2$	

distribution of persons in the various occupations does *not* fit the expected distributions as hypothesized by the sociologist. Since the χ^2 value is computed over all categories, a significant χ^2 value does not specify which categories have been major contributors to the statistical significance. To determine which of the categories are major contributors, the *standardized residual* is computed for each of the categories.[5] The **standardized residual** is defined as follows:

$$R = \frac{O - E}{\sqrt{E}} \qquad (21.3)$$

The standardized residuals for each of the categories in this example are found in Table 21.4. When a standardized residual for a category is greater than 2.00 (in absolute value), the researcher can conclude that it is a major contributor to the significant χ^2 value.[6] Notice that all the standardized residuals for this example are greater than 2.00; thus, all are major contributors to the significant χ^2 value.

The χ^2 value does not indicate where the statistical significance lies; this is determined by computing the standardized residuals.

Frequencies, Two-Sample Case and *K*-Sample Case: χ^2 Test of Homogeneity[7]

In the goodness-of-fit test, there was one variable that contained two or more categories. We can extend this analysis to more than one variable where each variable has two or more categories. For example, suppose the president in the consulting-policy example also wants to know if there is a difference between tenured and non-tenured faculty (Variable 1) in support of the policy (Variable 2). Also suppose the president is interested in whether there is a difference between tenured and non-tenured faculty (Variable 1) in their perception of the fringe benefits package most in need of re-evaluation (Variable 2), as well as in differences in perception of the fringe benefit package among faculty by rank (instructor, assistant professor, associate professor, and professor). The data are collected and entered into a file

---

[5] S. J. Haberman, "The Analysis of Residuals in Cross-classified Tables," *Biometrics* 29 (1984): 205–220, and Reynolds.

[6] Ibid.

[7] Many authors have called this test the *test of independence*. However, there is a subtle difference between the test of independence and the test of homogeneity. For a thorough discussion, see Marascuilo and McSweeney, pp. 134–137 and 196–197.

TABLE 21.5
Data from the Faculty Survey

ID	Tenure	Rank	Policy	Fringe Benefit
001	1	4	1	1
002	1	3	1	3
003	0	2	0	3
.	.	.	.	.
.	.	.	.	.
.	.	.	.	.
.	.	.	.	.
200	1	3	0	2

outlined in Table 21.5; the variables are defined and coded as follows:

ID — Identification number of the faculty member

Tenure — Yes (1)/No (0) indicating whether the faculty member is tenured

Rank — Academic rank of the faculty member:
(1) Instructor
(2) Assistant professor
(3) Associate professor
(4) Professor

Policy — Yes (1)/No (0) indicating whether the faculty member supports the consulting policy

Fringe Benefit — Fringe benefit most in need of re-evaluation:
Retirement (1)
Disability (2)
Hospitalization (3)
Major Medical (4)

There are three specific research questions to be addressed:

1. Is there a difference between tenured and non-tenured faculty members regarding the support of the consulting policy?

2. Do tenured and non-tenured select different fringe benefits most in need of re-evaluation?

3. Do faculty of different ranks select different fringe benefits most in need of re-evaluation?

Each of these questions can be answered using the χ^2 test of homogeneity.

TABLE 21.6
Data for University Consulting Example

	Support Policy	Do Not Support Policy	
Tenured	86	19	105
Non-tenured	82	13	95
Total	168	32	200

Research Question 1

Suppose we consider the random sample of 200 faculty as two different samples: one of size 105 selected from the tenured population and one of size 95 selected from the non-tenured population. The hypothetical data for this example are found in Table 21.6; these data indicate that 86 of the 105 tenured faculty support the new policy, and 82 of the 95 non-tenured faculty support it.

The observant reader will see that these data can be analyzed in several ways, depending on which way the null hypothesis is stated. We could state the null hypothesis in terms of no relationship between faculty status and support for the consulting policy. The phi (ϕ) coefficient could be computed using formula 20.3, and this coefficient could be tested for statistical significance using the general formula for hypothesis testing (formula 10.1). We could also test the null hypothesis that the proportion of tenured faculty supporting the new policy is equal to the proportion of non-tenured faculty supporting it (see Chapter 12). The sample proportions could be computed, and the difference between them could be tested for statistical significance using formula 12.2. In addition to these two methods, the χ^2 **test of homogeneity** can be used; the null hypothesis for this test is that there is no difference in the level of support between tenured and non-tenured faculty. It will be left as an exercise to show that the χ^2 value computed in the test of homogeneity equals the square of the *z* value (z^2) computed using formula 12.2.

Determination of the Expected Frequencies

In the one-sample case, the expected frequencies are determined by the statement of the null hypothesis. In our example, the percentage of persons in each occupational category was hypothesized a priori. These hypothesized percentages were then multiplied by the total number of persons in the sample to determine the expected frequency for each occupational category. For the test of homogeneity, the expected frequencies are computed based on the percentages in the marginal totals. For example, consider the data in Table 21.6. Of the 200 faculty members, 168, or 84 percent, indicated support for the consulting policy. If the null hypothesis (no difference between tenured and non-tenured faculty members) is true, then we would expect this percentage to be the same for both tenured and non-tenured faculty. Thus, the expected frequency for tenured faculty would be $0.84 \times 105 = 88.2$, and the expected frequency for non-tenured faculty would be $0.84 \times 95 = 79.8$. Similarly, 32, or 16 percent, of the faculty do not support the policy. Therefore, the expected frequency for tenured faculty would be $0.16 \times 105 = 16.8$, and the expected frequency for non-tenured faculty would be $0.16 \times 95 = 15.2$.

The most convenient way to calculate the expected frequency for each cell is to multiply the total row frequency (f_r) by the total column frequency (f_c) corresponding to the respective cell and then to divide this product by the total frequency (c).

$$\text{Expected frequency} = \frac{f_r \times f_c}{n} \tag{21.4}$$

The expected frequencies for each cell of the contigency table are calculated as follows and are found in the parentheses in Table 21.7.

	Support policy	*Do not support policy*
Tenured	$\dfrac{105 \times 168}{200} = 88.2$	$\dfrac{105 \times 32}{200} = 16.8$
Non-tenured	$\dfrac{95 \times 168}{200} = 79.8$	$\dfrac{95 \times 32}{200} = 15.2$

Notice that the sum of the expected frequencies for any row or column equals the respective row or column total. This can be a useful check of the calculations. Once the expected frequencies have been calculated, the χ^2 statistic is calculated using formula 21.1; this calculation is illustrated in Table 21.7.

TABLE 21.7
Data for Calculating χ^2 Statistic for the Consulting Example

	Support Policy	*Do Not Support Policy*	
Tenured	86(88.2)	19(16.8)	105
Non-tenured	82(79.8)	13(15.2)	95
	168	32	200

O	*E*	*O − E*	$(O - E)^2$	$(O - E)^2/E$
86	88.2	−2.2	4.84	0.055
82	79.8	2.2	4.84	0.061
19	16.8	2.2	4.84	0.288
13	15.2	−2.2	4.84	0.318
200	200	0		$0.722 = \chi^2$

In a contingency table, the expected frequencies are determined using the marginal totals. The expected frequency of the RC cell is determined by $f_r \times f_c/n$, that is, the product of the row and column frequencies divided by the sample size.

Determination of the Degrees of Freedom

For the one-sample case, it is relatively simple to define the degrees of freedom associated with the χ^2 statistic. The degrees of freedom are equal to the number of frequencies minus one $(K - 1)$. In our earlier example, there were five occupational categories; thus, there are $5 - 1 = 4$ degrees of freedom.

The degrees of freedom for the test of homogeneity are similarly defined. Consider the data in Table 21.7. Once we compute the expected frequency for tenured faculty that support the policy (88.2), the expected frequency for non-tenured faculty who support the policy is uniquely determined, since the total column frequency must be 168. That is, $168 - 88.2 = 79.2$. Similarly, once we compute the expected frequency for tenured faculty that support the policy (88.2), the expected frequency for tenured faculty who do not support the policy is uniquely determined; $105 - 88.2 = 16.8$. Now that the expected frequencies are determined for the first three cells of the contingency table, the expected frequency for the fourth cell can also be determined. Thus, for a 2×2 contingency, once the expected frequency for any one cell is determined, the expected frequencies for the remaining cells can be determined. The general formula for determining the degrees of freedom associated with the χ^2 statistic for any $R \times C$ contingency is as follows:

$$df = (\text{number of rows} - 1)(\text{number of columns} - 1)$$

$$= (R - 1)(C - 1) \tag{21.5}$$

For the $R \times C$ contingency table, there are $(R - 1)(C - 1)$ degrees of freedom associated with the χ^2 test of homogeneity.

The data for Research Question 1 concerning support for the new consulting policy are found in Table 21.7, along with the calculation of the χ^2 value using formula 21.1. For a 2×2 contingency table, the computational formula can be simplified to the following:

$$\chi^2 = \frac{n(AD - BC)^2}{(A + B)(C + D)(A + C)(B + D)} \tag{21.6}$$

where A, B, C, and D are the cell frequencies of the 2×2 contingency table illustrated in Table 21.7. For these data,

$$\chi^2 = \frac{200[(86)(13) - (19)(82)]^2}{(86 + 19)(82 + 13)(86 + 82)(19 + 13)}$$

$$= \frac{200(1{,}118 - 1{,}558)^2}{(105)(95)(168)(32)}$$

$$= 0.722$$

Notice that this χ^2 value is the same as the one computed using formula 21.1. Now consider these data in the context of the questions asked when testing the null hypothesis.

Step 1: State the hypotheses. The null hypothesis for this example is that there is no difference between tenured and non-tenured faculty concerning the new consulting policy. This null hypothesis is tested at $\alpha = .05$.

Step 2: Set the criterion for rejecting H_0. The test statistic to be computed for the 2×2 contingency table in this example is χ^2; either formula 21.1 or 21.6 can be used. The degrees of freedom associated with this test statistic are $(R - 1)(C - 1) = (2 - 1)(2 - 1) = 1$. Therefore, the critical value of the test statistic (χ^2_{cv}) is 3.841.

Step 3: Compute the test statistic. The expected cell frequencies were determined using formula 21.4 and are found in Table 21.7. Using formula 21.1 or 21.6, the χ^2 value is found to be 0.722.

Step 4: Interpret the results. Since the computed χ^2 value (0.722) does not exceed the critical value ($\chi^2_{cv} = 3.841$), the null hypothesis is not rejected, and the conclusion is that the opinions of the tenured and non-tenured faculty are the same *(homogeneous)* regarding the new consulting policy.

Research Question 2

The president of the university wants to know which part of the fringe-benefits program is most in need of re-evaluation: (1) retirement benefits, (2) disability benefits, (3) hospitalization benefits, or (4) major medical benefits. The two samples of faculty members (tenured and non-tenured) are asked to indicate which component is *most* in need of re-evaluation. These data are displayed in a 2×4 contingency table in Table 21.8. The numbers in parentheses are the expected frequencies for each of the cells. For these data, since the contingency table is larger than 2×2, neither the test of significance for the phi (ϕ) coefficient nor the test of two independent proportions can be used. The appropriate data analysis procedure is the χ^2 test of homogeneity.

TABLE 21.8
Data for Faculty Reevaluation of Fringe-Benefits Program

	Retirement	Disability	Hospitalization	Major Medical	
Tenured	68(47.25)	17(13.65)	8(27.30)	12(16.80)	105
Non-tenured	22(42.75)	9(12.35)	44(24.70)	20(15.20)	95
	90	26	52	32	200

The data in Table 21.8 will now be considered in the context of the questions that provide a logical procedure for testing the appropriate null hypothesis.

Step 1: State the hypothesis. The president of the university hypothesized that, in the population of tenured and non-tenured faculty members, there would be no difference in their responses concerning the re-evaluation of the fringe-benefits program. Assume that the level of significance (α) is set at .05.

Step 2: Set the criterion for rejecting H_0. The test statistic to be computed for the 2×4 contingency table in this example is χ^2. The degrees of freedom associated with this test statistic are $(R-1)(C-1) = (2-1)(4-1) = 3$. Therefore, the critical value of the test statistic (χ^2_{cv}) is 7.815.

Step 3: Compute the test statistic. The expected cell frequencies were determined and are found in Table 21.8. Using formula 21.1, the calculation of the χ^2 value is found in Table 21.9: $\chi^2 = 52.52$.

Step 4: Interpret the results. Since the computed χ^2 value (52.52) exceeds the critical value ($\chi^2_{cv} = 7.815$), the null hypothesis is rejected, and the conclusion is that the opinions of the tenured and non-tenured

TABLE 21.9
Calculation of χ^2 for Data in Table 21.8

O	E	O − E	$(O-E)^2$	$(O-E)^2/E$
68	47.25	20. 75	430.56	9. 11
22	42.75	−20. 75	430.56	10. 07
17	13.65	3. 35	11.22	0. 82
9	12.35	−3. 35	11.22	0. 91
8	27.30	−19. 30	372.49	13. 64
44	24.70	19. 30	372.49	15. 08
12	16.80	−4. 80	23.04	1. 37
20	15. 20	4. 80	23. 04	1. 52
200	200.00	0	—	$52.52 = \chi^2$

TABLE 21.10
Standardized Residuals for Fringe-Benefits Example

	Retirement	Disability	Hospitalization	Major Medical
Tenured	3.02	0.91	−3.69	−1.17
Non-tenured	−3.17	−0.95	3.88	1.23

faculty are *not* the same (*not homogeneous*) regarding the parts of the fringe-benefits program most in need of re-evaluation. As for the one-sample χ^2 goodness-of-fit test, the χ^2 value is computed over all cells, and a significant χ^2 value does not specify which cells have been major contributors to the statistically significant χ^2 value. This is done by computing the standardized residuals (R) for each of the cells; these residuals are found in Table 21.10. As before, when a standardized residual for a category is greater than 2.00 (in absolute value), the researcher can conclude that it is a major contributor to the significant χ^2 value.[8] Notice that the standardized residuals for cells 11 ($R_{11} = 3.02$), 21 ($R_{21} = -3.17$), 13 ($R_{13} = -3.69$), and 23 ($R_{23} = 3.88$) are greater than 2.00; thus, these are major contributors to the significant χ^2 value. These residuals indicate that, in comparing the observed frequencies with the expected frequencies, there were *more* tenured and *less* non-tenured faculty than expected who indicated that the retirement benefits needed reevaluation. Similarly, there were *less* tenured and *more* non-tenured faculty than expected who indicated that the hospitalization benefits needed re-evaluation.

Research Question 3

In the introduction to the χ^2 test of homogeneity, we described it as one of the more frequent uses of the χ^2 statistic. For Research Question 1, the χ^2 test of homogeneity was applied to a 2 × 2 contingency table. For Research Question 2, the test was applied to a 2 × 4 contingency table. Both illustrate a two-sample case for using this χ^2 statistic, i.e., the difference between groups on some categorical variable. The application of the χ^2 test of homogeneity for the K-sample case is the logical extension of that for the two-sample case. Rather than two groups, there are K groups; thus there is an $R \times C$ contingency table (where R = the number of rows and C = the number of columns).

[8] Marascuilo and McSweeney; Haberman; and Reynolds.

Consider Research Question 3, in which the university president wants to determine whether faculty of different academic ranks select different fringe benefits most in need of re-evaluation. The frequency data for this example are found in Table 21.11. Note that the expected frequencies for each of the cells (contained within the parentheses) were calculated using formula 21.4. The statistic was computed using formula 21.1; these calculations are shown in Table 21.12.

Step 1: State the hypothesis. The null hypothesis is that there will be no difference between faculty in the various ranks regarding their perception of the

TABLE 21.11
Frequencies for Faculty Indicating Fringe Benefit Most in Need of Re-evaluation

Rank	Retirement	Disability	Hospitalization	Major Medical	
Professor	35 (28.35)	8 (8.19)	10 (16.38)	10 (10.08)	63
Associate	30 (29.70)	8 (8.58)	20 (17.16)	8 (10.56)	66
Assistant	15 (19.35)	5 (5.59)	15 (11.18)	8 (6.68)	43
Instructor	10 (12.60)	5 (3.64)	7 (7.28)	6 (4.48)	28
	90	26	52	32	200

TABLE 21.12
Calculation of χ^2 for Research Question 3—Faculty Rank by Fringe Benefit Re-evaluation

O	E	$O - E$	$(O - E)^2$	$(O - E)^2/E$
35	28.35	6.65	44.22	1.56
30	29.70	0.30	0.09	0.00
15	19.35	−4.35	18.92	0.98
10	12.60	−2.60	6.76	0.54
8	8.19	−0.19	0.04	0.00
8	8.58	−0.58	0.34	0.04
5	5.59	−0.59	0.35	0.06
5	3.64	1.36	1.85	0.51
10	16.38	−6.38	40.70	2.49
20	17.16	2.84	8.07	0.47
15	11.18	3.82	14.59	1.31
7	7.28	−0.28	0.08	0.01
10	10.08	−0.08	0.01	0.00
8	10.56	−2.56	6.55	0.62
8	6.88	1.12	1.25	0.18
6	4.48	1.52	2.31	0.52
				9.29

fringe benefits that are most in need of re-evaluation. The .05 level of significance is used.

Step 2: Set the criterion for rejecting H_0. The test statistic for this example is computed using formula 21.1. The degrees of freedom associated with this test statistic are $(R - 1)(C - 1) = (4 - 1)(4 - 1) = 9$. Therefore, the critical value of this test statistic (χ^2_{CV}) is 16.919.

Step 3: Compute the test statistic. The expected cell frequencies were shown in Table 21.11; the computed value of χ^2 using formula 21.1 was shown in Table 21.12 to be 9.29.

Step 4: Interpret the results. Since the computed value of χ^2 (9.29) does *not* exceed the critical value (16.919), the null hypothesis is not rejected. The conclusion is that faculty members in the various ranks have similar perceptions about the components of the fringe benefit package most in need of re-evaluation. In other words, the faculty members in the various ranks are *homogeneous* in their perceptions. Note that since the null hypothesis was not rejected, there is no need to calculate the standardized residuals.

Small Expected Frequencies in Contingency Tables

The theoretical sampling distribution of χ^2 for 1 degree of freedom was illustrated in Figure 21.1. Notice that the distribution is continuous; there are no breaks in the continuity. However, when the *expected* frequencies in any of the cells of a 2×2 contingency table are small (less than 5), the sampling distribution of χ^2 for these data may depart substantially from continuity. Thus, the theoretical sampling distribution of χ^2 for 1 degree of freedom may poorly fit the data. For this situation, an adjustment, called the **Yates' correction for continuity**, has been suggested for application to these data.[9] However, based on a study by Camilli and Hopkins in 1978,[10] the Yates' correction for continuity is *not* recommended for the χ^2 test for the 2×2 table "since its use would result in an unnecessary loss of power" — that is, a tendency not to reject the null hypothesis when in fact it is false.

For contingency tables larger than 2×2, the lack of continuity in the χ^2 distribution resulting from small expected frequencies is of less consequence. However, when more than 20 percent of the cells have *expected* frequencies less than 5, or when one of the cells has no frequencies, the researcher is advised to combine adjacent rows or columns if this will not result in a distortion of the data.

[9]W.G. Cochran, "Some Methods for Strengthening the Common χ^2 Tests," *Biometrics* 10 (1954): 417–451.

[10]G. Camilli and K.D. Hopkins, "Applicability of Chi-Square to 2×2 Contingency Tables with Small Expected Frequencies," *Psychological Bulletin*, 85 (1978): 163–167.

Frequencies, Two-Sample Case: Dependent Samples

There is a χ^2 test for dependent samples involving frequencies called the **McNemar test** for significance of change. [11] This test may be used in pretest-posttest designs in which the same sample of subjects is categorized before and after some intervening treatment.

For example, suppose that a random sample of 38 assembly-line employees in a small manufacturing company is asked at two different times about their attitudes toward homeless people — once before and once after they have read several short magazine articles describing the characteristics, experiences, and values of typical homeless persons. The resulting data are found in Table 21.13. It is important to note the order of the data entries in this table. The cells marked A and D *must* be those indicating a change in response from pretest to posttest. The cells marked B and C *must* be those indicating no change in response from pretest to posttest. Now consider these data in the context of the steps for testing hypotheses.

Step 1: State the hypotheses. In this example, we are interested only in the cells of the contingency table that reflect a change of opinion about the homeless; namely, cells A and D. The specific null hypothesis is that, in the population, there will be an equal number of changes in both directions. This implies that, under the null hypothesis, the expected frequency in cell A will be equal to the expected frequency in cell D. This null hypothesis will be tested at $\alpha = .05$.

Step 2: Set the criterion for rejecting H_0. The test statistic for this example is again χ^2, and formula 21.1 can be applied. However, under this specific null hypothesis, the formula will be applied only to cells A and D. Further, because the expected frequencies for these two cells are

TABLE 21.13

Data for Computing χ^2 for Dependent Samples in the Attitudes-Toward-the-Homeless Example

		Before Reading Articles				
		Unfavorable		Favorable		
After Reading	*Favorable*	14	A	B	6	20
Articles	*Unfavorable*	16	C	D	2	18
		30			8	38

[11] Q. McNemar, *Psychological Statistics* (New York: Wiley, 1969), pp. 260–262.

hypothesized to be equal, the expected value for both cells is $(A + D)/2$. Therefore, formula 21.1 reduces to the following:

$$\chi^2 = \sum \frac{(O - E)^2}{E}$$

$$= \frac{\left(A - \dfrac{A + D}{2}\right)^2}{\dfrac{A + D}{2}} + \frac{\left(D - \dfrac{A + D}{2}\right)^2}{\dfrac{A + D}{2}}$$

$$= \frac{(A - D)^2}{A + D} \tag{21.7}$$

Since there are only two cells under consideration, once the expected frequency for one of the cells is determined, the expected frequency of the other is uniquely determined. Thus, there is 1 degree of freedom associated with this χ^2 test. From Table C.4, the critical value of χ^2 for 1 degree of freedom at $\alpha = .05$ is 3.841.

Step 3: Compute the test statistic. Using formula 21.7, the calculation of the χ^2 value for the data in Table 21.13 is

$$\chi^2 = \frac{(14 - 2)^2}{14 + 2}$$

$$= \frac{(12)^2}{16}$$

$$= 9.00$$

Step 4: Interpret the results. Since the computed χ^2 value (9.00) exceeds the critical value ($\chi^2_{cv} = 3.841$), the null hypothesis is rejected, and the conclusion is that the attitudes toward the homeless changed more in one direction than in the other. Inspection of the data in Table 21.13 indicates that more employees changed from unfavorable to favorable attitudes than the reverse.

In the McNemar test for significance of change, only cells *A* and *D* are considered. These are the cells that indicate a change in response from pretest to posttest.

Frequencies, Three-Sample Case: Dependent Samples

There are occasions when there are 3×3 contingency tables for dependent samples and the McNemar test will not do. A version of the Stuart-Maxwell statistic would be appropriate.[12] For example, suppose that someone was interested in whether people's perceived software needs for personal computers reflected their actual use. Imagine a study in which 100 randomly selected adults were asked, when purchasing their computers, whether they were most likely to use word-processing, spreadsheet, or database programs. One year later, the same persons were asked which type of software they used most often. The data for this example are found in Table 21.14.

The Stuart-Maxwell test, like the McNemar test, is focused on the cells that depict change. The Stuart-Maxwell test also uses the difference between row and column totals:

$$\bar{n}_{ij} = \frac{n_{ij} + n_{ji}}{2}$$

that is, $\bar{n}_{ij}$ is the mean of n_{ij} and n_{ji}, and

$$d_i = (\text{total for row } i - \text{total for column } j)$$

The formula for the Stuart-Maxwell test for a 3×3 contingency table of paired observations is:

$$\chi^2 = \frac{\bar{n}_{23}d_1^2 + \bar{n}_{13}d_2^2 + \bar{n}_{12}d_3^2}{2(\bar{n}_{12}\bar{n}_{13} + \bar{n}_{12}\bar{n}_{23} + \bar{n}_{13}\bar{n}_{23})} \qquad (21.8)$$

We can use the steps for testing hypotheses to see whether there is a difference between the anticipated use of computer software and the actual use.

Step 1: State the hypotheses. The null hypothesis for this example is that there is no difference between people's anticipated use of software and their actual use. This null hypothesis is tested using $\alpha = .05$.

Step 2: Set the criterion for rejecting H_0. The appropriate sampling distribution is the chi-square distribution, and because we are interested in the three pairs of cells that indicate change, there are 2 degrees of freedom. If we

TABLE 21.14
Anticipated and Actual Software Use

		Anticipated Use			
		Word Processing	Spread-sheet	Data-base	Total
Actual Use	Word Processing	30	4	5	39
	Spreadsheet	15	15	3	33
	Database	5	11	12	28
	Total	50	30	20	

[12]J. L. Fleiss, *Statistical Methods for Rates and Proportions*, 2d ed. (New York: Wiley, 1981).

use the .05 level of significance and there are 2 degrees of freedom, the critical value of chi-square, consulting Table C.4, is 5.991.

Step 3: Compute the test statistic. Using formula 21.8 and the data from Table 21.14, the calculated value of χ^2 is:

$$\chi^2 = \frac{\frac{3+11}{2}(39-50)^2 + \frac{5+5}{2}(33-30)^2 + \frac{4+15}{2}(28-20)^2}{2\left[\left(\frac{4+15}{2} \times \frac{5+5}{2}\right) + \left(\frac{4+15}{2} \times \frac{3+11}{2}\right) + \left(\frac{5+5}{2} \times \frac{3+11}{2}\right)\right]}$$

$$= 5.03$$

Step 4: Interpret the results. The calculated χ^2 was less than the critical value, so the null hypothesis cannot be rejected. The conclusion is that people's anticipated use of software and their actual use are similar.

The Stuart-Maxwell test, as presented here, is appropriate for 3 × 3 contingency tables where the data consist of pairs of scores.

Summary

This chapter focused on some nonparametric tests that are used with dependent variables measured on nominal scales (i.e. frequencies). These tests use the chi-square (χ^2) distribution and are analogous to the parametric tests presented in earlier chapters. These tests do not depend on assumptions of normality and homogeneity of variance, as do the parametric tests of the preceding chapters. The statistical tests that were described in this chapter are summarized in Table 21.15.

One-dimensional to k-dimensional statistical tests were introduced, and there were two-dimensional tests for independent samples as well as dependent samples. These methods for testing hypotheses are widely used in behavioral research because there are many possibilities for cross-tabulating pairs of variables in a data set. We would urge that you focus on a small number of well-conceived hypotheses in your research rather than blindly cross-tabulating all pairs of variables and then computing a bevy of chi-square statistics.

There are entire books on nonparametric statistics. We present only a few of these tests, those that you are most likely to encounter in the literature. Those in this chapter are for nominal data and use the chi-square distribution. Those in Chapter 22 are for ordinal data and have several different sampling distributions. Fortunately, the logic of hypothesis testing and the steps that are used when testing hypotheses are the same for nonparametric and parametric tests.

TABLE 21.15
Summary of Nonparametric Test Statistics for Nominal Data

Name of Test	Hypothesis Tested	Test Statistic	Sampling Distribution
One-sample case	H_0: Goodness-of-fit	$\chi^2 = \Sigma \dfrac{(O-E)^2}{E}$	χ^2 with $k-1$ degrees of freedom
Two-sample case (test of homogeneity)	H_0: Homogeneity	$\chi^2 = \dfrac{n(AD-BC)^2}{(A+B)(C+D)(A+C)(B+D)}$	χ^2 with 1 degree of freedom
k-sample case (test of homogeneity)	H_0: Homogeneity	$\chi^2 = \Sigma \dfrac{(O-E)^2}{E}$	χ^2 with $(r-1)(c-1)$ degrees of freedom
Two-sample case, dependent samples (McNemar test)	H_0: $A = D$	$\chi^2 = \dfrac{(A-D)^2}{A+D}$	χ^2 with 1 degree of freedom
Three-sample case, dependent samples (Stuart-Maxwell test)	H_0: $n_{ij} = n_{ji}$	$\chi^2 = \dfrac{\bar{n}_{23}d_1^2 + \bar{n}_{13}d_2^2 + \bar{n}_{12}d_3^2}{2(\bar{n}_{12}\bar{n}_{13} + \bar{n}_{12}\bar{n}_{23} + \bar{n}_{13}\bar{n}_{23})}$	χ^2 with 2 degrees of freedom

Exercises

For these exercises, use the following format:
1. State the hypotheses.
2. Set the criterion for rejecting H_0.
3. Compute the test statistic.
4. Interpret the results.

1. A research psychologist wants to investigate the impact of instructor feedback on mastery of a complex learning task. Four groups of ten students are selected to participate. One group receives only positive feedback, another only negative feedback. The third group receives both positive and negative feedback, and the fourth group receives no feedback. Using the four steps for hypothesis testing, carry out the χ^2 test of homogeneity at the .05 level of significance. Also compute the standardized residuals.

	Successful	*Unsuccessful*
Positive	6	4
Negative	4	6
Both	8	2
None	3	7

2. The chairperson of an organization's planning committee conjectures that membership preference for an upcoming national convention approximates the following distribution:

City	*Percentage Positive Responses*
Dallas	12%
Miami Beach	20%
New York	15%
Chicago	8%
Washington, D.C.	10%
San Francisco	18%
Other	17%

Preliminary responses are received from a random sample of 200 members. The results are as follows:

City	*No. Positive Responses*
Dallas	20
Miami Beach	45
New York	20
Chicago	10
Washington, D.C.	17
San Francisco	50
Other	38

Carry out the χ^2 goodness-of-fit test, using $\alpha = .05$.

*3. The management work force of a company was downsized from 150 to 120 employees. The age distributions before and after the downsizing are given below. Using the goodness-of-fit test, determine whether there is any evidence of age discrimination based upon these distributions. Use $\alpha = .05$ for testing the appropriate hypothesis. Hint: Do the frequencies after the downsizing "fit proportionally" with the frequencies before the downsizing?

	Age					
	Under 30	*30–39*	*40–49*	*50–59*	*60 & over*	*Total*
Before	15	60	35	28	12	150
After	15	46	28	20	11	120

4. The following data are obtained from a study designed to examine the relationship between marital status and entertainment preference. For the following data, carry out the test of homogeneity, using $\alpha = .05$. Also compute the standardized residuals.

	Individual	Small Group	Large Group
Single	18	4	3
Married	8	12	5
Separated/divorced	10	7	8
Widowed	6	15	4

5. Using the χ^2 value obtained in Exercise 4, compute both the C and the V coefficients using the formulas from Chapter 20.

6. Following the model developed by the League of Women Voters, a local high school conducts an issues forum. A random sample of 50 students is asked before and after the forum about their intention to vote on a specific issue. Using the following data, conduct the McNemar test for change at the .05 level of significance.

		Before Forum	
		No	Yes
After	Yes	16	11
Forum	No	17	6

7. Women in their third trimester of pregnancy were asked whether they intended to breast feed their babies, bottle feed them, or use a mixture of breast and bottle feeding. Six months after the birth of the babies, the mothers again were asked about feeding practices. Use the Stuart-Maxwell test, at the .05 level, to see whether the women fed their babies as they had intended.

		Actual		
		Breast	Bottle	Mixed
	Breast	20	20	20
Intended	Bottle	1	25	4
	Mixed	2	10	8

8. A school administrator wished to determine whether support for a new suspension policy differed among people according to their ages. He surveyed 20 people in each of five age groups. Use the goodness-of-fit test to determine whether support for the policy was uniform across the age categories. Use the .05 significance level.

Age Category	Number in Support of the Policy
20–29	10
30–39	12
40–49	15
50–59	19
60 and older	18

9. A group of sexually active, single college students was questioned about whether they always used condoms. They were asked the same question three months later, after attending workshops on "safer sex." Using the McNemar test, see whether the change in condom use was significant at the .05 level.

		Before	
		Did Not Use	Used
After	Used	24	46
	Did Not Use	6	4

10. A psychologist studying heroes asked each of 50 nine-year-old boys to name one hero figure in his life. Three categories (sports figures, fantasy figures, and family members) emerged. The same boys were asked the same question six years later. Using the Stuart-Maxwell test, compute and interpret the results.

		Age 15		
		Sports Figures	Fantasy Figures	Family Members
	Sports Figures	10	1	4
Age 9	Fantasy Figures	5	5	10
	Family Members	4	1	10

11. Consider the data in Exercise 4 of Chapter 20. Conduct the test of homogeneity and compute the standardized residuals. Compare the results and interpretation with the interpretation of the ϕ coefficient.

12. Men and women were asked which of the following four Olympic events were their favorite. The data below are the numbers of people who chose each event. Test the hypothesis that men and women had similar responses. Use the .01 level of significance.

	Gymnastics	Diving	Kayaking	Basketball
Women	45	27	13	30
Men	40	21	29	35

13. The following data were obtained in a community survey concerning recycling. Taxpayers were asked to respond to this statement:

Households that do not participate in the municipal recycling program should pay an additional tax.

	Strongly Disagree	Disagree	Uncertain	Agree	Strongly Agree
Homeowners	125	100	150	220	235
Renters	120	145	170	135	100

Test the hypothesis that homeowners and renters responded homogeneously to this statement. Use the .05 level of significance.

14. Again consider using the data in Exercise 4 of Chapter 20. Using the methods outlined in Chapter 11, conduct the two-sample test for independent proportions. Compare these results with the χ^2 test of homogeneity, illustrating that, for a 2×2 contingency table, $\chi^2 = z^2$.

15. Consider the data in Exercise 9 of Chapter 20. Conduct the test of homogeneity and compute the standardized residuals. Compare the results and interpretation with the interpretation of the λ and λ_Y.

22

Other Nonparametric Tests

C hapter 21 contained nonparametric statistics for situations in which the dependent variable was measured on a nominal scale. This chapter includes a number of nonparametric tests for instances where the dependent variable is measured on an ordinal scale. These tests are analogous to the parametric tests that were presented in Chapters 8 through 12, but they are distribution-free. That is, we do not have to be concerned about whether the distribution of the dependent variable for the population is a normal distribution. We will also not be concerned with the assumption of homogeneity of variance. The statistical tests that are included in this chapter are comparatively simple to compute, but they have a wide variety of research applications.

Ordinal Data: Two-Sample Case, Independent Samples

There are several nonparametric tests of significance[1] for the two-sample case with ordinal data, but only two of them will be discussed here.[2] They are the **median**

---

[1] There are several appropriate tests of significance for the one-sample case for ordinal data. Three of these tests are (1) the one-sample sign test, (2) the Mann-Kendall test for trends, and (3) the Kolmogorov-Smirnov one-sample test. These tests are discussed in Hollander and Wolfe, pp. 39–47.

[2] See Hollander and Wolfe.

test and the **Mann-Whitney U test**; both will be applied to the same example and data.

Suppose a sociologist is interested in the attitudes of income-producing and nonincome-producing women about equal rights for women. A sample of ten income-producing women (women with jobs outside the home, group 1) and a sample of ten nonincome-producing women (women working at home, group 2) are selected and administered an attitude instrument. The resulting data are found in Table 22.1. For the purposes of this example, assume the data to be ordinal.

The Median Test

Step 1: State the Hypotheses. The null hypothesis for the median test is that the two samples have been selected from populations with the same or a common median. Assume that we will test this null hypothesis against a nondirectional alternative at the .05 level of significance. The hypotheses would be $H_0: \text{Mdn}_1 = \text{Mdn}_2$ and $H_a: \text{Mdn}_1 \neq \text{Mdn}_2$.

Step 2: Set the Criterion for Rejecting H_0. The test statistic for the median test is χ^2. The first step is to develop the 2×2 contingency table from the data in Table 22.1. We must determine the **common median** for these data. This is done by combining the scores from both groups and listing them from low to high. Using the methods of Chapter 3, the median for the

TABLE 22.1
Data on Attitudes About Equal Rights for Women

Income-producing Women (Group 1)		Nonincome-producing Women (Group 2)	
Score	*Rank*	*Score*	*Rank*
19	3	16	1
22	5	18	2
28	8	21	4
32	11	26	6
34	13	27	7
37	14	29	9
40	17	31	10
42	18	33	12
43	19	38	15
46	20	39	16
	$128 = R_1$		$82 = R_2$

Median for both groups combined $= \dfrac{31 + 32}{2} = 31.5$

combined group is as follows:

$$Mdn = \frac{31 + 32}{2}$$

$$= 31.5$$

Now we categorize the data from Table 22.1 into groups that fall above and below this common median; the 2×2 contingency table is as follows:

	Group 1	Group 2	
Above Mdn	7	3	10
Below Mdn	3	7	10
	10	10	20

Notice that seven members of group 1 and three members of group 2 have attitude scores above the median for the combined group.

With a 2×2 contingency table, the associated degrees of freedom are $(R - 1)(C - 1) = 1$. Therefore, with $\alpha = .05$, the critical value of χ^2 is 3.841.

Step 3: Compute the Test Statistic. Using formula 21.6, we find

$$\chi^2 = \frac{n(AD - BC)^2}{(A + B)(C + D)(A + C)(B + D)}$$

$$= \frac{20(7 \times 7 - 3 \times 3)^2}{(7 + 3)(3 + 7)(7 + 3)(3 + 7)}$$

$$= \frac{20(49 - 9)^2}{(10)(10)(10)(10)}$$

$$= 3.20$$

Step 4: Interpret the Results. Since the observed value of χ^2 (3.20) does *not* exceed the critical value, the null hypothesis is *not* rejected. The conclusion is that there is no difference in the attitudes of income-producing and nonincome-producing women about equal rights for women.

The null hypothesis for the median test is that the populations from which the samples are selected have a common median.

The Mann-Whitney U Test

Step 1: State the Hypotheses. In the median test, the null hypothesis tested was that there is no difference between the medians of the two populations from which the samples were selected. In that case, the test statistic was sensitive only to the differences between the medians and did not take into consideration the total distributions of scores from the two groups. In contrast, the Mann-Whitney U test is sensitive to both the central tendency of the scores and the distribution of scores. The null hypothesis is thus stated in the more general terms; there is no difference in the *scores* of the populations from which the samples were selected. For our example, the null hypothesis is that there is no difference in attitude about equal rights for women between populations of income-producing and nonincome-producing women. Again, the null hypothesis is tested against the nondirectional alternative hypothesis at the .05 level of significance. Symbolically, H_0: Attitude$_1$ = Attitude$_2$ and H_a: Attitude$_1 \neq$ Attitude$_2$.

Step 2: Set the Criterion for Rejecting H_0. The calculation of the U statistic takes into consideration the central tendency, as well as the total distribution of scores for both groups, and is defined as the smaller of U_1 and U_2.

$$U_1 = n_1 n_2 + \frac{n_1(n_1 + 1)}{2} - R_1 \qquad (22.1)$$

$$U_2 = n_1 n_2 + \frac{n_2(n_2 + 1)}{2} - R_2$$

where

n_1 = number of observations in group 1
n_2 = number of observations in group 2
R_1 = sum of the ranks assigned to group 1
R_2 = sum of the ranks assigned to group 2

The sampling distribution of U is known and is used for testing hypotheses in the same way as the t distributions and the χ^2 distributions. The critical values for one-tailed and two-tailed tests for various significance levels are found in Table C.14. In using this table, U is the *smaller* of U_1 and U_2. In order to reject the null hypothesis at the .05 level for a two-tailed test (or at the .025 level for a one-tailed test), the computed U must be *less* than the table value. For this example, the sampling distribution of U for $n_1 = 10$ and $n_2 = 10$ is used to test the null hypothesis. The critical value from Table C.14 for this example is 24. Thus, the computed value of U must be *less* than 24 in order to reject the null hypothesis.

Step 3: Compute the Test Statistic. Before applying formula 22.1 to the data in Table 22.1, it is necessary to assign ranks to the scores for the combined group; these ranks are found in the table. U_1 and U_2 can then be computed.

$$U_1 = (10)(10) + \frac{(10)(10+1)}{2} - 128$$

$$= 100 + 55 - 128$$

$$= 27$$

$$U_2 = (10)(10) + \frac{(10)(10+1)}{2} - 82$$

$$= 100 + 55 - 82$$

$$= 73$$

Thus, since U_1 is the smaller of the two, $U = 27$. Notice that the group with the highest scores (group 1) and, correspondingly, the highest sum of ranks has the smaller value of U.

Step 4: *Interpret the Results.* Since the observed value of U (27) is *not less* than the critical value (24), the null hypothesis is *not* rejected. The conclusion is that there is no difference in the attitude about equal rights between income-producing and nonincome-producing women. Notice that this conclusion is the same as that reached with the median test. Although both tests are appropriate for these data, the Mann-Whitney U test is statistically more powerful and has been shown to be the better alternative to the two-sample t test for independent means. Since it is more sensitive and thus more likely to lead to the rejection of the null hypothesis when it is false, the authors recommend the use of the Mann-Whitney U test when the assumptions underlying the t test (normality and homogeneity of variance) cannot be adequately met.

Mann-Whitney U Test for Large Samples   When the size of the samples for both groups is greater than 20 ($n_1, n_2 > 20$), the sampling distribution of U approaches the normal distribution. The mean of this sampling distribution is given by:

$$\mu_U = \frac{n_1 n_2}{2} \tag{22.2}$$

where

n_1 = sample size for group 1

n_2 = sample size for group 2

and the standard deviation of the sampling distribution (the standard error of U) is given by:

$$\sigma_U = \sqrt{\frac{(n_1)(n_2)(n_1 + n_2 + 1)}{12}} \tag{22.3}$$

Thus, our general formula for the test statistic (discussed in Chapter 9) can be used to test the null hypothesis

$$z = \frac{U - \mu_U}{\sigma_U}$$

$$= \frac{U - \dfrac{n_1 n_2}{2}}{\sqrt{\dfrac{(n_1)(n_2)(n_1 + n_2 + 1)}{12}}} \qquad (22.4)$$

As usual, if a computed z value exceeds the critical value at a specified level of significance, the null hypothesis is rejected.

> The hypothesis tested using the Mann-Whitney U test is that two population distributions are the same for a specified variable. It is a statistically more powerful test than the median test.

Ordinal Data, *K*-Sample Case: Kruskal-Wallis One-Way Analysis of Variance

For ordinal data, the nonparametric analog to one-way analysis of variance is the **Kruskal-Wallis one-way analysis of variance**. The calculation of the test statistic (H) for the Kruskal-Wallis test is similar to the calculation of that for the Mann-Whitney U. Consider the following example: Suppose a researcher is interested in the differences among elementary, middle school, and high school teachers on an authoritarianism inventory. Random samples are selected from the three teacher populations, and these samples (of six, five, and six subjects, respectively) are administered the authoritarianism inventory. The resulting data are found in Table 22.2. For this scale, higher scores indicate a more authoritarian attitude.

Step 1: State the Hypotheses. The null hypothesis for the Kruskal-Wallis test is analogous to the null hypothesis for the one-way ANOVA. The null hypothesis for ANOVA is that there is no difference between the means of the K populations from which the samples were selected. The null hypothesis for the Kruskal-Wallis test, like that for the Mann-Whitney U test, is expressed in the more general terms — namely, that there is *no difference in the distribution of "scores" of the K populations*. For this example, the null hypothesis is that there is no difference between the authoritarianism scores of elementary, middle school, and high school teacher populations. The alternative hypothesis for the Kruskal-Wallis test is that at least two of the K populations or a combination of populations differ. The level of significance was set a priori at .05.

TABLE 22.2
Data for Elementary, Middle School, and High School Teachers on Authoritarianism Scale

Elementary Teachers		Middle School Teachers		High School Teachers	
Score	Rank	Score	Rank	Score	Rank
52	4	66	13	63	10
46	1	49	3	65	12
62	9	64	11	58	8
48	2	53	5	70	15
57	7	68	14	71	16
54	6			73	17
	$29 = R_1$		$46 = R_2$		$78 = R_3$

Step 2: Set the Criterion for Rejecting H_0. The test statistic for this example is the Kruskal-Wallis H; the general formula for computing H, given K samples, is

$$H = \frac{12}{N(N+1)} \sum_{k=1}^{K} \frac{R_k^2}{n_k} - 3(N+1) \qquad (22.5)$$

where

$N = \Sigma n_k$ = total number of observations

n_k = number of observations in the kth sample

R_k = sum of the ranks in the kth sample

The sampling distribution for H is the χ^2 distribution with $K - 1$ degrees of freedom, where K is the number of samples. For this example, since $K = 3$, there are 2 degrees of freedom, and the critical value of H (from Table C.4) for $\alpha = .05$ is 5.991.

Step 3: Compute the Test Statistic. For the data in Table 22.2, the Kruskal-Wallis H is

$$H = \frac{12}{(17)(17+1)} \left[\frac{(29)^2}{6} + \frac{(46)^2}{5} + \frac{(78)^2}{6} \right] - 3(17+1)$$

$$= \frac{2}{51}(1,557.37) - 54$$

$$= 7.86$$

Step 4: Interpret the Results. Since the calculated value of H is greater than the critical value, the researcher rejects the null hypothesis that there is no difference in the authoritarianism scores of elementary, middle-school, and high school teachers. Inspection of the data indicates that the high school teachers tend to have higher authoritarianism scores than both the middle- and elementary school teachers, whereas the middle-school teachers tend to have higher scores than the elementary school teachers.[3]

> The Kruskal-Wallis one-way analysis of variance for ranks is applicable for two or more independent samples. The null hypothesis tested is that the population distributions from which the samples were selected are the same.

Tied Ranks

For both the Mann-Whitney U test and the Kruskal-Wallis H test, there is a possibility of **tied ranks** in the data. If tied ranks do occur, a correction factor may be applied,[4] but this correction has only a minimal effect on the calculated value of U or H and is not presented here. However, if the number of tied ranks is excessive, the use of either the Mann-Whitney U test or the Kruskal-Wallis H test could be questionable.

Ordinal Data: Two-Sample Case, Dependent Samples

The nonparametric analog of the two-sample case with dependent samples for ordinal data is the **Wilcoxon matched-pairs signed-rank test**. This test is commonly used in designs that involve either matched pairs of subjects or pre- and posttests. Consider the following example. Suppose a psychologist is interested in the aggressive behavior of young adolescents with learning disabilities before and after a series of counseling sessions. A random sample of 12 adolescents is selected for the study. The measure of aggressive behavior is the sum of subjective ratings by five trained judges. The measures are taken before and after the treatment. The data for the study are found in Table 22.3.

Step 1: State the Hypotheses. As was the case for the Mann-Whitney U test and the Kruskal-Wallis H test, the null hypothesis is stated in general terms. In this example, the null hypothesis is that there is no difference in the aggressive behavior of populations of children with learning disabilities

---

[3]Multiple comparison procedures for the Kruskal-Wallis test are available and should be applied when the null hypothesis is rejected. For a discussion of these procedures, see Hollander and Wolfe, pp. 124–130.

[4]See S. Siegel, *Nonparametric Statistics for the Behavioral Sciences* (New York: Wiley, 1956), pp. 188–189.

TABLE 22.3
Data on the Aggressive Behavior of Children with Learning Disabilities

Child	Pretest Score	Posttest Score	Difference	Rank of Difference	Ranks with Less Frequent Sign
1	36	21	15	11	
2	23	24	−1	−1	1
3	48	36	12	10	
4	54	30	24	12	
5	40	32	8	7	
6	32	35	−3	−3	3
7	50	43	7	6	
8	44	40	4	4	
9	36	30	6	5	
10	29	27	2	2	
11	33	22	11	9	
12	45	36	9	8	
					$4 = T$

before and after a series of counseling sessions. The psychologist decides to test this null hypothesis against the directional alternative that the aggressive behavior will lessen after the counseling sessions. The level of significance is set at .01.

Step 2: Set the Criterion for Rejecting H_0. The test statistic for the Wilcoxon test is T; the procedure for determining T is discussed in the next step. The sampling distribution of T was developed by Wilcoxon; critical values for one-tailed and two-tailed tests for several significance levels are found in Table C.15. For this example, with 12 matched pairs and $\alpha = .01$, the critical value of T is 10. In order to reject the null hypothesis, the calculated value of T must be *less* than 10.

Step 3: Compute the Test Statistic. The Wilcoxon T is determined as follows:

1. Determine the difference between the pretest and posttest scores for each individual. These are difference scores.
2. Rank the absolute values of the difference scores, and then place the appropriate sign with the rank; if pretest score is larger than posttest score, the sign is positive. (For example, for the data of Table 22.3, the difference of −1 receives a rank of 1 but retains its minus sign, and so on.)
3. Sum the ranks with the less frequent sign. In the example, there are two negative and ten positive signs; therefore, we sum the ranks of the two negatives, getting $T = 4$.

Step 4: Interpret the Results. Since the calculated value of T (4) is *less* than the critical value (10), the null hypothesis is rejected. The psychologist concludes that the aggressive behavior of learning-disabled children can be reduced as a result of the counseling sessions.

The Wilcoxon test is used with matched pairs of observations and tests the null hypothesis of no difference in the matched populations. This test can be used in a pretest-posttest design.

The Wilcoxon Test for Larger Samples

When the size of the sample is large ($n > 25$), the sampling distribution of T has been shown to approximate the normal distribution. The mean of this sampling distribution is given by:

$$\mu_T = \frac{n(n + 1)}{4} \tag{22.6}$$

The standard deviation of the sampling distribution (the standard error of T) is given by:

$$\sigma_T = \sqrt{\frac{n(n + 1)(2n + 1)}{24}} \tag{22.7}$$

Thus, we can use the general formula for the test statistic (discussed in Chapter 10) to test the null hypothesis:

$$z = \frac{T - \mu_T}{\sigma_T}$$

$$= \frac{T - \dfrac{(n)(n + 1)}{4}}{\sqrt{\dfrac{n(n + 1)(2n + 1)}{24}}} \tag{22.8}$$

As usual, if the computed z value exceeds the critical value (z_{cv}) at a specified level of significance, the null hypothesis is rejected.

Summary

In this chapter, we have presented a selected number of nonparametric statistical tests that are the analogs of the parametric tests of Chapters 8 through 12. These tests of significance are used when the assumptions underlying the use of parametric tests cannot be met. Specifically, they should be considered when assumptions of normality and homogeneity of variance cannot reasonably be met.

The statistical tests described in this chapter are summarized in Table 22.4. The table includes the hypothesis tested, the computational formula of the test statistic, and the test statistic's underlying distribution.

TABLE 22.4
Summary of Nonparametric Statistics for Ordinal Data

Name of Test	Hypothesis Tested	Test Statistic	Sampling Distribution
Independent Samples			
Two-sample case, median test	H_0: $Mdn_1 = Mdn_2$	$$\chi^2 = \frac{n(AD - BC)^2}{(A + B)(C + D)(A + C)(B + D)}$$	χ^2 with 1 degree of freedom
Mann-Whitney U test (small samples)	H_0: Population distributions are the same	$U = $ Smaller of U_1 and U_2 $$U_1 = n_1 n_2 + \frac{n_1(n_1 + 1)}{2} - R_1$$ $$U_2 = n_1 n_2 + \frac{n_2(n_2 + 1)}{2} - R_2$$	Sampling distribution of U
Mann-Whitney U test (large samples)	H_0: Population distributions are the same	$$z = \frac{U - \frac{(n_1)(n_2)}{2}}{\sqrt{\frac{(n_1)(n_2)(n_1 + n_2 + 1)}{12}}}$$	Normal distribution
k-sample case, Kruskal-Wallis test	H_0: Population distributions are the same	$$H = \frac{12}{N(N + 1)} \sum \frac{R_k^2}{n_k} - 3(N + 1)$$	χ^2 with $K - 1$ degrees of freedom
Dependent Samples			
Wilcoxon test (small samples)	H_0: Population difference $= 0$	$T = $ Sum of ranks with signs of lesser frequency	Sampling distribution of T
Wilcoxon test (large samples)	H_0: Population difference $= 0$	$$z = \frac{T - \frac{n(n + 1)}{4}}{\sqrt{\frac{n(n + 1)(2n + 1)}{24}}}$$	Normal distribution

Several of the examples in this chapter involved relatively small sample sizes. These examples reflect research situations in the behavioral sciences in which large samples are not feasible. For example, the number of subjects available may be limited because subjects are atypical or because the measurement or treatment involved in the research is expensive or otherwise demanding of time and resources. For these situations, the parametric assumptions are difficult to justify. Since the nonparametric tests can be used when assumptions about the population distribution are not necessary, small samples can be used. However, it must be emphasized that statistical precision is enhanced with larger samples. For example, the assumption of normality in the population becomes less important as the sample size increases. Thus, small samples should not be used when larger ones are readily available.

In the final analysis, the statistical procedures used in any research study must be appropriate to the conditions of the research and the specific hypotheses. There is little point in stating hypotheses that require parametric procedures if the parametric assumptions cannot be met. On the other hand, we would not employ nonparametric procedures if parametric procedures were applicable. When the assumptions are tenable, the parametric procedures provide more information and can be used to test more complex hypotheses. The interaction in ANOVA is a good case in point. Thus, whether or not nonparametric procedures are "better" for a specific research study is a function of the conditions of the research and the hypotheses to be tested.

Exercises

1. A child psychologist wants to investigate the relationship between the gender of a child and the level of response to nonverbal communication cues. The researcher believes that females will miss fewer cues and thus receive lower scores, taking into consideration both the accuracy and depth of the interpretations. The following results were obtained for ten male and ten female children. Complete both the median test and the Mann-Whitney U Test, using $\alpha = .05$.

Male	Female
10	7
13	8
15	9
16	11
19	12
21	14
22	17
23	18
25	20
26	24

*2 A school district has three large senior high schools. The teacher evaluation system involves, among other activities, periodic observation of teacher performance.

One of the measures that can be obtained from the observation inventory is an indicator of the quality of the student-teacher interaction in the classroom. The quality score can range from zero to 35; the measurement scale is assumed to be ordinal. There is an interest in determining if the quality of interaction differs among subjects taught, i.e., Mathematics, Chemistry, History, and English. Five teachers are selected at random from each of the four subjects, observed in their teaching, and given a quality score. The data are found below. Carry out the Kruskal-Wallis one-way analysis of variance using $\alpha = .05$.

	Quality Scores				
Mathematics	15	17	26	27	24
Chemisty	21	18	23	13	25
History	30	14	12	22	28
English	16	29	19	20	33

3. A school psychologist investigates the relationship of preschool background to emotional adjustment during the first grade. A teacher within the system is asked to develop adjustment ratings for students based on some carefully designed criteria. It is assumed that the resulting data are ordinal in nature and that the scores must be ranked. Carry out the Kruskal-Wallis one-way analysis of variance, using $\alpha = .05$.

Home with parent	*Home with babysitter*	*Nursery school*	*Home with friend or relative*
42	37	47	31
35	40	49	44
39	32	34	38
50	33	46	
45		41	
48		43	
36			

4. The director of a human services agency wants to assess the impact of summer camp attendance on the self-confidence of mentally retarded youth. Two groups of 12 members each are selected for participation. Careful screening by the camp director makes possible a matched-pairs design. One child from each of the 12 pairs is assigned to the experimental treatment. The self-confidence index is assumed to be an ordinal measure, and the scores should be ranked before the analysis. Conduct the Wilcoxon matched-pairs signed-rank test, using $\alpha = .01$.

Control	*Experimental*		*Control*	*Experimental*
10	17		10	22
13	11		11	20
8	18		10	23
10	9		13	24
13	16		6	14
14	10		9	23

5. Faculty members at a southwestern college rated their dean on a scale that was purported to measure leadership effectiveness. The scores below are those of randomly selected tenured and non-tenured faculty. Using the .01 level of significance, compare the ratings of tenured and non-tenured faculty a) using the median test, and b) using the Mann-Whitney U test.

Rating of the Dean

Tenured faculty	Non-tenured faculty
82	94
68	42
63	50
90	55
77	88
74	45
60	48
85	40
86	66
80	64
92	58
75	89

6. An exercise physiologist compared the perceived exertion of women between the ages of 20 and 25 who exercised with three different pieces of equipment—a stationary bicycle, a treadmill, and free weights. Perceived exertion was rated on a scale of 1 to 50. Eighteen women, randomly divided into three groups, used the exercise equipment for 30 minutes before rating their perceived exertion. Use the Kruskal-Wallis test to see if there were differences among the three kinds of exercise. Use $\alpha = .05$.

Bicycle	Treadmill	Weights
33	42	24
38	35	30
31	40	26
44	46	34
41	37	36
28	39	22

7. Ten people suffering chronic pain were given a rating scale that measured the intensity of their pain. They were then given a drug (actually a placebo) and encouraging information concerning the effectiveness of the drug. After taking the drug for five days, they again rated the intensity of their pain. The ratings are listed below. Use the Wilcoxon matched-pairs signed-rank test to determine if the pretest scores differed significantly from the posttest scores. Use $\alpha = .05$.

Patient	Pretest	Posttest
1	48	35
2	27	29
3	36	29
4	44	21
5	22	24
6	35	30
7	29	27
8	48	37
9	25	28
10	32	24

8. A clinical psychologist was interested in whether people who feared different things had different levels of depression. She gave a depression questionnaire to persons whose major fears were a) being in a crowded room, b) spiders and snakes, and c) public speaking. The scores on the depression measure are ordinal and must be ranked. Test the hypothesis that the depression scores of the three populations are the same. Use the Kruskal-Wallis test, with $\alpha = .05$.

Crowded room	Spiders and snakes	Public speaking
34	28	52
47	33	46
40	35	50
42	30	49
43		44
		55

9. A sociologist wished to compare men and women with regard to their attitudes toward the homeless. Ten men and ten women were asked to rate the intensity of their feelings on a 1 (never think about it) to 30 (overwhelmingly concerned) scale. Complete the median test and the Mann-Whitney U Test, using the .05 level of significance.

Men	Women
6	20
18	13
8	17
7	4
2	20
5	16
15	19
22	23
14	28
25	24

10. A political scientist surveyed men and women in four age groups about how concerned they were about the national economy. The data are listed below; higher scores indicate greater concern.

a. Compare men and women using the median test.

b. Compare men and women using the Mann-Whitney U test.

c. Compare the different age groups using the Kruskal-Wallis test.

d. Interpret the result in a couple of sentences.

Person	Age	Gender	Concern about the economy
1	40	M	30
2	50	M	33
3	30	M	20
4	20	M	10
5	50	F	36
6	30	F	18
7	20	M	6
8	40	M	26
9	30	F	16
10	40	F	24
11	50	F	35
12	20	F	5
13	30	F	13
14	40	F	29
15	20	M	11
16	20	M	8
17	30	F	21
18	50	M	31
19	50	M	32
20	40	F	27

Appendix A
Glossary

Absolute value   The value of a number without regard to its algebraic sign.

Adjusted means   In ANCOVA, group means adjusted for the effects of the covariate.

Alternative hypothesis (H_a)   Possible outcomes not covered by the null hypothesis.

b coefficient   Regression coefficient in raw score form.

Backward solution   In multiple regression, all predictor variables are initially entered into the model and then are deleted one at a time when they do not contribute to regression.

Bar graph   A graph of a variable measured on the nominal scale; the heights of the bars represent the frequency of occurrence for the specific category.

Beta (β) coefficient   Regression coefficient in standard score form.

Between-groups variation   Variation between the group means and the mean of the total group; denoted s_B^2.

Binomial distribution   Distribution generated by taking the binomial $(X + Y)$ and raising it to the nth power $(X + Y)^n$.

Biserial correlation coefficient (r_b)   The correlation between two variables; one measured on an interval scale and the other a dichotomy with underlying continuity.

Box plot   A graphical summary of a set of scores that can illustrate both the central tendency and the variation.

Central limit theorem   A theorem which provides the mathematical basis for using the normal distribution as the sampling distribution of all sample means of a given sample size. The theorem states that this distribution of sample means (1) is normally distributed, (2) has a mean equal to μ, and (3) has a variance equal to σ^2/n.

Central tendency   A central point on the scale of measurement, between the extreme scores in the distribution, around which the scores are distributed.

Chi-square (χ^2) distribution   A family of distributions used as sampling distributions in both parametric and nonparametric tests of significance.

Class intervals   Intervals of scores in a frequency distribution derived by combining several scores into an interval, thus reducing the number of categories.

Cluster sampling   The random selection of clusters (groups of population members) rather than individual population members.

Coefficient of determination   The square of the correlation coefficient (r^2); a measure of the shared variance.

Combinations   The number of ways different subsets of events can be selected.

Comparisonwise error rate (α)   The probability of making a Type I error for each comparison.

Complex contract   A comparison of group means involving more than two groups in the comparison.

Compound event   An event that involves two or more single events that are not necessarily mutually exclusive.

Conditional distribution   A distribution of error scores around the respective predicted Y score for a given value of X.

Confidence interval   A range of values that we are confident contains the population parameter.

Constant   A characteristic that assumes the same value for all members of the group under study.

Contingency coefficient (Pearson's)   A correlation coefficient used to determine the degree of association between two variables in a contingency table.

Continuous variable   A variable that can take on any value in the measurement scale being used.

Correlated data   Data from subjects measured under more than one condition.

Correlation   The nature, or extent, of the relationship between two variables.

Correlation coefficient   An index of the relationships between two variables.

Covariance   The average of the cross-products of deviation scores.

Covariate   A variable, assumed to be related to the dependent variable, that is controlled through statistical analysis.

Criterion variable   In the process of prediciton, the variable for which scores are estimated on the basis of knowledge of the scores on the predictor variable.

Critical value   The value in the sampling distribution that represents the beginning of the region of rejection.

Cross-products   The multiplication of two scores X and Y for each subject.

Cumulative frequency distribution   The sum of the frequency scores in any class interval plus the frequencies of scores in all preceding class intervals on the scale of measurement.

Curvilinear relationship   A relationship between two variables when the points in a scatterplot tend to locate along a curved line rather than a straight line.

Data   Bits of information that are gathered on some characteristic of a group of individuals or objects under study.

Data curve   A set of data points connected by straight line segments, illustrating the relationship between two variables.

Degrees of freedom   The number of observations less the number of restrictions placed on them.

Dependent samples   Two or more samples of scores from a single population (or matched population) such that scores from one measurement are matched with scores on the other measurement, e.g., the same respondent measured under two conditions.

Dependent variable   The variable that is, or is presumed to be, the result of the manipulation of the independent variable.

Descriptive statistics   Procedures used for classifying and summarizing, or describing, data.

Deviation score   The difference between a given score and the mean, $x = (X - \mu)$.

Dichotomous variable   A discrete variable with only two classifications.

Directional or **one-tailed test**   Statistical test in which the region of rejection is located in one of the two tails of the sampling distribution.

Disordinal interaction   Condition that exists when lines in an interaction plot of cell means intersect within the plot.

Effect size   In determining the power of a test, the difference between the value specified in the H_0 and the value specified in the H_a.

Errors in prediction   The differences between actual Y scores and the predicted scores.

Eta (η) coefficient   The correlation between two variables measured on an interval scale that is an index of the nonlinear relationship between the variables.

Exact limits   The points that separate intervals in a frequency distribution.

Expected mean squares   Mathematical expressions consisting of parameters that describe the components of variance estimates.

Experimental control   When the independent variables are identified by the researcher as the primary variables under investigation.

Experimentwise error rate (α_E)   The probability of making a Type I error for a set of possible comparisons.

Exponent   The power to which the number is raised.

F **distribution**   A family of sampling distributions that require two degrees of freedom values for identifying the specific distribution.

F **ratio**   The ratio of two variance estimates.

Factor   Another term for an independent variable.

Factorial   A mathematical expression indicating the product of all integers 1 through n.

Factorial design   The design of a research study in which two or more independent variables are considered simultaneously.

Fisher *z* transformation   Formula for transforming the correlation coefficient such that the sampling distribution of the transformed correlation coefficient is the normal distribution.

Fixed-effects model   A two-way ANOVA in which the levels of both independent variables are selected by the researcher because they are of particular interest.

Forward solution   In multiple regression, a process for selecting predictor variables that are entered one at a time until the increase in R^2 is no longer statistically significant.

Frequency distribution   A tabulation of data that indicates the number of times given scores or groups of scores appear.

Frequency polygon   A graph on which the frequency of each class interval is plotted at the midpoint and the midpoints are connected with straight line segments.

Goodness-of-fit   A statistical test of whether or not the observed frequencies are a good fit to the expected frequencies

Graph   A pictorial representation of a set of data.

Histogram   A type of graph that depicts the frequencies of individual scores, or scores in a class interval, by the height of the bars.

Homogeneity   The extent to which the members of the group tend to be the same on the variables being investigated.

Homogeneity of regression   In ANCOVA, the assumption that the regression lines within each group are equal.

Homogeneity of variance   An assumption that the variances of the populations are equal.

Homoscedasticity   The assumption, in regression, that the standard deviations of all conditional distributions are equal.

Hypothesis   A conjecture about one or more population parameters.

Hypothesis testing   Determining whether some supposed value for an unknown population parameter is tenable or justifiable.

Independent samples   Samples selected from separate and distinct populations.

Independent variable   A variable that is controlled or manipulated by the researcher. A categorical variable used to form the groupings of observations.

Inferential statistics   Procedures for making generalizations about a population by studying a subset of the population, called a sample.

Interaction   Condition that exists when the effect of the levels of one independent variable is not the same across the levels of a second independent variable.

Intercept   The value on the Y axis when X equals 0.

Interquartile range (IQR)   The range of values between the 75th percentile (3rd quartile) and the 25th percentile (1st quartile).

Interval estimation   A process that produces a range of values for the parameter that are tenable for a given level of confidence.

Interval scale   A scale having distinctive and ordered categories with equal interval differences.

Kurtosis   The degree of peakedness in a symmetric distribution.

Laws of probability   Laws that describe the behavior of events given certain conditions.

Least-squares criterion   The condition that requires fitting a line to a scatterplot of points in such a way that the sum of the squared distances from the data points to the lines is a minimum.

Level of confidence   A probability that indicates the degree of confidence that the computed interval contains the parameter being estimated.

Level of significance (or alpha (α) level)   The probability of making a Type I error if H_0 is rejected.

Levels of the independent variable   Classification of groupings of observations.

Linear model   An estimation of the components of one score in the population.

Linear regression line   The mathematical equation of a straight line: $Y = bX + a$.

Linear relationship   A relationship represented on a scatterplot by a random scatter of points about a straight line.

Main effects   Differences on the dependent variable attributed to differences between levels of the independent variables.

Mean   The arithmetic average of the scores in a distribution.

Mean deviation   The average deviation from the mean for the scores in a distribution.

Mean square   A variance estimate found by dividing the sum of squares by the degrees of freedom.

Measurement   The process of assigning numbers to characteristics according to a defined rule.

Measures of association   Coefficients for contingency tables larger than 2×2.

Median   The point on the scale of measurement below which 50 percent of the scores fall.

Midpoint   The point on the scale of measurement that is halfway through the interval.

Mixed-effects model   A two-way ANOVA in which the levels of one independent variable are fixed and the levels of the other independent variable are random.

Mode   The most frequent score in a distribution of scores.

Multimodal distribution   A distribution in which two or more scores have the same frequency, which is also the greatest frequency.

Multiple correlation   The relationship between the criterion variable (Y) and multiple predictor variables (X_k).

Multiple linear regression   Predicting scores on the criterion variable (Y) from scores on multiple predictor variables (X_k).

Negative correlation   The relationship between two variables depicted by a pattern of points in a scatterplot that tends to run from the upper left to lower right.

Newman-Keuls method   A multiple comparison procedure designed to make all pairwise comparisons of means, which are ranked from low to high, while maintaining the Type I error rate at the predetermined α level.

Nominal scale   A measurement scale that simply classifies without ordering.

Nondirectional or **two-tailed test**   Test in which the region of rejection is located in both tails of the sampling distribution.

Nonparametric tests   Statistical tests of significance that require fewer assumptions than parameter tests.

Normal curve equivalent (NCE) score   A normalized standard score; the mean of the distribution of NCE scores is 50 and the standard deviation is 21.

Normal distribution   An underlying distribution based on the normal curve.

Normalized standard score   Scores that are transformed into percentiles and then into z scores using the properties of the normal distribution.

Null hypothesis (H_0)   A statement of no difference or no relationship.

Observed frequencies   The observed frequencies of occurrence in the categories of a contingency table.

Ogive   The graph of a cumulative percentage distribution.

One-tailed, or **directional, test**   Hypothesis testing when the null hypothesis is tested against a directional alternative hypothesis.

One-way ANOVA   An analysis that includes only one independent variable with more than two levels.

Ordinal interaction   Condition that exists when the lines in an interaction plot of cell means do not intersect within the plot.

Ordinal scale   A measurement scale that has distinctive and ordered categories.

Outlier   An unusual score in a distribution that is considered extreme and may warrant special consideration.

Pairwise comparisons   All possible comparisons when testing the differences between means for multiple groups.

Parameter   A descriptive measure of a population.

Parametric assumptions   Assumptions underlying the use of certain inferential statistical procedures: (1) the samples are random samples from defined populations; (2) the samples are independent; (3) the dependent variable is measured on at least an interval scale; (4) the dependent variable is normally distributed in the population; and (5) the population variances are equal (homogeneity of variance).

Parametric statistics   Procedures that require parametric assumptions.

Parametric tests   Statistical tests of hypotheses in which the null hypothesis includes a specified value for the population parameter and in which certain assumptions have been met (see parametric assumptions).

Part correlation   The relationship between two variables, X and Y, with the influence of a third variable, Z, removed from either X or Y.

Partial correlation   The relationship between two variables, X and Y, with the influence of a third variable, Z, removed from both.

Partitioning total variation   Partitioning the sum of squared deviations around the grand mean, $\Sigma (X - \mu)^2$.

Pearson product-moment correlation coefficient   The index of the linear relationship between two variables, called the Pearson r.

Percentile   The point in a distribution at or below which a given percentage of scores is located.

Percentile rank   A score that is transformed from the original measurement scale to a percentile scale.

Phi (ϕ) coefficient   The correlation between two variables, both of which are dichotomies.

Planned, or **a priori, comparisons**   Predetermined tests of differences between groups conducted in lieu of ANOVA.

Planned orthogonal contrasts   A set of complex comparisons of group means; all the comparisons are statistically independent.

Point-biserial correlation coefficient (r_{Pb})   The correlation between two variables; one is measured on an interval scale and the other is a dichotomy.

Point estimate   The sample statistic that represents the "best" estimate of the population parameter.

Pooled estimate of the population variance   The estimate of the population variance that takes into consideration not only the variances of the two samples, s_1^2 and s_2^2, but also the respective sample sizes, n_1 and n_2.

Population   All members of some defined group.

Positive correlation   The relationship between two variables depicted by a pattern of points in a scatterplot that tends to run from lower left to upper right.

Post hoc multiple-comparison tests   Follow-up methods used to determine which pair or pairs of means differ.

Power   The probability of rejecting the null hypothesis when it is false.

Prediction   The process of estimating scores on one variable (Y) from knowledge of scores on another variable (X).

Predictor variable   The known variable (X) that is used in the prediction process.

Probability   The ratio of the number of favorable outcomes to the total number of possible outcomes for an event.

Proportion   The fractional part of a group that possesses some specific characteristic.

Qualitative variables   Variables measured on the nominal or ordinal scale.

Quantitative variables   Variables measured on the interval or ratio scale.

Random-effects model   A two-way ANOVA in which the levels of both independent variables are randomly selected by the researcher from populations of levels.

Range   The number of units on the scale of measurement needed to include both the highest and lowest score.

Rank distribution   The organizing of scores to order them from highest to lowest.

Ratio scale   A scale having distinctive and ordered categories with equal intervals and a true zero point.

Real numbers   All positive and negative numbers (including fractions and decimals) from negative infinity to positive infinity.

Region of rejection   The area of the sampling distribution that represents those values of the sample statistic that are highly improbable if the null hypothesis is true.

Regression coefficient   A number indicating the slope of the regression line.

Regression constant   A number indicating the Y intercept of the regression line.

Repeated measures ANOVA   An ANOVA in which subjects are measured two or more times and the total variation is partitioned into three components: (1) variation among individuals; (2) variation among test occasions; and (3) residual variation.

Residual variation   Variation not due to either individuals or test occasions in repeated measures ANOVA.

Sample   A subset of a population.

Sample size   The number of observations in a sample.

Sampling distribution   The distribution of a statistic (e.g., sample mean) for all possible samples of a given sample size.

Sampling distribution of the mean   The distribution of sample means from all possible samples of a given sample size.

Sampling fraction   The ratio of the size of the sample to the size of the population.

Scatterplot   A graph that plots pairs of scores for each individual on the two variables.

Scheffé method   A multiple comparison procedure that involves computing an F value for each combination of means.

Signed numbers   Real numbers with the algebraic sign added.

Simple effects   Comparisons of differences between means for the levels of one independent variable within the levels of the second independent variable.

Simple random sample   A sample in which all population members have the same probability of being selected and the selection of each member is independent of the selection of all other members.

Skewed distribution   A frequency distribution which has many scores at one end of the scale of measurement and progressively fewer scores at the other end.

Skewness   The degree to which the majority of scores in a frequency distribution are located at one end of the scale of measurement with progressively fewer scores toward the opposite end of the scale.

Slope   The amount of change in Y that corresponds to a change of one unit in X.

Spearman rho (ρ) coefficient   A correlation coefficient used when the level of measurement for both variables is ordinal.

Spreadsheet   A two-dimensional array of data cells.

Standard deviation   Square root of the variance. A measure on the variation/dispersion of scores in a distribution.

Standard error of estimate   The standard deviation of the errors in prediction.

Standard error of the difference between independent proportions   The standard deviation of the sampling distribution of differences between two sample proportions.

Standard error of the difference between means   The standard deviation of the sampling distribution of differences between two sample means.

Standard error of the mean   The standard deviation of the sampling distribution of sample means.

Standard error of the regression coefficient   The standard deviation of the sampling distribution of regression coefficients.

Standard error of the sample proportion   The standard deviation of the sampling distribution of sample proportions.

Standard error of the transformed correlation coefficient   The standard deviation of the sampling distribution of z_r.

Standard normal distribution   A normal distribution in standard score form with mean of 0 and standard deviation of 1.0.

Standard score   A transformed score that indicates the number of standard deviations a corresponding raw score is above or below the mean.

Standardized residual   In a χ^2 test, an indicator of the major contributors to the significant χ^2 value.

Statistic   A descriptive measure of a sample.

Statistical control   When variance is controlled through statistical analyses.

Statistical estimation   The process of estimating a parameter from the corresponding sample statistic.

Statistical precision   The accuracy by which a statistic is tested or a parameter estimated; in general, the smaller the standard error, the greater the precision.

Statistical significance   The difference between the hypothesized population parameter and the corresponding sample statistic is said to be statistically significant when the probability that the difference occurred by chance is less than the significance level (α level).

Statistics   The entire body of mathematical theory and the procedures that are used to analyze data.

Stem-and-leaf displays   A graphical summary that illustrates the shape of a distribution of scores.

Stepwise solution   In multiple regression, predictor variables are added one at a time as in the forward solution; however, after each new variable is added, all the other variables are tested individually for significance as if they were added last.

Stratified random sampling   The selection of a sample in which the population is divided into subpopulations called strata, with all strata being represented in the sample.

Studentized range (Q) distributions   A family of distributions used in post hoc multiple comparison procedures.

Sum of squares   The sum of squared deviations for all scores in a distribution, e.g., $\Sigma(X - \mu)^2$.

Symmetric distribution   A distribution that is the same on either side of the median.

Systematic sampling   A procedure for selecting a probability sample in which every kth member of the population is selected and in which $1/k$ is the sampling fraction.

t distribution   A family of symmetric, bell-shaped distributions. As the sample size increases, the specific t distribution increasingly approximates the normal distribution.

T score   The scores in a transformed distribution with a mean of 50 and standard deviation of 10: $T = 10z + 50$.

Test of independence   A χ^2 test of signifcance used to determine whether the effects of one variable are independent of the effects of the second variable.

Test of simple effects   In two-way ANOVA, comparison of means on one factor at different levels of the second factor.

Test statistic   A standard score indicating the difference between the observed sample statistic and the hypothesized value for the population parameter.

Theoretical sampling distribution   Distribution based on information from a single sample in conjunction with mathematical theory.

Theorized or **expected frequencies**   The theoretical frequencies of occurrence in the categories of a contingency table that are determined from the null hypothesis.

Tied ranks   Ties among the ranks of ordinal data.

Total variation   Variation of all subjects around the mean for the total group.

Transformed scores   Raw scores that are transformed into a different distribution of scores with a new scale of measurement, with a predetermined mean and standard deviation.

Trend analysis   A set of complex comparisons of group means used to investigate the nature of the pattern of the quantitative levels of the independent variable.

Tukey method   A multiple comparison procedure designed to make all pairwise comparisons of means while maintaining the Type I error rate at the predetermined α level.

Tukey/Kramer (TK) method   A multiple comparison procedure designed to be used when the group sizes differ.

Two-tailed, or **nondirectional, test**   Hypothesis testing when the null hypothesis is tested against a nondirectional alternative hypothesis.

Type I error   Rejecting a null hypothesis when in fact it is true.

Type I error rate   The probability of making one or more Type I errors in a series of hypothesis tests.

Type II error   Retaining a null hypothesis when in fact it is false.

Unbiased estimate   An estimate for which the mean of all possible sample values, for a given sample size, equals the parameter being estimated.

Underlying distribution   The distribution of all possible outcomes of a particular event.

Uniform, or rectangular, distribution   A set of scores that are evenly distributed throughout the distribution.

Unit normal distribution   Normal distribution with mean equal to 0 and variance equal to 1.

Variable   A characteristic that can take on different values for different members of the group under study.

Variance   A measure of the variation/dispersion of scores in a distribution.

Variance estimates   Estimates of the variance in the population using the sample variance.

Variation   A quantitative measure of the extent to which the scores are dispersed throughout the distribution.

Weighted average score   A composite score generated by transforming individual scores into standard scores and then applying weights.

Weighted mean   The mean of a distribution of weighted scores.

Wilcoxon matched-pairs signed-rank test   Nonparametric test for ordinal data with dependent sample.

Within-groups variation   Variation among all subjects within a group; denoted s_w^2.

Yates' correction for continuity   Adjustment in the 2×2 χ^2 test when there are small expected frequencies.

z score   A standard score: $z = (X - \mu)/\sigma$.

Appendix B
Data Sets and Computer Exercises

In several chapters of this book, computer output is presented for examples that use the survey of high school seniors data set. We have used the Statistical Package for the Social Sciences for Windows on a PC (SPSS, 1996) to analyze the data for these examples. Although the examples and computer solutions illustrate how statistical software packages relieve researchers of the computational burden formerly associated with statistics, they do not relieve them of the responsibility for appropriate application and proper interpretation of the results.

Accompanying this text are the survey of high school seniors data and three additional data sets that can be used for practice in analyzing data using computers. The data sets are described in Part I of this Appendix using the coding procedures that were discussed in Chapter 2. Suggested exercises based on these data sets and corresponding to the statistical methods in many of the chapters are provided in Part II.

Part I: The Data Sets

HSB500.sav and HSB75.sav

The survey of high school seniors data set contains responses from 28,240 high school seniors. A random sample of 500 seniors was selected for the purposes of this book; a second sample of 75 seniors was also selected. Instructors may wish to use either or both data sets in creating additional exercises. The variables are:

GENDER   Gender of High School Senior
                        0 = Female
                        1 = Male

MAED and FAED   Education of Mother and Education of Father
                        2 = Less than High School
                        3 = High School Graduate
                        4 = Less than two years of Vocational Training
                        5 = More than two years of Vocational Training
                        6 = Less than two years of College
                        7 = More than two years of College
                        8 = College Graduate—Baccalaureate Degree
                        9 = Master's Degree
                        10 = M.D./Ph.D.

ALG1   Enrolled in Algebra 1
                        1 = Student has taken the course
                        0 = Student has not taken the course

ALG2   Enrolled in Algebra 2
                        1 = Student has taken the course
                        0 = Student has not taken the course

GEO   Enrolled in Geometry
                        1 = Student has taken the course
                        0 = Student has not taken the course

TRIG   Enrolled in Trigonometry
                        1 = Student has taken the course
                        0 = Student has not taken the course

CALC   Enrolled in Calculus
                        1 = Student has taken the course
                        0 = Student has not taken the course

GRADES	Student Report of High School Grades
	8 = Mostly A's
	7 = Half A's/Half B's
	6 = Mostly B's
	5 = Half B's/Half C's
	4 = Mostly C's
	3 = Half C's/Half D's
	2 = Mostly D's
	1 = Below D's

MATHGR	Student Report of Grades in Mathematics Courses
	1 = Mostly A's and B's
	0 = Otherwise

MATHACH — A 25-item Mathematics test with scores ranging from −8.33 to 25.

VISUAL — A 16-item test of Visualization in three dimensions, for determining how a three-dimensional object would look if its spatial position were changed; scores range from −4 to 16.

MOSAIC — A 56-item test involving the detection of relationships in patterns of tiles; scores range from −28 to 56.

ODE.sav

ODE.sav lists information from 94 school districts in northwest Ohio. The data come from the Ohio Department of Education. The variables are:

SCHOOLS	school district name
STUDENTS	number of students in the district
INCOME	median income in the district
PROPERTY	property valuation per pupil
WELFARE	percent of students from families on welfare
SALARY	average teacher salary
INSTRUCT	per pupil expenditures on instruction
ATTEND	average daily attendance expressed as a percent
PASS9th	percent of students passing all 9th grade state proficiency tests on the first opportunity
PASS4th	percent of students passing all 4th grade state proficiency tests on the first opportunity

HWJ100.sav

HWJ100.sav contains information on 100 applicants for management positions at a large corporation. The data come from a study of applicant profiles. The variables are:

JOBS	number of related jobs the applicant has held
EXPER	number of job-related unpaid experiences the applicant has had
VERBAL	score on a verbal aptitude test
QUANT	score on a quantitative aptitude test
INTVW	interviewer's rating of the applicant; a 1–10 rating with 10 as the highest rating
COLLEGE	rating of the quality of the undergraduate college the applicant attended; a 1–10 rating with 10 as the highest rating
GPA	college grade-point average
ACTIV	number of extracurricular activities participated in during college
SELFCONF	score on a measure of self-confidence
EXTROV	score on a measure of extroversion

CPS50.sav

CPS50.sav contains data from a study of 50 youth who had been in custody of Children's Protective Services and were emancipated upon reaching their 18th birthdays. Scores are available on the following variables.

ILS	Independent Living Scale
SCS	Self-Confidence Scale
SSI	Social Skills Inventory
AAS	Academic Aptitude Test
PAS	Personal Adjustment Scale
Age	1 = 18 years old, 2 = 19 years old, 3 = 20 years old
SAS	Social Adjustment Scale

Part II: Computer Exercises

Chapter 2

1. Do a frequency distribution for the percent of students who passed the 9th grade proficiency test (ODE.sav-pass9th).

2. Do a frequency distribution for the number of related jobs that the applicant has had (HWJ100.sav-jobs).

3. Do a histogram for the number of extracurricular activities participated in while in college (HWJ100.sav-activ).

Chapter 3

1. Find the mean and standard deviation of the average teacher salaries (ODE.sav-salary).

2. Find the range and variance of the quantitative aptitude scores of the applicants (HWJ100.sav-quant).

3. Convert the property valuation per pupil to standard scores (z-scores) and list them (ODE.sav-property)

Chapter 5

1. Determine the correlation between the average teacher's salary and per pupil expenditures on instruction (ODE.sav-salary and instruct).

2. Determine the correlation between the score on a verbal aptitude test and the score on a quantitative aptitude test (HWJ100.sav-verbal and quant).

3. Determine the correlation between the score on the Self-Confidence Scale and the score on the Social Skill Inventory (CPS50-SCS and SSI).

Chapter 6

1. Find the equation of the regression line between median income in the district and number of students in the district (ODE.sav-income and students).

2. Determine the prediction equation for predicting the percentage of students passing all 9th grade proficiency tests on the first opportunity from per pupil expenditures on instruction (ODE.sav-pass9th and instruct).

3. Determine the prediction equation for predicting college grade-point average from the score on a verbal aptitude test (HWJ100.sav-gpa and verbal).

Chapter 8

For the exercises in this chapter, assume that CPS50.sav consists of scores from a random sample of a large population.

1. Test the hypothesis that the population mean of the Independent Living Scale score (CPS50.sav-ILS) is 54.6. Use alpha = .05. Consider this a two-tailed test, assuming that the hypothesis was formulated before the sample was measured.

2. Test the hypothesis that the population mean of the Self-Confidence Scale score (CPS50.sav-SCS) is less than 52. Use alpha = .05.

3. Test the hypothesis that the population mean of the Social Skills Inventory (CPS50.sav-SSI) is 51. Use alpha = .05, and make the same assumption as for Exercise 1.

Chapter 11

For the exercises in this chapter, assume that ODE.sav consists of scores from a random sample of 94 school districts. Recode the variable, median income in the district (income) into a dichotomy, the 47 districts with the lowest median incomes (those less than $23.700), and the 47 districts with the highest median incomes (those greater than $23,700). Consider each of these 47 districts as a random sample of larger populations of districts. Designate those with the lowest median incomes, population 1; those with the highest median incomes, population 2.

1. Test the hypothesis that the mean, average teacher salary (ODE.sav-salary) is the same for the two populations. Use alpha = .05.

2. Test the hypothesis that the mean number of students per district (ODE.sav-students) is the same for the two poulations. Use alpha = .01.

3. Test the hypothesis that the mean amount spent per pupil on instruction (ODE.sav-instruct) in population 1 equals that spent in population 2. Use alpha = .05.

Chapter 14

For the exercises in this chapter, assume that CPS50.sav consists of scores from a random sample of a large population.

1. Complete a one-way ANOVA on the Self-Confidence Scale scores, with age group as the independent variable. Test the null hypothesis that the populations from which the age groups were selected have equal means. Use alpha = .05. (CPS50.sav-SCS and Age)

2. Repeat the procedures in Exercise 1 with Social Skills Inventory scores as the dependent variable (CPS50.sav-SSI and Age).

3. Repeat the procedures in Exercise 1 with Academic Aptitude Scale scores as the dependent variable (CPS50.sav-AAS and Age).

Chapter 15

1. For the ANOVA computed in Exercise 1 of Chapter 14, if statistical significance was attained, perform a Tukey method HSD test to determine which group means are significantly different.

2. For the ANOVA computed in Exercise 2 of Chapter 14, if statistical significance was attained, execute a Newman-Keuls method test to detemine which group means are significantly different.

3. For the ANOVA computed in Exercise 3 of Chapter 14, if statistical significance was attained, perform a Scheffé method test to determine which group means are significantly different.

Chapter 16

1. Recode the median income in the district into those above and below $23,700 and recode the number of students in the district into those less than 1000, 1000 to 2000, and those with more than 2000 students. Then run a 2×3 ANOVA with the dependent variable being "the percent of students who passed the 9th grade proficiency test on the first try" (ODE.sav-income, students, and pass9th).

2. Repeat the above analysis for the 4th grade proficiency tests (ODE.sav-income, students, and pass4th).

3. Recode the number of related jobs and the years of related experience to be 0, 1, or more than 1. Then run a 3×3 ANOVA with the interviewer's rating of the applicant as the dependent variable (HWJ100.sav-jobs, exper, and intvw).

Chapter 17

1. Find the equation for predicting verbal aptitude scores from quantitative aptitude scores and then test the hypothesis that $\beta = 0$ (HWJ100.sav-verbal and quant).

2. Find the equation for predicting the "percent of students who passed the 9th grade proficiency tests on the first try" from the average daily attendance. Then test the hypothesis that $\beta = 0$ (ODE.sav-pass9th and attend).

3. Find the equation for predicting the "percent of students who passed the 9th grade proficency tests on the first try" from per pupil expenditures on instruction. Then test the hypothesis that $\beta = 0$ (ODE.sav-pass9th and instruct).

Chapter 18

1. What percent of the variance in the scores on the Independent Living Scale is accounted for when the Personal Adjustment Scale, the Social Adjustment Scale, and the Social Skills Inventory are used together as predictors (CPS50.sav-ILS, PAS, SAS, and SSI)?

2. Find the equation for predicting per pupil instructional expenditures from the following set of variables: percent of families on welfare, property valuation per pupil, median income in the district, and the number of students in the district (ODE.sav-instruct, welfare, property, income, and students). Would you recommend keeping all of these variables in the prediction equation?

3. Is the number of extracurricular activities in college predictable from the combination of self-confidence, extroversion, and verbal aptitude (HWJ100.sav-activ, selfconf, extrov, and verbal)? Is the R_2 statistically significant?

Chapter 19

1. Complete a one-way ANCOVA on the Self-Confidence Scale scores, with age group as the independent variable and the Social Skills Scale as the covariate. Use alpha = .05. (CPS50.sav-SCS, Age, and SSI).

2. Complete a one-way ANCOVA on the Personal Adjustment Scale scores, with age group as the independent variable and the Social Adjustment Scale as the covariate. Use alpha = .05. (CPS50.sav-PAS, Age, and SAS).

3. Complete a one-way ANCOVA on the Independent Living Scale scores, with age group as the independent variable and the Social Skills Inventory as the covariate. Use alpha = .05. (CPS50.sav-ILS, Age, and SSI).

Chapter 21

1. Recode the ratings of the colleges and the interviews (HWJ100.sav-college and intvw) such that 1–3 are 1; 4–7 are 2; and 8–10 are 3. Then use chi-square to test whether the two ratings are independent.

2. Recode the median incomes as above or below $23,700 and recode the teachers' salary as above and below $32,700 (ODE.sav-income and salary). Then use chi-square to test whether the two ratings are independent.

3. Recode the number of related jobs and the years of related experience to be 0, 1, or more than 1 (HWJ100.sav-jobs and exper). Then use chi-square to test whether the two ratings are independent.

Appendix C
Tables

632

TABLE C.1
Areas under Standard Normal Curve for Values of z

z	Area between $\bar{X}$ and z	Area beyond z	Ordinate	z	Area between $\bar{X}$ and z	Area beyond z	Ordinate	z	Area between $\bar{X}$ and z	Area beyond z	Ordinate
0.00	.0000	.5000	.3989	0.40	.1554	.3446	.3683	0.80	.2881	.2119	.2897
0.01	.0040	.4960	.3989	0.41	.1591	.3409	.3668	0.81	.2910	.2090	.2874
0.02	.0080	.4920	.3989	0.42	.1628	.3372	.3653	0.82	.2939	.2061	.2850
0.03	.0120	.4880	.3988	0.43	.1664	.3336	.3637	0.83	.2967	.2033	.2827
0.04	.0160	.4840	.3986	0.44	.1700	.3300	.3621	0.84	.2995	.2005	.2803
0.05	.0199	.4801	.3984	0.45	.1736	.3264	.3605	0.85	.3023	.1977	.2780
0.06	.0239	.4761	.3982	0.46	.1772	.3228	.3589	0.86	.3051	.1949	.2756
0.07	.0279	.4721	.3980	0.47	.1808	.3192	.3572	0.87	.3078	.1922	.2732
0.08	.0319	.4681	.3977	0.48	.1844	.3156	.3555	0.88	.3106	.1894	.2709
0.09	.0359	.4641	.3973	0.49	.1879	.3121	.3538	0.89	.3133	.1867	.2685
0.10	.0398	.4602	.3970	0.50	.1915	.3085	.3521	0.90	.3159	.1841	.2661
0.11	.0438	.4562	.3965	0.51	.1950	.3050	.3503	0.91	.3186	.1814	.2637
0.12	.0478	.4522	.3961	0.52	.1985	.3015	.3485	0.92	.3212	.1788	.2613
0.13	.0517	.4483	.3956	0.53	.2019	.2981	.3467	0.93	.3238	.1762	.2589
0.14	.0557	.4443	.3951	0.54	.2054	.2946	.3448	0.94	.3264	.1736	.2565
0.15	.0596	.4404	.3945	0.55	.2088	.2912	.3429	0.95	.3289	.1711	.2541
0.16	.0636	.4364	.3939	0.56	.2123	.2877	.3410	0.96	.3315	.1685	.2516
0.17	.0675	.4325	.3932	0.57	.2157	.2843	.3391	0.97	.3340	.1660	.2492
0.18	.0714	.4286	.3925	0.58	.2190	.2810	.3372	0.98	.3365	.1635	.2468
0.19	.0753	.4247	.3918	0.59	.2224	.2776	.3352	0.99	.3389	.1611	.2444
0.20	.0793	.4207	.3910	0.60	.2257	.2743	.3332	1.00	.3413	.1587	.2420
0.21	.0832	.4168	.3902	0.61	.2291	.2709	.3312	1.01	.3438	.1562	.2396
0.22	.0871	.4129	.3894	0.62	.2324	.2676	.3292	1.02	.3461	.1539	.2371
0.23	.0910	.4090	.3885	0.63	.2357	.2643	.3271	1.03	.3485	.1515	.2347
0.24	.0948	.4052	.3876	0.64	.2389	.2611	.3251	1.04	.3508	.1492	.2323
0.25	.0987	.4013	.3867	0.65	.2422	.2578	.3230	1.05	.3531	.1469	.2299
0.26	.1026	.3974	.3857	0.66	.2454	.2546	.3209	1.06	.3554	.1446	.2275
0.27	.1064	.3936	.3847	0.67	.2486	.2514	.3187	1.07	.3577	.1423	.2251
0.28	.1103	.3897	.3836	0.68	.2517	.2483	.3166	1.08	.3599	.1401	.2227
0.29	.1141	.3859	.3825	0.69	.2549	.2451	.3144	1.09	.3627	.1379	.2203
0.30	.1179	.3821	.3814	0.70	.2580	.2420	.3123	1.10	.3643	.1357	.2179
0.31	.1217	.3783	.3802	0.71	.2611	.2389	.3101	1.11	.3665	.1335	.2155
0.32	.1255	.3745	.3790	0.72	.2642	.2358	.3079	1.12	.3686	.1314	.2131
0.33	.1293	.3707	.3778	0.73	.2673	.2327	.3056	1.13	.3708	.1292	.2107
0.34	.1331	.3669	.3765	0.74	.2704	.2296	.3034	1.14	.3729	.1271	.2083
0.35	.1368	.3632	.3752	0.75	.2734	.2266	.3011	1.15	.3749	.1251	.2059
0.36	.1406	.3594	.3739	0.76	.2764	.2236	.2989	1.16	.3770	.1230	.2036
0.37	.1443	.3557	.3725	0.77	.2794	.2206	.2966	1.17	.3790	.1210	.2012
0.38	.1480	.3520	.3712	0.78	.2823	.2177	.2943	1.18	.3810	.1190	.1989
0.39	.1517	.3483	.3697	0.79	.2852	.2148	.2920	1.19	.3830	.1170	.1965

Source: From table Ili of R. A. Fisher & F. Yates, *Statistical Tables for Biological, Agricultural and Medical Research*, 6th edition, (London: Longman Group, Ltd., 1974). Reprinted by permission of Addison-Wesley Longman, Ltd.

TABLE C.1 (cont.)

z	Area between $\overline{X}$ and z	Area beyond z	Ordinate	z	Area between $\overline{X}$ and z	Area beyond z	Ordinate	z	Area between $\overline{X}$ and z	Area beyond z	Ordinate
1.20	.3849	.1151	.1942	1.60	.4452	.0548	.1109	2.00	.4772	.0228	.0540
1.21	.3869	.1131	.1919	1.61	.4463	.0537	.1092	2.01	.4778	.0222	.0529
1.22	.3888	.1112	.1895	1.62	.4474	.0526	.1074	2.02	.4783	.0217	.0519
1.23	.3907	.1093	.1872	1.63	.4484	.0516	.1057	2.03	.4788	.0212	.0508
1.24	.3925	.1075	.1849	1.64	.4495	.0505	.1040	2.04	.4793	.0207	.0498
1.25	.3944	.1056	.1826	1.65	.4505	.0495	.1023	2.05	.4798	.0202	.0488
1.26	.3962	.1038	.1804	1.66	.4515	.0485	.1006	2.06	.4803	.0197	.0478
1.27	.3980	.1020	.1781	1.67	.4525	.0475	.0989	2.07	.4808	.0192	.0468
1.28	.3997	.1003	.1758	1.68	.4535	.0465	.0973	2.08	.4812	.0188	.0459
1.29	.4015	.0985	.1736	1.69	.4545	.0455	.0957	2.09	.4817	.0183	.0449
1.30	.4032	.0968	.1714	1.70	.4554	.0446	.0940	2.10	.4821	.0179	.0440
1.31	.4049	.0951	.1691	1.71	.4564	.0436	.0925	2.11	.4826	.0174	.0431
1.32	.4066	.0934	.1669	1.72	.4573	.0427	.0909	2.12	.4830	.0170	.0422
1.33	.4082	.0918	.1647	1.73	.4582	.0418	.0893	2.13	.4834	.0166	.0413
1.34	.4099	.0901	.1626	1.74	.4591	.0409	.0878	2.14	.4838	.0162	.0404
1.35	.4115	.0885	.1604	1.75	.4599	.0401	.0863	2.15	.4842	.0158	.0395
1.36	.4131	.0869	.1582	1.76	.4608	.0392	.0848	2.16	.4846	.0154	.0387
1.37	.4147	.0853	.1561	1.77	.4616	.0384	.0833	2.17	.4850	.0150	.0379
1.38	.4162	.0838	.1539	1.78	.4625	.0375	.0818	2.18	.4854	.0146	.0371
1.39	.4177	.0823	.1518	1.79	.4633	.0367	.0804	2.19	.4857	.0143	.0363
1.40	.4192	.0808	.1497	1.80	.4641	.0359	.0790	2.20	.4861	.0139	.0355
1.41	.4207	.0793	.1476	1.81	.4649	.0351	.0775	2.21	.4864	.0136	.0347
1.42	.4222	.0778	.1456	1.82	.4656	.0344	.0761	2.22	.4868	.0132	.0339
1.43	.4236	.0764	.1435	1.83	.4664	.0336	.0748	2.23	.4871	.0129	.0332
1.44	.4251	.0749	.1415	1.84	.4671	.0329	.0734	2.24	.4875	.0125	.0325
1.45	.4265	.0735	.1394	1.85	.4678	.0322	.0721	2.25	.4878	.0122	.0317
1.46	.4279	.0721	.1374	1.86	.4686	.0314	.0707	2.26	.4881	.0119	.0310
1.47	.4292	.0708	.1354	1.87	.4693	.0307	.0694	2.27	.4884	.0116	.0303
1.48	.4306	.0694	.1334	1.88	.4699	.0301	.0681	2.28	.4887	.0113	.0297
1.49	.4319	.0681	.1315	1.89	.4706	.0294	.0669	2.29	.4890	.0110	.0290
1.50	.4332	.0668	.1295	1.90	.4713	.0287	.0656	2.30	.4893	.0107	.0283
1.51	.4345	.0655	.1276	1.91	.4719	.0281	.0644	2.31	.4896	.0104	.0277
1.52	.4357	.0643	.1257	1.92	.4726	.0274	.0632	2.32	.4898	.0102	.0270
1.53	.4370	.0630	.1238	1.93	.4732	.0268	.0620	2.33	.4901	.0099	.0264
1.54	.4382	.0618	.1219	1.94	.4738	.0262	.0608	2.34	.4904	.0096	.0258
1.55	.4394	.0606	.1200	1.95	.4744	.0256	.0596	2.35	.4906	.0094	.0252
1.56	.4406	.0594	.1182	1.96	.4750	.0250	.0584	2.36	.4909	.0091	.0246
1.57	.4418	.0582	.1163	1.97	.4756	.0244	.0573	2.37	.4911	.0089	.0241
1.58	.4429	.0571	.1145	1.98	.4761	.0239	.0562	2.38	.4913	.0087	.0235
1.59	.4441	.0559	.1127	1.99	.4767	.0233	.0551	2.39	.4916	.0084	.0229

TABLE C.1 (cont.)

z	Area between $\overline{X}$ and z	Area beyond z	Ordinate	z	Area between $\overline{X}$ and z	Area beyond z	Ordinate	z	Area between $\overline{X}$ and z	Area beyond z	Ordinate
2.40	.4918	.0082	.0224	2.80	.4974	.0026	.0079	3.20	.4993	.0007	
2.41	.4920	.0080	.0219	2.81	.4975	.0025	.0077	3.21	.4993	.0007	
2.42	.4922	.0078	.0213	2.82	.4976	.0024	.0075	3.22	.4994	.0006	
2.43	.4925	.0075	.0208	2.83	.4977	.0023	.0073	3.23	.4994	.0006	
2.44	.4927	.0073	.0203	2.84	.4977	.0023	.0071	3.24	.4994	.0006	
2.45	.4929	.0071	.0198	2.85	.4978	.0022	.0069	3.25	.4994	.0006	
2.46	.4931	.0069	.0194	2.86	.4979	.0021	.0067	3.30	.4995	.0005	
2.47	.4932	.0068	.0189	2.87	.4979	.0021	.0065	3.35	.4996	.0004	
2.48	.4934	.0066	.0184	2.88	.4980	.0020	.0063	3.40	.4997	.0003	
2.49	.4936	.0064	.0180	2.89	.4981	.0019	.0061	3.45	.4997	.0003	
2.50	.4938	.0062	.0175	2.90	.4981	.0019	.0060	3.50	.4998	.0002	
2.51	.4940	.0060	.0171	2.91	.4982	.0018	.0058	3.60	.4998	.0002	
2.52	.4941	.0059	.0167	2.92	.4982	.0018	.0056	3.70	.4999	.0001	
2.53	.4943	.0057	.0163	2.93	.4983	.0017	.0055	3.80	.4999	.0001	
2.54	.4945	.0055	.0158	2.94	.4984	.0016	.0053	3.90	.49995	.00005	
2.55	.4946	.0054	.0154	2.95	.4984	.0016	.0051	4.00	.49997	.00003	
2.56	.4948	.0052	.0151	2.96	.4985	.0015	.0050				
2.57	.4949	.0051	.0147	2.97	.4985	.0015	.0048				
2.58	.4951	.0049	.0143	2.98	.4986	.0014	.0047				
2.59	.4952	.0048	.0139	2.99	.4986	.0014	.0046				
2.60	.4953	.0047	.0136	3.00	.4987	.0013	.0044				
2.61	.4955	.0045	.0132	3.01	.4987	.0013	.0033				
2.62	.4956	.0044	.0129	3.02	.4987	.0013	.0024				
2.63	.4957	.0043	.0126	3.03	.4988	.0012	.0017				
2.64	.4959	.0041	.0122	3.04	.4988	.0012	.0012				
2.65	.4960	.0040	.0119	3.05	.4989	.0011	.0009				
2.66	.4961	.0039	.0116	3.06	.4989	.0011	.0006				
2.67	.4962	.0038	.0113	3.07	.4989	.0011	.0004				
2.68	.4963	.0037	.0110	3.08	.4990	.0010	.0003				
2.69	.4964	.0036	.0107	3.09	.4990	.0010	.0002 *				
2.70	.4965	.0035	.0104	3.10	.4990	.0010					
2.71	.4966	.0034	.0101	3.11	.4991	.0009					
2.72	.4967	.0033	.0099	3.12	.4991	.0009					
2.73	.4968	.0032	.0096	3.13	.4991	.0009					
2.74	.4969	.0031	.0093	3.14	.4992	.0008					
2.75	.4970	.0030	.0091	3.15	.4992	.0008					
2.76	.4971	.0029	.0088	3.16	.4992	.0008					
2.77	.4972	.0028	.0086	3.17	.4992	.0008					
2.78	.4973	.0027	.0084	3.18	.4993	.0007					
2.79	.4974	.0026	.0081	3.19	.4993	.0007					

*For values of z greater than 3.09, the height of the curve is negligible and the value of the ordinate is close to zero.

TABLE C.2
Standard Scores (or Deviates) Corresponding to Divisions of the Area under the Normal Curve Into a Larger Proportion (*B*)

B The larger area	*z* Standard score	*B* The larger area	*z* Standard score	*B* The larger area	*z* Standard score
.500	.0000	.675	.4538	.850	1.0364
.505	.0125	.680	.4677	.855	1.0581
.510	.0251	.685	.4817	.860	1.0803
.515	.0376	.690	.4959	.865	1.1031
.520	.0502	.695	.5101	.870	1.1264
.525	.0627	.700	.5244	.875	1.1503
.530	.0753	.705	.5388	.880	1.1750
.535	.0878	.710	.5534	.885	1.2004
.540	.1004	.715	.5681	.890	1.2265
.545	.1130	.720	.5828	.895	1.2536
.550	.1257	.725	.5978	.900	1.2816
.555	.1383	.730	.6128	.905	1.3106
.560	.1510	.735	.6280	.910	1.3408
.565	.1637	.740	.6433	.915	1.3722
.570	.1764	.745	.6588	.920	1.4051
.575	.1891	.750	.6745	.925	1.4395
.580	.2019	.755	.6903	.930	1.4757
.585	.2147	.760	.7063	.935	1.5141
.590	.2275	.765	.7225	.940	1.5548
.595	.2404	.770	.7388	.945	1.5982
.600	.2533	.775	.7554	.950	1.6449
.605	.2663	.780	.7722	.955	1.6954
.610	.2793	.785	.7892	.960	1.7507
.615	.2924	.790	.8064	.965	1.8119
.620	.3055	.795	.8239	.970	1.8808
.625	.3186	.800	.8416	.975	1.9600
.630	.3319	.805	.8596	.980	2.0537
.635	.3451	.810	.8779	.985	2.1701
.640	.3585	.815	.8965	.990	2.3263
.645	.3719	.820	.9154	.995	2.5758
.650	.3853	.825	.9346	.996	2.6521
.655	.3989	.830	.9542	.997	2.7478
.660	.4125	.835	.9741	.998	2.8782
.665	.4261	.840	.9945	.999	3.0902
.670	.4399	.845	1.0152	.9995	3.2905

Source: From J. P. Guilford and B. Fruchter, *Fundamental Statistics in Psychology and Education*, 6th ed. (New York: McGraw-Hill, Inc., 1978). Copyright ©1978. Reproduced by permission of the publishers.

TABLE C.3
Critical Values of the *t* Distribution

	Level of significance for one-tailed test					
	.10	.05	.025	.01	.005	.0005
	Level of significance for two-tailed test					
df	.20	.10	.05	.02	.01	.001
1	3.078	6.314	12.706	31.821	63.657	636.619
2	1.886	2.920	4.303	6.965	9.925	31.598
3	1.638	2.353	3.182	4.541	5.841	12.941
4	1.533	2.132	2.776	3.747	4.604	8.610
5	1.476	2.015	2.571	3.365	4.032	6.859
6	1.440	1.943	2.447	3.143	3.707	5.959
7	1.415	1.895	2.365	2.998	3.499	5.405
8	1.397	1.860	2.306	2.896	3.355	5.041
9	1.383	1.833	2.262	2.821	3.250	4.781
10	1.372	1.812	2.228	2.764	3.169	4.587
11	1.363	1.796	2.201	2.718	3.106	4.437
12	1.356	1.782	2.179	2.681	3.055	4.318
13	1.350	1.771	2.160	2.650	3.012	4.221
14	1.345	1.761	2.145	2.624	2.977	4.140
15	1.341	1.753	2.131	2.602	2.947	4.073
16	1.337	1.746	2.120	2.583	2.921	4.015
17	1.333	1.740	2.110	2.567	2.898	3.965
18	1.330	1.734	2.101	2.552	2.878	3.922
19	1.328	1.729	2.093	2.539	2.861	3.883
20	1.325	1.725	2.086	2.528	2.845	3.850
21	1.323	1.721	2.080	2.518	2.831	3.819
22	1.321	1.717	2.074	2.508	2.819	3.792
23	1.319	1.714	2.069	2.500	2.807	3.767
24	1.318	1.711	2.064	2.492	2.797	3.745
25	1.316	1.708	2.060	2.485	2.787	3.725
26	1.315	1.706	2.056	2.479	2.779	3.707
27	1.314	1.703	2.052	2.473	2.771	3.690
28	1.313	1.701	2.048	2.467	2.763	3.674
29	1.311	1.699	2.045	2.462	2.756	3.659
30	1.310	1.697	2.042	2.457	2.750	3.646
40	1.303	1.684	2.021	2.423	2.704	3.551
60	1.296	1.671	2.000	2.390	2.660	3.460
120	1.289	1.658	1.980	2.358	2.617	3.373
∞	1.282	1.645	1.960	2.326	2.576	3.291

Source: From table III of R. A. Fisher & F. Yates, *Statistical Tables for Biological, Agricultural and Medical Research*, 6th edition, (London: Longman Group, Ltd., 1974). Reprinted by permission of Addison-Wesley Longman, Ltd.

TABLE C.4
Upper Percentage Points of the χ^2 Distribution

df	.99	.98	.95	.90	.80	.70	.50	.30	.20	.10	.05	.02	.01	.001
1	.0³157	.0³628	.00393	.0158	.0642	.148	.455	1.074	1.642	2.706	3.841	5.412	6.635	10.827
2	.0201	.0404	.103	.211	.446	.713	1.386	2.408	3.219	4.605	5.991	7.824	9.210	13.815
3	.115	.185	.352	.584	1.005	1.424	2.366	3.665	4.642	6.251	7.815	9.837	11.345	16.266
4	.297	.429	.711	1.064	1.649	2.195	3.357	4.878	5.989	7.779	9.488	11.668	13.277	18.467
5	.554	.752	1.145	1.610	2.343	3.000	4.351	6.064	7.289	9.236	11.070	13.388	15.086	20.515
6	.872	1.134	1.635	2.204	3.070	3.828	5.348	7.231	8.558	10.645	12.592	15.033	16.812	22.457
7	1.239	1.564	2.167	2.833	3.822	4.671	6.346	8.383	9.803	12.017	14.067	16.622	18.475	24.322
8	1.646	2.032	2.733	3.490	4.594	5.527	7.344	9.524	11.030	13.362	15.507	18.168	20.090	26.125
9	2.088	2.532	3.325	4.168	5.380	6.393	8.343	10.656	12.242	14.684	16.919	19.679	21.666	27.877
10	2.558	3.059	3.940	4.865	6.179	7.267	9.342	11.781	13.442	15.987	18.307	21.161	23.209	29.588
11	3.053	3.609	4.575	5.578	6.989	8.148	10.341	12.899	14.631	17.275	19.675	22.618	24.725	31.264
12	3.571	4.178	5.226	6.304	7.807	9.034	11.340	14.011	15.812	18.549	21.026	24.054	26.217	32.909
13	4.107	4.765	5.892	7.042	8.634	9.926	12.340	15.119	16.985	19.812	22.362	25.472	27.688	34.528
14	4.660	5.368	6.571	7.790	9.467	10.821	13.339	16.222	18.151	21.064	23.685	26.873	29.141	36.123
15	5.229	5.985	7.261	8.547	10.307	11.721	14.339	17.322	19.311	22.307	24.996	28.259	30.578	37.697
16	5.812	6.614	7.962	9.312	11.152	12.624	15.338	18.418	20.465	23.542	26.296	29.633	32.000	39.252
17	6.408	7.255	8.672	10.085	12.002	13.531	16.338	19.511	21.615	24.769	27.587	30.995	33.409	40.790
18	7.015	7.906	9.390	10.865	12.857	14.440	17.338	20.601	22.760	25.989	28.869	32.346	34.805	42.312
19	7.633	8.567	10.117	11.651	13.716	15.352	18.338	21.689	23.900	27.204	30.144	33.687	36.191	43.820
20	8.260	9.237	10.851	12.443	14.578	16.266	19.337	22.775	25.038	28.412	31.410	35.020	37.566	45.315
21	8.897	9.915	11.591	13.240	15.445	17.182	20.337	23.858	26.171	29.615	32.671	36.343	38.932	46.797
22	9.542	10.600	12.338	14.041	16.314	18.101	21.337	24.939	27.301	30.813	33.924	37.659	40.289	48.268
23	10.196	11.293	13.091	14.848	17.187	19.021	22.337	26.018	28.429	32.007	35.172	38.968	41.638	49.728
24	10.856	11.992	13.848	15.659	18.062	19.943	23.337	27.096	29.553	33.196	36.415	40.270	42.980	51.179
25	11.524	12.697	14.611	16.473	18.940	20.867	24.337	28.172	30.675	34.382	37.652	41.566	44.314	52.620
26	12.198	13.409	15.379	17.292	19.820	21.792	25.336	29.246	31.795	35.563	38.885	42.856	45.642	54.052
27	12.879	14.125	16.151	18.114	20.703	22.719	26.336	30.319	32.912	36.741	40.113	44.140	46.963	55.476
28	13.565	14.847	16.928	18.939	21.588	23.647	27.336	31.391	34.027	37.916	41.337	45.419	48.278	56.893
29	14.256	15.574	17.708	19.768	22.475	24.577	28.336	32.461	35.139	39.087	42.557	46.693	49.558	58.302
30	14.953	16.306	18.493	20.599	23.364	25.508	29.336	33.530	36.250	40.256	43.773	47.962	50.892	59.703

Source: From table IV of R. A. Fisher & F. Yates: *Statistical Tables for Biological, Agricultural and Medical Research,* 6th edition, (London: Longman Group, Ltd., 1974). Reprinted by permission of Addison-Wesley Longman, Ltd.

For df > 30, the expression $\sqrt{2\chi^2} - \sqrt{2df - 1}$ may be used as a normal deviate with unit variance.

TABLE C.5
Upper Percentage Points of the *F* Distribution

df for de-nomi-nator	α	|	1	2	3	4	5	6	7	8	9	10	11	12
							df for numerator							
1	.25		5.83	7.50	8.20	8.58	8.82	8.98	9.10	9.19	9.26	9.32	9.36	9.41
	.10		39.9	49.5	53.6	55.8	57.2	58.2	58.9	59.4	59.9	60.2	60.5	60.7
	.05		161	200	216	225	230	234	237	239	241	242	243	244
2	.25		2.57	3.00	3.15	3.23	3.28	3.31	3.34	3.35	3.37	3.38	3.39	3.39
	.10		8.53	9.00	9.16	9.24	9.29	9.33	9.35	9.37	9.38	9.39	9.40	9.41
	.05		18.5	19.0	19.2	19.2	19.3	19.3	19.4	19.4	19.4	19.4	19.4	19.4
	.01		98.5	99.0	99.2	99.2	99.3	99.3	99.4	99.4	99.4	99.4	99.4	99.4
3	.25		2.02	2.28	2.36	2.39	2.41	2.42	2.43	2.44	2.44	2.44	2.45	2.45
	.10		5.54	5.46	5.39	5.34	5.31	5.28	5.27	5.25	5.24	5.23	5.22	5.22
	.05		10.1	9.55	9.28	9.12	9.01	8.94	8.89	8.85	8.81	8.79	8.76	8.74
	.01		34.1	30.8	29.5	28.7	28.2	27.9	27.7	27.5	27.3	27.2	27.1	27.1
4	.25		1.81	2.00	2.05	2.06	2.07	2.08	2.08	2.08	2.08	2.08	2.08	2.08
	.10		4.54	4.32	4.19	4.11	4.05	4.01	3.98	3.95	3.94	3.92	3.91	3.90
	.05		7.71	6.94	6.59	6.39	6.26	6.16	6.09	6.04	6.00	5.96	5.94	5.91
	.01		21.2	18.0	16.7	16.0	15.5	15.2	15.0	14.8	14.7	14.5	14.4	14.4
5	.25		1.69	1.85	1.88	1.89	1.89	1.89	1.89	1.89	1.89	1.89	1.89	1.89
	.10		4.06	3.78	3.62	3.52	3.45	3.40	3.37	3.34	3.32	3.30	3.28	3.27
	.05		6.61	5.79	5.41	5.19	5.05	4.95	4.88	4.82	4.77	4.74	4.71	4.68
	.01		16.3	13.3	12.1	11.4	11.0	10.7	10.5	10.3	10.2	10.1	9.96	9.89
6	.25		1.62	1.76	1.78	1.79	1.79	1.78	1.78	1.78	1.77	1.77	1.77	1.77
	.10		3.78	3.46	3.29	3.18	3.11	3.05	3.01	2.98	2.96	2.94	2.92	2.90
	.05		5.99	5.14	4.76	4.53	4.39	4.28	4.21	4.15	4.10	4.06	4.03	4.00
	.01		13.7	10.9	9.78	9.15	8.75	8.47	8.26	8.10	7.98	7.87	7.79	7.72
7	.25		1.57	1.70	1.72	1.72	1.71	1.71	1.70	1.70	1.69	1.69	1.69	1.68
	.10		3.59	3.26	3.07	2.96	2.88	2.83	2.78	2.75	2.72	2.70	2.68	2.67
	.05		5.59	4.74	4.35	4.12	3.97	3.87	3.79	3.73	3.68	3.64	3.60	3.57
	.01		12.2	9.55	8.45	7.85	7.46	7.19	6.99	6.84	6.72	6.62	6.54	6.47
8	.25		1.54	1.66	1.67	1.66	1.66	1.65	1.64	1.64	1.63	1.63	1.63	1.62
	.10		3.46	3.11	2.92	2.81	2.73	2.67	2.62	2.59	2.56	2.54	2.52	2.50
	.05		5.32	4.46	4.07	3.84	3.69	3.58	3.50	3.44	3.39	3.35	3.31	3.28
	.01		11.3	8.65	7.59	7.01	6.63	6.37	6.18	6.03	5.91	5.81	5.73	5.67
9	.25		1.51	1.62	1.63	1.63	1.62	1.61	1.60	1.60	1.59	1.59	1.58	1.58
	.10		3.36	3.01	2.81	2.69	2.61	2.55	2.51	2.47	2.44	2.42	2.40	2.38
	.05		5.12	4.26	3.86	3.63	3.48	3.37	3.29	3.23	3.18	3.14	3.10	3.07
	.01		10.6	8.02	6.99	6.42	6.06	5.80	5.61	5.47	5.35	5.26	5.18	5.11
10	.25		1.49	1.60	1.60	1.59	1.59	1.58	1.57	1.56	1.56	1.55	1.55	1.54
	.10		3.29	2.92	2.73	2.61	2.52	2.46	2.41	2.38	2.35	2.32	2.30	2.28
	.05		4.96	4.10	3.71	3.48	3.33	3.22	3.14	3.07	3.02	2.98	2.94	2.91
	.01		10.0	7.56	6.55	5.99	5.64	5.39	5.20	5.06	4.94	4.85	4.77	4.71

Source: Abridged from Table 18 of E. Pearson and H. Hartley, *Biometrika Tables for Statisticians*, Vol. 1, 3rd ed. (Cambridge: Cambridge University Press, 1966). Reprinted with the permission of the Biometrika Trustees.

TABLE C.5 (cont.)

				df for numerator									α	df for denominator
15	20	24	30	40	50	60	100	120	200	500	∞		
9.49	9.58	9.63	9.67	9.71	9.74	9.76	9.78	9.80	9.82	9.84	9.85	.25	
61.2	61.7	62.0	62.3	62.5	62.7	62.8	63.0	63.1	63.2	63.3	63.3	.10	1
246	248	249	250	251	252	252	253	253	254	254	254	.05	
3.41	3.43	3.43	3.44	3.45	3.45	3.46	3.47	3.47	3.48	3.48	3.48	.25	
9.42	9.44	9.45	9.46	9.47	9.47	9.47	9.48	9.48	9.49	9.49	9.49	.10	2
19.4	19.4	19.5	19.5	19.5	19.5	19.5	19.5	19.5	19.5	19.5	19.5	.05	
99.4	99.4	99.5	99.5	99.5	99.5	99.5	99.5	99.5	99.5	99.5	99.5	.01	
2.46	2.46	2.46	2.47	2.47	2.47	2.47	2.47	2.47	2.47	2.47	2.47	.25	
5.20	5.18	5.18	5.17	5.16	5.15	5.15	5.14	5.14	5.14	5.14	5.13	.10	3
8.70	8.66	8.64	8.62	8.59	8.58	8.57	8.55	8.55	8.54	8.53	8.53	.05	
26.9	29.7	26.6	26.5	26.4	26.4	26.3	26.2	26.2	26.2	26.1	26.1	.01	
2.08	2.08	2.08	2.08	2.08	2.08	2.08	2.08	2.08	2.08	2.08	2.08	.25	
3.87	3.84	3.83	3.82	3.80	3.80	3.79	3.78	3.78	3.77	3.76	3.76	.10	4
5.86	5.80	5.77	5.75	5.72	5.70	5.69	5.66	5.66	5.65	5.64	5.63	.05	
14.2	14.0	13.9	13.8	13.7	13.7	13.7	13.6	13.6	13.5	13.5	13.5	.01	
1.89	1.88	1.88	1.88	1.88	1.88	1.87	1.87	1.87	1.87	1.87	1.87	.25	
3.24	3.21	3.19	3.17	3.16	3.15	3.14	3.13	3.12	3.12	3.11	3.10	.10	5
4.62	4.56	4.53	4.50	4.46	4.44	4.43	4.41	4.40	4.39	4.37	4.36	.05	
9.72	9.55	9.47	9.38	9.29	9.24	9.20	9.13	9.11	9.08	9.04	9.02	.01	
1.76	1.76	1.75	1.75	1.75	1.75	1.74	1.74	1.74	1.74	1.74	1.74	.25	
2.87	2.84	2.82	2.80	2.78	2.77	2.76	2.75	2.74	2.73	2.73	2.72	.10	6
3.94	3.87	3.84	3.81	3.77	3.75	3.74	3.71	3.70	3.69	3.68	3.67	.05	
7.56	7.40	7.31	7.23	7.14	7.09	7.06	6.99	6.97	6.93	6.90	6.88	.01	
1.68	1.67	1.67	1.66	1.66	1.66	1.65	1.65	1.65	1.65	1.65	1.65	.25	
2.63	2.59	2.58	2.56	2.54	2.52	2.51	2.50	2.49	2.48	2.48	2.47	.10	7
3.51	3.44	3.41	3.38	3.34	3.32	3.30	3.27	3.27	3.25	3.24	3.23	.05	
6.31	6.16	6.07	5.99	5.91	5.86	5.82	5.75	5.74	5.70	5.67	5.65	.01	
1.62	1.61	1.60	1.60	1.59	1.59	1.59	1.58	1.58	1.58	1.58	1.58	.25	
2.46	2.42	2.40	2.38	2.36	2.35	2.34	2.32	2.32	2.31	2.30	2.29	.10	8
3.22	3.15	3.12	3.08	3.04	3.02	3.01	2.97	2.97	2.95	2.94	2.93	.05	
5.52	5.36	5.28	5.20	5.12	5.07	5.03	4.96	4.95	4.91	4.88	4.86	.01	
1.57	1.56	1.56	1.55	1.55	1.54	1.54	1.53	1.53	1.53	1.53	1.53	.25	
2.34	2.30	2.28	2.25	2.23	2.22	2.21	2.19	2.18	2.17	2.17	2.16	.10	9
3.01	2.94	2.90	2.86	2.83	2.80	2.79	2.76	2.75	2.73	2.72	2.71	.05	
4.96	4.81	4.73	4.65	4.57	4.52	4.48	4.42	4.40	4.36	4.33	4.31	.01	
1.53	1.52	1.52	1.51	1.51	1.50	1.50	1.49	1.49	1.49	1.48	1.48	.25	
2.24	2.20	2.18	2.16	2.13	2.12	2.11	2.09	2.08	2.07	2.06	2.06	.10	10
2.85	2.77	2.74	2.70	2.66	2.64	2.62	2.59	2.58	2.56	2.55	2.54	.05	
4.56	4.41	4.33	4.25	4.17	4.12	4.08	4.01	4.00	3.96	3.93	3.91	.01	

TABLE C.5 (cont.)

df for denominator	α	\multicolumn{12}{c}{df for numerator}											
		1	2	3	4	5	6	7	8	9	10	11	12
11	.25	1.47	1.58	1.58	1.57	1.56	1.55	1.54	1.53	1.53	1.52	1.52	1.51
	.10	3.23	2.86	2.66	2.54	2.45	2.39	2.34	2.30	2.27	2.25	2.23	2.21
	.05	4.84	3.98	3.59	3.36	3.20	3.09	3.01	2.95	2.90	2.85	2.82	2.79
	.01	9.65	7.21	6.22	5.67	5.32	5.07	4.89	4.74	4.63	4.54	4.46	4.40
12	.25	1.46	1.56	1.56	1.55	1.54	1.53	1.52	1.51	1.51	1.50	1.50	1.49
	.10	3.18	2.81	2.61	2.48	2.39	2.33	2.28	2.24	2.21	2.19	2.17	2.15
	.05	4.75	3.89	3.49	3.26	3.11	3.00	2.91	2.85	2.80	2.75	2.72	2.69
	.01	9.33	6.93	5.95	5.41	5.06	4.82	4.64	4.50	4.39	4.30	4.22	4.16
13	.25	1.45	1.55	1.55	1.53	1.52	1.51	1.50	1.49	1.49	1.48	1.47	1.47
	.10	3.14	2.76	2.56	2.43	2.35	2.28	2.23	2.20	2.16	2.14	2.12	2.10
	.05	4.67	3.81	3.41	3.18	3.03	2.92	2.83	2.77	2.71	2.67	2.63	2.60
	.01	9.07	6.70	5.74	5.21	4.86	4.62	4.44	4.30	4.19	4.10	4.02	3.96
14	.25	1.44	1.53	1.53	1.52	1.51	1.50	1.49	1.48	1.47	1.46	1.46	1.45
	.10	3.10	2.73	2.52	2.39	2.31	2.24	2.19	2.15	2.12	2.10	2.08	2.05
	.05	4.60	3.74	3.34	3.11	2.96	2.85	2.76	2.70	2.65	2.60	2.57	2.53
	.01	8.86	6.51	5.56	5.04	4.69	4.46	4.28	4.14	4.03	3.94	3.86	3.80
15	.25	1.43	1.52	1.52	1.51	1.49	1.48	1.47	1.46	1.46	1.45	1.44	1.44
	.10	3.07	2.70	2.49	2.36	2.27	2.21	2.16	2.12	2.09	2.06	2.04	2.02
	.05	4.54	3.68	3.29	3.06	2.90	2.79	2.71	2.64	2.59	2.54	2.51	2.48
	.01	8.68	6.36	5.42	4.89	4.56	4.32	4.14	4.00	3.89	3.80	3.73	3.67
16	.25	1.42	1.51	1.51	1.50	1.48	1.47	1.46	1.45	1.44	1.44	1.44	1.43
	.10	3.05	2.67	2.46	2.33	2.24	2.18	2.13	2.09	2.06	2.03	2.01	1.99
	.05	4.49	3.63	3.24	3.01	2.85	2.74	2.66	2.59	2.54	2.49	2.46	2.42
	.01	8.53	6.23	5.29	4.77	4.44	4.20	4.03	3.89	3.78	3.69	3.62	3.55
17	.25	1.42	1.51	1.50	1.49	1.47	1.46	1.45	1.44	1.43	1.43	1.42	1.41
	.10	3.03	2.64	2.44	2.31	2.22	2.15	2.10	2.06	2.03	2.00	1.98	1.96
	.05	4.45	3.59	3.20	2.96	2.81	2.70	2.61	2.55	2.49	2.45	2.41	2.38
	.01	8.40	6.11	5.18	4.67	4.34	4.10	3.93	3.79	3.68	3.59	3.52	3.46
18	.25	1.41	1.50	1.49	1.48	1.46	1.45	1.44	1.43	1.42	1.42	1.41	1.40
	.10	3.01	2.62	2.42	2.29	2.20	2.13	2.08	2.04	2.00	1.98	1.96	1.93
	.05	4.41	3.55	3.16	2.93	2.77	2.66	2.58	2.51	2.46	2.41	2.37	2.34
	.01	8.29	6.01	5.09	4.58	4.25	4.01	3.84	3.71	3.60	3.51	3.43	3.37
19	.25	1.41	1.49	1.49	1.47	1.46	1.44	1.43	1.42	1.41	1.41	1.40	1.40
	.10	2.99	2.61	2.40	2.27	2.18	2.11	2.06	2.02	1.98	1.96	1.94	1.91
	.05	4.38	3.52	3.13	2.90	2.74	2.63	2.54	2.48	2.42	2.38	2.34	2.31
	.01	8.18	5.93	5.01	4.50	4.17	3.94	3.77	3.63	3.52	3.43	3.36	3.30
20	.25	1.40	1.49	1.48	1.46	1.45	1.44	1.43	1.42	1.41	1.40	1.39	1.39
	.10	2.97	2.59	2.38	2.25	2.16	2.09	2.04	2.00	1.96	1.94	1.92	1.89
	.05	4.35	3.49	3.10	2.87	2.71	2.60	2.51	2.45	2.39	2.35	2.31	2.28
	.01	8.10	5.85	4.94	4.43	4.10	3.87	3.70	3.56	3.46	3.37	3.29	3.23

TABLE C.5 (cont.)

15	20	24	30	40	50	60	100	120	200	500	∞	α	df for denominator
1.50	1.49	1.49	1.48	1.47	1.47	1.47	1.46	1.46	1.46	1.45	1.45	.25	
2.17	2.12	2.10	2.08	2.05	2.04	2.03	2.00	2.00	1.99	1.98	1.97	.10	11
2.72	2.65	2.61	2.57	2.53	2.51	2.49	2.46	2.45	2.43	2.42	2.40	.05	
4.25	4.10	4.02	3.94	3.86	3.81	3.78	3.71	3.69	3.66	3.62	3.60	.01	
1.48	1.47	1.46	1.45	1.45	1.44	1.44	1.43	1.43	1.43	1.42	1.42	.25	
2.10	2.06	2.04	2.01	1.99	1.97	1.96	1.94	1.93	1.92	1.91	1.90	.10	12
2.62	2.54	2.51	2.47	2.43	2.40	2.38	2.35	2.34	2.32	2.31	2.30	.05	
4.01	3.86	3.78	3.70	3.62	3.57	3.54	3.47	3.45	3.41	3.38	3.36	.01	
1.46	1.45	1.44	1.43	1.42	1.42	1.42	1.41	1.41	1.40	1.40	1.40	.25	
2.05	2.01	1.98	1.96	1.93	1.92	1.90	1.88	1.88	1.86	1.85	1.85	.10	13
2.53	2.46	2.42	2.38	2.34	2.31	2.30	2.26	2.25	2.23	2.22	2.21	.05	
3.82	3.66	3.59	3.51	3.43	3.38	3.34	3.27	3.25	3.22	3.19	3.17	.01	
1.44	1.43	1.42	1.41	1.41	1.40	1.40	1.39	1.39	1.39	1.38	1.38	.25	
2.01	1.96	1.94	1.91	1.89	1.87	1.86	1.83	1.83	1.82	1.80	1.80	.10	14
2.46	2.39	2.35	2.31	2.27	2.24	2.22	2.19	2.18	2.16	2.14	2.13	.05	
3.66	3.51	3.43	3.35	3.27	3.22	3.18	3.11	3.09	3.06	3.03	3.00	.01	
1.43	1.41	1.41	1.40	1.39	1.39	1.38	1.38	1.37	1.37	1.36	1.36	.25	
1.97	1.92	1.90	1.87	1.85	1.83	1.82	1.79	1.79	1.77	1.76	1.76	.10	15
2.40	2.33	2.29	2.25	2.20	2.18	2.16	2.12	2.11	2.10	2.08	2.07	.05	
3.52	3.37	3.29	3.21	3.13	3.08	3.05	2.98	2.96	2.92	2.89	2.87	.01	
1.41	1.40	1.39	1.38	1.37	1.37	1.36	1.36	1.35	1.35	1.34	1.34	.25	
1.94	1.89	1.87	1.84	1.81	1.79	1.78	1.76	1.75	1.74	1.73	1.72	.10	16
2.35	2.28	2.24	2.19	2.15	2.12	2.11	2.07	2.06	2.04	2.02	2.01	.05	
3.41	3.26	3.18	3.10	3.02	2.97	2.93	2.86	2.84	2.81	2.78	2.75	.01	
1.40	1.39	1.38	1.37	1.36	1.35	1.35	1.34	1.34	1.34	1.33	1.33	.25	
1.91	1.86	1.84	1.81	1.78	1.76	1.75	1.73	1.72	1.71	1.69	1.69	.10	17
2.31	2.23	2.19	2.15	2.10	2.08	2.06	2.02	2.01	1.99	1.97	1.96	.05	
3.31	3.16	3.08	3.00	2.92	2.87	2.83	2.76	2.75	2.71	2.68	2.65	.01	
1.39	1.38	1.37	1.36	1.35	1.34	1.34	1.33	1.33	1.32	1.32	1.32	.25	
1.89	1.84	1.81	1.78	1.75	1.74	1.72	1.70	1.69	1.68	1.67	1.66	.10	18
2.27	2.19	2.15	2.11	2.06	2.04	2.02	1.98	1.97	1.95	1.93	1.92	.05	
3.23	3.08	3.00	2.92	2.84	2.78	2.75	2.68	2.66	2.62	2.59	2.57	.01	
1.38	1.37	1.36	1.35	1.34	1.33	1.33	1.32	1.32	1.31	1.31	1.30	.25	
1.86	1.81	1.79	1.76	1.73	1.71	1.70	1.67	1.67	1.65	1.64	1.63	.10	19
2.23	2.16	2.11	2.07	2.03	2.00	1.98	1.94	1.93	1.91	1.89	1.88	.05	
3.15	3.00	2.92	2.84	2.76	2.71	2.67	2.60	2.58	2.55	2.51	2.49	.01	
1.37	1.36	1.35	1.34	1.33	1.33	1.32	1.31	1.31	1.30	1.30	1.29	.25	
1.84	1.79	1.77	1.74	1.71	1.69	1.68	1.65	1.64	1.63	1.62	1.61	.10	20
2.20	2.12	2.08	2.04	1.99	1.97	1.95	1.91	1.90	1.88	1.86	1.84	.05	
3.09	2.94	2.86	2.78	2.69	2.64	2.61	2.54	2.52	2.48	2.44	2.42	.01	

df for numerator

TABLE C.5 (cont.)

df for denominator	α	1	2	3	4	5	6	7	8	9	10	11	12
							df for numerator						
22	.25	1.40	1.48	1.47	1.45	1.44	1.42	1.41	1.40	1.39	1.39	1.38	1.37
	.10	2.95	2.56	2.35	2.22	2.13	2.06	2.01	1.97	1.93	1.90	1.88	1.86
	.05	4.30	3.44	3.05	2.82	2.66	2.55	2.46	2.40	2.34	2.30	2.26	2.23
	.01	7.95	5.72	4.82	4.31	3.99	3.76	3.59	3.45	3.35	3.26	3.18	3.12
24	.25	1.39	1.47	1.46	1.44	1.43	1.41	1.40	1.39	1.38	1.38	1.37	1.36
	.10	2.93	2.54	2.33	2.19	2.10	2.04	1.98	1.94	1.91	1.88	1.85	1.83
	.05	4.26	3.40	3.01	2.78	2.62	2.51	2.42	2.36	2.30	2.25	2.21	2.18
	.01	7.82	5.61	4.72	4.22	3.90	3.67	3.50	3.36	3.26	3.17	3.09	3.03
26	.25	1.38	1.46	1.45	1.44	1.42	1.41	1.39	1.38	1.37	1.37	1.36	1.35
	.10	2.91	2.52	2.31	2.17	2.08	2.01	1.96	1.92	1.88	1.86	1.84	1.81
	.05	4.23	3.37	2.98	2.74	2.59	2.47	2.39	2.32	2.27	2.22	2.18	2.15
	.01	7.72	5.53	4.64	4.14	3.82	3.59	3.42	3.29	3.18	3.09	3.02	2.96
28	.25	1.38	1.46	1.45	1.43	1.41	1.40	1.39	1.38	1.37	1.36	1.35	1.34
	.10	2.89	2.50	2.29	2.16	2.06	2.00	1.94	1.90	1.87	1.84	1.81	1.79
	.05	4.20	3.34	2.95	2.71	2.56	2.45	2.36	2.29	2.24	2.19	2.15	2.12
	.01	7.64	5.45	4.57	4.07	3.75	3.53	3.36	3.23	3.12	3.03	2.96	2.90
30	.25	1.38	1.45	1.44	1.42	1.41	1.39	1.38	1.37	1.36	1.35	1.35	1.34
	.10	2.88	2.49	2.28	2.14	2.05	1.98	1.93	1.88	1.85	1.82	1.79	1.77
	.05	4.17	3.32	2.92	2.69	2.53	2.42	2.33	2.27	2.21	2.16	2.13	2.09
	.01	7.56	5.39	4.51	4.02	3.70	3.47	3.30	3.17	3.07	2.98	2.91	2.84
40	.25	1.36	1.44	1.42	1.40	1.39	1.37	1.36	1.35	1.34	1.33	1.32	1.31
	.10	2.84	2.44	2.23	2.09	2.00	1.93	1.87	1.83	1.79	1.76	1.73	1.71
	.05	4.08	3.23	2.84	2.61	2.45	2.34	2.25	2.18	2.12	2.08	2.04	2.00
	.01	7.31	5.18	4.31	3.83	3.51	3.29	3.12	2.99	2.89	2.80	2.73	2.66
60	.25	1.35	1.42	1.41	1.38	1.37	1.35	1.33	1.32	1.31	1.30	1.29	1.29
	.10	2.79	2.39	2.18	2.04	1.95	1.87	1.82	1.77	1.74	1.71	1.68	1.66
	.05	4.00	3.15	2.76	2.53	2.37	2.25	2.17	2.10	2.04	1.99	1.95	1.92
	.01	7.08	4.98	4.13	3.65	3.34	3.12	2.95	2.82	2.72	2.63	2.56	2.50
120	.25	1.34	1.40	1.39	1.37	1.35	1.33	1.31	1.30	1.29	1.28	1.27	1.26
	.10	2.75	2.35	2.13	1.99	1.90	1.82	1.77	1.72	1.68	1.65	1.62	1.60
	.05	3.92	3.07	2.68	2.45	2.29	2.17	2.09	2.02	1.96	1.91	1.87	1.83
	.01	6.85	4.79	3.95	3.48	3.17	2.96	2.79	2.66	2.56	2.47	2.40	2.34
200	.25	1.33	1.39	1.38	1.36	1.34	1.32	1.31	1.29	1.28	1.27	1.26	1.25
	.10	2.73	2.33	2.11	1.97	1.88	1.80	1.75	1.70	1.66	1.63	1.60	1.57
	.05	3.89	3.04	2.65	2.42	2.26	2.14	2.06	1.98	1.93	1.88	1.84	1.80
	.01	6.76	4.71	3.88	3.41	3.11	2.89	2.73	2.60	2.50	2.41	2.34	2.27
∞	.25	1.32	1.39	1.37	1.35	1.33	1.31	1.29	1.28	1.27	1.25	1.24	1.24
	.10	2.71	2.30	2.08	1.94	1.85	1.77	1.72	1.67	1.63	1.60	1.57	1.55
	.05	3.84	3.00	2.60	2.37	2.21	2.10	2.01	1.94	1.88	1.83	1.79	1.75
	.01	6.63	4.61	3.78	3.32	3.02	2.80	2.64	2.51	2.41	2.32	2.25	2.18

TABLE C.5 (cont.)

15	20	24	30	40	50	60	100	120	200	500	∞	α	df for de-nomi-nator
						df for numerator							
1.36	1.34	1.33	1.32	1.31	1.31	1.30	1.30	1.30	1.29	1.29	1.28	.25	
1.81	1.76	1.73	1.70	1.67	1.65	1.64	1.61	1.60	1.59	1.58	1.57	.10	22
2.15	2.07	2.03	1.98	1.94	1.91	1.89	1.85	1.84	1.82	1.80	1.78	.05	
2.98	2.83	2.75	2.67	2.58	2.53	2.50	2.42	2.40	2.36	2.33	2.31	.01	
1.35	1.33	1.32	1.31	1.30	1.29	1.29	1.28	1.28	1.27	1.27	1.26	.25	
1.78	1.73	1.70	1.67	1.64	1.62	1.61	1.58	1.57	1.56	1.54	1.53	.10	24
2.11	2.03	1.98	1.94	1.89	1.86	1.84	1.80	1.79	1.77	1.75	1.73	.05	
2.89	2.74	2.66	2.58	2.49	2.44	2.40	2.33	2.31	2.27	2.24	2.21	.01	
1.34	1.32	1.31	1.30	1.29	1.28	1.28	1.26	1.26	1.26	1.25	1.25	.25	
1.76	1.71	1.68	1.65	1.61	1.59	1.58	1.55	1.54	1.53	1.51	1.50	.10	26
2.07	1.99	1.95	1.90	1.85	1.82	1.80	1.76	1.75	1.73	1.71	1.69	.05	
2.81	2.66	2.58	2.50	2.42	2.36	2.33	2.25	2.23	2.19	2.16	2.13	.01	
1.33	1.31	1.30	1.29	1.28	1.27	1.27	1.26	1.25	1.25	1.24	1.24	.25	
1.74	1.69	1.66	1.63	1.59	1.57	1.56	1.53	1.52	1.50	1.49	1.48	.10	28
2.04	1.96	1.91	1.87	1.82	1.79	1.77	1.73	1.71	1.69	1.67	1.65	.05	
2.75	2.60	2.52	2.44	2.35	2.30	2.26	2.19	2.17	2.13	2.09	2.06	.01	
1.32	1.30	1.29	1.28	1.27	1.26	1.26	1.25	1.24	1.24	1.23	1.23	.25	
1.74	1.67	1.64	1.61	1.57	1.55	1.54	1.51	1.50	1.48	1.47	1.46	.10	30
2.01	1.93	1.89	1.84	1.79	1.76	1.74	1.70	1.68	1.66	1.64	1.62	.05	
2.70	2.55	2.47	2.39	2.30	2.25	2.21	2.13	2.11	2.07	2.03	2.01	.01	
1.30	1.28	1.26	1.25	1.24	1.23	1.22	1.21	1.21	1.20	1.19	1.19	.25	
1.66	1.61	1.57	1.54	1.51	1.48	1.47	1.43	1.42	1.41	1.39	1.38	.10	40
1.92	1.84	1.79	1.74	1.69	1.66	1.64	1.59	1.58	1.55	1.53	1.51	.05	
2.52	2.37	2.29	2.20	2.11	2.06	2.02	1.94	1.92	1.87	1.83	1.80	.01	
1.27	1.25	1.24	1.22	1.21	1.20	1.19	1.17	1.17	1.16	1.15	1.15	.25	
1.60	1.54	1.51	1.48	1.44	1.41	1.40	1.36	1.35	1.33	1.31	1.29	.10	60
1.84	1.75	1.70	1.65	1.59	1.56	1.53	1.48	1.47	1.44	1.41	1.39	.05	
2.35	2.20	2.12	2.03	1.94	1.88	1.84	1.75	1.73	1.68	1.63	1.60	.01	
1.24	1.22	1.21	1.19	1.18	1.17	1.16	1.14	1.13	1.12	1.11	1.10	.25	
1.55	1.48	1.45	1.41	1.37	1.34	1.32	1.27	1.26	1.24	1.21	1.19	.10	120
1.75	1.66	1.61	1.55	1.50	1.46	1.43	1.37	1.35	1.32	1.28	1.25	.05	
2.19	2.03	1.95	1.86	1.76	1.70	1.66	1.56	1.53	1.48	1.42	1.38	.01	
1.23	1.21	1.20	1.18	1.16	1.14	1.12	1.11	1.10	1.09	1.08	1.06	.25	
1.52	1.46	1.42	1.38	1.34	1.31	1.28	1.24	1.22	1.20	1.17	1.14	.10	200
1.72	1.62	1.57	1.52	1.46	1.41	1.39	1.32	1.29	1.26	1.22	1.19	.05	
2.13	1.97	1.89	1.79	1.69	1.63	1.58	1.48	1.44	1.39	1.33	1.28	.01	
1.22	1.19	1.18	1.16	1.14	1.13	1.12	1.09	1.08	1.07	1.04	1.00	.25	
1.49	1.42	1.38	1.34	1.30	1.26	1.24	1.18	1.17	1.13	1.08	1.00	.10	∞
1.67	1.57	1.52	1.46	1.39	1.35	1.32	1.24	1.22	1.17	1.11	1.00	.05	
2.04	1.88	1.79	1.70	1.59	1.52	1.47	1.36	1.32	1.25	1.15	1.00	.01	

TABLE C.6
Transformation of r to z_r

r	z_r	r	z_r	r	z_r	r	z_r	r	z_r
.000	.000	.200	.203	.400	.424	.600	.693	.800	1.099
.000	.000	.200	.203	.400	.424	.600	.693	.800	1.099
.005	.005	.205	.208	.405	.430	.605	.701	.805	1.113
.010	.010	.210	.213	.410	.436	.610	.709	.810	1.127
.015	.015	.215	.218	.415	.442	.615	.717	.815	1.142
.020	.020	.220	.224	.420	.448	.620	.725	.820	1.157
.025	.025	.225	.229	.425	.454	.625	.733	.825	1.172
.030	.030	.230	.234	.430	.460	.630	.741	.830	1.188
.035	.035	.235	.239	.435	.466	.635	.750	.835	1.204
.040	.040	.240	.245	.440	.472	.640	.758	.840	1.221
.045	.045	.245	.250	.445	.478	.645	.767	.845	1.238
.050	.050	.250	.255	.450	.485	.650	.775	.850	1.256
.055	.055	.255	.261	.455	.491	.655	.784	.855	1.274
.060	.060	.260	.266	.460	.497	.660	.793	.860	1.293
.065	.065	.265	.271	.465	.504	.665	.802	.865	1.313
.070	.070	.270	.277	.470	.510	.670	.811	.870	1.333
.075	.075	.275	.282	.475	.517	.675	.820	.875	1.354
.080	.080	.280	.288	.480	.523	.680	.829	.880	1.376
.085	.085	.285	.293	.485	.530	.685	.838	.885	1.398
.090	.090	.290	.299	.490	.536	.690	.848	.890	1.422
.095	.095	.295	.304	.495	.543	.695	.858	.895	1.447
.100	.100	.300	.310	.500	.549	.700	.867	.900	1.472
.105	.105	.305	.315	.505	.556	.705	.877	.905	1.499
.110	.110	.310	.321	.510	.563	.710	.887	.910	1.528
.115	.116	.315	.326	.515	.570	.715	.897	.915	1.557
.120	.121	.320	.332	.520	.576	.720	.908	.920	1.589
.125	.126	.325	.337	.525	.583	.725	.918	.925	1.623
.130	.131	.330	.343	.530	.590	.730	.929	.930	1.658
.135	.136	.335	.348	.535	.597	.735	.940	.935	1.697
.140	.141	.340	.354	.540	.604	.740	.950	.940	1.738
.145	.146	.345	.360	.545	.611	.745	.962	.945	1.783
.150	.151	.350	.365	.550	.618	.750	.973	.950	1.832
.155	.156	.355	.371	.555	.626	.755	.984	.955	1.886
.160	.161	.360	.377	.560	.633	.760	.996	.960	1.946
.165	.167	.365	.383	.565	.640	.765	1.008	.965	2.014
.170	.172	.370	.388	.570	.648	.770	1.020	.970	2.092
.175	.177	.375	.394	.575	.655	.775	1.033	.975	2.185
.180	.182	.380	.400	.580	.662	.780	1.045	.980	2.298
.185	.187	.385	.406	.585	.670	.785	1.058	.985	2.443
.190	.192	.390	.412	.590	.678	.790	1.071	.990	2.647
.195	.198	.395	.418	.595	.685	.795	1.085	.995	2.994

TABLE C.7
Critical Values of the Correlation Coefficient

	Level of significance for one-tailed test					Level of significance for one-tailed test			
	.05	.025	.01	.005		.05	.025	.01	.005
	Level of significance for two-tailed test					Level of significance for two-tailed test			
df	.10	.05	.02	.01	df	.10	.05	.02	.01
1	.988	.997	.9995	.9999	21	.352	.413	.482	.526
2	.900	.950	.980	.990	22	.344	.404	.472	.515
3	.805	.878	.934	.959	23	.337	.396	.462	.505
4	.729	.811	.882	.917	24	.330	.388	.453	.496
5	.669	.754	.833	.874	25	.323	.381	.445	.487
6	.622	.707	.789	.834	26	.317	.374	.437	.479
7	.582	.666	.750	.798	27	.311	.367	.430	.471
8	.549	.632	.716	.765	28	.306	.361	.423	.463
9	.521	.602	.685	.735	29	.301	.355	.416	.456
10	.497	.576	.658	.708	30	.296	.349	.409	.449
11	.476	.553	.634	.684	35	.275	.325	.381	.418
12	.458	.532	.612	.661	40	.257	.304	.358	.393
13	.441	.514	.592	.641	45	.243	.288	.338	.372
14	.426	.497	.574	.623	50	.231	.273	.322	.354
15	.412	.482	.558	.606	60	.211	.250	.295	.325
16	.400	.468	.542	.590	70	.195	.232	.274	.303
17	.389	.456	.528	.575	80	.183	.217	.256	.283
18	.378	.444	.516	.561	90	.173	.205	.242	.267
19	.369	.433	.503	.549	100	.164	.195	.230	.254
20	.360	.423	.492	.537					

Source: From table VII, of R. A. Fisher & F. Yates, *Statistical Tables for Biological, Agricultural and Medical Research*, 6th edition, (London: Longman Group, Ltd., 1974). Reprinted by permission of Addison-Wesley Longman, Ltd.

TABLE C.8
Percentage Points of the Studentized Range

Error df	α	\multicolumn{10}{c}{r = number of means or number of steps between ordered means}									
		2	3	4	5	6	7	8	9	10	11
5	.05	3.64	4.60	5.22	5.67	6.03	6.33	6.58	6.80	6.99	7.17
	.01	5.70	6.98	7.80	8.42	8.91	9.32	9.67	9.97	10.24	10.48
6	.05	3.46	4.34	4.90	5.30	5.63	5.90	6.12	6.32	6.49	6.65
	.01	5.24	6.33	7.03	7.56	7.97	8.32	8.61	8.87	9.10	9.30
7	.05	3.34	4.16	4.68	5.06	5.36	5.61	5.82	6.00	6.16	6.30
	.01	4.95	5.92	6.54	7.01	7.37	7.68	7.94	8.17	8.37	8.55
8	.05	3.26	4.04	4.53	4.89	5.17	5.40	5.60	5.77	5.92	6.05
	.01	4.75	5.64	6.20	6.62	6.96	7.24	7.47	7.68	7.86	8.03
9	.05	3.20	3.95	4.41	4.76	5.02	5.24	5.43	5.59	5.74	5.87
	.01	4.60	5.43	5.96	6.35	6.66	6.91	7.13	7.33	7.49	7.65
10	.05	3.15	3.88	4.33	4.65	4.91	5.12	5.30	5.46	5.60	5.72
	.01	4.48	5.27	5.77	6.14	6.43	6.67	6.87	7.05	7.21	7.36
11	.05	3.11	3.82	4.26	4.57	4.82	5.03	5.20	5.35	5.49	5.61
	.01	4.39	5.15	5.62	5.97	6.25	6.48	6.67	6.84	6.99	7.13
12	.05	3.08	3.77	4.20	4.51	4.75	4.95	5.12	5.27	5.39	5.51
	.01	4.32	5.05	5.50	5.84	6.10	6.32	6.51	6.67	6.81	6.94
13	.05	3.06	3.73	4.15	4.45	4.69	4.88	5.05	5.19	5.32	5.43
	.01	4.26	4.96	5.40	5.73	5.98	6.19	6.37	6.53	6.67	6.79
14	.05	3.03	3.70	4.11	4.41	4.64	4.83	4.99	5.13	5.25	5.36
	.01	4.21	4.89	5.32	5.63	5.88	6.08	6.26	6.41	6.54	6.66
15	.05	3.01	3.67	4.08	4.37	4.59	4.78	4.94	5.08	5.20	5.31
	.01	4.17	4.84	5.25	5.56	5.80	5.99	6.16	6.31	6.44	6.55
16	.05	3.00	3.65	4.05	4.33	4.56	4.74	4.90	5.03	5.15	5.26
	.01	4.13	4.79	5.19	5.49	5.72	5.92	6.08	6.22	6.35	6.46
17	.05	2.98	3.63	4.02	4.30	4.52	4.70	4.86	4.99	5.11	5.21
	.01	4.10	4.74	5.14	5.43	5.66	5.85	6.01	6.15	6.27	6.38
18	.05	2.97	3.61	4.00	4.28	4.49	4.67	4.82	4.96	5.07	5.17
	.01	4.07	4.70	5.09	5.38	5.60	5.79	5.94	6.08	6.20	6.31
19	.05	2.96	3.59	3.98	4.25	4.47	4.65	4.79	4.92	5.04	5.14
	.01	4.05	4.67	5.05	5.33	5.55	5.73	5.89	6.02	6.14	6.25
20	.05	2.95	3.58	3.96	4.23	4.45	4.62	4.77	4.90	5.01	5.11
	.01	4.02	4.64	5.02	5.29	5.51	5.69	5.84	5.97	6.09	6.19
24	.05	2.92	3.53	3.90	4.17	4.37	4.54	4.68	4.81	4.92	5.01
	.01	3.96	4.55	4.91	5.17	5.37	5.54	5.69	5.81	5.92	6.02
30	.05	2.89	3.49	3.85	4.10	4.30	4.46	4.60	4.72	4.82	4.92
	.01	3.89	4.45	4.80	5.05	5.24	5.40	5.54	5.65	5.76	5.85
40	.05	2.86	3.44	3.79	4.04	4.23	4.39	4.52	4.63	4.73	4.82
	.01	3.82	4.37	4.70	4.93	5.11	5.26	5.39	5.50	5.60	5.69
60	.05	2.83	3.40	3.74	3.98	4.16	4.31	4.44	4.55	4.65	4.73
	.01	3.76	4.28	4.59	4.82	4.99	5.13	5.25	5.36	5.45	5.53
120	.05	2.80	3.36	3.68	3.92	4.10	4.24	4.36	4.47	4.56	4.64
	.01	3.70	4.20	4.50	4.71	4.87	5.01	5.12	5.21	5.30	5.37
∞	.05	2.77	3.31	3.63	3.86	4.03	4.17	4.29	4.39	4.47	4.55
	.01	3.64	4.12	4.40	4.60	4.76	4.88	4.99	5.08	5.16	5.23

TABLE C.8 (cont.)

| r = number of means or number of steps between ordered means | | | | | | | | | | Error |
12	13	14	15	16	17	18	19	20	α	df
7.32	7.47	7.60	7.72	7.83	7.93	8.03	8.12	8.21	.05	5
10.70	10.89	11.08	11.24	11.40	11.55	11.68	11.81	11.93	.01	
6.79	6.92	7.03	7.14	7.24	7.34	7.43	7.51	7.59	.05	6
9.48	9.65	9.81	9.95	10.08	10.21	10.32	10.43	10.54	.01	
6.43	6.55	6.66	6.76	6.85	6.94	7.02	7.1_			7
8.71	8.86	9.00	9.12	9.24	9.35	9.4_				
6.18	6.29	6.39	6.48	6.57	6.65	6.7_				8
8.18	8.31	8.44	8.55	8.66	8.76	8.8_				
5.98	6.09	6.19	6.28	6.36	6.44	6.5_	6.58	6.64	.05	9
7.78	7.91	8.03	8.13	8.23	8.33	8.41	8.49	8.57	.01	
5.83	5.93	6.03	6.11	6.19	6.27	6.34	6.40	6.47	.05	10
7.49	7.60	7.71	7.81	7.91	7.99	8.08	8.15	8.23	.01	
5.71	5.81	5.90	5.98	6.06	6.13	6.20	6.27	6.33	.05	11
7.25	7.36	7.46	7.56	7.65	7.73	7.81	7.88	7.95	.01	
5.61	5.71	5.80	5.88	5.95	6.02	6.09	6.15	6.21	.05	12
7.06	7.17	7.26	7.36	7.44	7.52	7.59	7.66	7.73	.01	
5.53	5.63	5.71	5.79	5.86	5.93	5.99	6.05	6.11	.05	13
6.90	7.01	7.10	7.19	7.27	7.35	7.42	7.48	7.55	.01	
5.46	5.55	5.64	5.71	5.79	5.85	5.91	5.97	6.03	.05	14
6.77	6.87	6.96	7.05	7.13	7.20	7.27	7.33	7.39	.01	
5.40	5.49	5.57	5.65	5.72	5.78	5.85	5.90	5.96	.05	15
6.66	6.76	6.84	6.93	7.00	7.07	7.14	7.20	7.26	.01	
5.35	5.44	5.52	5.59	5.66	5.73	5.79	5.84	5.90	.05	16
6.56	6.66	6.74	6.82	6.90	6.97	7.03	7.09	7.15	.01	
5.31	5.39	5.47	5.54	5.61	5.67	5.73	5.79	5.84	.05	17
6.48	6.57	6.66	6.73	6.81	6.87	6.94	7.00	7.05	.01	
5.27	5.35	5.43	5.50	5.57	5.63	5.69	5.74	5.79	.05	18
6.41	6.50	6.58	6.65	6.73	6.79	6.85	6.91	6.97	.01	
5.23	5.31	5.39	5.46	5.53	5.59	5.65	5.70	5.75	.05	19
6.34	6.43	6.51	6.58	6.65	6.72	6.78	6.84	6.89	.01	
5.20	5.28	5.36	5.43	5.49	5.55	5.61	5.66	5.71	.05	20
6.28	6.37	6.45	6.52	6.59	6.65	6.71	6.77	6.82	.01	
5.10	5.18	5.25	5.32	5.38	5.44	5.49	5.55	5.59	.05	24
6.11	6.19	6.26	6.33	6.39	6.45	6.51	6.56	6.61	.01	
5.00	5.08	5.15	5.21	5.27	5.33	5.38	5.43	5.47	.05	30
5.93	6.01	6.08	6.14	6.20	6.26	6.31	6.36	6.41	.01	
4.90	4.98	5.04	5.11	5.16	5.22	5.27	5.31	5.36	.05	40
5.76	5.83	5.90	5.96	6.02	6.07	6.12	6.16	6.21	.01	
4.81	4.88	4.94	5.00	5.06	5.11	5.15	5.20	5.24	.05	60
5.60	5.67	5.73	5.78	5.84	5.89	5.93	5.97	6.01	.01	
4.71	4.78	4.84	4.90	4.95	5.00	5.04	5.09	5.13	.05	120
5.44	5.50	5.56	5.61	5.66	5.71	5.75	5.79	5.83	.01	
4.62	4.68	4.74	4.80	4.85	4.89	4.93	4.97	5.01	.05	∞
5.29	5.35	5.40	5.45	5.49	5.54	5.57	5.61	5.65	.01	

TABLE C.9
Coefficients of Orthogonal Polynomials

k	Polynomial	Coefficients										$\sum c_{ij}^2$
3	Linear	−1	0	1								2
	Quadratic	1	−2	1								6
	Linear	−3	−1	1	3							20
4	Quadratic	1	−1	−1	1							4
	Cubic	−1	3	−3	1							20
	Linear	−2	−1	0	1	2						10
5	Quadratic	2	−1	−2	−1	2						14
	Cubic	−1	2	0	−2	1						10
	Quartic	1	−4	6	−4	1						70
	Linear	−5	−3	−1	1	3	5					70
6	Quadratic	5	−1	−4	−4	−1	5					84
	Cubic	−5	7	4	−4	−7	5					180
	Quartic	1	−3	2	2	−3	1					28
	Linear	−3	−2	−1	0	1	2	3				28
7	Quadratic	5	0	−3	−4	−3	0	5				84
	Cubic	−1	1	1	0	−1	−1	1				6
	Quartic	3	−7	1	6	1	−7	3				154
	Linear	−7	−5	−3	−1	1	3	5	7			168
	Quadratic	7	1	−3	−5	−5	−3	1	7			168
8	Cubic	−7	5	7	3	−3	−7	−5	7			264
	Quartic	7	−13	−3	9	9	−3	−13	7			616
	Quintic	−7	23	−17	−15	15	17	−23	7			2184
	Linear	−4	−3	−2	−1	0	1	2	3	4		60
	Quadratic	28	7	−8	−17	−20	−17	−8	7	28		2772
9	Cubic	−14	7	13	9	0	−9	−13	−7	14		990
	Quartic	14	−21	−11	9	18	9	−11	−21	14		2002
	Quintic	−4	11	−4	−9	0	9	4	−11	4		468
	Linear	−9	−7	−5	−3	−1	1	3	5	7	9	330
	Quadratic	6	2	−1	−3	−4	−4	−3	−1	2	6	132
10	Cubic	−42	14	35	31	12	−12	−31	−35	−14	42	8580
	Quartic	18	−22	−17	3	18	18	3	−17	−22	18	2860
	Quintic	−6	14	−1	−11	−6	6	11	1	−14	6	780

Source: From table XXIII of R. A. Fisher & F. Yates: *Statistical Tables for Biological, Agricultural and Medical Research*, 6th edition, (London: Longman Group, Ltd., 1974). Reprinted by permission of Addison-Wesley Longman, Ltd.

TABLE C.10
Binomial Coefficients

N	$\binom{N}{0}$	$\binom{N}{1}$	$\binom{N}{2}$	$\binom{N}{3}$	$\binom{N}{4}$	$\binom{N}{5}$	$\binom{N}{6}$	$\binom{N}{7}$	$\binom{N}{8}$	$\binom{N}{9}$	$\binom{N}{10}$
0	1										
1	1	1									
2	1	2	1								
3	1	3	3	1							
4	1	4	6	4	1						
5	1	5	10	10	5	1					
6	1	6	15	20	15	6	1				
7	1	7	21	35	35	21	7	1			
8	1	8	28	56	70	56	28	8	1		
9	1	9	36	84	126	126	84	36	9	1	
10	1	10	45	120	210	252	210	120	45	10	1
11	1	11	55	165	330	462	462	330	165	55	11
12	1	12	66	220	495	792	924	792	495	220	66
13	1	13	78	286	715	1287	1716	1716	1287	715	286
14	1	14	91	364	1001	2002	3003	3432	3003	2002	1001
15	1	15	105	455	1365	3003	5005	6435	6435	5005	3003
16	1	16	120	560	1820	4368	8008	11440	12870	11440	8008
17	1	17	136	680	2380	6188	12376	19448	24310	24310	19448
18	1	18	153	816	3060	8568	18564	31824	43758	48620	43758
19	1	19	171	969	3876	11628	27132	50388	75582	92378	92378
20	1	20	190	1140	4845	15504	38760	77520	125970	167960	184756

TABLE C.11
Sample Sizes for Interval Data Using One-Tailed Tests with Varying Effect Sizes and Levels of Power

	$\alpha = .05$					
	Power of the Statistical Test					
d	.75	.80	.85	.90	.95	.99
.10	538	619	719	857	1,083	1,577
.20	135	155	180	215	271	395
.30	62	71	82	98	121	176
.40	35	41	47	56	70	103
.50	23	27	31	37	45	66
.60	17	20	23	26	32	46
.70	13	15	18	21	24	35
.75	12	13	15	19	21	31
.80	11	12	14	17	19	28
.90	9	10	11	13	16	22
1.00	8	9	10	11	13	19
1.25	6	7	7	8	10	14
1.50	6	6	6	7	9	11

	$\alpha = .01$					
.10	901	1,004	1,131	1,302	1,577	2,165
.20	226	251	283	326	395	542
.30	101	112	126	145	176	241
.40	59	66	74	85	103	136
.50	39	43	48	55	66	91
.60	28	30	34	39	46	64
.70	21	23	26	29	35	48
.75	19	21	23	26	31	42
.80	17	19	21	23	27	37
.90	14	16	17	19	22	30
1.00	13	14	15	16	19	25
1.25	11	11	12	12	13	18
1.50	9	10	10	10	11	14

Source: From D. E. Hinkle, J. D. Oliver, and C. A. Hinkle, "How Large Should the Sample Part Be? Part II—The One-sample Case for Survey Research," *Educational and Psychological Measurement*, 45 (1985), pp. 271–280. Copyright ©1985. Reprinted by permission of Sage Publications, Inc.

TABLE C.11 (cont.)
Sample Sizes for Interval Data Using Two-Tailed Tests with
Varying Effect Sizes and Levels of Power

			$\alpha = .05$			
			Power of the Statistical Test			
d	*.75*	*.80*	*.85*	*.90*	*.95*	*.99*
.10	695	785	898	1,051	1,300	1,838
.20	174	197	225	263	325	459
.30	80	90	103	117	145	205
.40	46	52	58	68	84	119
.50	30	34	38	44	54	77
.60	22	24	27	32	38	54
.70	17	19	21	24	29	40
.75	15	16	18	21	26	36
.80	13	15	16	19	23	32
.90	11	12	14	15	19	26
1.00	10	11	12	13	16	22
1.25	8	9	9	10	11	15
1.50	7	8	9	9	10	12

			$\alpha = .01$			
.10	1,057	1,168	1,305	1,488	1,782	2,404
.20	265	292	327	372	446	601
.30	118	130	145	166	198	268
.40	70	77	86	97	116	151
.50	46	50	56	63	75	101
.60	33	36	40	45	53	71
.70	25	27	30	34	40	53
.75	22	24	27	30	35	46
.80	20	22	24	27	31	41
.90	17	18	20	22	25	34
1.00	15	15	17	18	21	28
1.25	12	12	13	14	15	19
1.50	11	11	12	12	13	16

TABLE C.12
Sample Sizes for Varying Numbers of Treatment Levels and Effect Sizes to be Detected

	Number of treatment levels (k)	Sample size per treatment level with differences to be detected between treatment levels ($C\sigma$) of:				
		$.5\sigma$	$.75\sigma$	1.0σ	1.25σ	1.5σ
		$\alpha = .05$				
Power = .75	2	55	26	15	10	7
	3	69	31	18	12	9
	4	79	35	21	14	10
	5	88	39	23	15	10
	6	95	42	24	16	11
	7	101	45	26	17	11
	8	107	48	27	18	12
		$\alpha = .01$				
	2	85	39	23	15	11
	3	103	46	27	18	13
	4	117	53	30	20	14
	5	130	58	33	21	15
	6	137	61	35	22	16
	7	143	64	37	23	17
	8	148	66	38	24	18
		$\alpha = .05$				
Power = .80	2	62	29	17	12	8
	3	78	35	20	14	10
	4	88	40	23	16	11
	5	96	43	25	17	12
	6	103	47	27	18	13
	7	110	50	29	19	13
	8	117	52	30	20	14
		$\alpha = .01$				
	2	93	43	25	17	12
	3	112	52	29	20	14
	4	124	56	32	22	16
	5	134	60	34	23	17
	6	144	64	36	25	18
	7	150	68	38	26	19
	8	156	71	40	27	19

Source: From D. E. Hinkle, J. D. Oliver, and C. A. Hinkle, "How Large Should the Sample Part Be? A Question with No Simple Answer? Or . . .," *Educational and Psychological Measurement*, 43 (1983), pp. 1051–1060. Copyright ©1983. Reprinted by permission of Sage Publications, Inc.

TABLE C.12 (cont.)

Number of treatment levels (k)	Sample size per treatment level with differences to be detected between treatment levels (Cσ) of:				
	.5σ	.75σ	1.0σ	1.25σ	1.5σ
	$\alpha = .05$				

Power = .85

	.5σ	.75σ	1.0σ	1.25σ	1.5σ
2	72	32	19	13	9
3	87	39	22	15	11
4	98	44	25	17	12
5	108	48	28	18	13
6	116	52	30	19	14
7	123	55	31	20	14
8	129	57	32	21	15

$\alpha = .01$

	.5σ	.75σ	1.0σ	1.25σ	1.5σ
2	103	48	28	18	13
3	124	56	32	21	15
4	138	62	35	23	17
5	150	67	38	25	18
6	162	72	41	27	19
7	169	76	44	29	19
8	176	79	46	30	20

$\alpha = .05$

Power = .90

	.5σ	.75σ	1.0σ	1.25σ	1.5σ
2	85	39	22	14	11
3	101	45	26	17	12
4	114	51	29	19	13
5	124	56	31	20	14
6	133	59	33	22	15
7	140	62	35	23	16
8	146	65	37	24	17

$\alpha = .01$

	.5σ	.75σ	1.0σ	1.25σ	1.5σ
2	120	55	32	21	15
3	142	64	37	24	17
4	158	72	41	26	19
5	170	77	44	28	20
6	180	81	46	30	21
7	187	84	48	32	22
8	192	87	50	33	23

TABLE C.12 (cont.)

Number of treatment levels (k)	Sample size per treatment level with differences to be detected between treatment levels ($C\sigma$) of:				
	$.5\sigma$	$.75\sigma$	1.0σ	1.25σ	1.5σ
	$\alpha = .05$				

Power = .95

k	$.5\sigma$	$.75\sigma$	1.0σ	1.25σ	1.5σ
2	104	47	27	18	13
3	124	56	32	21	15
4	139	63	36	23	16
5	149	67	39	25	17
6	158	71	41	27	18
7	167	75	43	28	19
8	176	79	45	29	20

$\alpha = .01$

k	$.5\sigma$	$.75\sigma$	1.0σ	1.25σ	1.5σ
2	144	64	37	24	18
3	167	76	43	28	20
4	183	82	47	31	22
5	197	88	50	33	24
6	207	93	53	35	25
7	217	98	56	36	26
8	226	102	59	37	27

$\alpha = .05$

Power = .99

k	$.5\sigma$	$.75\sigma$	1.0σ	1.25σ	1.5σ
2	148	67	38	25	17
3	173	77	44	28	20
4	192	85	48	31	22
5	208	93	52	34	24
6	218	98	55	36	25
7	227	102	58	38	26
8	235	105	60	39	27

$\alpha = .01$

k	$.5\sigma$	$.75\sigma$	1.0σ	1.25σ	1.5σ
2	194	87	49	32	23
3	222	100	57	37	26
4	240	109	62	40	29
5	258	117	67	43	31
6	274	122	70	45	32
7	284	127	72	47	33
8	294	133	74	49	34

TABLE C.13
Determination of r_{tet} for Various Values of bc/ad or ad/bc from a Four-Fold Contingency Table

r_{tet}	$\dfrac{bc}{ad}$ or $\dfrac{ad}{bc}$	r_{tet}	$\dfrac{bc}{ad}$ or $\dfrac{ad}{bc}$	r_{tet}	$\dfrac{bc}{ad}$ or $\dfrac{ad}{bc}$	r_{tet}	$\dfrac{bc}{ad}$ or $\dfrac{ad}{bc}$
0	1.000	.26	1.941–1.993	.51	4.068–4.205	.76	11.513–12.177
.01	1.013–1.039	.27	1.994–2.048	.52	4.206–4.351	.77	12.178–12.905
.02	1.040–1.066	.28	2.049–2.105	.53	4.352–4.503	.78	12.906–13.707
.03	1.067–1.093	.29	2.106–2.164	.54	4.504–4.662	.79	13.708–14.592
.04	1.094–1.122	.30	2.165–2.225	.55	4.663–4.830	.80	14.593–15.574
.05	1.123–1.151	.31	2.226–2.288	.56	4.831–5.007	.81	15.575–16.670
.06	1.152–1.180	.32	2.289–2.353	.57	5.008–5.192	.82	16.671–17.899
.07	1.181–1.211	.33	2.354–2.421	.58	5.193–5.388	.83	17.900–19.287
.08	1.212–1.242	.34	2.422–2.491	.59	5.389–5.595	.84	19.288–20.865
.09	1.243–1.275	.35	2.492–2.563	.60	5.596–5.813	.85	20.866–22.674
.10	1.276–1.308	.36	2.564–2.638	.61	5.814–6.043	.86	22.675–24.766
.11	1.309–1.342	.37	2.639–2.716	.62	6.044–6.288	.87	24.767–27.212
.12	1.343–1.377	.38	2.717–2.797	.63	6.289–6.547	.88	27.213–30.105
.13	1.378–1.413	.39	2.798–2.881	.64	6.548–6.822	.89	30.106–33.577
.14	1.414–1.450	.40	2.882–2.968	.65	6.823–7.115	.90	33.578–37.815
.15	1.451–1.488	.41	2.969–3.059	.66	7.116–7.428	.91	37.816–43.096
.16	1.489–1.528	.42	3.060–3.153	.67	7.429–7.761	.92	43.097–49.846
.17	1.529–1.568	.43	3.154–3.251	.68	7.762–8.117	.93	49.847–58.758
.18	1.569–1.610	.44	3.252–3.353	.69	8.118–8.499	.94	58.759–71.035
.19	1.611–1.653	.45	3.354–3.460	.70	8.500–8.910	.95	71.036–88.964
.20	1.654–1.697	.46	3.461–3.571	.71	8.911–9.351	.96	88.965–117.479
.21	1.698–1.743	.47	3.572–3.687	.72	9.352–9.828	.97	117.480–169.503
.22	1.744–1.790	.48	3.688–3.808	.73	9.829–10.344	.98	169.504–292.864
.23	1.791–1.838	.49	3.809–3.935	.74	10.345–10.903	.99	292.865–923.687
.24	1.839–1.888	.50	3.936–4.067	.75	10.904–11.512	1.00	923.688–∞
.25	1.889–1.940						

Source: From *Statistical Methods in Education and Psychology*, by G. V Glass and J. C. Stanley. Copyright ©1970 by Allyn and Bacon. Reprinted by permission.

Values in this table were calculated by Thomas O. Maguire.

If bc/ad is greater than 1, the value of r_{tet} is read directly from this table. If ad/bc is greater than 1, the table is entered with ad/bc and the value of r_{tet} is *negative*.

TABLE C.14
Quantiles of the Mann-Whitney Test Statistics

n	p	m = 2	3	4	5	6	7	8	9	10	11	12	13	14	15	16	17	18	19	20
	.001	0	0	0	0	0	0	0	0	0	0	0	0	0	0	0	0	0	0	0
	.005	0	0	0	0	0	0	0	0	0	0	0	0	0	0	0	0	0	1	1
2	.01	0	0	0	0	0	0	0	0	0	0	0	1	1	1	1	1	1	2	2
	.025	0	0	0	0	0	0	1	1	1	1	2	2	2	2	2	3	3	3	3
	.05	0	0	0	1	1	1	2	2	2	2	3	3	4	4	4	4	5	5	5
	.10	0	1	1	2	2	2	3	3	4	4	5	5	5	6	6	7	7	8	8
	.001	0	0	0	0	0	0	0	0	0	0	0	0	0	0	0	1	1	1	1
	.005	0	0	0	0	0	0	0	1	1	1	2	2	2	3	3	3	3	4	4
3	.01	0	0	0	0	0	1	1	2	2	2	3	3	3	4	4	5	5	5	6
	.025	0	0	0	1	2	2	3	3	4	4	5	5	6	6	7	7	8	8	9
	.05	0	1	1	2	3	3	4	5	5	6	6	7	8	8	9	10	10	11	12
	.10	1	2	2	3	4	5	6	6	7	8	9	10	11	11	12	13	14	15	16
	.001	0	0	0	0	0	0	0	0	1	1	1	2	2	2	3	3	4	4	4
	.005	0	0	0	0	1	1	2	2	3	3	4	4	5	6	6	7	7	8	9
4	.01	0	0	0	1	2	2	3	4	4	5	6	6	7	9	8	9	10	10	11
	.025	0	0	1	2	3	4	5	5	6	7	8	9	10	11	12	12	13	14	15
	.05	0	1	2	3	4	5	6	7	8	9	10	11	12	13	15	16	17	18	19
	.10	1	2	4	5	6	7	8	10	11	12	13	14	16	17	18	19	21	22	23
	.001	0	0	0	0	0	0	1	2	2	3	3	4	4	5	6	6	7	8	8
	.005	0	0	0	1	2	2	3	4	5	6	7	8	8	9	10	11	12	13	14
5	.01	0	0	1	2	3	4	5	6	7	8	9	10	11	12	13	14	15	16	17
	.025	0	1	2	3	4	6	7	8	9	10	12	13	14	15	16	18	19	20	21
	.05	1	2	3	5	6	7	9	10	12	13	14	16	17	19	20	21	23	24	26
	.10	2	3	5	6	8	9	11	13	14	16	18	19	21	23	24	26	28	29	31
	.001	0	0	0	0	0	0	2	3	4	5	5	6	7	8	9	10	11	12	13
	.005	0	0	1	2	3	4	5	6	7	8	10	11	12	13	14	16	17	18	19
6	.01	0	0	2	3	4	5	7	8	9	10	12	13	14	16	17	19	20	21	23
	.025	0	2	3	4	6	7	9	11	12	14	15	17	18	20	22	23	25	26	28
	.05	1	3	4	6	8	9	11	13	15	17	18	20	22	24	26	27	29	31	33
	.10	2	4	6	8	10	12	14	16	18	20	22	24	26	28	30	32	35	37	39
	.001	0	0	0	0	1	2	3	4	6	7	8	9	10	11	12	14	15	16	17
	.005	0	0	1	2	4	5	7	8	10	11	13	14	16	17	19	20	22	23	25
7	.01	0	1	2	4	5	7	8	10	12	13	15	17	18	20	22	24	25	27	29
	.025	0	2	4	6	7	9	11	13	15	17	19	21	23	25	27	29	31	33	35
	.05	1	3	5	7	9	12	14	16	18	20	22	25	27	29	31	34	36	38	40
	.10	2	5	7	9	12	14	17	19	22	24	27	29	32	34	37	39	42	44	47
	.001	0	0	0	1	2	3	5	6	7	9	10	12	13	15	16	18	19	21	22
	.005	0	0	2	3	5	7	8	10	12	14	16	18	19	21	23	25	27	29	31
8	.01	0	1	3	5	7	8	10	12	14	16	18	21	23	25	27	29	31	33	35
	.025	1	3	5	7	9	11	14	16	18	20	23	25	27	30	32	35	37	39	42
	.05	2	4	6	9	11	14	16	19	21	24	27	29	32	34	37	40	42	45	48
	.10	3	6	8	11	14	17	20	23	25	28	31	34	37	40	43	46	49	52	55

Critical regions correspond to values less than (or greater than) but not including the appropriate quantile.

TABLE C.14 (cont.)

n	p	m = 2	3	4	5	6	7	8	9	10	11	12	13	14	15	16	17	18	19	20
	.001	0	0	0	2	3	4	6	8	9	11	13	15	16	18	20	22	24	26	27
	.005	0	1	2	4	6	8	10	12	14	17	19	21	23	25	28	30	32	34	37
9	.01	0	2	4	6	8	10	12	15	17	19	22	24	27	29	32	34	37	39	41
	.025	1	3	5	8	11	13	16	18	21	24	27	29	32	35	38	40	43	46	49
	.05	2	5	7	10	13	16	19	22	25	28	31	34	37	40	43	46	49	52	55
	.10	3	6	10	13	16	19	23	26	29	32	36	39	42	46	49	53	56	59	63
	.001	0	0	1	2	4	6	7	9	11	13	15	18	20	22	24	26	28	30	33
	.005	0	1	3	5	7	10	12	14	17	19	22	25	27	30	32	35	38	40	43
10	.01	0	2	4	7	9	12	14	17	20	23	25	28	31	34	37	39	42	45	48
	.025	1	4	6	9	12	15	18	21	24	27	30	34	37	40	43	46	49	53	56
	.05	2	5	8	12	15	18	21	25	28	32	35	38	42	45	49	52	56	59	63
	.10	4	7	11	14	18	22	25	29	33	37	40	44	48	52	55	59	63	67	71
	.001	0	0	1	3	5	7	9	11	13	16	18	21	23	25	28	30	33	35	38
	.005	0	1	3	6	8	11	14	17	19	22	25	28	31	34	37	40	43	46	49
11	.01	0	2	5	8	10	13	16	19	23	26	29	32	35	38	42	45	48	51	54
	.025	1	4	7	10	14	17	20	24	27	31	34	38	41	45	48	52	56	59	63
	.05	2	6	9	13	17	20	24	28	32	35	39	43	47	51	55	58	62	66	70
	.10	4	8	12	16	20	24	28	32	37	41	45	49	53	58	62	66	70	74	79
	.001	0	0	1	3	5	8	10	13	15	18	21	24	26	29	32	35	38	41	43
	.005	0	2	4	7	10	13	16	19	22	25	28	32	35	38	42	45	48	52	55
12	.01	0	3	6	9	12	15	18	22	25	29	32	36	39	43	47	50	54	57	61
	.025	2	5	8	12	15	19	23	27	30	34	38	42	46	50	54	58	62	66	70
	.05	3	6	10	14	18	22	27	31	35	39	43	48	52	56	61	65	69	73	78
	.10	5	9	13	18	22	27	31	36	40	45	50	54	59	64	68	73	78	82	87
	.001	0	0	2	4	6	9	12	15	18	21	24	27	30	33	36	39	43	46	49
	.005	0	2	4	8	11	14	18	21	25	28	32	35	39	43	46	50	54	58	61
13	.01	1	3	6	10	13	17	21	24	28	32	36	40	44	48	52	56	60	64	68
	.025	2	5	9	13	17	21	25	29	34	38	42	46	51	55	60	64	68	73	77
	.05	3	7	11	16	20	25	29	34	38	43	48	52	57	62	66	71	76	81	85
	.10	5	10	14	19	24	29	34	39	44	49	54	59	64	69	75	80	85	90	95
	.001	0	0	2	4	7	10	13	16	20	23	26	30	33	37	40	44	47	51	55
	.005	0	2	5	8	12	16	19	23	27	31	35	39	43	47	51	55	59	64	68
14	.01	1	3	7	11	14	18	23	27	31	35	39	44	48	52	57	61	66	70	74
	.025	2	6	10	14	18	23	27	32	37	41	46	51	56	60	65	70	75	79	84
	.05	4	8	12	17	22	27	32	37	42	47	52	57	62	67	72	78	83	88	93
	.10	5	11	16	21	26	32	37	42	48	53	59	64	70	75	81	86	92	98	103
	.001	0	0	2	5	8	11	15	18	22	25	29	33	37	41	44	48	52	56	60
	.005	0	3	6	9	13	17	21	25	30	34	38	43	47	52	56	61	65	70	74
15	.01	1	4	8	12	16	20	25	29	34	38	43	48	52	57	62	67	71	76	81
	.025	2	6	11	15	20	25	30	35	40	45	50	55	60	65	71	76	81	86	91
	.05	4	8	13	19	24	29	34	40	45	51	56	62	67	73	78	84	89	95	101
	.10	6	11	17	23	28	34	40	46	52	58	64	69	75	81	87	93	99	105	111

TABLE C.14 (cont.)

n	p	m = 2	3	4	5	6	7	8	9	10	11	12	13	14	15	16	17	18	19	20
	.001	0	0	3	6	9	12	16	20	24	28	32	36	40	44	49	53	57	61	66
	.005	0	3	6	10	14	19	23	28	32	37	42	46	51	56	61	66	71	75	80
16	.01	1	4	8	13	17	22	27	32	37	42	47	52	57	62	67	72	77	83	88
	.025	2	7	12	16	22	27	32	38	43	48	54	60	65	71	76	82	87	93	99
	.05	4	9	15	20	26	31	37	43	49	55	61	66	72	78	84	90	96	102	108
	.10	6	12	18	24	30	37	43	49	55	62	68	75	81	87	94	100	107	113	120
	.001	0	1	3	6	10	14	18	22	26	30	35	39	44	48	53	58	62	67	71
	.005	0	3	7	11	16	20	25	30	35	40	45	50	55	61	66	71	76	82	87
17	.01	1	5	9	14	19	24	29	34	39	45	50	56	61	67	72	78	83	89	94
	.025	3	7	12	18	23	29	35	40	46	52	58	64	70	76	82	88	94	100	106
	.05	4	10	16	21	27	34	40	46	52	58	65	71	78	84	90	97	103	110	116
	.10	7	13	19	26	32	39	46	53	59	66	73	80	86	93	100	107	114	121	128
	.001	0	1	4	7	11	15	19	24	28	33	38	43	47	52	57	62	67	72	77
	.005	0	3	7	12	17	22	27	32	38	43	48	54	59	65	71	76	82	88	93
18	.01	1	5	10	15	20	25	31	37	42	48	54	60	66	71	77	83	89	95	101
	.025	3	8	13	19	25	31	37	43	49	56	62	68	75	81	87	94	100	107	113
	.05	5	10	17	23	29	36	42	49	56	62	69	76	83	89	96	103	110	117	124
	.10	7	14	21	28	35	42	49	56	63	70	78	85	92	99	107	114	121	129	136
	.001	0	1	4	8	12	16	21	26	30	35	41	46	51	56	61	67	72	78	83
	.005	1	4	8	13	18	23	29	34	40	46	52	58	64	70	75	82	88	94	100
19	.01	2	5	10	16	21	27	33	39	45	51	57	64	70	76	83	89	95	102	108
	.025	3	8	14	20	26	33	39	46	53	59	66	73	79	86	93	100	107	114	120
	.05	5	11	18	24	31	38	45	52	59	66	73	81	88	95	102	110	117	124	131
	.10	8	15	22	29	37	44	52	59	67	74	82	90	98	105	113	121	129	136	144
	.001	0	1	4	8	13	17	22	27	33	38	43	49	55	60	66	71	77	83	89
	.005	1	4	9	14	19	25	31	37	43	49	55	61	68	74	80	87	93	100	106
20	.01	2	6	11	17	23	29	35	41	48	54	61	68	74	81	88	94	101	108	115
	.025	3	9	15	21	28	35	42	49	56	63	70	77	84	91	99	106	113	120	128
	.05	5	12	19	26	33	40	48	55	63	70	78	85	93	101	108	116	124	131	139
	.10	8	16	23	31	39	47	55	63	71	79	87	95	103	111	120	128	136	144	152

TABLE C.15
Table of Critical values of *T* in the Wilcoxon Matched-Pairs Signed-Ranks Test

	Level of significance for one-tailed test		
	.025	.01	.005
	Level of significance for two-tailed test		
N	.05	.02	.01
6	0	—	—
7	2	0	—
8	4	2	0
9	6	3	2
10	8	5	3
11	11	7	5
12	14	10	7
13	17	13	10
14	21	16	13
15	25	20	16
16	30	24	20
17	35	28	23
18	40	33	28
19	46	38	32
20	52	43	38
21	59	49	43
22	66	56	49
23	73	62	55
24	81	69	61
25	89	77	68

Source: From F. Wilcoxon and R. A. Wilcox, *Some Rapid Approximate Statistical Procedures, Revised Edition*, (Pearl River, NY: Lederle Laboratories, 1964).

Appendix D
Answers to Exercises

Chapter 1

1. a. 64   **b.** 145   **c.** 99   **d.** 300   **e.** 86   **f.** 274   **g.** 13   **h.** 52

2. a. 50   **b.** −25   **c.** 19   **d.** 227

3. a. 343   **b.** 169   **c.** 8   **d.** 9   **e.** 3125   **f.** 121   **g.** 735

4. a. $(3 + 7)(8 + 1) = (10)(9)$ Adding numbers within parentheses

$\qquad\qquad\qquad\quad = 90$ Multiplying numbers

  b. $(3)(7) + (128/8) = (3)(7) + (16)$ Dividing within parentheses

$\qquad\qquad\qquad\qquad\quad = (21) + (16)$ Multiplying within parentheses

$\qquad\qquad\qquad\qquad\quad = 37$ Adding the numbers

  c. $\sqrt{169}(3 - 15) = \sqrt{169}(-12)$ Subtracting within parentheses

$\qquad\qquad\qquad\quad = \pm 13(-12)$ Square root can be either $-$ or $+$

$\qquad\qquad\qquad\quad = -156 \; or +156$ $(+13)(-12) = -156$ while

$\qquad\qquad\qquad\qquad\qquad\qquad\qquad\quad (-13)(-12) = +156$

  d. $3^2 + 5(6) - (16 + 3) = 3^2 + 5(6) - 19$ Adding within parentheses

$\qquad\qquad\qquad\qquad\qquad = 3^2 + 30 - 19$ Multiplying numbers

$\qquad\qquad\qquad\qquad\qquad = 9 + 30 - 19$ Squaring the number

$\qquad\qquad\qquad\qquad\qquad = 20$ Adding (subtracting) the numbers

5. $(5)(4) + (5)(6) + (5)(12) + (5)(2) + (5)(2) = 5(4 + 6 + 12 + 2 + 2)$

$\quad\; 20 \;\; + \;\; 30 \;\; + \quad 60 \;\; + \;\; 10 \;\; + \;\; 10 \;\; = 5(26)$

$\qquad\qquad\qquad\qquad\qquad\qquad\qquad\qquad\quad 130 = 130$

6. a. 63   **b.** 52   **c.** 30   **d.** 48   **e.** 44

7. a. continuous   **b.** discrete   **c.** discrete   **d.** discrete   **e.** continuous

8. a. ratio   **b.** ordinal   **c.** ordinal   **d.** ratio   **e.** ratio   **f.** nominal   **g.** ratio
h. nominal   **i.** interval   **j.** ratio

9. a. 5   **b.** -19 **c.** -2.56 **d.** -149.62 **e.** 84   **f.** -1.12 **g.** 3.51   **h.** 19.81
i. 0.547   **j.** -23.94 **k.** 43.75

10. a. 11.25   **b.** -525 **c.** 31.25

11. a. 20   **b.** 114

c. $(2+7+5+6)^2 \neq 2^2 + 7^2 + 5^2 + 6^2$
$$20^2 \neq 4 + 49 + 25 + 36$$
$$400 \neq 114$$

d. $(3)(2) + (3)(7) + (3)(5) + (3)(6) = 3(2+7+5+6)$
$$6 \;+\; 21 \;+\; 15 \;+\; 18 \;\; = 3(20)$$
$$60 = 60$$

Chapter 2

1. a. Frequencies for actual scores

Score	f
37	2
36	3
35	5
34	4
33	3
32	3
31	2
30	4
29	4
28	1

b. Frequencies within intervals

Interval	f
36.5–39.5	2
33.5–36.5	12
30.5–33.5	8
27.5–30.5	9

2. a. 10|2
9|88
8|0111223345556788
7|011122233344444455555667778999
6|6788

b. Class interval	f
100–104	1
95–99	2
90–94	0
85–89	7
80–84	9
75–79	14
70–74	15
65–69	4

d. Class interval	f
102–104	1
99–101	0
96–98	2
93–95	0
90–92	0
87–89	3
84–86	5
81–83	7
78–80	5
75–77	10
72–74	11
69–71	4
66–68	4

3. a. 8|0111223
7|122444677889
6|12445578
5|12445669
4|2222344667789
3|33455678899
2|8

b. Class interval	f
83–87	1
78–82	9
73–77	6
68–72	4
63–67	5
58–62	3
53–57	5
48–52	4
43–47	7
38–42	8
33–37	7
28–32	1

c. Class interval	f
80–89	7
70–79	12
60–69	8
50–59	8
40–49	13
30–39	11
20–29	1

d. The frequencies correspond to the number of leaves.

4. *Class interval*   *f*

Class interval	f
69–73	1
64–68	9
59–63	14
54–58	41
49–53	34
44–48	35
39–43	21
34–38	16
29–33	7
24–28	2

The peak of the frequency polygon is shifted to the right (higher scores) by approximately 10 units.

5. a.

Class Interval	Exact Limits	Midpoint	Group I				Group II			
			f	cf	%	c%	f	cf	%	c%
150–154	149.5–154.5	152	2	200	1.0	100.0	3	200	1.5	100.0
145–149	144.5–149.5	147	3	198	1.5	99.0	7	197	3.5	98.5
140–144	139.5–144.5	142	4	195	2.0	97.5	12	190	6.0	95.0
135–139	134.5–139.5	137	6	191	3.0	95.5	17	178	8.5	89.0
130–134	129.5–134.5	132	9	185	4.5	92.5	25	161	12.5	80.5
125–129	124.5–129.5	127	18	176	9.0	88.0	33	136	16.5	68.0
120–124	119.5–124.5	122	22	158	11.0	79.0	38	103	19.0	51.5
115–119	114.5–119.5	117	34	136	17.0	68.0	23	65	11.5	32.5
110–114	109.5–114.5	112	37	102	18.5	51.0	15	42	7.5	21.0
105–109	104.5–109.5	107	28	65	14.0	32.5	10	27	5.0	13.5
100–104	99.5–104.5	102	15	37	7.5	18.5	8	17	4.0	8.5
95–99	94.5–99.5	97	12	22	6.0	11.0	5	9	2.5	4.5
90–94	89.5–94.5	92	7	10	3.5	5.0	3	4	1.5	2.0
85–89	84.5–89.5	87	3	3	1.5	1.5	1	1	0.5	0.5

c. The frequency distribution for group I has a small positive skew. The frequency distribution for group II is more sharply peaked (kurtotic) than the frequency distribution for group I and has a more distinct negative skew. The frequency distribution for group II is also shifted to the right of the distribution of group I.

Chapter 3

1. a.

Class Interval	Exact Limits	f	cf	c%
3.8–4.0	3.75–4.05	4	120	100.00
3.5–3.7	3.45–3.75	8	116	96.67
3.2–3.4	3.15–3.45	15	108	90.00
2.9–3.1	2.85–3.15	18	93	77.50
2.6–2.8	2.55–2.85	20	75	62.50
2.3–2.5	2.25–2.55	17	55	45.83
2.0–2.2	1.95–2.25	12	38	31.67
1.7–1.9	1.65–1.95	12	26	21.67
1.4–1.6	1.35–1.65	10	14	11.67
1.1–1.3	1.05–1.35	4	4	3.33
0.8–1.0	0.75–1.05	0	0	0.00

b. $P_{10} = 1.35 + (8/10)(0.3) = 1.59$
$P_{45} = 2.25 + (16/17)(0.3) = 2.53$
$P_{60} = 2.55 + (17/20)(0.3) = 2.81$
$P_{95} = 3.45 + (6/8)(0.3) = 3.68$

c. $PR_{2.40} = 38.75$
$PR_{2.75} = 56.94$
$PR_{3.25} = 81.67$
$PR_{3.60} = 93.33$

3.

Class Interval	Exact Limits	f	cf	c%
69–73	68.5–73.5	1	180	100.00
64–68	63.5–68.5	9	179	99.44
59–63	58.5–63.5	14	170	94.44
54–58	53.5–58.5	41	156	86.67
49–53	48.5–53.5	34	115	63.89
44–48	43.5–48.5	35	81	45.00
39–43	38.5–43.5	21	46	25.56
34–38	33.5–38.5	16	25	13.89
29–33	28.5–33.5	7	9	5.00
24–28	23.5–28.5	2	2	1.11

a. $P_{15} = 38.98$
$P_{45} = 48.50$
$P_{80} = 57.04$

b. $PR_{67} = 97.94$
$PR_{40} = 17.39$
$PR_{34} = 5.89$

4. a. Max $= 4.00$
$P_{75} = 3.10$
Mdn $= 2.63$
$P_{25} = 2.05$
Min $= 1.10$

5. a.   I. Max $= 154$
$$P_{75} = 122.68$$
Mdn $= 114.23$
$$P_{25} = 106.82$$
Min $= 85$

    II. Max $= 154$
$$P_{75} = 132.30$$
Mdn $= 124.11$
$$P_{25} = 116.24$$
Min $= 85$

6. a. Max $= 104$ **c.** RUB $= 97.31$
$$P_{75} = 82.64$$ RLB $= 58.19$
Mdn $= 76.60$
$$P_{25} = 72.86$$
Min $= 66$

d. Potential outliers $= 102, 98, 98$

7. $\Sigma X = 312$ $\Sigma X^2 = 4182$ SS $= 288.24$
  a. $(\overline{X}) = 12.48$
    Median $= 13$
    Mode $= 14$

  b. Range $= 19 - 6 + 1 = 14$
$$s^2 = \frac{288.24}{24} = 12.01$$
$$s = 3.47$$

8. a. $\overline{X} = \dfrac{100}{10} = 10$
  b. $\Sigma(X - \overline{X}) = 0$

9. a. $\Sigma X = 140$ $\Sigma X^2 = 2110$ SS $= 150$ $\overline{X} = 14$ $s = \sqrt{\dfrac{150}{9}} = 4.08$

  b. $\Sigma X = 100$ $\Sigma X^2 = 1150$ SS $= 150$ $\overline{X} = 10$ $s = \sqrt{\dfrac{150}{9}} = 4.08$
  The mean increased by 4, and the standard deviation remained the same.

10. $\Sigma X = 3096$ $\Sigma X^2 = 243{,}366$ SS $= 3735.6$
  a. $\overline{X} = 77.4$
    Median $= 79.5$
  b. $s^2 = 95.78$ $s = 9.79$

11. Mean = 46   median = 43   mode = 40   skewed to right
Mean = 43   median = 43   mode = 43   symmetric
Mean = 40   median = 43   mode = 46   skewed to left

12. a. $\overline{X}_T = \dfrac{16{,}371.80}{234} = 69.96$ **b.** $\overline{X}_M = \dfrac{7339.20}{104} = 70.57$

c. $\overline{X}_O = \dfrac{8884.80}{111} = 80.04$ $\overline{X}_F = \dfrac{9032.60}{130} = 69.48$

$\overline{X}_C = \dfrac{7487.00}{123} = 60.87$

13. $\Sigma X = 2331$ $\Sigma X^2 = 182{,}389$ SS $= 1270.30$

$\overline{X} = 77.70$ $s = \sqrt{\dfrac{1270.30}{29}} = 6.62$

Range $= 89 - 65 + 1 = 25$

The range and standard deviation decrease whereas the mean remains stable.

14. $\Sigma(X - \overline{X})^2 = 54.40$
$\Sigma(X - 7)^2 = 56$

15. $X_8 = 37$

16. $z = 2.32$ Transformed score $(T) = 73.2$
$z = 1.84$ $T = 68.4$
$z = 0.00$ $T = 50.0$
$z = -0.37$ $T = 46.3$

17. a. $\overline{X} = 20.07$
$s = 4.86$

b.

X_i	z_i
24	0.81
25	1.01
14	−1.25
18	−0.43
24	0.81
20	−0.01
20	−0.01
12	−1.66
24	0.81
28	1.63
26	1.22
17	−0.63
15	−1.04
18	−0.43
16	−0.84

$\Sigma z_i = -0.01 \approx 0.00$

$\Sigma z_i^2 = 14.02 \approx 14.00$

c. $\bar{z} = \dfrac{\Sigma z_i}{n} = \dfrac{0.0}{15} = 0.0$

$s_z = \sqrt{\dfrac{14.00}{14}} = 1.00$

18. $\Sigma T = 749.9 \quad \Sigma T^2 = 38{,}890.59 \quad SS = 1400.59$

$\bar{T} = \dfrac{749.9}{15} = 49.99 \quad s = \sqrt{\dfrac{1400.59}{14}} = 10.00$

19. a. $\Sigma(X+5) = 376 \quad \Sigma(X+5)^2 = 9756 \quad SS = 330.93$

$\bar{X} = \dfrac{376}{15} = 25.07 \quad s = \sqrt{\dfrac{330.93}{14}} = 4.86$

$X_i' = X_i + 5$	z_i'
29	0.81
30	1.01
19	−1.25
23	−0.43
29	0.81
25	−0.01
25	−0.01
17	−1.66
29	0.81
33	1.63
31	1.22
22	−0.63
20	−1.04
23	−0.43
21	−0.84

The z scores for the new distributions are identical to those obtained for the original distribution.

20. a. Convert the score from the first distribution to a standard z score.

b. Transform the z score to the desired distribution using equation 4.6 in the text.

$X' = 10z + 100$

For the example, $X' = 115$. The equivalent standard score is 1.50 (since $\bar{X} = 100$ and $s = 10$). To find the corresponding raw score in the original distribution, use equation 4.6 again.

$X' = 4z + 18$

$= (4)(1.5) + 18$

$= 24$

Thus, the original raw score is 24.

21. Convert both scores to z scores using general formula and compare z scores.

$$z = \frac{X - \bar{X}}{s}$$

For John's mathematics score of 40,

$$z_{40} = \frac{40 - 35.2}{4.5} = \frac{4.8}{4.5} = +1.07$$

For John's humanities score of 115,

$$z_{115} = \frac{115 - 107.8}{9.6} = \frac{7.2}{9.6} = +0.75$$

Since John's z score for mathematics is higher than the z score for humanities, the best performance was on the mathematics test.

22. a. $z_C = .30$ **b.** Mathematics is John's strongest subject.
     $z_M = .63$
     $z_E = -.83$

23. z scores:

	GPA	SAT-V	SAT-M
Student 1	−0.30	−0.50	0.75
Student 2	1.36	1.50	1.25

Weighted composite scores: Student 1 = −0.19
                                 Student 2 =   1.39

Chapter 4

1. a. Convert raw score to z scores using general formula.

$$z_{87} = \frac{87 - 92}{11.5} = \frac{-5}{11.5} = -0.43$$

The raw score 87 is 0.43 standard deviation units *below* the mean.

$$z_{96} = \frac{96 - 92}{11.5} = \frac{4}{11.5} = +0.35$$

The raw score 96 is 0.35 standard deviation units *above* the mean.

b. For the scores 80 and 100, determine the z scores and find the area between these scores and the mean.

$$z_{80} = \frac{80 - 92}{11.5} = \frac{-12}{11.5} = -1.04 \qquad \text{area} = .3508$$

$$z_{100} = \frac{100 - 92}{11.5} = \frac{8}{11.5} = +0.70 \qquad \text{area} = .2580$$

Summing the two areas $(.3508 + .2580 = .6088)$

Multiplying $(.6088 \times 150) = 91.32$.

Thus, 60.88% of the scores (91.32 or 91) are between 80 and 100.

For the scores 95 and 100, determine the z scores and find the area between these scores and the mean.

$$z_{95} = \frac{95 - 92}{11.5} = \frac{3}{11.5} = +0.26 \qquad \text{area} = .1026$$

$$z_{100} = \frac{100 - 92}{11.5} = \frac{8}{11.5} = +0.70 \qquad \text{area} = .2580$$

Subtract the two areas $(.2508 - .1012 = .1554)$

Multiplying $(.1554 \times 150) = 23.31$.

Thus, 15.54% of the scores (23.31 or 23) are between 95 and 100.

c. Find the z score corresponding to the percentile and then convert to raw score. From Table C.2, the z score corresponding to 80th percentile is $+0.8416$. Transformed score for the 80th percentile,

$$X' = s'(z) + \overline{X'}$$

$$= 11.5(.8416) + 92$$

$$= 101.68 \text{ which is the 80th percentile}$$

2. a. $X = 76 \quad z = \quad 1.33$ 　　**b.** between 48 and 80 $= 158.76$
　　　$X = 38 \quad z = -1.83$ 　　　　　between 65 and 75 $= \quad 46.32$
　　　$X = 50 \quad z = -0.83$ 　　　　　between 34 and 52 $= \quad 47.28$

c. $>80 = 9.5$ 　　　　　　　　**d.** $<35 = 3.76$
　　$>60 = 100$ 　　　　　　　　　　$<50 = 40.66$
　　$>40 = 190.5$ 　　　　　　　　　$<75 = 178.88$

e. $P_{35} = 55.38$
　　$P_{80} = 70.10$
　　$PR_{55} = 33.72$
　　$PR_{70} = 79.67$

3. *Student* 　　*National* 　　*Large-city*

John 　　　66.28 　　　78.81
Mary 　　　20.33 　　　42.07

4. *Using Actual Scores* 　　　　　　*Using Whole Numbers*

A's	> 87.29	A's	≥ 87
B's	79.99–87.29	B's	80–86
C's	70.01–79.98	C's	70–79
D's	62.71–70.00	D's	63–69
F's	< 62.71	F's	≤ 62

5. $X = 60 \qquad NCE = 75.62$
　　$X = 43 \qquad NCE = 35.09$
　　$X = 33 \qquad NCE = 13.04$

6. $z = -2.67$ 　 Area below $= 0.0038$ or 0.38%

7. $< 46: Z = -1.5$　Area below $= 0.0668$ or 6.68%
　　$> 68: Z = 2.17$　Area above $= 0.015$ or 1.5%

8. a. $Z = -1.58$　Area below $= 0.0571$

　　b. $Z = 1.14$　Area above $= 0.1271$

　　c. $Z = -0.54$
　　　$X = (-0.54)(7.7) + 41.2 = 37.04 \approx 37$

Chapter 5

1. b.

X	Y	XY	X^2	Y^2
2	6	12	4	36
6	14	84	36	196
5	12	60	25	144
4	10	40	16	100
1	4	4	1	16
Σ　18	46	200	82	492

$$r = \frac{(5)(200) - (18)(46)}{\sqrt{[(5)(82) - (18)^2][(5)(492) - (46)^2]}}$$
$$= 172/172 = 1.0$$

2. a. Use the raw score formula for computing the Pearson r; summary data needed are:

$$\sum X = 119 \qquad \sum Y = 143 \qquad \sum XY = 1{,}761$$
$$\left(\sum X\right)^2 = 14{,}161 \qquad \left(\sum Y\right)^2 = 20{,}449 \qquad n = 10$$
$$\sum X^2 = 1{,}515 \qquad \sum Y^2 = 2{,}155$$

Applying the raw score formula for the Pearson r,

$$r = \frac{n \sum XY - \sum X \sum Y}{\sqrt{[n \sum X^2 - (\sum X)^2][n \sum Y^2 - (\sum Y)^2]}}$$

$$= \frac{10(1{,}761) - (119)(143)}{\sqrt{[10(1{,}515) - 14{,}161][10(2{,}155) - 20{,}449]}}$$

$$= \frac{17{,}610 - 17{,}017}{\sqrt{(989)(1{,}101)}} = \frac{593}{1{,}043.5} = 0.568$$

This Pearson r indicates that there is a positive relationship between the two tests, indicating that high performance on one test is associated with high scores on the other test.

b. The coefficient of determination, r^2, is defined as the proportion of variance in one variable that can be attributed to the variance in the second variable. For this example,

$$r^2 = (0.568)^2 = 0.323$$

3. $r = \sqrt{r^2} = \sqrt{.60} = \pm.77$

4. a. Amount of pesticide sprayed versus number of insects found a day later.

   b. Number of errors on one task versus number of errors on a second task in a dual-task procedure (both tasks attended to simultaneously).

5. a. $r = \dfrac{10(35,923) - (423)(838)}{\sqrt{[10(18,619) - (423)^2][10(70,592) - (838)^2]}} = .92$

   b. $\Sigma X = 250 \qquad \Sigma Y = 443 \qquad \Sigma XY = 22,206$
      $\Sigma X^2 = 12,570 \quad \Sigma Y^2 = 39,313$

     $r = \dfrac{5(22,206) - (250)(443)}{\sqrt{[5(12,570) - (250)^2][5(39,313) - (443)^2]}} = +.84$

   c. The correlation found in part b is somewhat lower than that found in part a. The difference is due to a more homogeneous group.

   d. $r_a^2 = .85$
      $r_b^2 = .71$

6. These data do exhibit a curvilinear trend: as years of marriage increase, level of marital satisfaction decreases. But this relationship reverses after about 10 years of marriage. Thus, the Pearson product-moment correlation coefficient will underestimate the relationship between the two variables. In fact, the obtained r here is quite low: $r = -.23$.

7. $\Sigma X = 11,296 \qquad \Sigma Y = 57.67 \qquad \Sigma XY = 32,814.94$
   $\Sigma X^2 = 6,456,890 \quad \Sigma Y^2 = 169.7709$

   a. $r = \dfrac{20(32,814.94) - (11,296)(57.67)}{\sqrt{[20(6,456,890) - (11,296)^2][20(169.7709) - (57.67)^2]}} = .47$

   b. $r^2 = .22$

8. $\Sigma X = 376 \qquad \Sigma Y = 161 \qquad \Sigma XY = 4,317$
   $\Sigma X^2 = 12,714 \quad \Sigma Y^2 = 2,213$

   $r = \dfrac{15(4,317) - (376)(161)}{\sqrt{[15(12,714) - (376)^2][15(2,213) - (161)^2]}} = .22$

   $r^2 = .05$

The correlation between the number of cigarettes smoked and IQ is positive, but very small. Only a very small proportion of variance is shared by the two variables (5 percent). Other factors are likely to be much more important in determining a person's scores on these two variables.

9. $r = -.58$

10. $s_{xy} = 17.86$

11. $\rho = .89$

12. $\rho = .88$

13. a. $r = -.48$

    b. There is a moderate negative correlation between anxiety and performance.

    c. r would remain the same.

    d. Correlation indicates association. There may or may not be a cause-and-effect relationship.

14. a. The correlation for all children between grades 2 and 6.

    b. The correlation for fourth graders with vocabulary scores above the school median.

Chapter 6

1. b. Use the raw score formula for computing both the regression coefficient and the regression constant; summary data needed are:

$$\sum X = 118 \qquad \sum Y = 104.3 \qquad \sum XY = 591.50$$
$$(\sum X)^2 = 13{,}924 \qquad n = 24$$
$$\sum X^2 = 704$$

Applying the raw score formula for the regression coeffecient,

$$b = \frac{n\sum XY - \sum X \sum Y}{[n\sum X^2 - (\sum X)^2]}$$

$$= \frac{24(591.50) - (118)(104.3)}{[24(704) - 13{,}924]} = \frac{1{,}888.6}{2{,}972} = 0.635$$

Applying the raw score formula for the regression constant,

$$a = \overline{Y} - b\overline{X}$$

$$= 4.344 - (0.635)(4.917) = 1.222$$

Therefore the regression equation is

$$\hat{Y} = 0.635X + 1.222$$

d. Substituting into the regression equation for $X = 5$ and $X = 7$:

$$\hat{Y} = 0.635(5) + 1.222 = 4.397$$

$$\hat{Y} = 0.635(7) + 1.222 = 5.667$$

e. In order to determine the standard error of estimate $s_{Y \cdot X}$, we must determine the Pearson r, which is 0.94, and the standard deviation of the predictor variable, s_Y, which is 1.57. Thus, $s_{Y \cdot X}$ is found as follows:

Using formula 6.11,

$$s_{Y\cdot X} = s_Y\sqrt{1 - r^2}\sqrt{(n-1)/(n-2)}$$

$$= 1.57\sqrt{(1 - (0.94)^2}\sqrt{(23)/(22)}$$

$$= 1.57\sqrt{(1 - 0.8836)}\sqrt{1.045} = 1.57\sqrt{.1164}\sqrt{1.045} = 0.548$$

Using formula 6.12,

$$s_{Y\cdot X} = s_Y\sqrt{1 - r^2}$$

$$= 1.57\sqrt{(1 - (0.94)^2)} = 1.57\sqrt{(1 - 0.8836)} = 1.57\sqrt{.1164} = 0.536$$

f. Since the correlation between the two tests is very high, $r = 0.94$, the standard error of estimate, $s_{Y\cdot X}$, is small. Thus, the errors in predicting the long test from the short test are very small.

2. b. $\hat{Y} = -2.951X + 13.648$

   d. $\hat{Y} = 4.795$

   e. *using 6.11:* $s_{Y\cdot X} = 3.048$

   using 6.12: $s_{Y\cdot X} = 2.937$

3. b. $\hat{Y} = 0.076X + 11.519$

   d. $\hat{Y} = 53.319$

   e. *using 6.11:* $s_{Y\cdot X} = 7.244$

   using 6.12: $s_{Y\cdot X} = 6.932$

4. a. $b = 0.853$

   $a = 29.351$

   $\hat{Y} = 0.853X + 29.351$

   b. $\hat{Y} = 67.736$

   c. *using 6.11:* $s_{Y\cdot X} = 9.474$

   using 6.12: $s_{Y\cdot X} = 9.216$

5. b. $\hat{Y} = 0.065X + 66.548$

   d. $\hat{Y} = 77.923$

   e. *using 6.11:* $s_{Y\cdot X} = 8.898$

   using 6.12: $s_{Y\cdot X} = 8.630$

6. a. $z_y = 0.64z_x$

   b.

$z_x = -1.25$	$z_{\hat{y}} = -0.80$
$z_x = -0.72$	$z_{\hat{y}} = -0.46$
$z_x = 0.00$	$z_{\hat{y}} = 0.00$
$z_x = 0.66$	$z_{\hat{y}} = 0.42$
$z_x = 1.64$	$z_{\hat{y}} = 1.05$

c.

$X = 45$	$z_x = 0.58$	$z_{\hat{y}} = 0.37$	*Note* : $\overline{X} = 39.80$
$X = 33$	$z_x = -0.76$	$z_{\hat{y}} = -0.49$	$s_x = 9.00$
$X = 68$	$z_x = 3.13$	$z_{\hat{y}} = 2.00$	

d. Since $z_y = (Y - \overline{Y})/s_y = (Y - 63.3)/12$, then $Y = 12z_y + 63.3$ (by algebraic manipulation). Thus,

$z_{\hat{y}} = 0.37$	$\hat{Y} = 67.7$	$\hat{Y} = 67.7$	Computed from the
$z_{\hat{y}} = -0.49$	$\hat{Y} = 57.4$	$\hat{Y} = 57.5$	regression equation
$z_{\hat{y}} = 2.00$	$\hat{Y} = 87.3$	$\hat{Y} = 87.4$	determined in

Exercise 4
$(\hat{Y} = 0.853X + 29.351)$

7. b. $\hat{Y} = -0.044X + 3.616$

   d. $\hat{Y} = 2.516$

   e. *using* 6.11: $s_{Y \cdot X} = 0.346$

     using 6.12: $s_{Y \cdot X} = 0.339$

8. a. $b = r(s_Y/s_X) = (.57)(5.1/6.3) = 0.462$

     $a = (\Sigma Y - b\Sigma X)/n = [568 - (.462)(922)]/35 = 4.058$

     $\hat{Y} = 0.462X + 4.058$

   b. $\hat{Y} = 19.304$

   c. *using* 6.11: $s_{Y \cdot X} = 4.255$

     using 6.12: $s_{Y \cdot X} = 4.192$

Chapter 7

1. a. $P(6) = .10$ **b.** $P(6 \text{ } or \text{ } 4) = .20$

   c. $P(\text{odd number}) = .50$ **d.** $P(\text{multiple of 3}) = .30$

2. a. $P(A \text{ } or \text{ } B) = 1.00$

   b. $P(A \text{ } or \text{ } C) = .60$

   c. $P(B \text{ } or \text{ } C) = .70$

3. a. $P(A \text{ } and \text{ } B) = .25$

   b. $P(A \text{ } and \text{ } C) = .15$

   c. $P(B \text{ } and \text{ } C) = .15$

4. a. $P(A \text{ } and \text{ } B) = .278$

5.

Outcome	f
6	1
7	1
8	2
9	2
10	2
11	1
12	1

Outcome	f
3	1
4	1
5	2
6	2
7	3
8	3
9	4
10	4
11	5
12	4
13	4
14	3
15	3
16	2
17	2
18	1
19	1

6. a. P(ace of spades) = .0192

   b. P(ace) = .0769

   c. P(heart) = .25

   d. P(ace of spaces *and* ace of hearts) = .00037(w/ replacement)

    = .00038 (w/o replacement)

   e. P(3 aces) = .00046 (w/ replacement)

    = .00018 (w/o replacement)

   f. P(ace, king, queen) = .00046 (w/ replacement)

    = .00048 (w/o replacement)

7. a. .04   **b.** .03   **c.** .10   **d.** .22

8. The first step in this exercise is to determine the total number of combinations of seven individuals taken three at a time. The general formula for combinations of N things taken n at a time is:

$$_NC_n = \frac{N!}{n!(N-n)!}$$

For this example,

$$_7C_3 = \frac{7!}{3!(7-3)!} = \frac{7!}{3!4!} = \frac{7 \times 6 \times 5 \times 4 \times 3 \times 2 \times 1}{(3 \times 2 \times 1)(4 \times 3 \times 2 \times 1)} = 35$$

Therefore, there are 35 different combinationss of 7 things taken 3 at a time.

Next, consider the possible combinations of males and females; they are

(1) 3 males–0 females

(2) 2 males–1 female

(3) 1 male–2 females

(4) 0 males–3 females

For (1)–3 males/0 females:

$$_4C_3 \times {_3}C_0 = \frac{4!}{3!1!} \times \frac{3!}{0!3!} = 4$$

For (2)–2 males/1 female:

$$_4C_2 \times {_3}C_1 = \frac{4!}{2!2!} \times \frac{3!}{1!2!} = 18$$

For (3)–1 male/2 females:

$$_4C_1 \times {_3}C_2 = \frac{4!}{1!3!} \times \frac{3!}{2!1!} = 12$$

For (4)–0 males/3 females:

$$_4C_0 \times {_3}C_3 = \frac{4!}{0!4!} \times \frac{3!}{3!0!} = 1$$

Note that only possibilities (3) and (4) illustrate 2 or more women on the committee. That is, $12 + 1 = 13$ possibilities out of the 35 possibilities exist for having 2 or 3 women on the committee. Therefore the probability would be:

$$P(2 \text{ or more women}) = 13/35 = 0.37$$

9. 190

10. a. Binomial distribution **b.** .375

11. a. $P(X < 100) = .2119$ **b.** $P(X > 150) = .1151$
c. $P(110 < X < 125) = .2347$ **d.** $P(X < 75) = .0359$
$P(X > 160) = .0548$

12. a. $P(\text{pass}) = .9429$ **b.** $P(\text{honors}) = .0823$
c. $P(> 80) = .4562$ **d.** $P(< 85) = .6736$

13. a. $N = 144$ $\overline{X} = 120$ $s_{\overline{X}} = 2.08$ **b.** $N = 400$ $\overline{X} = 120$ $s_{\overline{X}} = 1.25$

14. a. Number of committees with:

3 Professors	21
2 Professors	105
1 Professor	105
0 Professors	21
	352

b. Prob (0 Professors) $= .083$

c. Prob (3 Professors) $= .083$

15. a. $P(\overline{X} > 121.4) = .2514$ **b.** $P(\overline{X} < 118.2) = .1922$
c. $P(\overline{X} < 120.8) = .6480$ **d.** $P(\overline{X} > 119.4) = .6141$

16. a. $P(\overline{X} > 121.4 = .1314$ **b.** $P(\overline{X} < 118.2) = .0749$
c. $P(\overline{X} < 120.8) = .7389$ **d.** $P(\overline{X} > 119.4) = .6844$

17. a. $P(\overline{X} < \$35,000 = .2148$ **b.** $P(\overline{X} > \$35,700) = .1251$
c. $P(\$34,900 < \overline{X} < \$35,500) = .5834$
d. $P(\overline{X} > \$36,000) = .0233$

18. $P(\overline{X} > \$40,000) = .2578$
$P(\overline{X} < \$25,000) = .0764$

Chapter 8

1. *Step 1: State the Hypotheses.*
The null hypothesis is tested against the non-directional alternative hypothesis. For this example, the research hypothesis is that hyperactive children are different from normal children in creativity.

$$H_0: \mu = 50$$

$$H_a: \mu \neq 50$$

Step 2: Set the Criterion for Rejecting the H_0.
Since the standard deviation in the population (σ) is known, the sampling distribution of means for testing this hypothesis is the normal distribution. With $\alpha = .05$, the critical value of z for this two-tailed test is

$$z_{CV} = \pm 1.96$$

Step 3: Compute the Test Statistic.
First, the standard error of the mean is computed:

$$\sigma_{\overline{X}} = \sigma/\sqrt{n} = 16/\sqrt{50} = 2.26$$

Then the general formula is used to compute the test statistic:

$$z = \frac{\overline{X} - \mu}{\sigma_{\overline{X}}} = \frac{152.5 - 150}{2.26} = \frac{2.5}{2.26} = 0.93$$

Step 4: Decide about H_0.
Since the computed value for the test statistic ($z = 0.93$) does not exceed the critical value ($z_{CV} = \pm 1.96$), the null hypothesis is *not* rejected. The associated probability statement would be: The probability that the observed sample mean ($\overline{X} = 152.5$) would have occurred by chance, if the null hypothesis is true

$(H_0: \mu = 50)$, is greater than .05. By failing to reject the null hypothesis, we conclude that the population of hyperactive children does not differ from the normal population in terms of creativity.

2. 1. $H_0: \mu = 2.75$

   $H_a: \mu \neq 2.75$

2. $t_{cv} = \pm 1.96$

3. $s_{\bar{X}} = .0406$, $t = 2.4631$

4. Reject H_0; $p < .05$.

3. 1. $H_0: \mu = 2.75$

   $H_a: \mu \neq 2.75$

2. $t_{cv} = \pm 2.045$

3. $s_{\bar{X}} = .1187$, $t = 0.8425$

4. Do not reject H_0; $p > .05$.

4. a. $\sigma_{\bar{X}} = 0.80$

The distribution of the means of all possible samples of size 225.

b. 1. $H_0: \mu = 175$

   $H_a: \mu \neq 175$

2. $z_{cv} = \pm 1.96$

3. $z = -1.75$

4. Do not reject H_0; $p > .05$.

c. The probability of obtaining a sample mean of 173.6 or less, if the population mean is 175, is greater than .05.

d. 1. $H_0: \mu = 175$

   $H_a: \mu < 175$

2. $z_{cv} = -1.645$

3. $z = -1.75$

4. Reject H_0; $p < .05$.

e. No, it is possible to reject H_0 with a one-tailed test but not with a two-tailed test.

f. No, σ is still known.

g. A larger difference because the standard error of the mean would be greater than it was with $n = 225$.

5. a. $H_0: \mu = 162$

   $H_a: \mu \neq 162$

b. $t_{cv} = \pm 1.645$

c. $s_{\bar{X}} = 1.41$, $t = -1.42$

d. Do not reject H_0; $p > .10$.

6. a. right tail

b. right tail

c. left tail

d. left tail

7. If the null hypothesis is rejected at the .05 level of significance, then it may or may not be rejected at the .01 level of significance. However, if the null hypothesis is rejected at the .01 level of significance, then it will also be rejected at the .05 level of significance.

8. When the value of the test statistic falls in the region of rejection, the null hypothesis is rejected because the value of the test statistic is unlikely to have occurred by chance.

9. $P_{01} = -2.552$

$P_{05} = -1.734$

$P_{95} = 1.734$

$P_{98} = 2.101$

$P_{99} = 2.552$

10. The area beyond $t = -1.383$ is .100.

The area beyond $t = 3.250$ is .005.

The proportion between $1.000 - .105 = .895$.

11. 1. H_0: $\mu = 10$

H_a: $\mu > 10$

2. $t_{cv} = +2.65$

3. $s_{\bar{X}} = 0.72$, $t = 2.28$

4. Do not reject H_0; $p < .01$.

12. The probability that a sample mean of 85 or more would appear by chance, if the population mean is 80, is less than .05.

13. a. one-tailed test

b. .05 level of significance

c. $n = 444$

14. 1. H_0: $\mu = 1.6$

H_a: $\mu < 1.6$

2. $t_{cv} = -2.718$

3. $s_{\bar{X}} = 0.101$, $t = -1.069$

4. Do not reject H_0; $p > .01$.

15. a. ± 4.604

   b. 1.833

   c. 2.567

   d. ± 1.717

   e. 2.763

   f. 1.678

16. Because the critical value indicates $\alpha/2$ of the area for a non-directional test and α of the area for a directional test.

17. ± 1.711

Chapter 9

1. 1. $\bar{X} = 2.85 \quad s_{\bar{X}} = .0406$

   $\text{CI}_{90} = (2.7832, 2.9168)$
   $\text{CI}_{95} = (2.7704, 2.9296)$

2. $\bar{X} = 2.85 \quad s_{\bar{X}} = .1187$

   $\text{CI}_{95} = (2.6073, 3.0927)$

   $\text{CI}_{99} = (2.5229, 3.1771)$

3. The general formula for confidence intervals is used in this example

$$\text{CI} = \text{sample mean} \pm (\text{critical value})(\text{standard error of the mean})$$

First, the standard error of the mean is computed:

$$s_{\bar{X}} = s/\sqrt{n} = 8.48/\sqrt{64} = 1.06$$

Since the standard deviation in the population (σ) is unknown and we use the sample standard deviation (s) as an estimate, the sampling distribution of means for developing the confidence interval is the t distribution with $n - 1$ degrees of freedom. For this example, $n = 64$; thus, the sampling distribution would be the t distribution with $64 - 1 = 63$ degrees of freedom. Assuming the 95-percent confidence level, the critical value for this confidence interval is 2.00. Now, the general formula is used to compute the confidence interval:

$$CI_{95} = \overline{X} \pm (2.00)(1.06)$$
$$= 84.2 \pm (2.00)(1.06)$$
$$= 84.2 \pm 2.12$$
$$= (82.08, 86.32)$$

Conclusion: We are 95 percent confident that the interval from 82.08 to 86.32 contains the population mean (μ) for the performance of seventh grade boys in the district on the physical performance test.

4. a. 1. $H_0: \mu = 65$ $\overline{X} = 65.6$ $s_{\overline{X}} = 0.20$

        $H_a: \mu > 65$

    2. $df = 399$ $t_{cv} = +1.645$

    3. $t = 3.00$

    4. Reject H_0; $p < .05$.

   b. $CI_{95} = (65.21, 65.99)$

5. $\overline{X} = 7.5$ $s_{\overline{X}} = 1.72$

   $CI_{90} = (4.67, 10.33)$

   $CI_{99} = (3.07, 11.93)$

6. For $n = 150$ $s_{\overline{X}} = 0.19$

   $CI_{95} = (22.23, 22.97)$

   For $n = 10$ $s_{\overline{X}} = 0.74$

   $CI_{95} = (20.93, 24.27)$

7. $\overline{X} = 31.2$ $s_{\overline{X}} = 4.5$

   $CI_{90} = (23.46, 38.94)$

   $CI_{99} = (18.46, 43.94)$

8. For $t_{cv} = 1.96$ $n = 96$

   For $t_{cv} = 1.99$ $n = 99$

9. All of these statements speak of the probability of μ, not the probability that the interval spans μ.

10. $\overline{X} = 75$

   $s_{\overline{X}} = 2.55$

   $s = 38.27$

11. The midpoint of the interval is not $\overline{X}$.

12. a. $CI_{95} = (88.97, 101.03)$

b. We are 95 percent confident that the interval from 88.97 to 101.03 spans the population mean.

c. CI_{99} is larger than CI_{95}.

d. $n = 138$

13. Statistical precision refers to the accuracy of estimates, the width of the confidence interval. A larger sample size yields a smaller standard error of the mean and thus greater statistical precision.

Chapter 10

1. *Step 1: State the Hypotheses.*

The null hypothesis is tested against the directional alternative hypothesis. For this example, the research hypothesis is that 65 percent of the union membership favors the set of salary and benefits demands. This hypothesis will be rejected only if significantly more than 65 percent of the membership favors the demands. Symbolically,

$$H_0: P = 0.65$$

$$H_a: P > 0.65$$

Step 2: Set the Criterion for Rejecting the H_0.

The sampling distribution of proportions for testing this hypothesis is the normal distribution. Using $\alpha = .10$, the critical value of z for this one-tailed test is

$$z_{cv} = +1.282$$

Step 3: Compute the Test Statistic.

First, the proportion, p, and the standard error of the proportion are computed:

$$p = 134/200 = 0.67$$

$$s_p = \sqrt{\frac{PQ}{n}} = \sqrt{\frac{(0.65)(0.35)}{200}} = 0.034$$

Then the general formula is used to compute the test statistic:

$$z = \frac{p - P}{s_p} = \frac{0.67 - 0.65}{0.034} = \frac{0.02}{0.034} = +0.588$$

Step 4: Decide about H_0.

Since the computed value for the test statistic ($z = +0.588$) does not exceed the critical value ($z_{cv} = +1.282$), the null hypothesis is *not* rejected. The associated probability statement would be: the probability that the observed sample proportion ($p = 0.67$) would have occurred by chance, if the null hypothesis is true (H_0: $P = 0.65$), is greater than .10. By failing to reject the null hypotheses, we conclude that the population of union members supporting the salary and benefits demands does not significantly exceed 0.65.

2. a. 1. H_0: $\rho = 0$
   H_a: $\rho \neq 0$
   2. $z_{cv} = \pm 1.96$
   3. $z_r = 0.400$ $s_{z_r} = 0.15$
      $z = 2.67$
   4. Reject H_0; $p < .05$.
   c. $CI_{95} = (.11, .69)$ for z_r
      $(.11, .60)$ for r

b. 1. H_0: $\rho = .35$
   H_a: $\rho \neq .35$
   2. $z_{cv} = \pm 1.96$
   3. $z = 0.23$
   4. Do not reject H_0; $p > .05$.

3. a. *Using 10.5:*
   1. H_0: $\rho = 0$
      H_a: $\rho \neq 0$
   2. $z_{cv} = \pm 1.96$
   3. $z_r = 0.523$ $s_{z_r} = 0.19$
      $z = 2.75$
   4. Reject H_0; $p < .05$.

   Using 10.7:
   1. H_0: $\rho = 0$
      H_a: $\rho \neq 0$
   2. For df $= 28$, $t_{cv} = \pm 2.048$.
   3. $t = 2.90$
   4. Reject H_0; $p < .05$.

b. $CI_{95} = (.15, .89)$ for z_r
      $(.15, .71)$ for r

4. a. 1. H_0: $P = .70$
   H_a: $P \neq .70$
   2. $z_{cv} = \pm 2.576$
   3. $p = 0.6$; $s_p = 0.03$
      $z = -3.33$
   4. Reject H_0; $p < .01$.

b. $CI_{99} = (.52, .68)$

5. a. 1. H_0: $P = .50$
   H_a: $P > .50$
   2. $z_{cv} = +1.282$
   3. $p = 0.60$; $s_p = 0.03$
      $z = 3.33$
   4. Reject H_0; $p < .10$.

b. $CI_{90} = (.55, .65)$

6. a. $r = .77$

b. 1. H_0: $\rho = 0$

H_a: $\rho > 0$

2. For df $= 7$, $t_{cv} = \pm 1.895$

3. $t = 3.19$

4. Reject H_0; $p < .05$.

c. $CI_{95} = (.22, 1.82)$ for z_r

$= (.22, .95)$ for r

7. a. 1. H_0: $P = .60$ **b.** $CI_{95} = (.64, .76)$

H_a: $P > .60$

2. $z_{cv} = \pm 1.645$

3. $z = 3.33$

4. Reject H_0; $p < .05$.

8. a. $r_{cv} = .325$. Do not reject H_0; $p > .05$.

b. $r_{cv} = .232$. Reject H_0; $p < .05$.

c. $r_{cv} = .393$

d. $r_{cv} = .358$

e. $r_{cv} = .418$. Reject H_0; $p < .01$.

f. $r_{cv} = .352$; $p < .10$.

$r_{cv} = .413$; $p > .05$.

The region of rejection is smaller for $\alpha = .05$ than for $\alpha = .10$.

9. In a one-tailed test, the entire region of rejection is in one tail. Therefore, the critical value is closer to the hypothesized value of the parameter than it would be in a two-tailed test.

10. a. $CI_{95} = (.780, .456)$ for z_r

$CI_{95} = (.65, .43)$ for r.

b. The interval does not contain the sample correlation.

c. $n \approx 686$

d. No, interval width is a function of n, not r.

e. CI_{90} could be used instead of CI_{95}.

11. a. $CI_{99} = (.27, .43)$

b. 8235 to 13,115 students

c. No, only z_{cv} changes.

$CI_{90} = (.30, .40)$ or 9150 to 12,200 students. The .90 level of confidence is adequate.

Chapter 11

1. *Step 1: State the Hypotheses.*
For this study, the null hypothesis would be that, in the population, there is no difference in the effects of the two drugs, A and B, on the hyperactivity of laboratory rats. For the alternative hypothesis, there is no reason to assume directionality; therefore, the null and alternative hypotheses would be:

$$H_0: \mu_A = \mu_B$$

$$H_a: \mu_A \neq \mu_B$$

Step 2: Set the Criterion for Rejecting H_0.
Note that the assumption is made that the two population variances are homogenous; therefore, the test of statistical significance will be based upon the pooled estimate for the population variance (see below). The general formula will be used to test the null hypothesis. The sampling distribution of the differences in sample means, used for testing the null hypothesis, is the t distribution with $n_A + n_B - 2$ degrees of freedom, i.e., $18 + 24 - 2 = 40$ df. The level of significance for this study is set a priori at $\alpha = .05$, the critical value of t for this two-tailed test is

$$t_{cv} = \pm 2.021$$

Step 3: Compute the Test Statistic.
Thr first step is to compute the pooled estimate for population variance (s^2) using formula 11.5; the second step is to compute the standard eerror of the differences ($s_{\overline{X}_1 - \overline{X}_2}$) using formula 11.1. The pooled estimate is:

$$s^2 = \frac{(n_A - 1)(s_A^2) + (n_B - 1)(s_B^2)}{n_A + n_B - 2}$$

$$= \frac{(18 - 1)(12.25) + (24 - 1)(10.24)}{18 + 24 - 2} = \frac{208.25 + 235.52}{40} = 11.09$$

The estimated standard error is:

$$s_{\overline{X}_1 - \overline{X}_2} = \sqrt{s^2(1/n_A + 1/n_B)} = \sqrt{11.09(1/18 + 1/24)} = 1.04$$

Thus, the test statistic is:

$$t = \frac{(\overline{X}_1 - \overline{X}_2) - (0)}{s_{\overline{X}_1 - \overline{X}_2}}$$

$$= \frac{(75.6 - 72.8) - (0)}{1.04} = \frac{2.8 - 0}{1.04} = 2.69$$

Step 4: Decide about H_0.
Since the computed value for the test statistic ($t = 2.69$) exceeds the critical value ($t_{cv} = \pm 2.021$), the null hypothesis is rejected. The associated probability

statement would be: the probability that the observed difference in sample means $(\overline{X}_1 - \overline{X}_2 = 2.8)$ would have occurred by chance, if the null hypothesis is true $(H_0: \mu_A = \mu_B)$, is less than .05. By rejecting the null hypotheses, we conclude that the two drugs have differing effects upon hyperactivity in laboratory rats. If the drugs were to reduce the level of hyperactivity, Drug B appears to be more effective.

Step 5: Construct the Confidence Interval.
For this study, the CI_{95} is computed as:

$$CI_{95} = (\overline{X}_1 - \overline{X}_2) \pm (t_{cv})(s_{\overline{X}_1 - \overline{X}_2})$$
$$= 2.8 \pm (2.021)(1.04)$$
$$= 2.8 \pm 2.10$$
$$= (0.70, 4.90)$$

We conclude with 95 percent confidence that the interval from 0.70 to 4.90 contains the difference between the population means. Note that this interval does not contain zero; this is consistent with the previous result in that the null hypothesis was rejected, i.e., there is a difference between the population means.

2. a. 1. $H_0: \mu_1 = \mu_2$ **b.** $CI_{90} = (-2.44, -1.16)$
 $H_a: \mu_1 \neq \mu_2$
 **2.** $t_{cv} = \pm 1.645$
 **3.** $s^2 = 11.50$ $s_{\overline{X}_1 - \overline{X}_2} = 0.39$
 $t = -4.62$
 **4.** Reject H_0; $p < .10$.

3. a. 1. $H_0: \mu_1 = \mu_2$ **b.** $CI_{99} = (-4.31, 2.71)$
 $H_a: \mu_1 \neq \mu_2$
 **2.** $t_{cv} = \pm 2.878$
 **3.** $s^2 = 7.42$ $s_{\overline{X}_1 - \overline{X}_2} = 1.22$
 $t = -0.66$
 **4.** Do not reject H_0; $p > .01$.

4. a. 1. $H_0: \mu_1 = \mu_2$
　　$H_a: \mu_1 \neq \mu_2$
　Using pooled estimate:
　2. $t_{cv} = \pm 1.746$
　3. $s^2 = 20.86$　$s_{\bar{X}_1-\bar{X}_2} = 2.17$
　　$t = -1.95$
　4. Reject H_0; $p < .10$.
　Using Cochran and Cox/Satterthwaite:
　2. df $= 13.312$　$t_{cv} = \pm 1.771$
　3. $s_{\bar{X}_1-\bar{X}_2} = 2.22$　$t = -1.91$
　4. Reject H_0; $p < .10$.

b. $CI_{90} = (-8.02, -0.44)$

5. a. 1. $H_0: \delta = 0$
　　$H_a: \delta \neq 0$
　2. $t_{cv} = \pm 2.365$
　3. $\Sigma d = 10$　$\Sigma d^2 = 82$　$\bar{d} = 1.25$
　　$s_d = 3.15$　$s_{\bar{d}} = 1.11$
　　$t = 1.13$
　4. Do not reject H_0; $p > .05$.

b. $CI_{95} = (-1.38, 3.88)$

6. a. 1. $H_0: \sigma_1^2 = \sigma_2^2$
　　$H_a: \sigma_1^2 \neq \sigma_2^2$
　2. $F_{cv} = 1.90$
　3. $F = 1.05$
　4. Do not reject H_0.
　c. $CI_{95} = (-11.80, 3.80)$

b. 1. $H_0: \mu_1 - \mu_2 = 0$
　　$H_a: \mu_1 - \mu_2 < 0$
　2. $t_{cv} = -1.678$
　3. $s^2 = 214.53$　$s_{\bar{X}_1-\bar{X}_2} = 3.88$
　　$t = -1.03$
　4. Do not reject H_0; $p > .05$.

7. a. 1. $H_0: \delta = 0$
　　$H_a: \delta \neq 0$
　2. $t_{cv} = \pm 1.833$
　3. $\Sigma d = 14$　$\Sigma d^2 = 38$　$\bar{d} = 1.4$
　　$s_d = 1.43$　$s_{\bar{d}} = 0.45$
　　$t = 3.11$
　4. Reject H_0; $p < .10$.

b. $CI_{90} = (0.58, 2.22)$

8. a. 1. $H_0: \sigma_1^2 = \sigma_2^2$
　　$H_a: \sigma_1^2 \neq \sigma_2^2$
　2. $F_{cv} = 0.418,\ 2.940$
　3. $F = 3.50$
　4. Reject H_0.
　b. 1. $H_0: \sigma_1^2 = \sigma_2^2$
　　$H_a: \sigma_1^2 \neq \sigma_2^2$
　2. $F_{cv} = 0.485,\ 1.860$
　3. $F = 0.50$
　4. Fail to reject H_0.

c. 1. $H_0: \sigma_1^2 = \sigma_2^2$
 $H_a: \sigma_1^2 \neq \sigma_2^2$
 2. $F_{cv} = 0.543, \; 2.160$
 3. $F = 3.58$
 4. Reject H_0.

d. 1. $H_0: \sigma_1^2 = \sigma_2^2$
 $H_a: \sigma_1^2 \neq \sigma_2^2$
 2. $F_{cv} = 0.633, \; 1.510$
 3. $F = 0.64$
 4. Fail to reject H_0.

9. The probability of obtaining an F ratio of 3.58 by chance, if the population variances are equal, is less than .10.
The probability of obtaining an F ratio of 0.64 by chance, if the population variances are equal, is greater than .10.

10. $F_{cv} = 1.70$
 $F_{cv} = 2.13$

11. a. 1. $H_0: \sigma_1^2 = \sigma_2^2$ $\qquad$ b. s^2 pooled $= 211.01$
 $H_a: \sigma_1^2 \neq \sigma_2^2$
 2. $F_{cv} = 1.64$
 3. $F = 1.10$
 4. Fail to reject H_0.

c. 1. $H_0: \mu_1 = \mu_2$ $\qquad$ d. $CI_{95} = (1.73, 14.07)$
 $H_a: \mu_1 \neq \mu_2$
 2. $t_{cv} = \pm 1.665$
 3. $t = 2.58$
 4. Reject H_0.

e. Using $\alpha = .10$, we would conclude that the population means differ. The confidence interval indicates that we are 95 percent confident that the interval from 1.73 to 14.07 spans the true difference between the population means.

12. a. They do not consist of pairs of scores. Each person's selection and performance is independent of those of others.

b. 1. $H_0: \sigma_1^2 = \sigma_2^2$
 $H_a: \sigma_1^2 \neq \sigma_2^2$
 2. $F_{cv} = 0.474, \; 1.75$
 3. $F = 2.50$
 4. Reject H_0.

c. $s_{\bar{X}_1 - \bar{X}_2} = 2.57$
$t = -2.41$
$df = 16.65$
$t_{cv} = \pm 2.120$
Reject H_0.

d. $CI_{95} = (-11.65, -0.75)$

e. We can conclude that the population variances are not equal, the population means are not equal, and that we are 95 percent confident that the interval from -11.65 to -0.75 spans the true difference between population means.

Chapter 12

1. *Step 1: State the Hypotheses.*
For this study, the null hypothesis would be that there is no difference between the levels of satisfaction for citizens from two diverse and geographically separated cities. The level of satisfaction was defined as the proportion of citizens responding "Yes" to the question. For the alternative hypothesis, there is no reason to assume directionality; therefore, the null and alternative hypotheses would be:

$$H_0: P_1 = P_2$$

$$H_a: P_1 \neq P_2$$

Step 2: Set the Criterion for Rejecting H_0.
The general formula will be used to test the null hypothesis. The sampling distribution of the differences in sample proportions, used for testing the null hypothesis, is the normal distribution. With $\alpha = .05$, the critical value of z for this two-tailed test is

$$z_{cv} = \pm 1.96$$

Step 3: Compute the Test Statistic.
The first step is to compute the pooled value for p and q; the second step is to compute the standard error of the differences $(s_{p_1 - p_2})$ using formula 12.3. The pooled value for p is:

$$p = \frac{f_1 + f_2}{n_1 + n_2}$$

$$= \frac{122 + 84}{200 + 150} = \frac{206}{350} = 0.59$$

$$q = 1 - p = 0.41$$

The standard error of the differences is:

$$s_{p_1-p_2} = \sqrt{(p)(q)(1/n_1 + 1/n_2)}$$

$$= \sqrt{(0.59)(0.41)(1/200 + 1/150)} = 0.053$$

Thus, the test statistic is:

$$z = \frac{(p_1 - p_2) - (0)}{s_{p_1-p_2}}$$

$$= \frac{(0.61 - 0.56) - (0)}{0.053} = \frac{0.05 - 0}{0.053} = 0.94$$

Step 4: Decide about H_0.
Since the computed value for the test statistic ($z = 0.94$) does not exceed the critical value ($z_{cv} = \pm 1.96$), the null hypothesis is *not* rejected. The associated probability statement would be: the probability that the observed difference in sample proportions ($p_1 - p_2 = .05$) would have occurred by chance, if the null hypothesis is true ($H_0: P_1 = P_2$), is greater than .05. By failing to reject the null hypotheses, we conclude that there is no difference in the level of satisfaction between citizens from the two cities.

Step 5: Construct the Confidence Interval.
For this study, the CI_{95} is computed as:

$$CI_{95} = (p_1 - p_2) \pm (z_{cv})(s_{P_1-P_2})$$

$$= 0.05 \pm (1.96)(0.053)$$

$$= 0.05 \pm .104$$

$$= (-0.054, 0.154)$$

We conclude with 95 percent confidence that the interval from -0.054 to 0.154 contains the difference between the population proportions. Note that this interval does contain zero; this is consistent with the previous result in that the null hypothesis was not rejected, i.e., there is no difference between the population proportions.

2. a. 1. $H_0: \rho_1 = \rho_2$
  $H_a: \rho_1 > \rho_2$
2. $z_{cv} = +1.645$
3. $z_{r1} = 0.523$ $z_{r2} = 0.224$ $s_{z_{r1}-z_{r2}} = 0.20$
  $z = 1.50$
4. Do not reject H_0; $p > .05$.

b. $CI_{95} = (-.09, .69)$ for z_r
  $(-.09, .60)$ for r

c. For $r_1 = .48$ $t = 3.87$ $t_{cv} = 2.009$
Reject H_0: $\rho_1 = 0$; $p < .05$.
For $r_2 = .22$ $t = 1.51$ $t_{cv} = \pm 2.015$
Do not reject H_0; $\rho_2 = 0$; $p > .05$.

3. a. 1. H_0: $\rho_1 = \rho_2$
  H_a: $\rho_1 \neq \rho_2$
2. $z_{cv} = \pm 1.645$
3. $z = 1.06$
4. Fail to reject H_0.

b. $CI_{90} = (-.076, .354)$ for z_r
  $(-.076, .340)$ for r

c. The probability that an observed difference between two sample correlations as great as this, would have occurred by chance if the population correlations were equal, is greater than .10. We are 90 percent confident that the interval from $-.076$ to $.340$ spans the true difference between the population correlation coefficients.

d. $CI_{95} = (-.118, .396)$ for z_r
  $(-.117, .377)$ for r

4. a. 1. H_0: $\rho_1 = \rho_2$
  H_a: $\rho_1 > \rho_2$
2. $z_{cv} = +1.645$
3. $z_{r1} = 0.662$ $z_{r2} = 0.590$ $s_{z_{r1}-z_{r2}} = 0.17$
  $z = 0.42$
4. Do not reject H_0; $p > .05$.

b. $CI_{95} = (-0.26, 0.40)$ for z_r
  $(-.26, .38)$ for r

5. a. 1. H_0: $P_1 = P_2$
  H_a: $P_1 \neq P_2$
2. $z_{cv} = \pm 1.96$
3. $p = 0.347$ $q = 0.653$ $s_{p_1-p_2} = 0.045$
  $P_1 = 0.40$ $P_2 = 0.30$
  $z = 2.222$
4. Reject H_0; $p < .05$.

b. $CI_{95} = (0.012, 0.188)$
c. 1. $CI_{95} = (.333, .467)$
  2. $CI_{95} = (.241, .359)$

6. a. 1. $H_0: P_1 = P_2$
$H_a: P_1 > P_2$
2. $z_{cv} = +1.645$
3. $p = 0.49 \quad q = 0.51 \quad s_{p_1-p_2} = 0.054$
$p_1 = 0.54 \quad p_2 = 0.46$
$z = 1.48$
4. Do not reject H_0; $p > .05$.

b. $CI_{95} = (-.026, .186)$

7. a. 1. $H_0: \delta_P = 0$
$H_a: \delta_P \neq 0$
2. $z_{cv} = \pm 1.96$
3. $p_1 = 0.30 \quad p_2 = 0.50 \quad s_{p_1-p_2} = 0.09$
$z = -2.22$
4. Reject H_0; $p < .05$.

b. $CI_{95} = (-.38, -.02)$

8. a. 1. $H_0: P_1 = P_2$
$H_a: P_1 \neq P_2$
2. $z_{cv} = \pm 1.96$
3. $p = 0.69 \quad q = 0.31 \quad s_{p_1-p_2} = 0.059 \quad p_1 = 0.75 \quad p_2 = 0.66$
$z = 1.525$
4. Do not reject H_0; $p > .05$.

b. $CI_{95} = (-.026, .206)$

c. There is no difference in the effectiveness of the drugs.

The probability is greater than .05 that the observed difference in the percentage of patients who reported relief would have occurred by chance if the drugs were equally effective.

We are 95 percent confident that the true difference between the population percentages is spanned by the interval from $-.05$ to .23.

9. a. 1. $H_0: \delta_P = 0$
$H_a: \delta_P \neq 0$
2. $z_{cv} = \pm 1.96$
3. $p_1 = 0.538 \quad p_2 = 0.475 \quad s_{p_1-p_2} = 0.054$
$z = 1.167$
4. Do not reject H_0; $p > .05$.

b. $CI_{95} = (-0.043, 0.169)$

Chapter 13

1. a. For $n = 144$, $\sigma_{\overline{X}} = 400$
   One-tailed test ($\alpha = .05$), $z = 0.15$, power $= 0.4404$
   Two-tailed test ($\alpha = .05$), $z = 0.46$ and -3.46, power $= 0.3228$

b. For $n = 144$, $\sigma_{\overline{X}} = 400$
   For $\alpha = .01$, $z = 0.83$, power $= .2033$
   For $\alpha = .10$, $z = -0.22$, power $= .5871$

c. For $n = 400$, $\sigma_{\overline{X}} = 240$
   For $\alpha = .05$, $z = -.86$, power $= .8051$

d. For $n = 144$, $\sigma_{\overline{X}} = 400$
   For $\alpha = .05$, $z = -0.86$, power $= .8051$

2. a. For H_a: $\mu = \mu_0 + 1\sigma_{\overline{X}}$, $z = 0.65$, power $= .2578$
   b. For H_a: $\mu = \mu_0 + 2\sigma_{\overline{X}}$, $z = -0.36$, power $= .6406$
   c. For H_a: $\mu = \mu_0 + 3\sigma_{\overline{X}}$, $z = -1.36$, power $= .9131$

3. The null hypothesis for this exercise is H_0: $\mu = 100$; the alternative hypothesis is H_a: $\mu = 105$. For a sample size greater than 120, the normal distribution can be used as the sampling distribution. The means for both sampling distributions are defined in the null and alternative hypotheses; the standard error of the mean for both distributions is,

$$s_{\overline{X}} = s/\sqrt{n} = 24/\sqrt{128} = 2.12$$

The two sampling distributions are illustrated below.

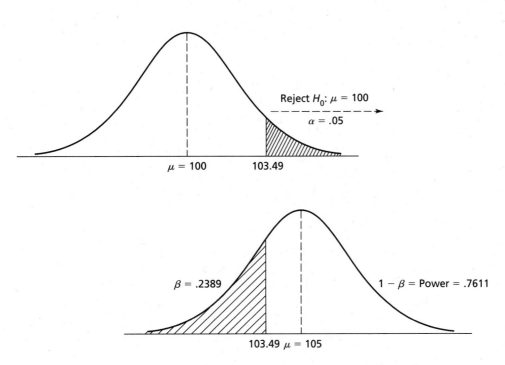

To determine the power of this test of H_0 against H_a, we will assume a one-tailed test and $\alpha = .05$. The first step is to determine the region of rejection in the sampling distribution of the H_0. Note that in the illustration, this critical value is 1.645 standard errors above the mean, i.e., $\mu + 1.645s_X = 100 + 1.645(2.12) = 103.49$. Now we project this point onto the sampling distribution of H_a, and compute the z score associated with the score in this distribution.

$$z = \frac{103.49 - 105}{2.12} = \frac{-1.51}{2.12} = -0.71$$

From Table C.1 in Appendix C we find that .2389 of the area is below the z score of -0.71 (see illustration), and .7611 of the area is about this z score. Thus, $\beta = .2389$ and the power $= .7611$.

Now consider the same exercise except for a sample size of 256. Again, assuming a one-tailed test and $\alpha = .05$, we will follow the same steps to determine the power. However, the illustration is not included. The standard error of the mean for the sampling distributions is now based upon $n = 256$,

$$s_{\overline{X}} = s/\sqrt{n} = 24/\sqrt{256} = 1.50$$

Thus, this critical value is 1.645 standard errors above the mean in the sampling distribution of H_0, i.e., $\mu + 1.645s_{\overline{X}} = 100 + 1.645(1.5) = 102.47$. Now consider how this point would be projected onto the sampling distribution of H_a, and compute the z score associated with the score in that distribution.

$$z = \frac{102.47 - 105}{1.50} = \frac{-2.31}{1.50} = -1.68$$

From Table C.1 in Appendix C, we find that .0455 of the area is below the z score of -1.69, and .9545 of the area is about this z score. Thus, $\beta = .0455$ and the power $= .9545$. Thus, by increasing the sample size from 128 to 256 (and holding all other conditions the same), we increase the power of the test from .7611 to .9545.

4. a. One-tailed test ($\alpha = .05$), $z = -0.59$, power $= .7224$
Two-tailed test ($\alpha = .05$), $z = -0.28$ and -4.19, power $= .6103$

b. For $\alpha = .01$, $z = 0.09$, power $= 0.4641$
For $\alpha = .10$, $z = -0.95$, power $= .8289$

c. For $n_1 = n_2 = 250$, $s_{\bar{X}_1 - \bar{X}_2} = 1.414$
For $\alpha = .05$, $z = -1.89$, power $= 0.9706$

d. For $\alpha = .05$, $z = 0.30$, power $= .3821$

5. a. For $n = 140$, $\sigma_{\bar{X}} = 1.06$
One-tailed test ($\alpha = .05$), $z = 0.25$, power $= .5987$
Two-tailed test ($\alpha = .05$), $z = 3.85$ and -0.08, power $= .4681$

b. For $\alpha = .01$, $z = -0.44$, power $= 0.3300$
For $\alpha = .10$, $z = 0.60$, power $= 0.7257$

c. For $n = 220$, $\sigma_{\bar{X}} = 0.84$
For $\alpha = .05$, $z = 0.74$, power $= .7704$

d. For $\alpha = .05$, $z = -0.23$, power $= .4090$

6. a. One-tailed test ($\alpha = .05$), $z = -0.19$, power $= .5753$
Two-tailed test ($\alpha = .05$), $z = 0.12$ and -3.80, power $= .4522$

b. For $\alpha = .01$, $z = 0.49$, power $= 0.312$
For $\alpha = .10$, $z = -0.56$, power $= .7123$

c. For $n_1 = n_2 = 120$, $\sigma_{\bar{X}_1 - \bar{X}_2} = 3.23$
$\alpha = .05$, $z = .10$, power $= .4602$

d. For $\alpha = .05$, $z = .79$, power $= .2148$

7. a. $n = 395.52$ or 396, $n = 501.76$ or 502

b. $n = 98.88$ or 99

c. $n = 1009.28$ or 1010, $n = 1140.48$ or 1141

d. $n = 252.32$ or 253

8. a. $n = 19.78$ or 20, $n = 25.09$ or 26

b. $n = 54.26$ or 55

c. $n = 50.46$ or 51, $n = 57.02$ or 58

d. $n = 123.32$ or 124

9. a. $n = 154.5$ or 155, $n = 196.00$ or 196

   b. $n = 24.72$ or 25

   c. $n = 394.25$ or 395, $n = 445.50$ or 446

   d. $n = 63.08$ or 64

10. a. $n = 206.00$ or 206, $n = 261.33$ or 262

   b. $n = 98.00$ or 98

   c. $n = 525.67$ or 526, $n = 594.00$ or 594

   d. $n = 222.75$ or 223

Chapter 14

1.

n_k	10	9	8	$N = 27$
T_k	109	131	137	$\Sigma T_k = 377$
$\overline{X}_k$	10.9	14.6	17.1	$T^2/N = 5264.04$
ΣX_{ik}^2	1339	2113	2529	$\Sigma\Sigma X_{ik}^2 = 5981$
T_k^2/n_k	1188.10	1906.78	2346.13	$\Sigma(T_k^2/n_k) = 5441.01$

a. $H_0: \mu_1 = \mu_2 = \mu_3$
   $H_a: \mu_i \neq \mu_k$ for some i, k

b. For df $= 2, 24$ and $\alpha = .05$, $F_{cv} = 3.40$

c.

Source	SS	df	MS	F	F_{cv}
Between	176.97	2	88.49	3.93	3.40
Within	539.99	24	22.50		
Total	716.96	26			

d. $\omega^2 = 0.18$

e. Reject H_0; $p < .05$.

2. *Step 1: State the Hypotheses.*
For this study, the null hypothesis is that there is no difference in solving the concept attainment problem for the respondents assigned to the various solution strategies.

$$H_0: \mu_A = \mu_B = \mu_C = \mu_D = \mu_E$$
$$H_a: \mu_i \neq \mu_k \quad \text{for some } i, k$$

Step 2: Set the Criterion for Rejecting H_0.

The test statistic for one-way ANOVA is the F ratio defined as MS_B/MS_W; the sampling distribution of the F ratio is the F distribution with $K - 1$ and $N - K$ degrees of freedom. In this study, there are $K - 1 = 5 - 1 = 4$ df associated with MS_B and $N - K = 65 - 5 = 60$ df associated with MS_W. Thus, the critical value for F with 4 and 60 degrees of freedom, from Table C.5 and for $\alpha = .05$, is $F_{cv} = 2.53$. The null hypothesis will be rejected if the calculated value of F exceeds the critical value.

Step 3: Complete the Summary ANOVA and Compute the Test Statistic.

The partial Summary ANOVA was given. It is only necessary to determine SS_W, enter the degrees of freedom values, compute the MS_B and MS_W, and then compute the F ratio.

$$SS_W = SS_T - SS_B = 347.45 - 105.41 = 242.04$$

$$MS_B = SS_B/df_B = 105.41/4 = 26.35$$

$$MS_W = SS_W/df_W = 242.04/60 = 4.03$$

$$F = \frac{MS_B}{MS_W} = \frac{26.35}{4.03} = 6.54$$

Summary ANOVA

Source	SS	df	MS	F	F_{cv}
Between	105.42	4	26.35	6.54	2.53
Within	242.04	60	4.03		
Total	347.45	64			

For this study, the calculated value of $F(6.54)$ exceeds the critical value ($F_{cv} = 2.53$); therefore, the null hypothesis, H_0: $\mu_A = \mu_B = \mu_C = \mu_D = \mu_E$, would be rejected. The probability that the observed differences would have occurred by chance if this null hypothesis were true (i.e., the population means are equal) is less than .05 (the level of significance set a priori).

Step 4: Compute ω^2.

ω^2 is defined as a measure of the strength of the association between the independent and dependent variables.

$$\omega^2 = \frac{SS_B - (K - 1)MS_W}{SS_B + MS_W} = \frac{105.41 - 4(4.03)}{347.45 + 4.03} = 0.25$$

Step 5: Interpret the Results.

The conclusion is that not all the population means are equal; the five solution strategies are not equally effective in solving the concept attainment problem.

At this point, we do no know which means differ significantly. This determination will be made using the techniques developed in Chapter 15. However, it appears that Strategy B is the most effective and Strategy D is the least effective. The ω^2 of 0.25 indicates that 25 percent of the variation in time to solve the concept attainment problem can be accounted for by the difference in the strategies.

3.

n_k	7	7	4	3	5
T_k	324	317	295	88	262
$\overline{X}_k$	46.29	45.29	73.75	29.33	52.40
ΣX_{ik}^2	15,764	15,127	21,997	2726	14,060
T_k^2/n_k	14,996.57	14,355.57	21,756.25	2581.33	13,728.80

$$N = 26$$
$$\Sigma T_k = 1286$$
$$T^2/N = 63,607.54$$
$$\Sigma\Sigma X_{ik}^2 = 69,674$$
$$\Sigma(T_k^2/n_k) = 67,418.52$$

a. $H_0: \mu_1 = \mu_2 = \mu_3 = \mu_4 = \mu_5$
  $H_a: \mu_i \neq \mu_k$ for some i, k
b. For df = 4, 21 and $\alpha = .05$, $F_{cv} = 2.84$

c.

Source	SS	df	MS	F	F_{cv}
Between	3810.98	4	952.75	8.87	2.84
Within	2255.48	21	107.40		
Total	6066.46	25			

d. $\omega^2 = 0.55$
e. Reject H_0; $p < .05$.

4.

n_k	11	11	$N = 22$
T_k	331	520	$\Sigma T_k = 851$
$\overline{X}_k$	30.09	47.27	$T^2/N = 32,918.23$
ΣX_{ik}^2	10,423	25,314	$\Sigma\Sigma X_{ik}^2 = 35,737$
T_k^2/n_k	9960.09	24,581.82	$\Sigma(T_k^2/n_k) = 34,541.91$

a. $H_0: \mu_1 = \mu_2$
  $H_a: \mu_1 \neq \mu_2$
b. For df = 1, 20 and $\alpha = .05$, $F_{cv} = 4.35$

c.

Source	SS	df	MS	F	F_{cv}
Between	1623.68	1	1623.68	27.17	4.35
Within	1195.09	20	59.75		
Total	2818.77	21			

d. $\omega^2 = 0.54$

e. Reject H_0; $p < .05$.

5. a. $H_0: \mu_1 = \mu_2$
   $H_a: \mu_1 \neq \mu_2$

b. For 20 df, $t_{cv} = \pm 2.086$

c. $s^2 = 59.76 \quad s_{\bar{X}_1 - \bar{X}_2} = 3.30 \quad t = -5.21$

d. Reject H_0: $p < .05$.

e. $t^2 = F$
   $(-5.21)^2 = 27.14 \approx 27.17$ (rounding error)
   $t_{cv}^2 = F_{cv}$
   $(2.086)^2 = 4.35$

6. a. $H_0: \mu_1 = \mu_2 = \mu_3 = \mu_4 = \mu_5$
   $H_a: \mu_i \neq \mu_k$ for some i, k

b. For df $= 4$, 35, $\alpha = .01$, $F_{cv} = 3.91$

c.

Source	SS	df	MS	F	F_{cv}
Between	95.80	4	23.95	6.54	3.91
Within	128.10	35	3.66		
Total	223.90	39			

d. $\omega^2 = 0.36$

e. Reject H_0; $p < .05$.

7. *Chapter 11:* $\quad s^2 = \dfrac{SS_1 + SS_2}{df_1 + df_2} = \dfrac{\Sigma(SS_k)}{\Sigma(df_k)}$

Chapter 14: $\quad MS_W = \dfrac{\Sigma(SS_k)}{\Sigma(df_k)}$

Both are pooled estimates of population variance.

8.

n_k	6	7	5	4	$N = 22$
T_k	420	553	350	332	$\Sigma T_k = 1655$
$\overline{X}_k$	70	79	70	83	$T^2/N = 124,501.14$
ΣX_{ik}^2	29,608	44,075	24,860	27,730	$\Sigma\Sigma X_{ik}^2 = 126,273$
T_k^2/n_k	29,400	43,687	24,500	27,556	$\Sigma(T_k^2/n_k) = 125,143$

a. $H_0: \mu_1 = \mu_2 = \mu_3 = \mu_4$
$H_a: \mu_i \neq \mu_k$ for some i, k

b. For df $= 3, 18$ and $\alpha = .05$, $F_{cv} = 3.16$

c.

Source	SS	df	MS	F	F_{cv}
Between	641.86	3	213.95	3.41	3.16
Within	1130.00	18	62.78		
Total	1771.86	21			

d. $\omega^2 = 0.25$

e. Reject H_0; $p < .05$.

9. a. $k(k-1)/2 = (4)(3)/2 = 6$ t tests

b. $\alpha_E = 1 - (1 - .05)^6 = 0.26$

10.

n_k	7	7	7	$N = 21$
T_k	122	120	198	$\Sigma T_k = 440$
$\overline{X}_k$	17.43	17.14	28.29	$T^2/N = 9219.05$
ΣX_{ik}^2	2190	2146	5688	$\Sigma\Sigma X_{ik}^2 = 10,042$
T_k^2/n_k	2126.29	2057.14	5600.57	$\Sigma(T_k^2/n_k) = 9784.00$

a. $H_0: \mu_1 = \mu_2 = \mu_3$
$H_a: \mu_i \neq \mu_k$ for some i, k

b. For df $= 2, 18$ and $\alpha = .05$, $F_{cv} = 3.55$

c.

Source	SS	df	MS	F	F_{cv}
Between	564.95	2	282.48	19.71	3.55
Within	258.00	18	14.33		
Total	822.95	20			

d. $\omega^2 = 0.64$

e. Reject H_0; $p < .05$.

11. $SS_B = \Sigma n_k(\bar{X}_k - \bar{X})^2 = 1026.18$
$SS_W = \Sigma(n_k - 1)s_k^2 = 262.37$

a. $H_0: \mu_1 = \mu_2 = \mu_3$
$H_a: \mu_i \neq \mu_k$ for some i, k

b. For df = 2, 24 and $\alpha = .01$, $F_{cv} = 5.61$

c.

Source	SS	df	MS	F	F_{cv}
Between	1026.18	2	513.09	46.94	5.61
Within	262.37	24	10.93		
Total	1288.55	26			

d. $\omega^2 = 0.77$

e. Reject H_0; $p < .05$.

12. a. $n = 23$ **b.** $n = 34$ **c.** $n = 36$ **d.** $n = 53$

13. a. $n = 35$ **b.** $n = 17$ **c.** $n = 76$ **d.** $n = 37$

14.

n	6	6	6	6	6
T_k	38	39	47	57	63
$\bar{X}_k$	6.33	6.50	7.83	9.50	10.50
ΣX_{ik}^2	244	267	393	561	703
T_k^2/n	240.67	253.50	368.17	541.50	661.50

$n = 30$	T_i	T_i^2/k
$\Sigma T_k = 244$	45	405.00
$T^2/N = 1984.53$	34	231.20
$\Sigma\Sigma X_{ik}^2 = 2168$	51	520.20
$\Sigma(T_k^2/n) = 2065.34$	27	145.80
	44	387.20
	43	369.80
	$\Sigma(T_i^2/k) =$	2059.20

a. $H_0: \mu_1 = \mu_2 = \mu_3 = \mu_4 = \mu_5$
$H_a: \mu_i \neq \mu_k$ for some i, k

b. For df = 4, 20 and $\alpha = .01$, $F_{cv} = 4.43$

c.

Source	SS	df	MS	F	F_{cv}
Occasions	80.81	4	20.20	14.43	4.43
Individuals	74.67	5	14.93		
Residual	27.99	20	1.40		
Total	183.47	29			

15.

n	8	8	8	$N = 24$
T_k	23.4	30.0	37.3	$\Sigma T_k = 90.7$
$\overline{X}_k$	2.93	3.75	4.66	$T^2/N = 342.77$
ΣX_{ik}^2	69.90	115.36	176.17	$\Sigma\Sigma X_{ik}^2 = 361.43$
T_k^2/n	68.45	112.50	173.91	$\Sigma(T_k^2/n) = 354.86$

T_i	T_i^2/k
10.5	36.75
11.7	45.63
12.4	51.25
14.4	69.12
10.0	33.33
9.4	29.45
11.3	42.56
11.0	40.33

$\Sigma(T_i^2/k) = 348.42$

a. $H_0\colon \mu_1 = \mu_2 = \mu_3$
$H_a\colon \mu_i \neq \mu_k$ for some $i,\ k$

b. For df $= 2, 14$ and $\alpha = .05,\ F_{cv} = 3.74$

c.

Source	SS	df	MS	F	F_{cv}
Occasions	12.09	2	6.05	86.43	3.74
Individuals	5.65	7	0.81		
Residual	0.92	14	0.07		
Total	18.66	23			

16.

n	10	10	10	10	$N = 40$
T_k	48	41	28.5	18	$T = 135.50$
$\overline{X}_k$	4.80	4.10	2.85	1.80	$T^2/N = 459.01$
ΣX_{ik}^2	249	178	94.25	36.50	$\Sigma\Sigma X_{ik}^2 = 557.75$
T_k^2/n	230.40	168.10	81.23	32.40	$\Sigma(T_k^2/n) = 512.13$

T_i	T_i^2/k
14.5	52.56
10.0	25.00
10.0	25.00
17.0	72.25
16.0	64.00
9.0	20.25
12.5	39.06
13.0	42.25
13.5	45.56
20.0	100.00

$\Sigma(T_i^2/k) = 485.93$

a. H_0: $\mu_1 = \mu_2 = \mu_3 = \mu_4$
H_a: $\mu_i \neq \mu_k$ for some i, k

b. For df $= 3, 27$ and $\alpha = .05$, $F_{cv} = 2.97$

c.

Source	SS	df	MS	F	F_{cv}
Occasions	53.12	3	17.71	25.67	2.97
Individuals	26.92	9	2.99		
Residual	18.70	27	0.69		
Total	98.74	39			

Chapter 15

1.

n_k	8	8	8	8
T_k	583	608	649	569
$\overline{X}_k$	72.88	76.00	81.13	71.13
ΣX_{ik}^2	42,531	46,304	52,751	40,587
T_k^2/n_k	42,486.13	46,208.00	52,650.13	40,470.13

$$N = 32$$
$$\Sigma T_k = 2409$$
$$T^2/N = 181,352.53$$
$$\Sigma\Sigma X_{ik}^2 = 182,173$$
$$\Sigma(T_k^2/n_k) = 181,814.39$$

a. $H_0: \mu_1 = \mu_2 = \mu_3 = \mu_4$
$H_a: \mu_i \neq \mu_k$ for some $i, \ k$

b. For df $= 3, 28$ and $\alpha = .05, \ F_{cv} = 2.95$

c.

Source	SS	df	MS	F	F_{cv}
Between	461.86	3	153.95	12.02	2.95
Within	358.61	28	12.81		
Total	820.47	31			

d. $\omega^2 = 0.51$

e. *Tukey method:*

Group 4		Q		
Group 1	1.38		
Group 2	3.83	2.46	
Group 3	7.87*	6.50*	4.04*

$^*p < .05 \ (Q_{cv} = 3.86$ for $r = 4, \ \mathrm{df} = 28)$

f. *Newman-Keuls procedure:*

	Q_{cv}		
2.90		
3.50	2.90	
3.86	3.50	2.90

Group 4		Q		
Group 1	1.38		
Group 2	3.83*	2.46	
Group 3	7.87*	6.50*	4.04*

$^*p < .05$

2. For the Tukey method, the null hypothesis tested for each pairwise comparison is

$$H_0: \mu_i = \mu_k \quad \text{for some } i, k$$

That is, each pair of population means is equal. The test statistic for the Tukey method is defined as Q. When the groups are equal in size, Q is defined as follows:

$$Q = \frac{\overline{X}_i - \overline{X}_k}{\sqrt{MS_W/n}}$$

For this example, $\sqrt{MS_W/n} = \sqrt{4.03/13} = 0.56$.

The critical value of Q for this example is determined from Table C.8 in Appendix C; for $r = 5$ and $df_W = 60$, $Q_{cv} = 3.98$. If the observed Q exceeds the critical value, the hypothesis for the pairwise comparison is rejected.

The first step (although not necessary) is to rank the group means from low to high. Then the differences between the pairs of means are determined and the Q values are computed. For the comparison of Group B with Group A,

$$Q = \frac{2.40 - 2.23}{0.56} = \frac{0.17}{0.56} = 0.30$$

The remainder of the Q values would be computed in the same manner.

Group	$\overline{X}_i$	$(\overline{X}_i - \overline{X}_k)$				Q			
Group B	2.23								
Group A	2.40	0.17				0.30			
Group C	3.79	1.56	1.39			2.79	2.48		
Group E	4.61	2.38	2.21	0.82		4.25*	3.95	1.46	
Group D	5.91	3.68	3.51	2.12	1.30	6.57*	6.27*	3.79	2.32

*$p < .05$ $Q_{cv} = 3.68$ for df $= 60$

For this example, note that the calculated values of Q that exceed the critical value are marked with an *. The Q value of 4.25 indicates a statistically significant difference between the means of Group E and Group B; the Q value of 6.57 indicates a statistically significant difference between Group D and Group B; and the Q value of 6.27 indicates a statistically significant difference between the means of Group A and Group D. In terms of strategy effectiveness, the shorter time to completion, the more effective the strategy. Therefore, we conclude that

Strategy B is more effective than both Strategies D and E;
Strategy A is more effective than Strategy D.

3. a. *Tukey method:*

	Q	
B		
A	0.20	
C	7.80*	7.59*

*$p < .05$ ($Q_{cv} = 3.61$ for $r = 3$ and df $= 18$)

b. *Newman-Keuls procedure:*

$$Q_{cv}$$

2.97	
3.61	2.97

c. *B*

A	0.20	
C	7.80*	7.59*

$$*p < .05$$

4. a. *Tukey/Kramer method:*

Denominator	A vs. B	1.54
	A vs. C	1.59
	B vs. C	1.64

b. *A*

$$Q$$

B	2.38	
C	3.87*	1.58

$*p < .05$ ($Q_{cv} = 3.53$ for $r = 3$ and df $= 24$)

5. a. *Tukey/Kramer method:*

Denominator	4 vs. 2	5.02
	4 vs. 1	5.02
	4 vs. 5	5.33
	4 vs. 3	5.58
	2 vs. 1	3.95
	2 vs. 5	4.27
	2 vs. 3	4.58
	1 vs. 5	4.27
	1 vs. 3	4.58
	5 vs. 3	4.92

b. *4*

$$Q$$

2	3.18			
1	3.38	.25		
5	4.33*	1.67	1.43	
3	7.96*	6.21*	6.00*	4.34*

$*p < .05$ ($Q_{cv} = 4.22$ for $r = 5$ and df $= 21$)

6. a. $H_0: (-1)\mu_1 + (0)\mu_2 + (1)\mu_3 + (0)\mu_4 = 0$
$F = 27.16$
For $F_{cv} = 8.22$, reject H_0; $p < .05$.

b. $H_0: (1)\mu_1 + (0)\mu_2 + (1)\mu_3 + (-2)\mu_4 = 0$
$F = 0.48$
For $F_{cv} = 8.22$, do not reject H_0; $p > .05$.

c. $H_0: (1)\mu_1 + (-3)\mu_2 + (1)\mu_3 + (1)\mu_4 = 0$
$F = 84.25$
For $F_{cv} = 8.22$, reject H_0; $p < .05$.

7. a. C: 1 −1 0 0
$F = 97.79^*$

b. C: 0 0 1 −1
$F = 10.29^*$

c. C: 1 1 −1 −1
$F = 1.79$

8. $\alpha_e = .01/6 = .00167$

9. $F_{linear} = 285.20^*$
$F_{quadratic} = 81.88^*$
$F_{cubic} = 1.00$
F_{cv} for df $= 1, 44$ at $\alpha = .01$ is 7.24.

10. *Tukey/Kramer method:*

Denominator		1 vs. 3	3.41
		1 vs. 2	3.12
		1 vs. 4	3.63
		3 vs. 2	3.27
		3 vs. 4	3.76
		2 vs. 4	3.50

Q

0.00		
2.88	2.75	
3.58	3.46	1.14

$^*p < .05$ ($Q_{cv} = 4.00$ for $r = 4$ and df $= 18$

11.

Linear	−3	−1	1	3
Quadratic	1	−1	−1	1
Cubic	−1	3	−3	1

$F_{linear} = 0.00$
$F_{quadratic} = 26.85^*$
$F_{cubic} = 9.17^*$
For df $= 1, 28$ and $\alpha = .01$, $F_{cv} = 7.64$

Chapter 16

For all problems, the general form of hypotheses is

row effect $\quad H_0: \mu_{1.} = \mu_{2.} = \cdots = \mu_{j.}$

column effect $\quad H_0: \mu_{.1} = \mu_{.2} = \cdots = \mu_{.k}$

interaction $\quad H_0:$ all $(\mu_{jk} - \mu_{j.} - \mu_{.k} + \mu)$

1. b. Rows $\quad$ df $= 1, 30 \quad$ and $\quad \alpha = .05 \quad F_{cv} = 4.17$

$\quad$ Columns $\quad$ df $= 2, 30 \quad$ and $\quad \alpha = .05 \quad F_{cv} = 3.32$

$\quad$ Interaction $\quad$ df $= 2, 30 \quad$ and $\quad \alpha = .05 \quad F_{cv} = 3.32$

c. $\quad \Sigma T_{j.}^2 = 801,037$

$$\Sigma T_{.k}^2 = 538,929$$

$$\Sigma T_{jk}^2 = 270,757$$

$$T = 1,265$$

$$\Sigma\Sigma\Sigma X_{ijk}^2 = 45,437$$

Source	SS	df	MS	F	F_{cv}
Rows	51.37	1	51.37	4.96*	4.17
Columns	460.06	2	230.03	22.20*	3.32
Interaction	164.05	2	82.03	7.92*	3.32
Within	310.83	30	10.36		
Total	986.31	35			

d. $\omega_{rows}^2 = 0.04$

$\quad \omega_{columns}^2 = 0.44$

$\quad \omega_{interaction}^2 = 0.14$

2. *Step 1: State the Hypotheses.*
For two-way ANOVA, there are three null hypotheses to be tested; one for the row effect (Gender), one for the column effect (Drug Dosage), and one for the interaction (Gender × Drug Dosage). For the row effect,

$$H_0: \mu_{M.} = \mu_{F.}$$

$$H_a: \mu_{M.} \neq \mu_{F.}$$

For the column effect,

$$H_0: \mu_{.1} = \mu_{.2} = \mu_{.3}$$

$$H_a: \mu_{.i} \neq \mu_{.k} \quad \text{for some } i, k$$

For the interaction effect,

$$H_0: \text{all}(\mu_{jk} - \mu_{j.} - \mu_{.k} + \mu) = 0$$

$$H_a: \text{all}(\mu_{jk} - \mu_{j.} - \mu_{.k} + \mu) \neq 0$$

Step 2: Set the Criterion for Rejecting H_0.
The test statistic for all three hypotheses is the F ratio; the sampling distribution for these test statistics in the F distribution with appropriate degrees of freedom. The degrees of freedom associated with the main effects, the interaction, and the within-cell mean squares are:

Gender	$(J - 1) = (2 - 1) = 1$
Drug Dosage	$(K - 1) = (3 - 1) = 2$
Interaction (Gender × Drug Dosage)	$(J - 1)(K - 1) = (2 - 1)(3 - 1)$
	$= 2$
Within Cell	$JK(n - 1) = (2)(3)(8 - 1) = 42$

With $\alpha = .05$, the critical values of the F ratios for test of the main effects and the interaction are (see Table C.5 in Appendix C):

Gender	df $= 1, 42$	$F_{cv} = 4.07$
Drug Dosage	df $= 2, 42$	$F_{cv} = 3.23$
Interaction (Gender × Drug Dosage)	df $= 2, 42$	$F_{cv} = 3.23$

Step 3: Compute the ANOVA and the Test Statistics.
The SS for Interaction (SS_{GD}) is the only Sum of Squares that is unknown. Since the SS_T is the sum of the other sum of squares terms, the SS_{GD} is found by subtraction,

$$\begin{aligned}
SS_{GD} &= SS_T - (SS_G + SS_D + SS_W) \\
&= 76,226.8 - (14,832.5 + 17,120.6 + 41,685.3) \\
&= 76,226.8 - 73,638.4 \\
&= 2,588.4
\end{aligned}$$

The Summary ANOVA can now be completed as follows:

	Summary ANOVA				
Source	SS	df	MS	F	F_{cv}
Gender	14,832.5	1	14,832.5	14.94*	4.07
Drug dosage	17,120.6	2	8,560.3	8.62*	3.23
Gender × Dosage	2,588.4	2	1,294.2	1.30	3.23
Within cell	41,685.3	42	992.5		
Total	76,226.8	47			

Step 4: Compute ω^2.

ω^2 is defined as a measure of the strength of the association between the independent and dependent variables.

$$\omega^2 = \frac{SS_{effect} - (df_{effect})MS_W}{SS_T + MS_W}$$

For the row effect (Gender)

$$\omega_{G^2} = \frac{14,832.5 - 1(992.5)}{76,226.8 + 992.5} = 0.18$$

For the column effect (Drug Dosage)

$$\omega_{D^2} = \frac{17,120.6 - 2(992.5)}{76,226.8 + 992.5} = 0.25$$

Step 5: Interpret the Results.

Note that the F ratio for the row effect (Gender: $F = 14.94$) and the column effect (Drug Dosage: $F = 8.62$) exceeded the respective critical values. Thus, the hypotheses for the row and column effects are rejected. The associated probability statement for each would be: the probability that the observed differences among the means (row or column, respectively) would have occurred by chance, if the null hypothesis is true, is less than .05. The conclusion is that not all the means in the respective populations are equal. The ω^2 for Gender ($\omega_{G^2} = 0.18$) implies that 18 percent of the variation in the performance can be attributed to Gender; the ω^2 for Drug Dosage ($\omega_{D^2} = 0.25$) implies that 25 percent of the variation in the performance can be attributed to Drug Dosage.

For Gender, since there are only two levels of this independent variable, we conclude that the performance of male rats across all levels of drug dosage was greater than the performance for female rats. For Drug Dosage, since there are more than two levels of the independent variable, we do not know which groups differ significantly and we must conduct a post hoc multiple procedure (Tukey). The data for this procedure are found below. The test statistic for this post hoc procedure is:

$$Q = \frac{\overline{X}_i - \overline{X}_k}{\sqrt{MS_W/n'}}$$

where $\sqrt{MS_W/n'} = \sqrt{992.5/16} = 7.88$

The critical value of Q for this example is determined from Table C.8; for $r = 3$ and $df_W = 42$, $Q_{cv} = 3.44$. If the observed Q exceeds the critical value, the hypothesis for the pairwise comparison is rejected.

The first step (although not necessary) is to rank the group means from low to high. Then the differences between the pairs of means are determined and the Q values are computed. For the comparison of Dosage 2 with Dosage 1,

$$Q = \frac{70.8 - 45.4}{7.88} = \frac{25.4}{7.88} = 3.22$$

Drug Dosage	$\bar{X}_i$	$\bar{X}_i - \bar{X}_k$		Q	
Dosage 1	45.4				
Dosage 2	70.8	25.4		3.22	
Dosage 3	90.9	45.5	20.1	5.77*	2.55

$*p < .05 \quad Q_{cv} = 3.44$ for df $= 42$

As can be seen from the table, the sample mean for Drug Dosage 3 differs significantly from the sample mean for Drug Dosage 1. We would conclude that the rats under Drug Dosage 3 had a higher mean on the performance than did the rats under Drug Dosage 1.

Since the F ratio for the interaction effect (Gender $\times$ Drug Dosage) was not statistically significant, the conclusion is that there was no difference in the performance of the male and female rats across the levels of Drug Dosage. Thus, there would be no need to plot the interaction means nor to conduct a test of simple effects.

3. b. Rows   df $= 1, 24$ and $\alpha = .05$ $F_{cv} = 4.26$
       Columns   df $= 3, 24$ and $\alpha = .05$ $F_{cv} = 3.01$
       Interaction   df $= 3, 24$ and $\alpha = .05$ $F_{cv} = 3.01$

c.
$$\Sigma T_{j\cdot}^2 = 660,105$$
$$\Sigma T_{\cdot k}^2 = 333,871$$
$$\Sigma\Sigma T_{jk}^2 = 167,301$$
$$T = 1,149$$
$$\Sigma\Sigma\Sigma X_{ijk}^2 = 42,065$$

Source	SS	df	MS	F	F_{cv}
Rows	0.28	1	0.28	0.03	4.26
Columns	477.60	3	159.20	15.94*	3.01
Interaction	91.09	3	30.36	3.04*	3.01
Within	239.75	24	9.99		
Total	808.72	31			

d. ω_{rows}^2 not computed
$$\omega_{columns}^2 = 0.55$$
$$\omega_{interaction}^2 = 0.08$$

4. b.

Rows	For df $= 1, 42$	and	$\alpha = .05$	$F_{cv} = 4.07$
Columns	For df $= 2, 42$	and	$\alpha = .05$	$F_{cv} = 3.22$
Interaction	For df $= 2, 42$	and	$\alpha = .05$	$F_{cv} = 3.22$

c.

Source	SS	df	MS	F	F_{cv}
Rows	97.85	1	97.85	4.45*	4.07
Columns	8.47	2	4.24	0.19	3.22
Interaction	412.84	2	206.42	9.38*	3.22
Within	923.80	42	22.00		
Total	1442.96	47			

d. $\omega^2_{rows} = 0.05$

$\omega^2_{columns}$ not computed

$\omega^2_{interaction} = 0.25$

5. b.

Rows	df $= 1, 16$	and	$\alpha = .05$	$F_{cv} = 4.49$
Columns	df $= 1, 16$	and	$\alpha = .05$	$F_{cv} = 4.49$
Interaction	df $= 1, 16$	and	$\alpha = .05$	$F_{cv} = 4.49$

c.

$$\Sigma T_{j.}^2 = 64,213$$

$$\Sigma T_{.k}^2 = 63,233$$

$$\Sigma T_{jk}^2 = 32,307$$

$$T = 355$$

$$\Sigma\Sigma\Sigma X_{ijk}^2 = 6,551$$

Source	SS	df	MS	F	F_{cv}
Rows	120.05	1	120.05	21.44*	4.49
Columns	22.05	1	22.05	3.94	4.49
Interaction	18.05	1	18.05	3.22	4.49
Within	89.60	16	5.60		
Total	249.75	19			

d. $\omega^2_{rows} = 0.45$

6. b.

Rows	df $= 2, 27$	and	$\alpha = .05$	$F_{cv} = 3.35$
Columns	df $= 2, 27$	and	$\alpha = .05$	$F_{cv} = 3.35$
Interaction	df $= 4, 27$	and	$\alpha = .05$	$F_{cv} = 2.73$

c. $\Sigma T_{j\cdot}^2 = 302,904$

$\Sigma T_{\cdot k}^2 = 282,186$

$\Sigma\Sigma T_{jk}^2 = 101,346$

$T = \quad 920$

$\Sigma\Sigma\Sigma X_{ijk}^2 = \quad 25,580$

Source	SS	df	MS	F	F_{cv}
Rows	1730.89	2	864.45	95.95*	3.35
Columns	4.39	2	2.20	0.24	3.35
Interaction	90.11	4	22.53	2.50	2.73
Within	243.50	27	9.02		
Total	2068.89	35			

d. $\omega_{\text{rows}}^2 = 0.82$

7. b.

Rows	$df = 1, 24$	and	$\alpha = .05$	$F_{cv} = 4.26$
Columns	$df = 2, 24$	and	$\alpha = .05$	$F_{cv} = 3.40$
Interaction	$df = 2, 24$	and	$\alpha = .05$	$F_{cv} = 3.40$

c. $\Sigma T_{j\cdot}^2 = 116,404$

$\Sigma T_{\cdot k}^2 = \quad 78,062$

$\Sigma\Sigma T_{jk}^2 = \quad 39,132$

$T = \quad 482$

$\Sigma\Sigma\Sigma X_{ijk}^2 = \quad 7,926$

Source	SS	df	MS	F	F_{cv}
Rows	16.14	1	16.14	3.89	4.26
Columns	62.07	2	31.04	7.48*	3.40
Interaction	4.06	2	2.03	0.49	3.40
Within	99.60	24	4.15		
Total	181.87	29			

d. $\omega_{\text{columns}}^2 = 0.29$

8. b.

Rows	$df = 1, 36$	and	$\alpha = .05$	$F_{cv} = 4.11$
Columns	$df = 1, 36$	and	$\alpha = .05$	$F_{cv} = 4.11$
Interaction	$df = 1, 36$	and	$\alpha = .05$	$F_{cv} = 4.11$

c. $\Sigma T_{j\cdot}^2 = 107{,}552$
$\Sigma T_{\cdot k}^2 = 110{,}152$
$\Sigma\Sigma T_{jk}^2 = 58{,}776$
$T = 448$
$\Sigma\Sigma\Sigma X_{ijk}^2 = 6{,}036$

Source	SS	df	MS	F	F_{cv}
Rows	360.00	1	360.00	81.82*	4.11
Columns	490.00	1	490.00	111.36*	4.11
Interaction	10.00	1	10.00	2.27	4.11
Within	158.40	36	4.40		
Total	1018.40	39			

d. $\omega_{\text{rows}}^2 = 0.35$
$\omega_{\text{columns}} = 0.48$

9. **a.** 7 observations per cell
   b. 20 observations per cell
   c. 15 observations per cell
   d. 44 observations per cell

10. **a.** 8 observations per cell
    b. 20 observations per cell
    c. 16 observations per cell
    d. 41 observations per cell

Chapter 17

1. **b.** $\hat{Y} = -0.238X + 36.84$
   d. For $X = 43$, $\hat{Y} = 26.604$
   e. $s_{Y\cdot X} = 4.072$ (formula 6.11)
   $s_{Y\cdot X} = 3.986$ (formula 6.12)
   f. For $X = 28$, $\hat{Y} = 30.176$
   $z = -1.271$, $p = .898$
   g. For $X = 33$, $s_{\hat{Y}} = 4.192$ $\text{CI}_{95} = (20.313, 37.659)$
   h. $r = -.429$ $t = -2.278$

Since t_{cv} for 23 df $= \pm 2.069$, reject H_0; $p < .05$.

i. $s_b = 0.1045 \qquad t = -2.277$

Since t_{cv} for 23 df $= \pm 2.069$, reject H_0; $p < .05$.

2. a. From the regression equation, $\hat{Y} = 0.635X + 1.22$,

For $X = 5$, $\hat{Y} = 4.40$

Consider the conditional distribution of Y scores for $X = 5$; this distribution is normally distributed with mean, $\hat{Y} = 4.40$, and standard deviation, $s_{Y \cdot X} = 0.56$.

Therefore, to find the percentage of Y scores of 5 (or greater) given $X = 5$, we use the general formula for a standard score.

$$z = \frac{Y - \hat{Y}}{s_{Y \cdot X}} = \frac{5 - 4.40}{0.56} = 1.07$$

From Table C.1 in Appendix C, the area to the right of a z score of 1.07 is .1423. Therefore, the proportion of Y scores of 5 (or greater) given X score of 5 is .1423.

b. For $X = 4$, $\hat{Y} = 3.76$

The standard error of estimate for this score is defined by

$$s_{\hat{Y}} = s_{Y \cdot X} \sqrt{1 + 1/n + \frac{(X - \overline{X})^2}{SS_X}}$$

Using the data from this example,

$$s_{\hat{Y}} = 0.56 \sqrt{1 + 1/24 + \frac{(4 - 4.92)^2}{123.83}} = 0.57$$

Using the general formula for the confidence interval, with $t_{cv} = 2.074$ for 22 df,

$$\text{CI}_{95} = \hat{Y} \pm (t_{cv})(s_{\hat{Y}})$$

$$= 3.76 \pm (2.074)(0.57)$$

$$= 3.76 \pm 1.18$$

$$= (2.58, 4.94)$$

c. *Step 1: State the Hypotheses.*

The null hypothesis is that the regression coefficient in the population (β) equals zero. We will test this null hypothesis against the nondirectional alternative hypothesis and set the level of significance at $\alpha = .05$. Symbolically,

$$H_0: \beta = 0$$

$$H_a: \beta \neq 0$$

Step 2: Set the Criterion for Rejecting H_0.
The general formula for the test statistic is used to test this null hypothesis.

$$t = \frac{\text{statistic} - \text{parameter}}{\text{Standard error of the statistic}}$$

The sampling distribution for this test statistic is the t distribtion with $n-2$ degreees of freedom. In this example, there are $n-2 = 24-2 = 22$ degrees. Thus, the critical values from Table C.3 for $\alpha = .05$ are $t_{cv} = \pm 2.074$. If the calculated value for the test statistic exceeds the critical values, the null hypothesis would be rejected.

Step 3: Compute the Test Statistic.
The test statistic for this example is as follows:

$$t = \frac{b - \beta}{s_b}$$

where

$$s_b = \frac{s_{Y \cdot X}}{\sqrt{SS_X}} = \frac{.56}{\sqrt{123.8}} = 0.05$$

Therefore,

$$t = \frac{0.635 - 0}{0.05} = 12.70$$

Step 4: Interpret the Results.
Since the observed value of the test statistic ($t = 12.70$) exceeds the critical value ($t_{cv} = \pm 2.074$), the null hypothesis H_0: $\beta = 0$ would be rejected. The associated probability statement is that the probability that we would observe a regression coefficient of 0.635 by chance if the null hypothesis is true is less than .05. The conclusion is that the scores on the short test are a statistically significant predictor of scores on the long test.

3. a. $\hat{Y} = 0.782X + 3.842$

   b. For $X = 13.5$, $\hat{Y} = 14.399$

   c. $s_{Y \cdot X} = 3.129$ (formula 6.11)

      $s_{Y \cdot X} = 3.116$ (formula 6.12)

   d. For $X = 15$, $\hat{Y} = 15.57$

      $z = 0.936$, $p = .1677$

   e. For $X = 13$, $s_{\hat{y}} = 3.144$ $CI_{95} = (10.127, 22.581)$

   f. For $r = .67$, $t = 9.805$; reject H_0; $p < .05$.

   g. $s_b = .0797$ $t = 9.812$ reject H_0; $p < .05$.

4. a. $\hat{Y} = 0.853X + 29.35$

    For $X = 45$, $\hat{Y} = 67.635$

$s_{Y \cdot X} = 9.465, z = -0.289, p = .2344$

b. For $X = 35$, $s_{\hat{y}} = 9.768$, $CI_{95} = (38.682, 79.728)$

c. For $r = .64$, $t = 3.534$; reject H_0, $p < .05$.

d. $S_b = 0.241$, $t = 3.539$; reject H_0, $p < .05$.

5. a. $\hat{Y} = 0.065X + 66.548$

For $X = 170$, $\hat{Y} = 77.598$

$s_{Y \cdot X} = 8.898, z = 0.27$ and $0.832, p = .1909$

b. For $X = 195$, $s_{\hat{y}} = 9.156$, $CI_{95} = (59.812, 98.635)$

c. For $r = .425$, $t = 1.878$; do not reject H_0, $p > .05$.

d. $S_b = 0.035$, $t = 1.857$; do not reject H_0, $p > .05$.

6. a. $\hat{Y} = -0.044X + 3.616$

For $X = 20$, $\hat{Y} = 2.736$

$s_{Y \cdot X} = 0.346, z = -0.682$ and $0.185, p = .3257$

b. For $X = 23$, $s_{\hat{y}} = 0.355$, $CI_{95} = (1.87, 3.34)$

c. For $r = -.641$, $t = -4.005$; reject H_0, $p < .05$.

d. $s_b = 0.011$, $t = -4.000$; reject H_0, $p < .05$

Chapter 18

1. *Step 1: Determine the Regression Model.*
From the data given, the raw score regression equation is:

$$\hat{Y} = 0.050583X_1 + 0.555965X_2 + 0.212757X_3 + 10.759157$$

The standard score regression equation is:

$$z_{\hat{y}} = 0.043450z_1 + 0.557349z_2 + 0.160511z_3$$

Step 2: Determine R and R^2.
When all three variables are entered into the regression model, the multiple correlation coefficent (R) and the coefficient of determination (R^2) are:

$$R = 0.68588$$

$$R^2 = 0.47043$$

The R^2 indicates that over 47 percent of the variance in the Independent Living Scale scores can be attributed to the variance of the combined predictor variables.

Step 3: Determine Whether the Multiple R Is Statistically Significant.
In the first step, we will test the mulitple R for statistical significance when all

the variables are included in the model. For the first test, the null hypothesis is $H_0: R_{pop} = 0$; the test statistic was defined as follows:

$$F = \frac{R^2/k}{(1 - R^2)/(N - k - 1)} = \frac{0.47043/3}{(1 - 0.47043)/(50 - 3 - 1)} = 13.62077$$

The sampling distribution for the test statistic is the F distribution with k and $N - k - 1$ degrees of freedom. For this example, $k = 3$ and $N - k - 1 = 50 - 3 - 1 = 46$. Thus, the appropriate F distribution would be for 3 and 46 degrees of freedom. With $\alpha = .05$, the critical value of F from Table C.5 is $F_{cv} = 2.81$. Since the computed value (13.62077) exceeds the critical value (2.81), the null hypothesis, $H_0: R_{pop} = 0$, is rejected. The associated probability statement is: the probability that the observed $R = 0.47043$ would have occurred by chance, if in fact the null hypothesis is true, is less than .05. Note the "Signif of F" value in the printout is .0000. As before, this implies that the actual probability is less than .0001.

Step 4: Determine the Significance of the Predictor Variables.
For this step, the question now becomes whether or not all three predictor variables contribute to the regression model. This question involves testing the null hypothesis $H_0: \beta_i = 0$ for each of the predictor variables. The test statistic for this hypothesis is:

$$t = \frac{b_i}{s_{bi}}$$

For testing the null hypothesis for each predictor variable using this test statistic, the sampling distribtuion is the t distribution for $N - k_1 - 1$ degrees of freedom. Assuming $\alpha = .05$ and for df $= 50 - 3 - 1 = 46$, the critical value $t_{cv} = \pm 2.01$. From the data given in the example, t values and the "Sig T" values are:

$t_1 = 0.294$ Sig T $= .7700$
$t_2 = 3.885$ Sig T $= .0003$
$t_3 = 1.194$ Sig T $= .2387$

Step 5: Interpret the Results.
Note that the t value for Variable 2 (Social Skills Inventory-SSI) exceeded the critical value; however, the t values for Variable 1 (Self-Confidence Scale-SCS) and for Variable 3 (Academic Aptitude Scale) did not. Thus, we can conclude that the one variable (SSI) regression model is as effective as the three variable model. An inspection of the correlation of the matrix shows why this is true. While X2 has the highest correlation with the Y ($r = 0.6689$), both X1 and X3 are highly correlated with X2 ($r = 0.6318$ and $r = 0.5237$, respectively).

2. a. $z_{\hat{y}} = 0.956z_1 - 0.141z_2$ $\hat{Y} = 0.982X_1 - 0.167X_2 + 1.973$
  b. $R = 0.857$, $R^2 = 0.734$
  c. For df $= 2$, 12, $F_{cv} = 3.89$ $F = 16.682$; reject H_0, $p < .05$

d. $s_{b_1} = 0.226$, $t = 4.345$; reject H_0, $p < .05$. $s_{b_2} = 0.261$, $t = -0.640$; do not reject H_0, $p > .05$.

e. $r_{y1\cdot2} = .782$ $r_{y2\cdot1} = -.182$ $r_{y(1\cdot2)} = .644$ $r_{y(2\cdot1)} = -.095$.

3. a. $z_{\hat{Y}} = 0.543z_1 - 0.482z_2$ $\hat{Y} = 0.769X_1 - 1.024X_2 + 15.131$

b. $R = 0.789$, $R^2 = 0.622$

c. For df = 2, 97 and $\alpha = .05$, $F_{cv} = 2.71$ $F = 77.750$, reject H_0, $p < .05$.

d. $s_{b_1} = 0.089$, $t = 8.640$; reject H_0, $p < .05$. $s_{b_2} = 0.134$, $t = -7.642$; reject H_0, $p < .05$.

e. $r_{y1\cdot2} = .656$ $r_{y2\cdot1} = -.611$ $r_{y(1\cdot2)} = .535$ $r_{y(2\cdot1)} = -.475$

4. a. $z_{\hat{Y}} = 0.056z_1 + 0.585z_2$

b. $R = 0.613$, $R^2 = 0.376$

c. For df = 2, 97 and $\alpha = .05$, $F_{cv} = 2.71$ $F = 31.33$; reject H_0, $p < .05$.

d. $r_{y1\cdot2} = .064$ $r_{y2\cdot1} = .550$ $r_{y(1\cdot2)} = .050$ $r_{y(2\cdot1)} = .522$.

5.

	I	II	III	
n	6	6	6	$N = 18$
T_k	50	74	65	$\Sigma T_k = 189$
$\overline{X}_k$	8.33	12.33	10.833	$T^2/N = 1984.500$
ΣX_{ik}^2	430	950	729	$\Sigma\Sigma X_{ik}^2 = 2109$
T_k^2/n_k	416.667	912.667	704.167	$\Sigma(T_k^2/n_k) = 2033.501$
s	1.633	2.733	2.229	
s^2	2.667	7.467	4.967	

Source	SS	df	MS	F	F_{cv}
Between	49.001	2	24.501	4.868	3.68
Within	75.499	15	5.033		
Total	124.50	17			

X_1	X_2	Y
$n = 18$	$n = 18$	$n = 18$
$\overline{X}_1 = 0.333$	$\overline{X}_2 = 0.33$	$\overline{Y} = 10.5$
$\Sigma X_1 = 6$	$\Sigma X_2 = 6$	$\Sigma Y = 189$
$\Sigma X_1^2 = 6$	$\Sigma X_2^2 = 6$	$\Sigma Y^2 = 2109$
$\Sigma X_1 X_2 = 0$	$\Sigma X_1 Y = 50$	$\Sigma X_2 Y = 74$
$s_1 = 0.485$	$s_2 = 0.485$	$s_{\hat{y}} = 2.706$
$r_{12} = -.50$	$r_{y1} = -.583$	$r_{y2} = .493$
$\beta_1 = -0.448$	$\beta_2 = 0.268$	$R^2 = 0.393$

$F = 4.925$; reject H_0, $p < .05$.

6. a. $z_{\hat{y}} = -0.169z_1 + 0.807z_2$ $\hat{Y} = -0.049X_1 + 1.158X_2 + 11.378$

b. $R = 0.938$, $R^2 = 0.880$

c. For df = 2, 12 and $\alpha = .05$, $F_{cv} = 3.89$ $F = 44.00$; reject H_0, $p < .05$.

d. $s_{b_1} = 0.045$, $t = -1.089$; do not reject H_0, $p > .05$. $s_{b_2} = 0.212$, $t = 5.462$; reject H_0, $p < .05$.

e. $r_{y1\cdot2} = -.312$ $r_{y2\cdot1} = .844$ $r_{y(1\cdot2)} = -.114$ $r_{y(2\cdot1)} = .544$

7. a. $z_{\hat{y}} = 0.650z_1 + 0.521z_2$ $\hat{Y} = 2.461X_1 + 1.624X_2 - 12.520$

b. $R = 0.922$, $R^2 = 0.850$

c. For df = 2, 97 and $\alpha = .05$, $F_{cv} = 2.71$ $F = 212.500$; reject H_0, $p < .05$.

d. $s_{b_1} = 0.152$, $t = 16.191$; reject H_0, $p < .05$. $s_{b_2} = 0.126$, $t = 12.889$; reject H_0, $p < .05$.

e. $r_{Y1\cdot2} = .853$ $r_{Y2\cdot1} = .794$ $r_{Y(1\cdot2)} = .633$ $r_{Y(2\cdot1)} = .507$

8. a. Adjusted $R^2 = .63$

b. The unadjusted R^2 accounts for 73% of the variance, while the adjusted R^2 accounts for 63%.

c. $F_{cv} = 2.66$, $F = 5.500$, reject H_0, $p < .05$.

d. These six predictors, when used in combination, are moderately strong and statistically significant predictors of coronary artery disease.

9. $F_{cv} = 3.97$, $F = 7.190$, reject H_0, $p < .05$.

a. The difference was significant at $\alpha = .05$.

b. Yes, the prediction would be more accurate.

10. $F_{cv} = 3.92$, $F = 17.680$, reject H_0, $p < .05$.

a. The increase in R was significant.

b. It is not possible to tell from the given information.

11. a. Adjusted $R^2 = 1.333$

b. R^2 cannot exceed 1.00. The error comes from having more predictors (k) than observations (n).

Chapter 19

1. a. ANOVA summary table: Verbal fluency (dependent variable)

Source	SS	df	MS	F	F_{cv}
Between	153.27	2	76.64	3.09	3.36
Within	670.60	27	24.84		
Total	823.87	29			

ANOVA summary table: General ability (covariate)

Source	SS	df	MS	F	F_{cv}
Between	23.40	2	11.70	1.59	3.36
Within	198.90	27	7.37		
Total	222.30	29			

b. $SS_{hreg} = 408.02$, $F = 0.22$, $p > .05$

c. ANCOVA summary table

Source	SS	df	MS	F	F_{cv}
Covariate	247.16	1	247.16	15.46	4.23
Between	160.94	2	80.47	5.03*	3.37
Within	415.77	26	15.99		
Total	823.87	29			

*$p < .05$

d. Adjusted means: Group 1 = 46.73 $MS''_W = 16.95$

Group 2 = 51.54

Group 3 = 46.53

e. Tukey Post Hoc Tests

	$\overline{Y}'_i$	$\overline{Y}'_i - \overline{Y}'_j$		Q	
$\overline{Y}'_3$	46.53	—	—	—	—
$\overline{Y}'_1$	46.73	.20	—	0.15	—
$\overline{Y}'_2$	51.54	5.01	4.81	38.5*	3.70*

$Q_{cv} = 3.50$ *$p < .05$

2. In a first step, consider the ANOVA for the covariate (Social Adjustment Scale-SAS). Given the data for the example, the F ratio ($F = 2.220$) was not statistically significant. Therefore, the null hypotheses, H_0: $\mu_1 = \mu_2 = \mu_3$, for the covariate was not rejected.

Step 1: State the Hypotheses.
The null hypothesis for ANCOVA is that there are not differences in the population means on the Independent Living Scale scores after they have been adjusted for the Social Adjustment Scale scores. The null and alternative hypotheses are:

$$H_0: \mu'_1 = \mu'_2 = \mu'_3$$

$$H_a: \mu'_i \neq \mu'_k \quad \text{for some } i, k$$

Step 2: Set the Criterion for Rejecting H_0.
The test statistic for this one-way ANCOVA is the F ratio defined as the ratio of MS_B and MS_W, both of which have been adjusted for the covariate, i.e.,

$$F = \frac{MS'_B}{MS'_W}$$

The sampling distribution for this F ratio is the F distribution for $K - 1$ and $N - K - 1$ degrees of freedom. For this example, there are $K - 1 = 3 - 1 = 2$ degrees of freedom associated with the MS'_B and $N - K - 1 = 50 - 3 - 1 = 46$ degrees of freedom associated with the MS'_W. Thus, assuming $\alpha = .05$, the critical value of F from Table C.5, for 2 and 46 degrees of freedom, is $F_{cv} = 3.20$.

Step 3: Compute the Test Statistic.
In ANCOVA, two test statistics are computed: one for testing the significance of the relationship between the covariate and the dependent variable; and one for testing the null hypothesis identified in Step 1, H_0: $\mu'_1 = \mu'_2 = \mu'_3 = \mu'_4$. For the first test statistic, the null hypotheis was that there is no relationship between the covariate and the dependent variable, i.e., H_0: $\rho = 0$. The test statistic is given by

$$F = \frac{MS'_{cov}}{MS'_W}$$

For this example,

$$F = \frac{831.943}{50.081} = 16.612$$

The critical value of F for this test statistic is identified in the F distribution for 1 and $N - K - 1 = 50 - 3 - 1 = 46$ degrees of freedom. Assuming $\alpha = .05$ and using Table C.5, $F_{cv} = 4.05$. Since the calculated F value ($F = 16.612$) exceeds the critical value ($F_{cv} = 4.05$), the null hypothesis of no relationship between the SAS scores and the ILS scores is rejected. The conclusion is that since the three groups did not differ on the SAS scores and since there is a statistically significant relationship between the SAS scores and the ILS scores, the variable SAS is an appropriate covariate.

The second test statistic was defined in Step 2,

$$F = \frac{MS'_B}{MS'_W}$$

and is used to test the null hypothesis defined in Step 1, $H_0: \mu'_1 = \mu'_2 = \mu'_3$. Note the data in the Summary ANCOVA,

$$F = \frac{922.174}{50.081} = 18.414$$

Step 4: Interpret the Results.
Since the calculated F value ($F = 18.414$) exceeds the critical value ($F_{cv} = 3.20$), the null hypothesis, $H_0: \mu'_1 = \mu'_2 = \mu'_3$ is rejected. The conclusion is that the adjusted means in the population are not equal. The associated probability statement would be: the probability that the observed differences among the adjusted means would have occurred by chance, if the null hypothesis is true, is less than .05. Note in the printout, "Sig of F" equals .000, which implies that the probability of such an occurrence is actually less than .001.

In order to identify which adjusted means differ, it is necessary to conduct the Tukey procedure for adjusted means. The adjusted means for the groups are computed using the data from the Multiple Classification Analysis:

Group 1 (20 year olds) $\overline{Y}'_1 = 51.60 + 10.36 = 61.96$
Group 2 (19 year olds) $\overline{Y}'_2 = 51.60 + 1.53 = 53.13$
Group 3 (18 year olds) $\overline{Y}'_3 = 51.60 - 6.70 = 44.90$

The Tukey procedure is conducted using the Tukey/Kramer (TK) formula for calculating the Q test statistic.

$$Q = \frac{\overline{Y}'_i - \overline{Y}'_k}{\sqrt{MS''_W \frac{1/n_i + 1/n_k}{2}}}$$

Note that it is necessary to use the MS''_W that has been adjusted the second time (see formula 19.4). Using the data given, we compute MS''_W to be

$$MS''_W = MS'_W \left(1 + \frac{SS_{B(X)}}{(K-1)SS_{W(X)}} \right)$$

$$= 50.081 \left(1 + \frac{148.120}{(2)1567.800} \right) = 52.447$$

The data for the Tukey procedure are below; the Q for the difference between Group 2 (19 year olds) and Group 1 (20 year olds) is calculated as follows:

$$Q = \frac{61.96 - 53.13}{\sqrt{52.447 \frac{1/20+1/10}{2}}} = \frac{8.83}{1.98} = 4.46$$

Group	$\bar{Y}_i'$	$\bar{Y}_i' - \bar{Y}_k'$		Q	
18 yr Olds	44.90				
19 yr Olds	53.13	8.23		5.08*	
20 yr Olds	61.96	17.06	8.83	8.62*	4.46*

*$p < .05$ $Q_{cv} = 3.43$ for df $= 46$

From the data, we conclude that the means for the groups differ significantly from each other, i.e., the 20 year olds had a significantly higher adjusted mean on the Independent Living Scale (ILS) than did the 19 year olds, which in turn had a significantly higher adjusted mean than did the 18 year olds.

3. a. **ANOVA summary table: Rate of gain (dependent variable)**

Source	SS	df	MS	F	F_{cv}
Between	126.76	2	63.38	17.13*	3.47
Within	77.74	21	3.70		
Total	204.50	23			

*$p < .001$

ANOVA summary table: Initial weight (covariate)

Source	SS	df	MS	F	F_{cv}
Between	19.08	2	9.54	3.22	3.47
Within	62.25	21	2.96		
Total	81.33	23			

b. $SS_{hreg} = 23.14$, $F = 1.28$, $p > .05$

c. ANCOVA summary table

Source	SS	df	MS	F	F_{cv}
Covariate	134.97	1	134.97	102.25*	4.35
Between	43.10	2	21.55	16.33*	3.49
Within	26.43	20	1.32		
Total	204.50	23			

*$p < .05$

d. Adjusted means: Group 1 = 18.35 $MS''_W = 1.500$
Group 2 = 20.40
Group 3 = 22.02

e. Tukey Post Hoc Tests

	$\overline{Y}'_i$	$\overline{Y}'_i - \overline{Y}'_j$		Q	
$\overline{Y}'_1$	18.33	—	—	—	—
$\overline{Y}'_2$	20.40	2.05	—	4.66*	—
$\overline{Y}'_3$	22.02	3.67	1.62	8.34*	3.68*

$Q_{cv} = 3.58$ *$p < .05$

4. a. ANOVA summary table: Problem solving (dependent variable)

Source	SS	df	MS	F	F_{cv}
Between	170.38	2	85.19	2.16	3.55
Within	709.43	18	39.41		
Total	879.81	20			

ANOVA summary table: Creativity (covariate)

Source	SS	df	MS	F	F_{cv}
Between	87.24	2	43.62	1.49	3.55
Within	525.43	18	29.19		
Total	612.67	20			

b. $SS_{hreg} = 254.86$, $F = 1.27$, $p > .05$

c. ANCOVA summary table

Source	SS	df	MS	F	F_{cv}
Covariate	351.92	1	351.92	20.08*	4.45
Between	229.93	2	114.97	6.56*	3.59
Within	297.96	17	17.53		
Total	879.81	20			

*$p < .05$

d. Adjusted means: Group 1 = 153.29 $MS''_W = 18.93$
Group 2 = 150.72
Group 3 = 145.27

e. Tukey Post Hoc tests

	$\overline{Y}'_j$	$\overline{Y}'_j - \overline{Y}'_k$		Q	
$\overline{Y}'_3$	145.27	—	—	—	—
$\overline{Y}'_2$	150.86	5.59	—	3.41	—
$\overline{Y}'_1$	153.29	8.02	2.43	4.89*	1.48

$Q_{cv} = 3.63$ *$p < .05$

Chapter 20

1. $r_{Pb} = -.81$

2. $\phi = -.22$

3. $r_{tet} = -.56$

4. $\phi = .35$

5. $r_{rb} = .28$

6. $r_{Pb} = -.83$

7. The assumption is that, even thought the internship is graded pass/fail, the performance during the internship has underlying continuity. So the appropriate correlation coefficient is the biserial coefficient; the formula is

$$r_b = \left(\frac{\overline{Y}_p - \overline{Y}_f}{\sigma_Y}\right)\left(\frac{pq}{Y}\right)$$

For this example, $\overline{Y}_p = 121.8$; $\overline{Y}_f = 105.6$; $\sigma_Y = 18.2$; $p = 0.62$; and $q = 0.38$. Y is the ordinate of the unit normal curve at the point of the division between the p and q proportions under the curve. Looking for the z score for the 62 percentile, we find from Table C.2 in Appendix C, $z = 0.3055$. Looking up the corresponding ordinate from Table C.1 and interpolating, we find $Y = .3808$. Therefore,

$$r_b = \left(\frac{121.8 - 105.6}{18.2}\right)\left(\frac{(.62)(.38)}{.3808}\right) = 0.55$$

This coefficient indicates that there is a moderate positive relationship between scores on the academic aptitude test taken at the beginning of the program and later success in the intership.

8. $r_b = .77$

9. $\lambda = .17$

   $\lambda_y = .25$

10. $\eta = .97$

11. $\lambda = .09$

12. $r_b = -.66$

13. $r_{rb} = .05$

Chapter 21

1. 1. No difference in success for the four feedback groups.
   2. For df $= 3$ and $\alpha = .05$, $\chi^2_{cv} = 7.815$
   3. $\chi^2 = 5.92$
   4. Do not reject H_0; $p > .05$.

   No standardized residuals are computed, since H_0 was not rejected.

2. 1. Membership preferences match a specified distribution.
   2. For df $= 6$ and $\alpha = .05$, $\chi^2_{cv} = 12.592$
   3. $\chi^2 = 13.24$
   4. Reject H_0; $p < .05$.

Standardized residuals	
City	*R*
Dallas	−0.82
Miami Beach	0.79
New York	0.61
Chicago	−1.50
Washington, D.C.	−0.67
San Francisco	2.33
Other	0.69

3. *Step 1: State the Hypotheses.*
The null hypothesis is that the observed frequencies in the downsized workforce do not differ from the expected frequencies; the expected frequencies are based upon the proportions of frequencies in the workforce prior to the downsizing.

Step 2: Set the Criterion for Rejecting H_0.
The test statistic for the above hypothesis is χ^2. The sampling distribution of this test statistic is the χ^2 distribution with $k - 1$ degrees of freedom, where k is the number of categories under investigation. The critical value of χ^2 for this example with $\alpha = .05$ and for $k - 1 = 5 - 1 = 4$ degrees of freedom, is found in Table C.4 to be 9.488.

Step 3: Compute the Test Statistic.
In this goodness-of-fit test, the first step is to determine the proportion of frequencies in each category before the downsizing:

$15/150 = .10$

$60/150 = .40$

$35/150 = .2333$

$28/150 = .1867$

$12/150 = .08$

Using these proportions, we compute the expected frequencies for the downsized workforce:

$.10 \times 120 = 12$

$.40 \times 120 = 48$

$.2333 \times 120 = 28$

$.1867 \times 120 = 22.4$

$.08 \times 120 = 9.6$

Now, we compute the χ^2 statistic:

$$\chi^2 = \sum \frac{(O - E)^2}{E}$$

$$= \frac{(15 - 12)^2}{12} + \frac{(46 - 48)^2}{48} + \frac{(28 - 28)^2}{28} + \frac{(20 - 22.4)^2}{22.4} + \frac{(11 - 9.6)^2}{9.6}$$

$$= .75 + .0833 + 0 + .2571 + .2042 = 1.2946$$

Step 4: Interpret the Results.
Since the calculated value of χ^2 (1.2946) does not exceed the critical value (9.488), we can conclude that the proportion of frequencies found in the downsized workforce does not depart significantly from the expected frequencies determined from the proportion of frequencies observed before downsizing.

4. 1. No difference in entertainment preference among individuals of different marital status.
   2. For df $= 6$ and $\alpha = .05$, $\chi^2_{cv} = 12.592$
   3. $\chi^2 = 18.39$
   4. Reject H_0; $p < .05$.

	Individual	*Standardized residuals* Small group	Large group
Single	2.32*	−1.79	−0.89
Married	−0.77	0.81	0.00
Separated/divorced	0.15	−0.81	1.34
Widowed	−1.39	1.79	−0.45

5. $C = 0.40$
   $V = 0.30$

6. 1. No change in voting intentions after forum.
   2. For df $= 1$ and $\alpha = .05$, $\chi^2_{cv} = 3.841$
   3. $\chi^2 = 4.55$
   4. Reject H_0; $p < .05$.

7. 1. No difference in women's actual and intended feeding practice.
   2. $\chi^2_{cv} = 5.991$
   3. $\chi^2 = 33.78$
   4. Reject H_0, $p < .05$.

8. 1. No difference in attitudes toward the policy.
   2. $\chi^2_{cv} = 9.488$
   3. $\chi^2 = 3.97$
   4. Fail to reject H_0, $p > .05$.

9. 1. No difference in behavior before and after intervention.
   2. $\chi^2_{cv} = 3.841$
   3. $\chi^2 = 14.29$
   4. Reject H_0, $p < .05$.

10. 1. Heroes the same at different ages.
   2. $\chi^2_{cv} = 5.991$

3. $\chi^2 = 9.47$

4. Reject H_0, $p < .05$.

11. 1. No difference in perceived life satisfaction for single and married graduate students.

    2. For df $= 1$ and $\alpha = .05$, $\chi^2_{cv} = 3.841$

    3. $\chi^2 = 14.44$

    4. Reject H_0; $p < .05$.

Standardized residuals
Satisfaction

	High	Low
Single	−1.81	2.06
Married	1.73	−1.98

12. 1. No difference in the patterns of choices for women and men.

    2. For df $= 3$ and $\alpha = .01$, $\chi^2_{cv} = 11.345$

    3. $\chi^2 = 7.12$

    4. Do not reject H_0; $p > .05$.

13. 1. No difference in the response patterns of choices for homeowners and renters.

    2. For df $= 4$ and $\alpha = .05$, $\chi^2_{cv} = 9.488$

    3. $\chi^2 = 68.04$

    4. Reject H_0; $p < .05$.

14. 1. H_0: $P_1 = P_2$

        H_a: $P_1 \neq P_2$

    2. $z_{cv} = \pm 1.96$

    3. $p = 0.57$ $q = 0.43$ $s_{p_1 - p_2} = .10$

    4. $z = -3.50$

    5. $\chi^2 = 14.43$

15. 1. No difference in attitude as a function of geographic location.

    2. For df $= 8$ and $\alpha = .05$, $\chi^2_{cv} = 15.507$

    3. $\chi^2 = 435.20$

    4. Reject H_0; $p < .05$.

| | | Standardized residuals | | | |
	Rural	Small suburban	Large suburban	Small urban	Large urban
Legalized	−8.03	−3.57	8.32	−1.99	4.20
Misdemeanor	−2.94	1.06	−1.82	2.74	0.29
Felony	12.13	3.14	−7.93	−0.29	−5.16

Chapter 22

1. 1. H_0: No difference in response to nonverbal communication cues for male and female children

H_0: $\text{Cues}_F < \text{Cues}_M$

Median test:

2. For df $= 1$ and $\alpha = .05$ (one-tailed test), $\chi^2_{cv} = 2.706$

3. $\chi^2 = 0.80$

4. Do not reject H_0; $p > .05$.

Mann-Whitney U test:

2. For $\alpha = .05$ (one-tailed test, $U_{cv} = 28$)

3. $U = 25$

4. Reject H_0; $p < .05$.

2. *Step 1: State the Hypotheses.*
The null hypothesis for the Kruskal-Wallis test in this example is that, in the population, there is no difference in the scores on the scale of student-teacher interaction. The alternative hypothesis is that at least two of the populations or combination of the populations differ.

Step 2: Set the Criterion for Rejecting H_0.
The test statistic for the Kruskal-Wallis test is

$$H = \frac{12}{N(N+1)} \sum \frac{R_{K^2}}{n_K} - 3(N+1)$$

The sampling distribution for H is the χ^2 distribution with $K - 1$ degrees of freedom, where K is the number of groups in the study. For this example, with $\alpha = .05$ and for $K - 1 = 4 - 1 = 3$ df, the critical value of H is 7.815.

Step 3: Compute the Test Statistic.
In the Kruskal-Wallis test, the first step is to rank all the scores regardless of the group and then to sum the ranks within the groups. Below are the scores and the corresponding ranks.

Mathematics	Chemistry	History	English
17(6)	13(2)	12(1)	16(5)
15(4)	18(7)	14(3)	19(8)
26(15)	21(10)	28(17)	20(9)
24(13)	23(12)	22(11)	33(20)
27(16)	25(14)	30(19)	29(18)

R_K 54   45   51   60

Now we compute the H statistic.

$$H = \frac{12}{20(21)}\left(\frac{(54)^2}{5} + \frac{(45)^2}{5} + \frac{(51)^2}{5} + \frac{(60)^2}{5}\right) - 3(21)$$

$$= 63.668 - 63 = .668$$

Step 4: Interpret the Results.
Since the calculated value of H (.73) does not exceed the critical value (7.815), we cannot reject the null hypothesis of no differences among the different populations of teachers on scores of student-teacher interaction. We conclude no difference between Mathematics, Chemistry, History, and English teachers.

3. 1. No difference in emotional adjustment as a function of preschool background.

  2. For df = 3 and $\alpha = .05$, $\chi^2_{cv} = 7.815$

  3. $H = 5.46$

  4. Do not reject H_0; $p > .05$.

4. 1. No difference in self-confidence due to summer camp attendance.

  2. $T_{cv} = 10$

  3. $T = 7$

  4. Reject H_0; $p < .01$.

5. 1. H_0: No difference between tenured and untenured faculty.

  Median test:

  2. $\chi^2_{cv} = 6.635$

  3. $\chi^2 = 6.00$

  4. Fail to reject H_0, $p > .01$.

  Mann-Whitney U test:

  2. $U_{cv} = 28$

  3. $U = 36$

  4. Fail to reject H_0, $p > .01$

6. 1. No difference in perceived exertion rates for women using a bicycle, a treadmill, or weights.

2. $\chi^2_{cv} = 5.991$

3. $H = 8.05$

4. Reject H_0, $p < .05$.

7. 1. Pretest scores and posted scores do not differ.

2. $T_{cv} = 8$

3. $T = 8$

4. Fail to reject H_0, $p > .05$.

8. 1. The depression scores for people who fear a crowded room, spiders and snakes, or public speaking are the same.

2. $\chi^2_{cv} = 5.991$

3. $H = 10.88$

4. Reject H_0, $p < .05$.

9. 1. No difference between men and women.

Median test:

2. $\chi^2_{cv} = 3.841$

3. $\chi^2 = 3.20$

4. Fail to reject H_0, $p > .05$.

Mann-Whitney U test:

2. $U_{cv} = 24$

3. $U = 28$

4. Fail to reject H_0, $p > .05$.

10. a. 1. H_0: Men and women are similar in terms of their concern about the economy.

Median test:

2. $\chi^2_{cv} = 3.841$

3. $\chi^2 = 0.00$

4. Fail to reject H_0, $p > .05$.

b. Mann-Whitney U test:

2. $U_{cv} = 24$

3. $U = 54$

4. Fail to reject H_0, $p > .05$.

c. 1. People of different ages are similar in terms of their concern about the economy.

2. $\chi^2_{cv} = 7.815$

3. $H = 17.86$

4. Reject H_0, $p < .05$.

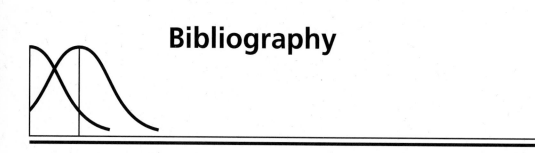

Bibliography

Anderson, N. H. (1961) Scales and Statistics: Parametric and Nonparametric. *Psychological Bulletin, 58,* 305–316.

Bartlett, M. S. (1937) Properties of Sufficiency and Statistical Tests. *Proceedings of the Royal Society of London, 160, series A,* 268–282.

Camilli, G., and Hopkins, K. D. (1978) Applicability of Chi-Square to 2×2 Contingency Tables with Small Expected Cell Frequencies. *Psychological Bulletin, 85, no. 1,* 163–167.

Carlson, J. E., and Trimm, N. H. (1974) Analysis of Nonorthogonal Fixed Effects Designs. *Psychological Bulletin, 81,* 563–570.

Carroll, J. B. (1961) The Nature of the Data, or How to Choose a Correlation Coefficient. *Psychometrika, 26,* 347–372.

Chase, L. J., and Tucker, R. K. Statistical Power: Derivation, Development and Data-analytic Implication. *The Psychological Record, 26,* 472–486.

Cochran, W. G. (1954) Some Methods for Strengthening the Common χ^2 Tests. *Biometrics, 10,* 417–451.

Cochran, W. G., and Cox, G. M. (1957) *Experimental Designs.* New York: Wiley.

Cohen, J. (1965) Some Statistical Issues in Psychological Research. In B. B. Wolman (Ed.), *Handbook of Clinical Psychology,* pp. 95–121. New York: McGraw-Hill.

Cohen, J. (1977) *Statistical Power Analysis for the Behavioral Sciences* (2nd ed.). New York: Academic Press.

Cook, T. D., and Campbell, D. T. (1979) *Quasi-experimentation: Design Analysis Issues for Field Settings.* Chicago: Rand McNally.

Cureton, E. E. (1956) Rank-Biserial Correlation. *Psychometrika, 21,* 287–290.

Cureton, E. E. (1968) Rank-Biserial Correlation—When Ties Are Present. *Educational and Psychological Measurement, 28,* 77–79.

Draper, N. R., and Smith, H. (1966) *Applied Regression Analysis.* New York: Wiley.

Elashoff, J. D. (1969) Analysis of Covariance: A Delicate Instrument. *American Educational Research Journal, 6,* 383–402.

Feldt, L. S. (1973) What Size Sample for Methods/Materials Experiments. *Journal of Educational Measurement, 10,* 221–226.

Fisher, R. A. (1915) Frequency Distributions of Values of the Correlation Coefficient in Samples of an Infinitely Large Population. *Biometrika, 10,* 507–521.

Gaito, J. (1980) Measurement Scales and Statistics: Resurgence of an Old Misconception. *Psychological Bulletin, 87,* 564–567.

Games, Paul A., and Klare, George R. (1967) *Elementary Statistics: Data Analysis for the Behavioral Sciences.* New York: McGraw-Hill.

Glass, G. V (1966) Note on Rank-Biserial Correlation. *Educational and Psychological Measurement, 26,* 623–631.

Glass, G. V et al. (1972) Consequences of Failure to Meet the Assumptions Underlying the Use of Analysis of Variance and Covariance. *Review of Educational Research, 42, no. 3,* 237–288.

Glass, G. V, and Hakstian, A. R. (1969) Measures of Association in Comparative Experiments: Their Development and Interpretation. *American Educational Research Journal, 6,* 403–414.

Guilford, J. P., and Fruchter, B. (1978) *Fundamental Statistics in Psychology and Education* (6th ed.). New York: McGraw-Hill.

Haberman, S. J. (1973) The Analysis of Residuals in Cross-classified Tables. *Biometrics, 29,* 205–220.

Hancock, G. R., and Klockers, A. J. (1996) The Quest for α: Developments in Multiple Comparison Procedures in the Quarter Century Since Games (1971), *Review of Educational Research, 66,* 269–306.

Hays, W. (1981) *Statistics.* New York: Holt.

Hinkle, D. E., and Oliver, J. D. (1983) How Large Should the Sample Be? A Question with No Simple Answer? Or ...? *Educational and Psychological Measurement, 43,* 1051–1060.

Hinkle, D. E., Oliver, J. D., and Hinkle, C. A. (1985) How Large Should the Sample Be? Part II—The One-sample Case for Survey Research. *Educational and Psychological Measurement, 45,* 271–280.

Hinkle, D. E., Wiersma, W., and Jurs, S. G. (1979) *Applied Statistics for the Behavioral Sciences.* Boston: Houghton Mifflin.

Hollander, M., and Wolfe, D. A. (1973) *Nonparametric Statistical Methods.* New York: Wiley.

Hollingshead, A. B. (1949) *Elmtown's Youth: The Impact of Social Classes on Adolescents.* New York: Wiley.

Horowitz, L. (1974) *Elements of Statistics for Psychology and Education.* New York: McGraw-Hill.

Huitema, B. E. (1980) *The Analysis of Covariance.* New York: Wiley.

Kaiser, H. F. (1960) Review of *Measurement and Statistics*, by Virginia Senders. *Psychometrika, 25,* 411–413.

Keppel, G. (1982) *Design and Analysis: A Researcher's Handbook* (2nd ed.). Englewood Cliffs, N.J.: Prentice-Hall.

Kerlinger, F. N. (1986) *Foundations of Behavioral Research.* New York: Holt.

Kirk, R. E. (1982) *Experimental Design: Procedures for the Behavioral Sciences.* Belmont, Calif.: Wadsworth.

Kish, L. (1967) *Survey Sampling.* New York: Wiley.

Lancaster, H. O., and Hamden, M. A. (1964) Estimates of the Correlation Coefficient in Contingency Tables with Possibly Nonmetrical Characters. *Psychometrika, 29,* 383–391.

Liebetrau, A. M. (1983) *Measures of Association.* Beverly Hills, Calif.: Sage Publications.

Lord, F. M. (1953) On the Statistical Treatment of Football Numbers. *American Psychologist, 8,* 750–751.

Lord, F. M. (1969) Statistical Adjustments When Comparing Pre-existing Groups. *Psychological Bulletin, 58,* 336–337.

Marascuilo, L. A., and McSweeney, M. (1977) *Nonparametric and Distribution-Free Methods for the Social Sciences.* Monterey, Calif.: Brooks/Cole.

Maxwell, S. E., Camp, C. J., and Avery, R. D. (1981) Measures of Strength of Association: A Comparative Examination. *Journal of Applied Psychology, 66,* 525–534.

McNemar, Q. (1969) *Psychological Statistics* (4th ed.). New York: Wiley.

National Opinion Research Center. (1980) *High School and Beyond; Information for Users: Base Year (1980) Data.* Chicago: National Opinion Research Center.

Ostle, B., and Mensing, R. W. (1975) *Statistics in Research.* Ames: University of Iowa Press.

Overall, J. E., and Spiegal, D. K. (1969) Concerning Least Squares Analysis of Experimental Data. *Psychological Bulletin, 72,* 311–322.

Pearson, E. S., and Hartley, H. O. (1966) *Biometrika Tables for Statisticians* (3rd ed., vol. 1). Cambridge, Eng.: Cambridge University Press.

Pedhazur, E. J. (1982) *Multiple Regression in Behavioral Research.* New York: Holt.

Resnikoff, G. J., and Lieberman, G. J. (1957) *Tables of the Non-central Distributions.* Stanford, Calif.: Stanford University Press.

Reynolds, H. T. (1984) *Analysis of Nominal Data.* Beverly Hills, Calif.: Sage Publications.

Satterthwaite, F. W. (1946) An Approximate Distribution of Estimates of Variance Components. *Biometrics Bulletin, 2,* 110–114.

Seber, G. A. F. (1977) *Linear Regression Analysis.* New York: Wiley.

Siegel, S. (1956) *Nonparametric Statistics for the Behavioral Sciences.* New York: McGraw-Hill.

Stevens, S. S. (ed). (1951) Mathematics, Measurement, and Psychophysics. In *Handbook of Experimental Psychology*. New York: Wiley.

Stevens, S. S. (1968) Measurement, Statistics, and the Schempiric View. *Science, 101,* 849–856.

Stoline, M. R. (1981) The Status of Multiple Comparisons: Simultaneous Estimation of All Pairwise Comparisons in One-Way ANOVA Designs. *American Statistician, 35,* 134–141.

Sudman, S. (1976) *Applied Sampling.* New York: Academic Press.

Index

TABLE 16.17
Summary of Computational Procedures for ANOVA, Two-Way Classification

	Row Effect	Column Effect	Interaction Effect	Within Cell	Total
Sum of Squares	$nK\sum_{j=1}^{J}(\bar{X}_{j\cdot}-\bar{X})^2$	$nJ\sum_{k=1}^{K}(\bar{X}_{\cdot k}-\bar{X})^2$	$n\sum_{k=1}^{K}\sum_{j=1}^{J}(\bar{X}_{jk}-\bar{X}_{j\cdot}-\bar{X}_{\cdot k}+\bar{X})^2$	$\sum_{k=1}^{K}\sum_{j=1}^{J}\sum_{i=1}^{n}(X_{ijk}-\bar{X}_{jk})^2$	$\sum_{k=1}^{K}\sum_{j=1}^{J}\sum_{i=1}^{n}(X_{ijk}-\bar{X})^2$
Computational Formula	$\dfrac{1}{nK}\sum_{j=1}^{J}T_{j\cdot}^2-\dfrac{T^2}{N}$	$\dfrac{1}{nJ}\sum_{k=1}^{K}T_{\cdot k}^2-\dfrac{T^2}{N}$	$\dfrac{1}{n}\sum_{k=1}^{K}\sum_{j=1}^{J}T_{jk}^2-\dfrac{1}{nK}\sum_{j=1}^{J}T_{j\cdot}^2$ $-\dfrac{1}{nJ}\sum_{k=1}^{K}T_{\cdot k}^2+\dfrac{T^2}{N}$	$\sum_{k=1}^{K}\sum_{j=1}^{J}\sum_{i=1}^{n}X_{ijk}^2$ $-\dfrac{1}{n}\sum_{k=1}^{K}\sum_{j=1}^{J}T_{jk}^2$	$\sum_{k=1}^{K}\sum_{j=1}^{J}\sum_{i=1}^{n}X_{ijk}^2-\dfrac{T^2}{N}$
Degrees of Freedom	$J-1$	$K-1$	$(J-1)(K-1)$	$JK(n-1)$	$N-1$
Mean Square	$\dfrac{SS_J}{J-1}$	$\dfrac{SS_K}{K-1}$	$\dfrac{SS_{JK}}{(J-1)(K-1)}$	$\dfrac{SS_W}{JK(n-1)}$	
E(MS) Fixed-Effects Model	$\sigma_e^2+\dfrac{nK\sum_{j=1}^{J}\alpha_{j\cdot}^2}{J-1}$	$\sigma_e^2+\dfrac{nJ\sum_{k=1}^{K}\beta_{\cdot k}^2}{K-1}$	$\sigma_e^2+\dfrac{n\sum_{k=1}^{K}\sum_{j=1}^{J}(\alpha\beta)_{jk}^2}{(J-1)(K-1)}$	σ_e^2	
E(MS) Mixed-Effects Model (Rows Fixed, Columns Random)	$\sigma_e^2+n\sigma_{jk}^2+\dfrac{nK\sum_{j=1}^{J}\alpha_{j\cdot}^2}{J-1}$	$\sigma_e^2+nJ\sigma_k^2$	$\sigma_e^2+n\sigma_{jk}^2$	σ_e^2	
E(MS) Random-Effects Model	$\sigma_e^2+n\sigma_{jk}^2+nK\sigma_j^2$	$\sigma_e^2+n\sigma_{jk}^2$ $+nJ\sigma_k^2$	$\sigma_e^2+n\sigma_{jk}^2$	σ_e^2	